Lecture Notes in Artificial Intelligence 928

Subseries of Lecture Notes in Computer Science
Edited by J. G. Carbonell and J. Siekmann

Lecture Notes in Computer Science

Edited by G. Goos, J. Hartmanis and J. van Leeuwen

Springer

Berlin
Heidelberg
New York
Barcelona
Budapest
Hong Kong
London
Milan
Paris
Tokyo

V.W. Marek A. Nerode M. Truszczyński (Eds.)

Logic Programming and Nonmonotonic Reasoning

Third International Conference, LPNMR '95
Lexington, KY, USA, June 26-28, 1995
Proceedings

 Springer

Series Editors

Jaime G. Carbonell
School of Computer Science
Carnegie Mellon University
Pittsburgh, PA 15213-3891, USA

Jörg Siekmann
University of Saarland
German Research Center for Artificial Intelligence (DFKI)
Stuhlsatzenhausweg 3, D-66123 Saarbrücken, Germany

Volume Editors

V. Wiktor Marek
Miroslaw Truszczyński
Department of Computer Science, University of Kentucky
Lexington, KY 40506, USA

Anil Nerode
Mathematical Sciences Institute, Cornell University
Ithaca, NY 14853, USA

CR Subject Classification (1991): I.2.3-4, F4.1, D.1.6

1991 Mathematics Subject Classification: 03B60, 03B70

ISBN 3-540-59487-6 Springer-Verlag Berlin Heidelberg New York

CIP data applied for

© Springer-Verlag Berlin Heidelberg 1995
Printed in Germany

Typesetting: Camera ready by author
SPIN: 10486224   06/3142 – 543210 – Printed on acid-free paper

Preface

This volume consists of refereed papers presented at the Third International Conference on Logic Programming and Non-monotonic Reasoning, LPNMR'95, in Lexington, KY. The conference was sponsored by the Association for Logic Programming.

LPNMR'95 is the third meeting in a series. The two previous meetings were held in Washington, DC, in July 1991, and in Lisbon, Portugal, in June 1993. LPNMR conferences are designed to promote research in the areas on the border between Logic Programming and Logical Foundations of Artificial Intelligence. By bringing together researchers in both disciplines they facilitate cooperation and cross-fertilization of ideas. Cross-disciplinary character of the conference is further stressed by the individuals who will deliver plenary presentations. These are: Hector Levesque, Alberto Martelli and Raymond Smullyan.

Many people contributed to the success of LPNMR'95. Diane Mier proved invaluable in attending to all local organizational matters in Lexington. Wilson Kone and his staff at Mathematical Science Institute, Cornell University, helped in timely execution of the review process. Special thanks are due to the Program Committee, and to additional reviewers listed elsewhere in this volume, for complete unbiased evaluations of submitted papers. We thank Howard Blair for coordinating electronic discussions of the reviews.

The conference was financially supported by the College of Engineering, the Office of Research and Graduate Studies, and the Center for Computational Sciences of the University of Kentucky, and by the Mathematical Science Institute (an ARO Center for Excellence), Cornell University.

March 1995

V. Wiktor Marek, Lexington, KY
Anil Nerode, Ithaca, NY
Miroslaw Truszczynski, Lexington, KY

Program Committee

K.R. Apt
H.A. Blair
P.M. Dung
M. Gelfond
G. Gottlob
A. Kakas
V. Lifschitz
V.W. Marek
A. Nerode
L. Pereira
T. Przymusinski
Y. Sagiv
V.S. Subrahmanian
M. Truszczynski
D. Warren

Reviewers

J.J. Alferes
C. Baral
P. Cholewinski
C.V. Damásio
Y. Dimopoulos
F. Dushin
T. Eiter
E. Marchiori
N. McCain
A. Mikitiuk
G. Schwarz
F. Teusink
H. Turner

Table of Contents

Complexity Results for Abductive Logic Programming 1
T. Eiter, G. Gottlob, N. Leone

A Terminological Interpretation of (Abductive) Logic Programming 15
M. Denecker

Abduction over 3-valued Extended Logic Programs 29
C.V. Damásio, L.M. Pereira

On Logical Constraints in Logic Programming 43
V.W. Marek, A. Nerode, J.B. Remmel

An Operator for Composing Deductive Data Bases with Theories
of Constraints 57
D. Aquilino, P. Asirelli, C. Renso, F. Turini

Update Rules in Datalog Programs 71
M. Halfeld Ferrari Alves, D. Laurent, N. Spyratos

Characterizations of the Stable Semantics by Partial Evaluation 85
S. Brass, J. Dix

Game Characterizations of Logic Program Properties 99
H.A. Blair

Computing the Well-Founded Semantics Faster 113
K.A. Berman, J.S. Schlipf, J.V. Franco

Loop Checking and the Well-Founded Semantics 127
V. Lifschitz, N. McCain, T.C. Przymusinski, R.F. Stärk

Annotated Revision Specification Programs 143
M. Fitting

Update by Means of Inference Rules 156
T.C. Przymusinski, H. Turner

A Sphere World Semantics for Default Reasoning 175
J.C.P. da Silva, S.R.M. Veloso

Revision by Communication 189
C. Witteveen, W. van der Hoek

Hypothetical Updates, Priority and Inconsistency in a Logic
Programming Language 203
D. Gabbay, L. Giordano, A. Martelli, N. Olivetti

Situation Calculus Specifications for Event Calculus Logic Programs 217
R. Miller

On the Extension of Logic Programming with Negation through
Uniform Proofs 231
L.-Y. Yuan, J.-H. You

Default Consequence Relations as a Logical Framework for
Logic Programs 245
A. Bochman

Skeptical Rational Extensions 259
A. Mikitiuk, M. Truszczyński

Reasoning with Stratified Default Theories 273
P. Cholewiński

Incremental Methods for Optimizing Partial Instantiation 287
R.T. Ng, X. Tian

A Transformation of Propositional Prolog Programs into Classical Logic ..302
R.F. Stärk

Nonmonotonic Inheritance, Argumentation and Logic Programming 316
P.M. Dung, T.C. Son

An Abductive Framework for Extended Logic Programming 330
A. Brogi, E. Lamma, P. Mancarella, P. Mello

Embedding Circumscriptive Theories in General Disjunctive Programs ... 344
C. Sakama, K. Inoue

Stable Classes and Operator Pairs for Disjunctive Programs 358
J. Kalinski

Nonmonotonicity and Answer Set Inference 372
D. Pearce

Trans-Epistemic Semantics for Logic Programs 388
A. Rajasekar

Computing the Acceptability Semantics 401
F. Toni, A.C. Kakas

Author Index 417

Complexity Results for Abductive Logic Programming

Thomas Eiter[1]   Georg Gottlob[1]   Nicola Leone[2] *

[1] Christian Doppler Lab for Expert Systems, Information Systems Department,
TU Vienna, A-1040 Wien, Paniglgasse 16, Austria.
email: (eiter|gottlob)@dbai.tuwien.ac.at
[2] Istituto per la Sistemistica e l'Informatica – C.N.R., c/o DEIS – UNICAL,
87036 Rende, Italy. email: nik@si.deis.unical.it

Abstract. In this paper, we argue that logic programming semantics can be more meaningful for abductive reasoning than classical inference by providing examples from the area of knowledge representation and reasoning. The main part of the paper addresses the issue of the computational complexity of the principal decisional problems in abductive reasoning, which are: Given an instance of an abduction problem (i) does the problem have solution (i.e., an explanation); (ii) does a given hypothesis belong to some explanation; and (iii) does a given hypothesis belong to all explanations. These problems are investigated here for the stable model semantics of normal logic programs.

1   Introduction

During the last years, there has been increasing interest in abduction in different areas of computer science. Given the observation of some facts, abduction aims at concluding the presence of other facts, from which, together with an underlying theory, the observed facts can be explained, i.e., deductively derived; roughly speaking, abduction amounts to an inverse of modus ponens. It has been recognized that abduction is an important principle of common-sense reasoning, and that abduction has fruitful applications in a number of areas such diverse as model-based diagnosis, speech recognition, maintenance of database views, and vision, cf. [17, 8].

In the past, most research on abduction concerned abduction from classical logic theories. However, the use of logic programming to perform abductive reasoning can be more appropriate in several application domains. For instance, consider the following scenario. Assume that it is Saturday and is known that Joe goes fishing on Saturdays if it's not raining. This may be represented by the following theory T:

$$T = \{ \, go\_fishing \leftarrow is\_saturday \wedge \neg rains \; ; \qquad is\_saturday \leftarrow \; \}$$

---

* This author was partially supported by the Italian National Research Council under grant 224.07.4/24.07.12; the main part of his work has been carried out while he was visiting the Christian Doppler Lab for Expert System.

Now you observe that Joe is not out for fishing. Intuitively, from this observation we conclude that it rains (i.e, we abduce *rains*), for otherwise Joe would be out for fishing. Nevertheless, under classical inference, the fact *rains* is not an explanation of $\neg go\_fishing$, as $T \cup \{rains\} \not\models \neg go\_fishing$ (neither can one find any explanation). On the contrary, if we adopt the semantics of logic programming, then, according with the intuition, we obtain that *rains* is an explanation of $\neg go\_fishing$, as it is entailed by $T \cup \{rains\}$.

In the context of logic programming, abduction has been first investigated by Eshghi and Kowalski [10] and Kakas and Mancarella [18] and, during the recent years, common interest in the subject has been growing rapidly [4, 20, 18, 17, 7, 6, 31], also for the observation that, compared to deduction, this kind of reasoning has some advantages for dealing with incomplete information [6, 2].

The above logic program is stratifiable [1], and thus has a generally accepted clear semantics; different proposals for the semantics of general logic programs exist. In this paper, we define a general abduction model for logic programming, where the inference operator (i.e., the semantics to be applied on the program) is not fixed *a priori* but can be specified by the user. In our opinion, this endows our model with a great flexibility, as the appropriate semantics can be chosen for each application domain. For instance, abduction with brave inference seems more appropriate for diagnosis, while the cautious inference operator appears well-suited for planning (see Section 3).

Roughly speaking, we describe abduction from logic programs as follows: Given a logic program LP formalizing a particular application domain, a set M of literals describing some manifestations, a set H of atoms containing possible individual hypotheses, and an inference operator $\models$ defining the semantics of the program, find an *explanation* (or *solution*) for M, i.e., a suitable set $S \subseteq H$ such that $LP \cup S \models M$.

For instance, the above example (Joe's Saturday business) can be represented as an abductive problem where

$$LP = \{ \quad go\_fishing \leftarrow is\_saturday \wedge \mathbf{not}\, rains, \quad is\_saturday \leftarrow \quad \},$$

$M = \{\neg go\_fishing\}$, $H = \{rains\}$, and the inference operator is $\models_{wf}$ (entailment under well-founded semantics). Then, $S = \{rains\}$ is a solution.

For another example, imagine an electrical circuit consisting of a simple stove with two hot-plates wired in parallel and a control light, which is on if at least one of the plates is on operation. Each plate has a fuse, and it is known that one of them can not stand much current and will melt if the current gets high, but it is not known which one. Consider the following disjunctive program P:

$$
\begin{aligned}
melted\_fuse_1 \vee melted\_fuse_2 &\leftarrow high\_current \\
light\_off &\leftarrow melted\_fuse_1 \wedge melted\_fuse_2 \\
light\_off &\leftarrow power\_failure \\
light\_off &\leftarrow broken\_bulb \\
burns\_plate_1 &\leftarrow \mathbf{not}\, melted\_fuse_1 \wedge \mathbf{not}\, power\_failure \\
burns\_plate_2 &\leftarrow \mathbf{not}\, melted\_fuse_2 \wedge \mathbf{not}\, power\_failure
\end{aligned}
$$

The first rule states that a fuse will melt on high current. The second through fourth rule describe situations under which the control light is off, namely, if both fuses are melted, the power fails, or the bulb is broken. The last two rules state that a hot-plate burns if there is no power failure and the fuse is not melted. [3]

For example, given the observation *light_off*, under the stable model semantics we can abduce *power_failure* as an explanation under both brave and cautious inference, i.e., $S = \{power\_failure\}$ is a solution for both the abductive problems $\mathcal{P} = \langle H, M, P, \models_{st}^{b} \rangle$ and $\mathcal{P}' = \langle H, M, P, \models_{st}^{c} \rangle$ where $H = \{power\_failure,$ *broken_bulb, high_current*$\}$, $M = \{light\_off\}$, and $\models_{st}^{b}$ and $\models_{st}^{c}$ denotes the brave and cautious inference operator, respectively. Indeed, $P \cup \{power\_failure\}$ has the single stable model $M_1 = \{power\_failure, light\_off\}$. Also *broken_bulb* is an explanation of *light_off*, while *high_current* is not.

Given the additional observation $\neg burns\_plate_1$ (i.e., the manifestations $M' = \{light\_off, \neg burns\_plate_1\}$), we still have *power_failure* as an explanation, but no longer *bulb_broken*. Under brave inference, however, $S' = \{bulb\_broken,$ *high_current*$\}$ is an explanation (i.e., S' is a solution for $\langle H, M', P, \models_{st}^{b} \rangle$); this is reasonable in some sense. Notice that *power_failure* is not an explanation under classical inference, since $P \cup \{power\_failure\} \not\models \neg burns\_plate_1$.

Finally, assume the observations *light_off*, $burns\_plate_1$, and $burns\_plate_2$ are made. The only explanation is *bulb_broken* (under cautious as well as brave inference), which is intuitive. Under classical inference, no explanation exists.

Usually, different abductive solutions may be regarded not as equally appealing. Following Occam's principle of parsimony, one is often willing to prefer as simplest solutions as possible. In particular, solutions should be nonredundant, i.e., an acceptable solution must not contain another solution properly. The property of subset-minimality is the most commonly applied criterion for acceptable abductive solutions, cf. [27, 32, 20]. For example, for the observation *light_off* in the above example, $\{bulb\_broken, power\_failure\}$ would be a solution (under both variants of inference). The simpler solution *bulb_broken* (or *power_failure* as well) is clearly preferable.

In the context of abduction, three main decision problems arise: (1) to determine whether an explanation for the given manifestations *exists* at all; (2) to determine whether an individual hypothesis $h \in H$ is *relevant*, i.e., it is part of at least one acceptable explanation; (3) to determine whether an individual hypothesis is *necessary*, i.e., it occurs in all acceptable explanations.

For instance, in the circuit example we have: (i) there exists a solution for the manifestation *light_off*, (ii) the atoms *bulb_broken* and *power_failure* are relevant for some solution, (iii) none of them (as well as any other hypothesis) is necessary. The complexity of the problems (1)–(3) will be analyzed.

This paper presents some results of [9], whose contribution to abductive logic programming is twofold. First, a general abductive model for propositional logic programs is defined, in which also some previously proposed abductive framework can be easily modeled. Second, the computational complexity of abductive problems under several semantics is determined: (i) the well-founded semantics

---

[3] To keep the example simple, we refrain from modeling switches for the hot-plates.

for normal logic programs [33], (ii) the stable model semantics for normal logic programs [12], (iii) the stable model semantics for DLPs [28], and (iv) the minimal model semantics for DLPs [25]. For each semantics we analyze in [9] the complexity of relevance, necessity and existence of abductive solutions, by considering also the case where solutions must satisfy some minimality criterion. Here, we focus on the stable model semantics for normal logic programs.

As discussed in the full paper [9], the abduction model naturally extends to the case of first order logic programs without functions. Moreover, the data-complexity of abduction from logic programs with variables exactly coincides to the complexity of abduction from propositional logic programs.

2 Preliminaries and Notation

Logic Programming We assume familiarity with the basic concepts of logic programming [21].

Let V be a set of of propositional variables. Each $a \in V$ is a *positive literal* (*atom*), and $\neg a$ is a *negative literal*. A *program clause* (or *rule*) r is

$$a \leftarrow b_1 \wedge \cdots \wedge b_k \wedge \mathbf{not}\, b_{k+1} \wedge \cdots \wedge \mathbf{not}\, b_m, \qquad m \geq 0,$$

where $a, b_1, \cdots, b_m$ are atoms; a is called the *head* of r, while the conjunction $b_1 \wedge \cdots \wedge b_k \wedge \mathbf{not}\, b_{k+1} \wedge \cdots \wedge \mathbf{not}\, b_m$ is its *body*. $\mathcal{C}_V$ represents the set of all clauses. A (propositional) *logic program* is a finite subset LP of $\mathcal{C}_V$.

The well-founded semantics assigns to each logic program LP a unique partial (3-valued) interpretation (i.e., a consistent set of literals over V), $WF(LP)$, called the *well-founded model of LP* [33]. Accordingly, it infers that a literal Q is true in LP, denoted $LP \models_{wf} Q$, iff it is true in the well-founded model (i.e., $Q \in WF(LP)$). This extends to inference of a set of literals as usual.

Given a logic program LP and an interpretation I of LP, I is called a *stable model* of LP iff $I = lm(LP^I)$ (see [12] for details). The collection of all stable models of LP is denoted by STM(LP). The modalities of brave and cautious reasoning are used to handle multiple stable models of a logic program.

Brave reasoning (or *credulous reasoning*) infers that a literal Q is true in LP (denoted $LP \models_{st}^{b} Q$) iff Q is true w.r.t. M for some $M \in STM(LP)$.

Cautious reasoning (or *skeptical reasoning*) infers that a literal Q is true in LP (denoted $LP \models_{st}^{c} Q$) iff (*i*) Q is true w.r.t. M for all $M \in STM(LP)$ and (*ii*) $STM(LP) \neq \emptyset$. [4]

The inferences $\models_{st}^{b}$ and $\models_{st}^{c}$ extend to sets of literals as usual.

A *disjunctive logic program* (DLP; cf.[22]) is a logic program with rules

$$a_1 \vee \cdots \vee a_n \leftarrow b_1 \wedge \cdots \wedge b_k \wedge \mathbf{not}\, b_{k+1} \wedge \cdots \wedge \mathbf{not}\, b_m, \qquad n \geq 1, \; m \geq 0.$$

The set of all minimal models for LP, denoted by MM(LP), has been adopted in [25] as the intended meaning of a positive (i.e., **not**-free) DLP LP. For general

---

[4] Note that (*ii*), i.e., consistency of LP under stable semantics, is usually not requested in the definition of cautious consequence; see also next section.

DLPs, the disjunctive stable model semantics [28] is widely acknowledged (cf. [22] for other proposals). An interpretation I is called a *(disj.) stable model* of a DLP LP iff I is a minimal model of LP^I [28], where LP^I is the Gelfond-Lifschitz transform of LP. The set of all stable models of LP is denoted by STM(LP).

As for normal logic programs, $\models_{st}^b$ and $\models_{st}^c$ denote the inference operators of brave and cautious reasoning of stable semantics, respectively. Similarly, $\models_{mm}^b$ and $\models_{mm}^c$ denote the corresponding operators for minimal model semantics.

Complexity Theory For NP-completeness and complexity theory, cf. [26]. The classes Σ_k^P and Π_k^P of the polynomial hierarchy are defined as follows:

$$\Sigma_0^P = \Pi_0^P = P \quad \text{and for all } k > 1, \quad \Sigma_k^P = NP^{\Sigma_{k-1}^P}, \quad \Pi_k^P = co\text{-}\Sigma_k^P.$$

In particular, NP $= \Sigma_1^P$ and co-NP $= \Pi_1^P$. The class D_k^P, which is defined as the class of problems that consist of the conjunction of two (independent) problems from Σ_k^P and Π_k^P, respectively, is considered to be further restricted in computational power. Notice that for all $k \geq 1$, clearly $\Sigma_k^P \subseteq D_k^P \subseteq \Sigma_{k+1}^P$; both inclusions are widely conjectured to be strict.

3 Abduction Model

In this section, we describe our formal model for abduction from logic programs and state the main decisional reasoning tasks for abductive reasoning. The abduction model is defined here for propositional programs; the generalization to logic programs with variables is given in the full paper [9].

Definition 1. Let V be a set of propositional atoms. A logic programming abduction problem (*LPAP*) $\mathcal{P}$ over V consists of a tuple $\langle H, M, LP, \models \rangle$, where $H \subseteq V$ is a finite set of hypotheses, $M \subseteq V \cup \{\neg v \mid v \in V\}$ is a finite set of manifestations, $LP \subseteq \mathcal{C}_V$ is a logic program, and $\models$ is an inference operator. □

Throughout the paper, V will be implicitly assumed as the set $V_{\mathcal{P}}$ of propositional atoms appearing in $\mathcal{P}$.

The definition of *LPAP* is extended by allowing in $\langle H, M, LP, \models \rangle$ that LP be a disjunctive logic program and $\models$ be an inference operator for DLP; we refer to this extended notion as *disjunctive LPAP*. All concepts and notions defined for *LPAP*s are naturally extended to disjunctive *LPAP*s.

Definition 2. Let $\mathcal{P} = \langle H, M, LP, \models \rangle$ be a *LPAP*, and let $S \subseteq H$. Then, S is a solution (or explanation) to $\mathcal{P}$ iff $LP \cup S \models M$. *Sol*($\mathcal{P}$) denotes the set of all solutions to $\mathcal{P}$. □

In this paper we consider abductive problems for the inference operators $\models_{wf}$, $\models_{st}^b$, and $\models_{st}^c$. Note that, in the definition of cautious abduction, we have not used the typical stable cautious operator $\models_{st}^{c1}$ (a literal is true if it is true in all stable models), but the inference operator $\models_{st}^c$ which requires additionally that the program has some stable model. The reason for that is the need to guarantee that $LP \cup S$ is "consistent".

Example 1. Consider the *LPAP* $\mathcal{P} = \langle\{d,b\},\{c\}, LP, \models_{st}^{c1}\rangle$, where $LP = \{a \leftarrow b \wedge \text{not } a , c \leftarrow d \}$. One reasonably expects that the observation of c abduces d. However, $LP \cup \{b\}$ has no stable model; hence, $LP \cup \{b\} \models_{st}^{c1} c$ and thus $\{b\}$ is a solution to $\mathcal{P}$. This is not the case if the $\models_{st}^{c}$ is used. Indeed, $\{b\}$ is not a solution to $\mathcal{P}' = \langle\{d,b\},\{c\}, LP, \models_{st}^{c}\rangle$, as $\models_{st}^{c}$ requires the consistency of the program.

The explicit introduction of the inference operator in the definition of abduction problem endows our abduction model with a great flexibility, since the appropriate semantics can be adopted on each application domain. This is particularly important in logic programming, as no semantics is generally accepted as the best one for *all* logic programs; rather, it is acknowledged that the right semantics must be chosen on the particular domain.

We briefly comment on the applicability on different domains of brave and cautious reasoning, respectively. We argue that brave inference is well-suited for diagnosis, while cautious inference is adequate for planning. In fact, consider a system represented by a logic program LP with stable model semantics. If the system is solicited by some input, represented by adding a set A of atoms to LP, then each stable model of $LP \cup A$ is a possible evolution of the system, i.e., each stable model represents a possible reaction of the system to A.

Diagnosis consists, loosely speaking, in deriving from an observed system state (characterized by the truth of a set F of facts), a suitable input A which caused this evolution. Now, since *each* stable model of $LP \cup A$ is a possible evolution of the system with input A, we can assert that A is a possible input that caused F if $LP \cup A \models_{st}^{b} F$. Thus, diagnostic problems can be naturally represented by abductive problems with brave inferences.

Suppose now that we want that the system is in a certain state (described by a set F of facts), and we have to determine the "right" input that enforces this state of the system (planning). In this case it is not sufficient to choose an input A such that F is true in some possible evolution of the system; rather, we look for an input A such that F is true in *all* possible evolutions, as we want to be sure that the system reacts in that particular way. In other words, we look for A such that $LP \cup A \models_{st}^{c} F$. Hence, planning activities can be represented by abductive problems with cautious inferences.

The following properties of a hypothesis in a *LPAP* $\mathcal{P}$ are of natural interest with respect to computing abductive solutions.

Definition 3. Let $\mathcal{P} = \langle H, M, LP, \models \rangle$ be a *LPAP* and $h \in H$. Then, h is relevant for $\mathcal{P}$ iff $h \in S$ for some $S \in Sol(\mathcal{P})$, and h is necessary for $\mathcal{P}$ iff $h \in S$ for every $S \in Sol(\mathcal{P})$.

We refer to the property opposite to necessity also as dispensability (cf. [16]). Now, the main decisional problems in abductive reasoning are as follows: Given a *LPAP* $\mathcal{P} = \langle H, M, LP, \models \rangle$,

1. does there exist a solution for $\mathcal{P}$? (*Consistency*)
2. is a given hypothesis $h \in H$ relevant for $\mathcal{P}$, i.e., does h contribute to some solution of $\mathcal{P}$? (*Relevance*)

3. is a given hypothesis $h \in H$ necessary for $\mathcal{P}$, i.e., is h contained in all solutions of $\mathcal{P}$? (*Necessity*)

It is common in abductive reasoning to prune the set of all solutions and to focus, guided by some principle of solution preference, on a set of preferred solutions. The most important preference is redundancy-free solutions, i.e., solutions which do not contain any other solution properly (cf. [27, 32, 20]). We refer to such solutions of a *LPAP* $\mathcal{P}$ as $\subseteq$-solutions and denote their collection by $Sol_\subseteq(\mathcal{P})$. [5] Relevance and necessity of a hypothesis for a *LPAP* $\mathcal{P}$ with respect to $Sol_\subseteq(\mathcal{P})$ rather than $Sol(\mathcal{P})$, are referred to as $\subseteq$-relevance and $\subseteq$-necessity.

It easy to see that h is $\subseteq$-necessary for a *LPAP* $\mathcal{P}$ iff h is necessary for $\mathcal{P}$. Thus, we do not deal explicitly with $\subseteq$-Necessity in our complexity analysis.

4 Complexity Results

we give here detailed proof sketches only of the complexity results of abduction under stable cautious semantics, which present the highest complexities and appear to be the most interesting. The other complexity results will be reported and discussed in Section 6; the detailed proofs of these results can be found in the full paper [9].

We fix some helpful notation. For a Boolean formula E in CNF (resp. DNF), we denote by $\overline{E}$ the DNF (resp. CNF) expression obtained by applying De Morgan's rule to $\neg E$ and cancelling all double negations. For instance, given $E = (x_1 \vee \neg x_2) \wedge x_3$, $\overline{E} = (\neg x_1 \wedge x_2) \vee \neg x_3$. Furthermore, given a Boolean expression E in CNF, we define $transf(E)$ as follows:

$$transf(E) = \{ abs \leftarrow b_1 \wedge \cdots \wedge b_k \wedge \mathbf{not}\, a_1 \wedge \cdots \wedge \mathbf{not}\, a_l$$
$$\mid a_1 \vee \cdots \vee a_l \vee \neg b_1 \vee \cdots \vee \neg b_k \in E \ (k, l \geq 0) \}$$

For a Boolean expression E in DNF, define $transf(E) = transf(\overline{E})$. Moreover, given a set Y of propositional variables, denote by Y' the set $\{y' \mid y \in Y\}$.

Given a set $Z = \{z_1, \ldots, z_n\}$ of propositional atoms we define $clos(Z)$ to be the following set of rules:

$$clos(Z) = \{z_i \leftarrow \mathbf{not}\, z_i' \ , \ z_i' \leftarrow \mathbf{not}\, z_i \mid i = 1, \ldots, n\}.$$

Let E be a Boolean CNF expression whose set of atoms is denoted by $\mathrm{Var}(E)$ and let $Z \subseteq \mathrm{Var}(Z)$. Consider $LP(E, Z) = transf(E) \cup clos(Z)$. Note that $LP(E, Z)$ is not stratified when $Z \neq \emptyset$.

Intuitively, z_i' represents the complement of z_i: z_i is true in a stable model M of LP iff z_i' is false in M, and conversely. The rules in $clos(Z)$ "give support" to the atoms in Z. For instance, given the expression $E = z_1 \vee z_2 \vee \neg z_3$, the logic program $LP(E, Z) = transf(\{z_1 \vee z_2 \vee \neg z_3\}) \cup clos(\{z_1, z_2, z_3\})$ is:

$$abs \leftarrow z_3 \wedge \mathbf{not}\, z_1 \wedge \mathbf{not}\, z_2, \quad z_i \leftarrow \mathbf{not}\, z_i', \quad z_i' \leftarrow \mathbf{not}\, z_i, \quad (1 \leq i \leq 3)$$

---

[5] In the full version of the paper we deal also with preference of solutions of smallest cardinality, cf. [27].

It is easy to see that there is a one-to-one correspondence between the interpretations of E and the stable models of LP, where the models of E correspond to the stable models of LP in which *abs* is false. Such a correspondence can be profitably used to check the validity of certain Quantified Boolean Formulas (QBFs). [6] For instance, the formula $\exists Z E$ is valid iff $LP \models_{st}^b \neg abs$. The formula $\forall Z E$ is valid iff $LP(E, Z) \models_{st}^c \neg abs$.

Let us first consider solution verification for *LPAP*s that use inference $\models_{st}^c$. In the light of the results from [29], it is unlikely that this problem polynomial.

Proposition 4. *Let* $\mathcal{P} = \langle H, M, LP, \models_{st}^c \rangle$ *be a LPAP. Deciding if* $S \subseteq H$ *fulfills* $S \in Sol(\mathcal{P})$ *is* D_1^P*-complete.*

Proof. D_1^P*-Membership.* The two problems (*p1*) Does $LP \cup S$ have a stable model? and (*p2*) Does each stable model N of $LP \cup S$ contain M? are by [3, 23] and [24, 30] in NP and co-NP, respectively. Thus, deciding $S \in Sol(\mathcal{P})$ can be reduced to deciding the logical conjunction of *p1* and *p2*, which implies that the problem lies in D_1^P.
D_1^P*-Hardness.* As shown in [29], given a propositional logic program LP and atom q, deciding whether $LP \models_{st}^c q$ is D_1^P-hard. This trivially reduces to deciding if the empty set is a solution of the *LPAP* $\mathcal{P} = \langle \emptyset, \{q\}, LP, \models_{st}^c \rangle$. □

Next, we consider consistency checking for *LPAP*s.

Theorem 5. *To decide if* $Sol(\mathcal{P}) \neq \emptyset$ *for a given LPAP* $\mathcal{P} = \langle H, M, LP, \models_{st}^c \rangle$ *is* Σ_2^P*-complete.*

Proof. (Sketch) Σ_2^P*-Membership.* We guess $S \subseteq H$. Then, by virtue of Proposition 4, we can verify that $S \in Sol(\mathcal{P})$ by two calls to an oracle in NP.
Σ_2^P*-Hardness.* We give a transformation from deciding whether a QBF $\Phi = \exists X \forall Y E$, $X = \{x_1, \ldots, x_n\}$, $Y = \{y_1, \ldots, y_m\}$ is valid. We assume wlog that E is in DNF with at most 3 literals in each disjunct.

Consider $\mathcal{P} = \langle X, \{abs\}, LP, \models_{st}^c \rangle$, where $LP = transf(\overline{E}) \cup clos(Y)$. For each $X1 \subseteq X$, there is a one-to-one correspondence between the stable models of $LP \cup X1$ and the possible truth assignments for $y_1, \cdots, y_m$, where the truth value of each x_i in the stable models is determined by membership in $X1$. A truth assignment to Y satisfies E iff *abs* is true in the corresponding stable model.

It holds that Φ is valid iff there exists $X1 \in Sol(\mathcal{P})$, i.e., $Sol(\mathcal{P}) \neq \emptyset$ iff Φ is valid. Thus, deciding $Sol(\mathcal{P}) \neq \emptyset$ for $\mathcal{P} = \langle H, M, LP, \models_{st}^c \rangle$ is Σ_2^P-hard. □

For deciding relevance, we obtain the same complexity result.

Theorem 6. *Deciding if a given hypothesis is relevant for a LPAP* $\mathcal{P} = \langle H, M, LP, \models_{st}^c \rangle$ *is* Σ_2^P*-complete.*

---

[6] A QBF is a formula $Q_1 X_1 Q_2 X_2 \cdots Q_k X_k E$, $k \geq 1$, where E is a Boolean expression whose atoms are from pairwise disjoint nonempty sets of variables $X_1, \ldots, X_k$, and the Q_i's are alternating quantifiers from $\{\exists, \forall\}$, for all $i = 1, \ldots, k$.

Proof. (Sketch) Σ_2^P-*Membership*. To prove that $h \in H$ is relevant, we can guess S, and verify that: $S \in Sol(\mathcal{P})$ and $h \in S$. By Proposition 4, verifying a guess for $S \in Sol(\mathcal{P})$ can be done by two calls to an oracle in NP.

Σ_2^P-*Hardness* We describe a transformation from deciding $Sol(\mathcal{P}) \neq \emptyset$.

Given a *LPAP* $\mathcal{P} = \langle H, M, LP, \models_{st}^c \rangle$ consider the *LPAP* $\mathcal{P}' = \langle H \cup \{q\}, M \cup \{q\}, LP \rangle, \models_{st}^c \rangle$, where q is a new variable. Since q does not appear in LP, every solution S' of $\mathcal{P}'$ must contain q. Clearly, for each $S \subseteq H$, $S \in Sol(\mathcal{P})$ iff $S \cup \{q\} \in Sol(\mathcal{P}')$. From Theorem 5, deciding $Sol(\mathcal{P}) \neq \emptyset$ is Σ_2^P-hard; hence, deciding if a given hypothesis is relevant for a *LPAP* $\mathcal{P}$ is Σ_2^P-hard as well. $\square$

Next, we consider the problem of deciding necessity.

Theorem 7. *Deciding if a given hypothesis is necessary for a given LPAP* $\mathcal{P} = \langle H, M, LP, \models_{st}^c \rangle$ *is* Π_2^P-*complete.*

Proof. (Sketch) Π_2^P-*Membership*. The proof that the complementary problem, deciding if the hypothesis is dispensable, is in Σ_2^P, is similar to the one for Σ_2^P-membership of Theorem 6.

Π_2^P-*Hardness*. Transformation of deciding whether a *LPAP* has no cautious solution. Let $\mathcal{P} = \langle X, \{abs\}, LP, \models_{st}^c \rangle$ be as in the proof of Theorem 5. Define $\mathcal{P}' = \langle X \cup \{w\}, \{abs\}, LP', \models_{st}^c \rangle$, where w is a new propositional variable and $LP' = LP \cup \{abs \leftarrow x_1 \wedge \cdots \wedge x_n \wedge w \}$. Notice that $X \cup \{w\}$ is a solution to $\mathcal{P}'$. For each $X1 \subseteq X$, X_1 is solution of $\mathcal{P}'$ iff it is a solution of $\mathcal{P}$. Hence, w is necessary for $\mathcal{P}'$ iff $Sol(\mathcal{P}) = \emptyset$. Hence, the result follows. $\square$

If we restrict solutions to those minimal under inclusion, we face an increase in complexity for Relevance. In fact, we encounter a problem complete for a class of the third level of the polynomial hierarchy,

Theorem 8. *Deciding if a given hypothesis is* $\subseteq$-*relevant for a LPAP* $\mathcal{P} = \langle H, M, LP, \models_{st}^c \rangle$ *is* Σ_3^P-*complete.*

Proof. (Sketch) Σ_3^P-*Membership*. Let $\mathcal{P} = \langle H, M, LP, \models_{st}^c \rangle$ be a *LPAP*. Given $E \in Sol(\mathcal{P})$, deciding if E is $\subseteq$ minimal for $\mathcal{P}$, i.e., deciding if $E \in Sol_\subseteq(\mathcal{P})$, is in Π_2^P. Indeed, the complementary problem, $E \notin Sol(\mathcal{P})$, is decidable as follows. Guess $E' \subseteq E$ and check that $E' \in Sol(\mathcal{P})$; by Proposition 4 this is done by single calls to an NP and a co-NP oracle. Thus the problem is in Σ_2^P. As a consequence, deciding if $E \in Sol_\subseteq(\mathcal{P})$ is in Π_2^P. Now, to decide if $h \in H$ is $\subseteq$-relevant for $\mathcal{P}$, we can proceed as follows. Guess $E \subseteq H$, verify by a single call to a Π_2^P oracle that $E \in Sol_\subseteq(\mathcal{P})$, and check in polynomial time that $h \in E$.

Σ_3^P-*Hardness*. Transformation from deciding if a QBF $\Phi = \exists X \forall Y \exists Z E$ is valid. We assume wlog that E is in CNF and has at most 3 literals per clause.

Let $X = \{x_1, \ldots, x_{n_x}\}$, $Y = \{y_1, \ldots, y_{n_y}\}$, and $Z = \{z_1, \ldots, z_{n_z}\}$. Take atoms $X' = \{x'_1, \ldots, x'_{n_x}\}$, $Q = \{q_1, \ldots, q_{n_x}\}$ and w, where the q_i and w are new variables, and define a logic program LP as follows:

$$LP = transf(E) \cup \{abs \leftarrow y_1 \wedge \cdots \wedge y_{n_y} \wedge w \} \cup$$
$$\{abs \leftarrow x_i \wedge x'_i \,, q_i \leftarrow x_i \,, q_i \leftarrow x'_i \mid i = 1 \ldots, n_x\} \cup clos(Z).$$

Consider the $LPAP$ $\mathcal{P} = \langle H, Q \cup \{abs\}, LP, \models_{st}^{c} \rangle$, where $H = X \cup X' \cup Y \cup \{w\}$. It holds that Φ is valid iff w is $\subseteq$-relevant for $\mathcal{P}$. Moreover, $\mathcal{P}$ is clearly constructible in polynomial time from Φ. Thus, the result follows. □

5   Related Work

In this section we briefly overview some works on abductive logic programming.

The best known definition of abductive logic programming is due to Kakas and Mancarella. In [18], they define abduction from (normal) logic programs under the brave modality of the stable model semantics, allowing that integrity constraints (given by propositional clauses) can be imposed on stable models. In our framework, their proposal can be represented by $LPAP$s of the form $\langle H, M, LP, \models_{st}^{b} \rangle$, where LP is a logic program. Note that integrity constraints can be easily simulated by proper rules and manifestations[7] (of course, we could extend our framework with integrity constraints). The various abductive problems for [18] have exactly the computational complexity of the $LPAP$s $\langle H, M, LP, \models_{st}^{b} \rangle$ (cf. Tables 1–2).

Abductive problems $\langle H, M, LP, \models \rangle$ where LP is a hierarchical (i.e., non recursive) program and $\models$ is the entailment operator based on Clark's program completion are studied in [4]. The relation of abduction to negation as failure has been investigated in [10, 13, 15]. Under some syntactic conditions, a one-to-one correspondence between the solutions of an abductive problem and the answer sets of an extended logic program obtained from it has been proven. In [2], a notion of *entailment* for abductive logic programs is defined. In some sense that approach is similar to $LPAP$s of the form $\langle H, M, LP, \models_{st}^{c} \rangle$, but it imposes stronger conditions. Critical overviews of several approaches to abductive logic programming have been given in [17, 14].

To our knowledge, not much work has been done on the side of complexity of abductive logic programming. The effective computation of abductive explanations has been addressed in several works [7, 5, 10, 15, 19], where suitable extensions of the $SLDNF$ procedure have been designed for abduction (these procedures return an abductive explanation of the given query).

In [31] the complexity of the existence of solutions for abduction from general and disjcuntive logic programs for the $\models_{st}^{b}$ (stable brave) is derived. The authors also analyze abduction under *possible model* semantics. A detailed analysis of the complexity of abduction from classical propositional theories is presented in [8]. A subclass of Horn theories for which finding a $\subseteq$-minimal solution is tractable is identified in [11]. Note that the results of [8] do not apply to the abductive framework described in this paper. In particular, even in the case of Horn programs, where the semantics of classical logic and logic programming coincide, the abductive framework of [8] is less expressive, and the corresponding reasoning tasks have different complexity. In the former framework, Necessity is polyno-

---

[7] E.g., a constraint $\neg(a_1 \wedge \ldots \wedge a_n)$ can be simulated by a rule $abs \leftarrow a_1 \wedge \ldots \wedge a_n$ and adding $\neg abs$ to the manifestations, where abs is a new propositional atom.

mial, while in the latter, this problem is co-NP-complete. This stems from the fact that the framework in [8] excludes negative literals from the manifestations.

Finally, the use of abductive logic programming for dealing with incomplete information has been discussed in [6, 2].

6 Overview of Results

Tables 1–4 summarize the complexities of the main decisional problems (consistency, relevance of a hypothesis, and necessity of a hypothesis) that emerge in abductive logic programming. We have considered the most acknowledged proposals for the semantics of logic programs, namely the well-founded semantics and the stable model semantics (in the usual variants of brave and cautious inference), and the stable as well as the minimal model semantics for disjunctive programs. Moreover, we have taken into account the commonly applied criteria for reducing the space of all solutions by means of a preference ($\preceq$) to those which are minimal with respect to inclusion ($\subseteq$). [8]

The complexity results cover the lower end of the polynomial hierarchy up to Σ_4^P. Each entry represents completeness for the respective class; in particular, P means logspace-completeness for P. (Notice that all polynomial time transformations in this paper are computable in logarithmic space; in fact, all results can be proven under logspace-reductions as well). The results reflect the conceptual complexity of combining the different common semantics and solution preference criteria. They show that abduction from logic programming, a topic of increasing interest, provides a rich variety of complete problems for a number of slots within the polynomial hierarchy.

Briefly, the following conclusions can be drawn. Abduction from logic programs has the same complexity for well-founded inference ($\models_{wf}$) and the brave variant of stable inference ($\models_{st}^b$), excepting the verification of a given solution, which is polynomial under the former and NP-complete under the latter. Notice that the hardness results for well-founded inference hold for stratified programs and, even more restrictive, for negation-free programs (i.e., Horn programs) as well. Since the semantics of Horn programs is clear, this shows that, loosely speaking, the well-founded semantics does not increase the inherent complexity of abduction from logic programs, which makes it attractive.

For the appropriate cautious variant of stable inference (cautious inference $\models_{st}^c$), the complexity is in each cased precisely one level higher in the polynomial hierarchy than for $\models_{st}^b$, again excepting solution verification, which has marginally higher complexity; this is noticeable since usually, the brave and the cautious variant of a reasoning task in a nonmonotonic formalism are of dual complexity (typically Σ_2^P and Π_2^P, respectively). The effect of the solution preference criteria on the complexity meets the intuitive expectations. Imposing

---

[8] Moreover, solutions of minimal cardinality (cf. [27]) are treated in the full paper, where completeness results for the complexity classes $\Delta_2^P[O(\log n)]$ and $\Delta_3^P[O(\log n)]$ from the refined polynomial hierarchy (see [26]) are established.

Table 1. Complexity of solution verification and consistency checking

LPAP $\mathcal{P} = \langle H, M, LP, \models \rangle$	Deciding $S \in SOL(\mathcal{P})$	Deciding $Sol(\mathcal{P}) \neq \emptyset$
well-founded inference ($\models_{wf}$)	P	NP
stable brave inference ($\models_{st}^{b}$)	NP	NP
stable cautious inference ($\models_{st}^{c}$)	D_1^P	Σ_2^P

Table 2. Complexity of deciding relevance and necessity

Logic Programming Abduction	$\preceq$-Relevance of $h \in H$		$\preceq$-Necessity of $h \in H$	
LPAP $\mathcal{P} = \langle H, M, LP, \models \rangle$	$= (no\ pref.)$	$\subseteq$	$= (no\ pref.)$	$\subseteq$
well-founded inf. ($\models_{wf}$)	NP	Σ_2^P	co-NP	co-NP
stable brave inf. ($\models_{st}^{b}$)	NP	Σ_2^P	co-NP	co-NP
stable cautious inf. ($\models_{st}^{c}$)	Σ_2^P	Σ_3^P	Π_2^P	Π_2^P

Table 3. Complexity of solution verification and consistency for disjunctive LPAPs

Disj. LPAP $\mathcal{P} = \langle H, M, LP, \models \rangle$	Deciding $S \in SOL(\mathcal{P})$	Deciding $Sol(\mathcal{P}) \neq \emptyset$
stable brave inference ($\models_{st}^{b}$)	Σ_2^P	Σ_2^P
stable cautious inference ($\models_{st}^{c}$)	D_2^P	Σ_3^P

Table 4. Complexity of deciding relevance and necessity for disjunctive LPAPs

Disj. Logic Programming Abduction	$\preceq$-Relevance of $h \in H$		$\preceq$-Necessity of $h \in H$	
Disj. LPAP $\mathcal{P} = \langle H, M, LP, \models \rangle$	$= (no\ pref.)$	$\subseteq$	$= (no\ pref.)$	$\subseteq$
stable brave inference ($\models_{st}^{b}$)	Σ_2^P	Σ_3^P	Π_2^P	Π_2^P
stable cautious inference ($\models_{st}^{c}$)	Σ_3^P	Σ_4^P	Π_3^P	Π_3^P

subset-minimality ($\subseteq$) leads in general to an increase by a full level in the polynomial hierarchy. It is worth mentioning that for abduction using classical semantics, imposing $\subseteq$ does not lead to a complexity increase, cf. [8].

The main complexity results for abduction from propositional disjunctive logic programs (DLPs) are summarized in Tables 3 and 4. We have that the complexity of a disjunctive *LPAP* under stable model semantics is exactly one level higher in the polynomial hierarchy than in the case of the corresponding (normal) *LPAP*. This is intuitively explained as follows. The complexity of checking whether an interpretation of a logic program is a stable model grows from polynomial time to co-NP-completeness when disjunction is allowed. This complexity increase causes the complexity of verifying an abductive solution under stable brave and stable cautious inference operators to grow from NP and D_1^P to Σ_2^P and D_2^P, respectively. Consequently, the complexity of all other abductive problems increases by one level in the polynomial hierarchy.

As is shown in the full paper, all results for disjunctive logic programming abduction remain valid for the restriction to positive (i.e., **not**-free) disjunctive logic programs. Since the stable semantics coincides with the minimal model semantics for such programs, we thus obtain for the minimal model inference operators $\models_{mm}^b$, and $\models_{mm}^c$ the same complexities as for $\models_{st}^b$ and $\models_{st}^c$, respectively. This is also intuitive, since checking whether an interpretation is a stable model of a DLP reduces efficiently to deciding whether it is a minimal model of a **not**-free DLP, which is co-NP-complete.

The results for the case of propositional programs carry over to the data complexity in the case of function-free first order logic programs (cf. [9]). The combined complexity parallels the data complexity with the corresponding exponential analogues (EXP for P, NEXP for NP etc).

References

1. K. Apt, H. Blair, and A. Walker. Towards a Theory of Declarative Knowledge. In J. Minker, editor, *Foundations of Deductive Databases and Logic Programming*, pages 89–148. Morgan Kaufman, Washington DC, 1988.
2. C. Baral and M. Gelfond. Logic Programming and Knowledge Representation *Journal of Logic Programming*, 19-20:72-144, 1994.
3. N. Bidoit and C. Froidevaux. General Logic Databases and Programs: Default Semantics and Stratification. *Information and Computation*, 19:15–54, 1991.
4. L. Console, D. Theseider Dupré, and P. Torasso. On the Relationship Between Abduction and Deduction. *Journal of Logic and Computation*, 1(5):661–690, 1991.
5. M. Denecker and D. De Schreye. SLDNFA: an abductive procedure for normal abductive logic programs. In *Proc. JICSLP-92*, pp. 686–700, 1992.
6. M. Denecker and D. De Schreye. Representing incomplete knowledge in abductive logic programming. In *Proc. ILPS-93*, pp. 147–163, 1993.
7. P. Dung. Negation as Hypotheses: An Abductive Foundation for Logic Programming. In *Proc. ICLP-91*. MIT Press, 1991.
8. T. Eiter and G. Gottlob. The Complexity of Logic-Based Abduction. *JACM*, 1995. Extended abstract Proc. *STACS-93*, LNCS 665, 70–79.

9. T. Eiter, G. Gottlob and N. Leone. Abduction from Logic Programs: Semantics and Complexity. *TR 94-72 Christian Doppler Lab for Expert Systems*, 1994.

10. K. Eshghi and R. Kowalski. Abduction Compared with Negation as Failure. In *Proc. ICLP-89*, pp. 234–254. MIT Press, 1989.

11. K. Eshghi. A Tractable Class of Abduction Problems. In *Proc. IJCAI-93*, pp. 3–8.

12. M. Gelfond and V. Lifschitz. The Stable Model Semantics for Logic Programming. In *Proc. Fifth Conference and Symposium on Logic Programming*, B. Kowalski and R.Bowen (eds), pp. 1070–1080, Cambridge Mass., 1988. MIT Press.

13. K. Inoue. Extended logic programs with default assumptions. In *Proc. ICLP-91*, pp. 490–504.

14. K. Inoue. Studies on Abductive and Nonmonotonic Reasoning. PhD thesis, Kyoto University, 1992.

15. N. Iwayama and K. Satoh. Computing abduction using the TMS. In *Proc. ICLP-91*, pp. 505–518, 1991.

16. J. Josephson, B. Chandrasekaran, J. J. W. Smith, and M. Tanner. A Mechanism for Forming Composite Explanatory Hypotheses. *IEEE TSMC-17*:445–454, 1987.

17. A. Kakas and R. Kowalski and F. Toni. Abductive logic programming. *Journal of Logic and Computation*, 2(6):719–771, 1992.

18. A. Kakas and P. Mancarella. Generalized Stable Models: a Semantics for Abduction. In *Proc. ECAI-90*, pp. 385–391, 1990.

19. A. Kakas and P. Mancarella. Stable theories for logic programs. In *Proc. ISLP-91*, pp. 88–100, 1991.

20. K. Konolige. Abduction versus closure in causal theories. *Artificial Intelligence*, 53:255–272, 1992.

21. J. LLoyd. *Foundations of Logic Programming*. Springer, Berlin, 1984.

22. J. Lobo, J. Minker, and A. Rajasekar. *Foundations of Disjunctive Logic Programming*. MIT Press, 1992.

23. W. Marek and M. Truszczyński. Autoepistemic Logic. *JACM*, 38:588–619, 1991.

24. W. Marek and M. Truszczyński. Computing Intersection of Autoepistemic Expansions. In A. Nerode, W. Marek, and V. Subrahmanian, editors, *Proc. LPNMR-91*, pp. 37–50, Washington DC, July 1991. MIT Press.

25. J. Minker. On Indefinite Data Bases and the Closed World Assumption. In *Proc. of the 6th Conference on Automated Deduction (CADE-82)*, pp. 292–308, 1982.

26. C. H. Papadimitriou. *Computational Complexity*. Addison-Wesley, 1994.

27. Y. Peng and J. Reggia. *Abductive Inference Models for Diagnostic Problem Solving*. Springer, 1990.

28. T. Przymusinski. Stable Semantics for Disjunctive Programs. *New Generation Computing*, 9:401–424, 1991.

29. D. Saccà. The Expressive Power of Stable Models for DATALOG Queries with Negation. H. Blair, W. Marek, A. Nerode, and J. Remmel (eds), *Proc. of the Second Workshop on Structural Complexity and Recursion-Theoretic Methods in Logic Programming*, pp. 150–162. Cornell University, Math. Sciences Institute, 1993.

30. J. Schlipf. The Expressive Powers of Logic Programming Semantics. *Proc. ACM PODS-90*, pp. 196–204. Extended version to appear in *JCSS*.

31. C. Sakama and K. Inoue. On the Equivalence between Disjunctive and Abductive Logic Programs. In *Proc. ICLP-94*, pp. 88–100, 1994.

32. B. Selman and H. J. Levesque. Abductive and Default Reasoning: A Computational Core. In *Proc. AAAI-90*, pp. 343–348, July 1990.

33. A. Van Gelder, K. Ross, and J. Schlipf. The well-founded semantics for general logic programs. *Journal of ACM*, 38(3):620–650, 1991.

A Terminological Interpretation of (Abductive) Logic Programming

Marc Denecker*

Department of Computing, K.U.Leuven,
Celestijnenlaan 200A, B-3001 Heverlee, Belgium.
e-mail : marcd@cs.kuleuven.ac.be

Abstract. The logic program formalism is commonly viewed as a modal or default logic. In this paper, we propose an alternative interpretation of the formalism as a terminological logic. A terminological logic is designed to represent two different forms of knowledge. A TBox represents definitions for a set of concepts. An ABox represents the *assertional knowledge* of the expert. In our interpretation, a logic program is a TBox providing definitions for all predicates; this interpretation is present already in Clark's completion semantics. We extend the logic program formalism such that some predicates can be left undefined and use classical logic as the language for the ABox. The resulting logic can be seen as an alternative interpretation of abductive logic program formalism. We study the expressivity of the formalism for representing uncertainty by proposing solutions for problems in temporal reasoning, with null values and open domain knowledge.

1 Introduction

The logic program formalism is commonly viewed as a modal, autoepistemic logic or, closely related, as a default logic. In these different interpretations the connective *not* is interpreted as a modal connective (*not p* means *it not believed that p*) or as a default connective (*not p* means *it can be assumed by default ¬p*). In this paper, we investigate a complementary view on logic programming which relates this formalism to another important class of logics in AI and knowledge representation, the terminological logics. At the origin of the terminological languages lies the observation that an expert's knowledge consists of a *terminological component* and an *assertional component* [1]. The terminological component, described in the TBox, consists of the definitions of the *technical vocabulary* of the expert. The role of the assertional component, described in the ABox is intimately tied to the representation of uncertainty: when a number of concepts cannot be defined, then other, less precise information may be available which can be represented as a set of *assertions*.

The view that we investigate in this paper is that a logic program is a TBox: it gives definitions to the predicates. The interpretation of a logic program as a

---

* supported by the research fund of the K.U.Leuven

formalism for representing definitions is already present in Clark's work on completion semantics [2]. Under Clark's interpretation, a program consists of a set of possibly empty definitions for all predicates; predicates are defined by enumerating exhaustively the cases in which they are true. Under this interpretation, a problem with the use of logic programs for knowledge representation, is that an expert needs to provide definitions for all predicates. In many applications, such complete information is not, available. A natural solution is to extend the logic program formalism such that a subset of the predicates can be defined while other predicates can be left undefined. One straightforward way to represent such an *open logic program*, as called in the sequel, is as a tuple (P, D), or P^D, which consists of a set P of program clauses and a set D of predicates which are defined in P; D contains all predicates which occur in the head of a program clause of P and possibly some other predicates with an empty definition. In the sequel, predicates in D are called *defined predicates*; other predicates are called *open predicates*. Note that in this formalism, one can distinguish predicates p which have an empty definition $(p \in D)$ and predicates which have no definition $(p \notin D)$. Just as in terminological logics, there is need for an ABox to represent assertional knowledge on open predicates; one can use classical first-order logic (FOL) for this. We thus obtain a new type of theory, called OLP-FOL theory or open theory, which consists of two components: an open logic program P^D and a set of FOL axioms T.

The interpretation of a logic program under the modal and default view on one hand and the terminological view on the other hand is fundamentally different. This appears very clearly in the following example. Consider the following logic program:

$$P = \{ \ dead(X) :\text{-} \ not \ alive(X) \ \}$$

Under the auto-epistemic interpretation, the program clause reads as: *if it is not believed that X is alive then X is dead*. P does not contain any information about the aliveness of, say John. Therefore, it is not believed that John is alive, and therefore, it may be derived that he is dead.

Under the terminological view, the set P can be embedded in two open logic programs $P^{\{dead, alive\}}$ and $P^{\{dead\}}$. Both open programs contain the same definition of *dead* which reads as: *X is dead iff X is not alive*. The first open logic program provides full information on alive: nobody is alive; hence, everybody is dead. The second open logic program contains no information on who is alive. Hence, this program does not allow to determine whether or not John is dead. Note that in $P^{\{dead\}}$ there is no evidence that John is alive; equivalently, it is not believed that John is alive. Yet the conclusion that he is dead cannot be drawn from $P^{\{dead\}}$. This shows that *not p* in OLP-FOL cannot be interpreted as *it is not believed that p* or *there is no evidence that p*. The *not* connective in OLP-FOL has the same *objective* modality as classical negation ¬ in FOL. This has important consequences for the negation as (finite) failure principle. Given an open logic program which defines all predicates, negation as (finite) failure acts as a solid inference rule to derive objective negative information. However, the inference rule cannot be applied for open logic programs. For example, the goal

$\leftarrow alive(john)$ finitely fails in $P^{\{dead\}}$; yet, $\neg alive(john)$ may not be derived.

OLP-FOL does not contain default rules nor modal connectives. Yet, the formalism is nonmonotonic. Extending a definition by adding a new positive case is a nonmonotonic operation. E.g. a logic program $\{p(a):\text{-}\}$ entails $\neg p(b)$. After adding $p(b):\text{-}$, it entails $p(b)$. OLP-FOL can be seen as an alternative interpretation of abductive logic programming [12]. The formal objects in OLP-FOL and the abductive logic program formalism (in the sequel: ALP) are the same: both contain tuples with a set of program clauses, an equivalent partition of the predicates in two subsets and a set of FOL formulas. Yet, OLP-FOL and ALP interpret these objects in different ways.

First, abductive logic programming is seen as the study of abduction in the context of logic programming. In contrast, we view OLP-FOL as a declarative logic in its own right. In [7], we investigated the use of other computational paradigms such as deduction and satisfiability checking on an equal basis as abduction. Second, the role of the formal model semantics is different in ALP and OLP. In ALP, a model semantics is meant as a specification of what abductive solutions must be computed [3] [13] [21]. In contrast, we look at model semantics as a formalisation of the declarative and descriptive meaning of an OLP-FOL theory. Third, the role of the FOL component in ALP and OLP seems different. In [12], a FOL formula is interpreted as an *integrity constraint*. The role of an integrity constraint is determined by its relation to abduction and can be classified either under the *theoremhood view*, the *consistency view* or *epistemic view*. On the other hand, OLP-FOL is an integration of two language components on equal basis. The FOL axioms in an OLP-FOL theory have the same role and meaning as in a FOL theory and are not bound to a specific role in any form of computation.

The content of this paper is the following. In section 2, we define the syntax of OLP-FOL and introduce the semantical primitives. Based on these, section 3 gives a semantics for the logic. Section 4 gives a number of well-chosen examples which clarify the role of OLP-FOL for knowledge representation. The paper closes with a discussion of related work and a conclusion. The proofs of the theorems are omitted and can be found in [4] and [6].

2 Preliminaries.

An alphabet Σ, terms, atoms, literals and formulae based on Σ are defined as usual. As usual, a free variable in a formula is not bound by a quantifier; a closed or ground formula does not contain free variables. Substitutions and variable assignments are defined as usual. A program clause here corresponds to a *general program clause* as defined in [18] except that we use the operator " :- " instead of "$\leftarrow$" in order to distinguish between program clauses and FOL implications.

An open logic program P^D is a pair of a set P of program clauses and a subset D of predicates of Σ, called *defined*, such that $=\notin D$ and all predicates in the head of program clauses in P are contained in D. The non-equality predicates in $\Sigma \setminus D$ are called *open*. A closed logic program is an open logic program which defines all non-equality predicates. The *definition* of a defined predicate p in P is the set of clauses with p in the head. This set may be empty. An OLP-FOL theory $\mathcal{T}$, also called open theory, is a tuple (P^D, T) with an open logic program P^D and a set of FOL axioms T.

In the next section we define a model semantics for OLP-FOL. The model semantics is based on the notion of a 3-valued Σ-interpretation (for a definition see e.g. [16]). It is essentially a domain, a correspondence between functors and functions on the domain and a correspondence between predicate symbols and 3-valued relations on the domain. With a Σ-interpretation I, a truth function $\mathcal{H}_I$ can be associated, which maps closed formulas to truth values, as usual.

3 The semantics of OLP-FOL.

The formal declarative semantics of an OLP-FOL theory is given by the class of its models. In the model theory of OLP-FOL, a model intuitively represents a possible state of the world. Both the terminological component and the assertional component of an OLP-FOL theory can be characterised by the way they impose restrictions on the possible states of the problem domain. Therefore, though OLP-FOL is an amalgamation of two different languages, it has a conceptually coherent semantics.

The semantics of the FOL-language is defined as usual except that 3-valued interpretations are allowed. Given is a Σ-interpretation I, with associated truth function $\mathcal{H}_I$. This function maps closed formulas to truth values and is defined as usual.

Definition 1. I satisfies a closed FOL formula F iff $\mathcal{H}_I(F) = \mathbf{t}$. I satisfies a FOL theory T iff for all $F \in T : \mathcal{H}_I(F) = \mathbf{t}$.

We denote this by $I \models F$ and $I \models T$.

For the semantics of the open logic program language, we follow more or less Clark's approach. Our approach extends both the 2-valued completion semantics for abduction in logic programming of [3] and the 3-valued completion semantics for closed logic programs of [16]. An open logic program P^D defines a set $comp_3(P^D)$, consisting of the *completed definitions* of the defined predicates of P^D [16]. As in [16], the classical equivalence operator $\leftrightarrow$ is replaced by a new operator $\Leftrightarrow$. The truth function $\mathcal{H}_I$ associated to some interpretation I is extended for this logical connective as given by the following table:

$$\mathcal{H}_I(E_1 \Leftrightarrow E_2) = \mathbf{t} \quad \text{iff} \quad \mathcal{H}_I(E_1) = \mathcal{H}_I(E_2)$$
$$\mathcal{H}_I(E_1 \Leftrightarrow E_2) = \mathbf{f} \quad \text{iff} \quad \mathcal{H}_I(E_1) \neq \mathcal{H}_I(E_2)$$

For two-valued interpretations, the meaning of the normal logical equivalence $\leftrightarrow$ and $\Leftrightarrow$ is the same. The use of $\Leftrightarrow$ is restricted to completed definitions and may not occur in the FOL component of an open theory.

An integration of a logic programming formalism and FOL raises a problem concerning the role of equality. Logic programming formalisms subsume virtually always the theory of Free Equality (denoted $FEQ(\Sigma)$), also called Clark's Equality theory [2]. These axioms lack in FOL. Because computational techniques in logic programming such as negation as failure rely on Free Equality, we define OLP-FOL as a free equality logic. For a definition of $FEQ(\Sigma)$, we refer to [2].

Using the above definitions, the declarative semantics of an OLP-FOL theory is defined as follows:

Definition 2. A Σ-interpretation I is a Σ-model of an OLP-FOL theory (P^D, T) iff I is 2-valued on all open predicates and I satisfies $T \cup FEQ(\Sigma) \cup comp_3(P^D)$.

Given is an OLP-FOL theory T and a FOL formula F based on Σ.

Definition 3. T is consistent iff there exists a Σ-model of T. Otherwise T is inconsistent. T entails F iff for each Σ-model M of T, $M \models F$. F is satisfiable with T iff there is a Σ-model of T which satisfies F.

The OLP-FOL logic is an integration of logic programming and classical logic. The integration of FOL in OLP-FOL is subject to one important restriction: the embedding of a FOL theory T will be equivalence preserving only if T entails the theory of Free Equality.

Theorem 4. *A closed logic program P based on Σ under Kunen's completion semantics [16] is logically equivalent with the OLP-FOL theory (P^D, ϕ) with $D = \Sigma \setminus \{=\}$.*

A FOL theory T based on Σ which entails $\mathrm{FEQ}(\Sigma)$ is logically equivalent with the OLP-FOL theory (ϕ^ϕ, T).

In principle, the free equality in OLP-FOL does not restrict the expressivity in OLP-FOL compared to FOL. N-ary functors for which the axioms in FEQ do not hold can always be replaced by n+1-ary predicates.

The above semantics raises a number of questions. Why 3-valued logic and how to interpret **u**? Why are the open predicates 2-valued?

The OLP component is to be used for definitions. However, due to the generality of the OLP-formalism, one can easily construct senseless definitions. Consider the following definition for p:

$$\{ \ p :- \neg p \ \}$$

Under a 2-valued completion semantics, an open logic program containing such definition would be inconsistent. The use of **u** is a more permissive solution to deal with badly constructed definitions. In 3-valued completion semantics, such a program remains consistent; yet the badly defined facts will have truth value **u** in all or a subclass of the models. At the same time, this shows that **u** is an error condition and not a truth value. **u** should be interpreted as *badly defined*. A 3-valued interpretation for an open predicate would not make sense under this interpretation of **u**: open predicates have no definition and cannot be badly defined. It should be stressed that many of the applications described in the following section critically rely on the 2-valuedness of open predicates.

For the purposes of this paper, one interesting advantage of using the 3-valued completion semantics is that it is the weakest semantics known for the (O)LP-formalism. In [4], the following theorem is proven:

Theorem 5. *If M is a model of P^D wrt (2-valued completion semantics [2] [3]) (generalised stable semantics [13]) (generalised well-founded semantics [21]) (justification semantics [5]) then M is a model of P^D wrt 3-valued completion semantics.*

As a consequence, the 3-valued completion semantics induces the weakest entailment relation $\models$: if an open theory entails F according to the 3-valued completion semantics then also wrt to the other semantics. In intuitive terms: the declarative meaning of an open logic program under the 3-valued completion semantics is a safe approximation of its meaning under these other semantics. As a consequence, many of the applications of OLP-FOL under 3-valued completion semantics described in this paper are preserved under the stronger semantics.

4 Knowledge Representation in OLP-FOL.

Crucial for knowledge representation in OLP-FOL is to distinguish assertional knowledge and *terminological* or *definitional* knowledge. Below, we explore this distinction in a number of problems. The problems include different forms of uncertainty: uncertainty on predicates, null values and uncertainty on the domain of discourse. First, we further explore the difference between the modal and the terminological interpretation.

4.1 A modal knowledge example.

Recall the example in section 1. The following definition for *dead*:

$dead(X) :- \neg alive(X)$

in an open program is interpreted as the phrase *X is dead iff X is not alive*. In contrast, the program clause in Extended Logic Programming (ELP):

$dead(X) \leftarrow not\ alive(X)$

under the modal interpretation is read as *X is dead if it is not believed that X is alive*. The program clause with strong negation:

$dead(X) \leftarrow \neg alive(X)$

under the modal interpretation is read as *X is dead if it is believed*[2] *that X is not alive.*

None of the representations in ELP have the same meaning as the representation in OLP-FOL. The *not* and $\neg$ of ELP have both a different modality than $\neg$ in FOL and in OLP-FOL. In the above example, only the OLP-FOL representation correctly represents our knowledge on the relationship between dead and alive. In

---

[2] Though [10] interprets the explicit negation $\neg$ in ELP as classical negation, various mappings of extended logic programs to autoepistemic logic have shown that it is more natural to interpret $\neg p$ as *it is believed that p is not true* [17].

other examples, the modal interpretation is the correct one. We recall an example from [10].

In ELP, one can represent the behaviour of a dare-devil who crosses a railroad if he has no evidence that there is a train coming by the following program clause:

$$cross \leftarrow not\ train$$

Under the modal interpretation, the logic program P_1 consisting of the above program clause contains no information whether there is a train or not, so *cross* should be true. Under answer set semantics this is correctly modelled: the unique answer set of P_1 is $\{cross\}$.

The behaviour of a more careful person who crosses if he believes that no train is coming can be represented using strong negation:

$$cross \leftarrow \neg train$$

The logic program P_2 consisting only of the above program clause contains no information whether there is a train or not, so *cross* should not be true. Under answer set semantics this is correctly modelled: the unique answer set of P_2 is $\{\}$.

Compare this with the OLP approach. Consider the set $P = \{cross :- \neg train\}$. The open logic program $P^{\{cross\}}$ contains no information on whether the train comes or not. The open logic program $P^{\{cross, train\}}$ contains full information on *train* because *train* has an empty definition. Both programs asserts that whether a person knows that a train is coming or not, he crosses iff there is no train coming. Clearly, neither of these OLP programs correspond to P_1 or P_2.

In our opinion, this example shows clearly the different modalities of *not* and $\neg$ in Extended Logic Programming (ELP) and of $\neg$ in classical logic and OLP-FOL. It shows also that both interpretations are of value. We believe that OLP-FOL is not suited to represent modal knowledge such as in the train example, while the modal interpretation is not suited to represent definitional information.

4.2 Temporal uncertainty.

The representation of uncertainty has been investigated most intensively in the context of temporal domains. Here we present solutions for a number of benchmark problems using an OLP-FOL version of event calculus [15]. The event calculus provides a more natural representation for reasoning on the *actual time-line*, compared to situation calculus which needs to be extended to reason on an actual time-line [22] [14]. The solutions of the benchmark problems below are similar to the abductive logic programs in [8]. However, [8] ignores the declarative aspects of the representation and focusses on the abductive reasoning.

A fundamental assumption in event calculus is that the state of the world at any point in time is determined completely by the initial state and the effects of actions in the past. Intuitively, a property P is true at a certain moment T if it is initiated by some earlier event and is not terminated since then. This knowledge can be interpreted as a definition for the predicate $holds\_at(P, T)$.

$holds\_at(P,T) :\text{-} happens(E), E < T, initiates(E,P), \neg clipped(E,P,T)$
$holds\_at(P,T) :\text{-} initially(P), \neg clipped(P,T)$
$clipped(E,P,T) :\text{-} happens(C), E < C < T, terminates(C,P)$
$clipped(P,T) :\text{-} happens(C), C < T, terminates(C,P)$

$clipped(E,P,T)$ is an auxiliary predicate which intuitively means that the property P is terminated during the interval $]E,T[$. $clipped(P,T)$ means that the property P is terminated before T. Interesting is that this definition for $holds\_at$ shows that *definitional knowledge* is not necessarily *terminological knowledge*: the definition for $holds\_at$ can hardly be seen as the definition of *technical vocabulary* of an expert. Here, this definition is used to represent an empirical observation relating the state of fluents with the occurrence of past actions.

In [8], we defend the view of event calculus as a linear time theory, in contrast to situation calculus which has branching time. That $<$ is a linear order is formally assured by adding the theory of linear order as FOL assertions:

$$X < Y \wedge Y < Z \rightarrow X < Z \qquad \text{transitivity}$$
$$\leftarrow X < Y, Y < X \qquad \text{asymmetry}$$
$$happens(X) \wedge happens(Y) \rightarrow X < Y \vee X = Y \vee Y > X \quad \text{linearity}$$

For simplicity, we assume here that two events cannot happen simultaneously.

The definitions of $holds\_at$ and $clipped$ rely on the predicates *initially, happens, $<$, initiates* and *terminates*, which describe the initial situation, the events, their order and their effects. Depending on the given information, different subsets of these predicates can be defined. We illustrate this in a number of well-known variants of the notorious Turkey shooting problem. In all these problems the effects of load, shoot and wait actions can be represented as definitions for *initiates* and *terminates*:

$initiates(E, loaded) :\text{-} act(E, loading)$
$terminates(E, loaded) :\text{-} act(E, shooting)$
$terminates(E, alive) :\text{-} act(E, shooting), holds\_at(loaded, E)$

A first problem is the Russian Turkey shooting problem, in which an indeterminate event of spinning the gun's chamber takes place instead of waiting. The effect of this event is that it potentially unloads the gun. A theory representing this problem should be satisfiable with both $holds\_at(alive, e4)$ and $\neg holds\_at(alive, e4)$.

The indeterminacy of the spinning action complicates the formulation of a definition for *terminates*. One possibility is to leave the predicate open and to represent all knowledge on it as FOL assertions. But it is significantly simpler, clearer and more concise to give a definition of *terminates* in which a new open predicate *good_luck*/1 is used which captures the indeterminacy. The open logic program T_{RTS} consists of the above program clauses augmented with the clause:

$terminates(E, loaded) :\text{-} act(E, spinning), holds\_at(loaded, E), good\_luck(E)$

The other predicates *initially*/1, *happens*/1, $< /2$, *act*/2 can be defined by exhaustive enumeration:

initially(alive), *happens(e1)*, *happens(e2)*, *happens(e3)*, *happens(e4)*, *act(e1, load)*, *act(e2, spin)*, *act(e3, shoot)* and the transitive closure of $e1 < e2 < e3 < e4$.

Both *holds_at(alive, e4)* and ¬*holds_at(alive, e4)* are satisfiable with $\mathcal{T}_{RTS}$. Note that the axioms of linear order are entailed by the definition of $<$ and can be dropped from $\mathcal{T}_{RTS}$.

Another variant is the Murder Mystery, in which a shooting event *e1* followed by a waiting event *e2* occur and the turkey is found dead at a next event *e3*. The initial state is unknown. A desired conclusion is that if the turkey is initially alive then the gun is initially loaded.

For this problem, all predicates except *initially/1* can be defined. The definitions of *holds_at*, *initiates*, *terminates* are given above. The predicates *happens*, $<$ and *act* can be defined by exhaustively describing the scenario:

happens(e1), *happens(e2)*, *happens(e3)*, *act(e1, shoot)*, *act(e2, wait)*, and the transitive closure of $e1 < e2 < e3$.

The knowledge that the turkey is dead at *e3* is formulated as a simple FOL assertion ¬*holds_at(alive, e3)*. The resulting theory $\mathcal{T}_{MM}$ provably entails[3]:

$$initially(loaded) \lor \neg initially(alive)$$

A variant of the murder mystery illustrates nicely the distinction between *definitional knowledge* and *assertional knowledge* and how these forms of knowledge are properly represented in OLP-FOL. Assume that the same events happen as in the murder mystery but that the turkey is found alive instead of dead at *e3*. Desired conclusions here are that initially the turkey is alive and that the gun is unloaded. The question now is: should we add the fact *holds_at(alive, e3)* to the open logic program or as a FOL axiom?

Adding the atom to the open logic program boils down to extending the definition of *holds_at*. It is clear that event calculus' basic assumption -the state is determined by initial state and effects of past actions- is not falsified in this example, so there seems no a priori reason why the definition should be changed. Stronger even, adding the atom as a program clause would alter the definition in a way which really contradicts our intuition: according to the extended definition, it would be possible that the fluent *alive* can originate at *e3* as by deus ex machina, without being caused by past actions[4]. This is evidenced by the behaviour of the corresponding program: the open logic program consisting of the program clauses of $\mathcal{T}_{MM}$ augmented with the program clause *holds_at(alive, e3):-* is satisfiable with ¬*initially(alive)*.

The correct representation is the theory $\mathcal{T}_{MM'}$ obtained from $\mathcal{T}_{MM}$ by substituting *holds_at(alive, e3)* for ¬*holds_at(alive, e3)* as a FOL assertion. $\mathcal{T}_{MM'}$ entails *initially(alive)* $\land$ ¬*initially(loaded)*.

In the previous examples, the theory of linear order is entailed by the program and can be dropped safely. The theory of linear order comes into play when there

---

[3] Note that again, axioms of the theory of linear order can be dropped from $\mathcal{T}_{MM}$.
[4] Note here the non-monotonicity of the concept of a definition.

is uncertainty about the events and/or their order. Take a scenario in which the gun is initially loaded (with one bullet) and the turkey is alive; at $e3$ the gun is fired. Before $e3$, two different events $e1$ and $e2$ take place at which the gun's chamber is turned for 180 degrees. The order of $e1$ and $e2$ is unknown. Note that no matter which is the order of $e1$ and $e2$, the gun's chamber is turned for 360 degrees, so it is loaded when the shooting event takes place. Therefore, a desired conclusion in this problem is that after the shooting the turkey is dead.

To represent this problem, a new action *turn* is introduced. When the gun is loaded, then *turn* removes the bullet from the barrel. The resulting state is called *opposite*. Again turning moves the bullet in the barrel and loads the gun. The definitions of *initiate* and *terminates* in $\mathcal{T}_{TS}$ are extended with:

$terminates(E, loaded) :- act(E, turn), holds\_at(loaded, E)$
$terminates(E, opposite) :- act(E, turn), holds\_at(opposite, E)$
$initiates(E, loaded) :- act(E, turn), holds\_at(opposite, E)$
$initiates(E, opposite) :- act(E, turn), holds\_at(loaded, E)$

The predicates *initially, happens, act* are defined by enumeration:

$initially(alive), initially(loaded), happens(e1), happens(e2), happens(e3),$
$happens(e4), act(e1, turn), act(e2, turn), act(e3, shoot).$

The remaining problem is the representation of the knowledge on $<$. It is tempting to add the partial order of known atoms of $<$ as a definition for $<$ to the open logic program. This approach for the representation of this form of uncertainty is proposed in [15]. The resulting logic program P_{LTTS}[5] contains definitions for all predicates. Note that the OLP-FOL theory consisting of P_{LTTS} augmented with the theory of linear order is inconsistent: P_{LTTS} violates the linearity constraint since it entails $\neg e1 < e2 \land \neg e2 < e1 \land \neg e1 = e2$. In [19], examples were given where this approach failed. Also in this case the approach fails: P_{LTTS} entails $holds\_at(alive, e4)$ and Prolog can be used to prove it. In more detail, one easily verifies that P_{LTTS} entails $holds\_at(loaded, e1)$ and $holds\_at(loaded, e2)$. Therefore, both $e1$ and $e2$ unload the gun, so that $\neg holds\_at(loaded, e3)$ is entailed. Since the gun is unloaded at $e3$, the shooting has no effect and the turkey remains alive.

A correct solution can be represented in OLP-FOL. Since there is uncertainty about $<$, this predicate cannot be defined. The correct theory $\mathcal{T}_{LTTS}$ is obtained from P_{LTTS} by dropping the definition of $<$ and adding the theory of linear order and the following atoms as FOL assertions:

$e1 < e3, e2 < e3, e3 < e4$[6]

$\mathcal{T}_{LTTS}$ correctly entails $\neg holds\_at(alive, e4)$.

Interesting applications of uncertainty on the events and their order in event calculus are planning problems. The goal of planning is to find an ordered set of

---

[5] LTTS stands for "Load Turn Turn Shoot".
[6] Note that this set of facts is not transitive. Due to the presence of the theory of linear order, additional facts such as $e1 < e4$ are entailed.

events which produces a given desired final state Q; the events are a priori unknown: in an OLP-FOL theory T representing a planning domain, $happens/1, <$ $/2$ and $act/2$ have no definition. An abductive procedure can be used to find a set Δ of definitions for these predicates such that the theory $T + \Delta \models Q$ [9].

4.3 Moore's example: a null value.

Moore gives the following example [20]. Three blocks, a, b and c, are arranged as shown:

The colour of a is green, c is blue and the colour of b is unknown. A desired conclusion is that there is a green block adjacent to a non-green block[7].

For simplicity, we assume that a, b and c are the only blocks and green and blue are the only colours. Due to this assumption, *block*, *colour* and *next* can be defined by exhaustive enumeration:

$$block(a), block(b), block(c), colour(blue), colour(green), next(a, b), next(b, c)$$

The predicate *adjacent* can be defined as well.

$$adjacent(X, Y) :- next(X, Y)$$
$$adjacent(X, Y) :- next(Y, X)$$

The fact that the colour of b is unknown poses a problem to define *has_colour*. An elegant solution would be to represent the colour of b by a *null value cb* of type *colour*. [23] formalises null values in classical logic. This approach exploits the fact that in classical logic, a constant represents some unique object of unknown identity. Unfortunately, this approach does not work in OLP-FOL due to the fact that OLP-FOL is a free equality logic. The reason is clear: the FOL assertion *colour(cb)* and the definition of *colour* entail that *cb* is either *green* or *blue* which is excluded by the free equality. However, there is a general and elegant way to simulate null values in a free equality logic. We introduce a new open predicate *colour_of_b/1* which -intuitively- represents the singleton $\{cb\}$. Using this predicate, *has_colour* can be defined as follows:

$$has\_colour(a, green) :-$$
$$has\_colour(c, blue) :-$$
$$has\_colour(b, C) :- colour\_of\_b(C)$$

The assertional knowledge that *colour_of_b/1* contains a unique *null value* of type *colour* is represented by the FOL assertions[8]:

---

[7] Moore's aim was to defend logic by showing that the desired conclusion can be obtained only by a reasoning by 2 cases (either b is green or not), a form of reasoning formalised only in logic.

[8] $\exists! X.F[X]$ is the standard notation for $\exists X.F[X] \wedge \forall X_1, X_2.F[X_1] \wedge F[X_2] \rightarrow X_1 = X_2$.

$\exists!C.colour\_of\_b(C)$
$\forall C.colour\_of\_b(C) \rightarrow colour(C)$

The result is an OLP-FOL theory T_{Moore} with one open predicate $colour\_of\_b/1$. The theory provably entails that there is a block which is not green adjacent to a green block.

4.4 The third man: an open domain problem.

The scenario of this simple problem is as follows. Bob is found murdered in his cottage while being visited by John. Suicide is excluded and John seems to have an alibi but which is hard to verify. The conclusions that can be derived in this domain clearly depend on whether it is known or not that Bob and John were the only persons in the cottage. Only if this knowledge is given, it can be concluded that John has a false alibi and is the murderer.

The following open theory T_{3M} represents this scenario:

$D = \{killed, alibi\}$
$P = \{\quad alibi(john)\,:\text{-}$
$\qquad killed(X, bob)\,:\text{-}\,murderer(X)\quad\}$
$T = \{\quad \exists X.\neg X = bob \wedge killed(X, bob)$
$\qquad \forall X.murderer(X) \wedge alibi(X) \rightarrow false\_alibi(X)\quad\}$

Open predicates are $murderer$ and $false\_alibi$.

Just as classical logic, OLP-FOL is an *open domain logic*: the domain closure assumption (DCA) is not a priori imposed. The theory T_{3M} has models in which bob and $john$ are the only domain elements. In these models, $false\_alibi(john)$ holds. There are other models in which other domain elements appear and where both facts are false. In these models, $murderer(x)$ is true for some domain element x not represented by $john$.

The example clarifies the role of general interpretations versus Herbrand interpretations. OLP-FOL (just like FOL) does not impose the DCA, due to the fact that general non-Herbrand interpretations are allowed. A logic with a model semantics based on Herbrand interpretations entails the DCA automatically. In many applications, these axioms are naturally satisfied in the problem domain; in other applications, they are not. From the knowledge representation point of view, basing OLP-FOL on Herbrand interpretations would restrict the expressivity in an undesirable way. If in the example, it is known that Bob and John are the only persons, then this must be represented explicitly by adding the DCA[9]:

$\forall X.X = bob \vee X = john$

The resulting theory entails $murderer(john)$.

An interesting class of domains where the domain closure assumption cannot be made are planning domains in the context of event calculus. In such domains, the events are a priori unknown and hence, the DCA cannot be added. Such domains are less easily described in a logic with a Herbrand-based semantics.

---

[9] It is well-known that when Σ contains functors with arity > 0, the DCA can not be represented correctly in FOL. As was shown in [5], the DCA can be represented correctly in OLP-FOL under stronger semantics than the 3-valued completion semantics.

5 Conclusions and future work.

The above examples show the expressivity of OLP-FOL for knowledge representation. The elegance of the above presented solutions may easily mislead. Many of the temporal reasoning problems in section 4.2 are considered as difficult knowledge representation problems and have been used to show problems with several temporal reasoning proposals. The representation of null values and open domain knowledge is still an open question in logic programming and, in the case of ELP, seems to require substantial changes to the semantics [24] [11].

An interesting issue which falls outside the scope of this paper is the representation of *inductive definitions* in the OLP-FOL formalism. OLP-FOL (under completion semantics) nor FOL are suited to represent inductive definitions; in ongoing work we showed that under stronger semantics such as justifications semantics [5], inductive definitions can be correctly represented in an open logic program. As a consequence, OLP-FOL under justification semantics significantly extends the FOL expressivity. Another interesting issue is how theorem provers and problem solvers for OLP-FOL can be developed. In [6], we show how existing abductive extensions of SLDNF resolution can be used for reasoning on OLP-FOL theories, not only to solve abductive problems but also for deductive and satisfiability checking problems.

To conclude, the OLP-FOL logic provides an alternative interpretation of (abductive) logic programming. The auto-epistemic and default interpretation on one hand and the terminological interpretation on the other hand are different interpretations which assign a different modality to negation in logic programs. In OLP-FOL, negation has the objective modality of negation in classical logic. As a consequence, it makes no sense to add *classical negation* to OLP-FOL. On the contrary, it would be interesting to add modal forms of negation, as found in the auto-epistemic interpretation. Such an extension would allow to represent also modal problems (e.g. the train example in section 4.1) and would combine the representational power of ELP and OLP-FOL. The problems of such an integration don't seem insurmountable. Last but not least, OLP-FOL is a declarative integration of logic programming and classical logic and opens the possibility of cross fertilisation between both fields.

Acknowledgements

I thank Danny De Schreye and Kristof Van Belleghem for many valuable discussions. Danny, Kristof and Marion Mircheva gave justified comments on earlier drafts. Hector Levesque pointed me to the analogy in the motivations for OLP-FOL and terminological languages.

References

1. R. J. Brachman and H.J. Levesque. Competence in Knowledge Representation. In *Proc. of the National Conference on Artificial Intelligence*, pages 189–192, 1982.
2. K.L. Clark. Negation as failure. In H. Gallaire and J. Minker, editors, *Logic and Databases*, pages 293–322. Plenum Press, 1978.

3. L. Console, D. Theseider Dupre, and P. Torasso. On the relationship between abduction and deduction. *Journal of Logic and Computation*, 1(5):661–690, 1991.

4. M. Denecker. *Knowledge Representation and Reasoning in Incomplete Logic Programming*. PhD thesis, Department of Computer Science, K.U.Leuven, 1993.

5. M. Denecker and D. De Schreye. Justification semantics: a unifying framework for the semantics of logic programs. In *Proc. of the Logic Programming and Nonmonotonic Reasoning Workshop*, pages 365–379, 1993.

6. M. Denecker and D. De Schreye. A terminological interpretation of (Abductive) Logic Programming. Draft, K.U.Leuven, 1994.

7. M. Denecker and D. De Schreye. Representing Incomplete Knowledge in Abductive Logic Programming. *Journal of Logic and Computation, to appear*, 1995.

8. M. Denecker, L. Missiaen, and M. Bruynooghe. Temporal reasoning with abductive event calculus. In *Proc. of the European Conference on Artificial Intelligence*, 1992.

9. K. Eshghi. Abductive planning with Event Calculus. In R.A. Kowalski and K.A. Bowen, editors, *Proc. of the International Conference on Logic Programming*, 1988.

10. M. Gelfond and V. Lifschitz. Logic Programs with Classical Negation. In D.H.D. Warren and P. Szeredi, editors, *Proc. of the 7th International Conference on Logic Programming 90*, page 579. MIT press, 1990.

11. M. Gelfond and H. Przymusinska. Reasoning in Open Domains . In *Proc. of Logic Programming and nonmonotonic reasoning workshop*, page 397, 1993.

12. A. C. Kakas, R.A. Kowalski, and F. Toni. Abductive Logic Programming. *Journal of Logic and Computation*, 2(6):719–770, 1993.

13. A.C. Kakas and P. Mancarella. Generalised stable models: a semantics for abduction. In *Proc. of the European Conference on Artificial Intelligence*, 1990.

14. R. Kowalski and F. Sadri. The Situation Calculus and Event Calculus Compared. In *Proc. of International Logic Programming Symposium*. MIT Press, 1994.

15. R.A. Kowalski and M. Sergot. A logic-based calculus of events. *New Generation Computing*, 4(4):319–340, 1986.

16. K. Kunen. Negation in Logic Programming. *Journal of Logic Programming*, 4:231–245, 1989.

17. V. Lifschitz and G. Schwarz. Extended Logic Programs as Autoepistemic Theories. In L. Pereira and A. Nerode, editors, *Proc. of the Logic Programming and Nonmonotonic Reasoning Workshop*, 1993.

18. J.W. Lloyd. *Foundations of Logic Programming*. Springer-Verlag, 1987.

19. L.R. Missiaen. *Localized abductive planning with the event calculus*. PhD thesis, Department of Computer Science, K.U.Leuven, 1991.

20. R.C. Moore. The role of logic in knowledge representation and commonsense reasoning. In *Proc. of AAAI-82*, pages 428–433, 1982.

21. L.M. Pereira, J.N. Aparicio, and J.J. Alferes. Hypothetical Reasoning with Well Founded Semantics. In B. Mayoh, editor, *Proc. of the 3th Scandinavian Conference on AI*. IOS Press, 1991.

22. J. Pinto and R.Reiter. Temporal Reasoning in Logic Programming: A Case for the Situation Calculus. In *Proc. of the International Conference on Logic Programming*, pages 203–221, 1993.

23. R. Reiter. On integrity constraints. In M. Vardi, editor, *Proc. of Conf. on Theoretical Aspects o Reasoning about Knowledge*, pages 97–111, 1988.

24. B. Traylor and M. Gelfond. Representing Null Values in Logic Programming. In *Proc. of the ILPS'93 workshop on Logic Programming with Incomplete Information*, pages 35–47, 1993.

Abduction over 3-valued extended logic programs

Carlos Viegas Damásio, Luís Moniz Pereira*

CRIA, Uninova and DCS, U. Nova de Lisboa,
2825 Monte da Caparica, Portugal
{cd|lmp}@fct.unl.pt

1 Introduction

Default negation was introduced so that negative conclusions could be drawn from logic programs. A first attempt to characterize this specific form of Closed World Assumption (CWA) [25] was the completion semantics [9].

This form of negation is capable of dealing with incomplete information via its non-monotonicity. However, completion semantics is inconsistent for a large class of normal logic programs and introduces undesirable tautologies. A spate of semantics proposals were set forth to solve these problems, from the late eighties onwards, including the well-founded semantics (WFS) [13], which addresses semantically non-terminating top-down computations by assigning such computations the truth value of "false" or "undefined", thereby giving semantics to every program. The well-founded semantics deals only with normal programs and thus provides no mechanism for explicitly declaring the falsity of literals. This can be a serious limitation.

Logic programming semantics has now been extended with a new explicit form of negation, beside the other implicit (or default) negation typical of Logic Programming.

Different semantics for such extended programs with a ¬-negation have appeared (e.g. [4, 15, 20, 24, 29, 30]). The generalization for extended programs of the WFS defined in [20], *WFSX*, adopted as the base semantics in this paper, provides a representational expressivity that captures a wide variety of logical reasoning forms [6, 16, 22].

The two forms of negation, default and explicit, are not unrelated: the "coherence principle" stipulates that the latter entails the former. Of course, introducing explicit negation requires, in addition, dealing with contradiction. Default negation and the revising of such default assumptions in the face of contradiction (or integrity constraint violation) are both non-monotonic reasoning mechanisms available in logic programming.

One major contribution of this paper is that of tackling contradiction, with some generality, and in particular within our choice semantics of extended logic programs, *WFSX*. Although such issues can be appproached from a belief revision point of view, they can likewise, and equivalently, be thought of within

---

* We thank Esprit BR project Compulog 2 (no. 6810), and JNICT for their support.

an abductive framework. Indeed, a significant result of this paper is to define such three-valued abductive framework, *directly* over three-valued extended logic programming semantics.

Like [28], we too allow for contradiction to appear and show moreover how it can be removed by freely changing the truth-values (among "true", "false", and "undefined") of a subset of pre-designated *revisable* literals: the abducibles. To do so, we start by introducing a paraconsistent version of *WFSX*, *WFSX$_p$*, that permits reasoning with contradictory programs. Consistency is regained by revising the truth value of the revisables that lend "support" to integrity constraint violating literals, which are detected in *WFSX$_p$*.

A notion of global minimal revision (improving on previous work in the literature) is defined as well, which establishes a closeness relation between the several abductive solutions, thereby defining minimality of abduction too. Forthwith, the changes required to achieve program revision can be enforced by introducing or deleting program rules, for the abducible literals only, without loss in generality.

Subsequently, we show how other abductive frameworks relate to our own (namely Generalised Stable Models (GSM) [18]). We also demonstrate our framework can be used to characterize various logic programming semantics, namely Stable Models [14], Stationary Models [24], Well Founded Semantics [13], Preferred extensions [11], Extended Partial Stable Models [1] and Answer Sets [15].

A final result of the paper is that of proving that GSM and the abductive framework defined here have the same expressive power.

2 Paraconsistent *WFSX*

In this section we recall the language of logic programs extended with explicit negation (or extended logic programs for short) and define the paraconsistent version of well founded semantics with explicit negation [20]. An extended program is a set of rules of the form $L_0 \leftarrow L_1, \ldots, L_m, not\ L_{m+1}, \ldots, not\ L_n$ $(0 \leq m \leq n)$ where each L_i is an objective literal. An objective literal is either an atom A or its explicit negation $\neg A$. In the sequel, we also use $\neg$ to denote complementary literal wrt. the explicit negation, so that $\neg\neg A = A$. The set of all objective literals of a program P is called the extended Herbrand base of P and denoted by $\mathcal{H}(P)$. The symbol not stands for negation by default. $not\ L$ is called a default literal. Literals are either objective or default literals. By $not\ \{a_1, \ldots, a_n, \ldots\}$ we mean $\{not\ a_1, \ldots, not\ a_n, \ldots\}$. An interpretation of an extended program P is denoted by $T \cup not\ F$, where T and F are subsets of $\mathcal{H}(P)$. Objective literals in T are said to be *true* in I, objective literals in F *false by default* in I, and in $\mathcal{H}(P) - I$ *undefined* in I.

WFSX follows from WFS for normal programs plus the coherence requirement relating the two forms of negation: "For any objective literal L, if $\neg L$ is entailed by the semantics then $not\ L$ is also entailed.

As mentioned in the introduction, we're also interested in detecting contradiction occurrences, in order to remove it. To remove a contradiction, as we shall see in detail in section 4, we need first to detect it and, more importantly, we need to know what are the "reasons" or "causes" that lead to contradiction, i.e. which assumptions made are responsible for contradiction by "supporting" it, meaning that if they were withdrawn then contradiction would be dispelled.

To determine such "causes" a paraconsistent *WFSX* (or *WFSX$_p$* for short) is required. The main idea is to calculate, always in keeping with coherence, all consequences of the program, even those leading to contradiction, as well as those arising from contradiction. For examples see [3].

Here we do not argue that *WFSX$_p$* is in fact a *"semantics"*. Instead we view *WFSX$_p$* as a tool needed for detecting contradiction and its causes, when *WFSX* is not able to assign meaning to a program. This tool is then used as the basis for removing contradiction where possible. The semantics assigned to the original program is defined as the *WFSX* of the program obtained after removing contradiction.

Since the purpose of *WFSX$_p$* is to detect contradiction on the basis of *WFSX*, every principle that enforces the truth-value of literals in *WFSX* must also be enforced in *WFSX$_p$*. Thus, coherence must be kept in *WFSX$_p$* too.

We will define *WFSX$_p$* in terms of alternating fixpoints of Gelfond–Lifschitz Γ–like operators [14, 15]. For more details and relationships see [1, 3]. In definition 1 every literal $\neg A$ is treated as a new literal, say $\neg\_A$, unrelated to A.

Definition 1 The Γ–operator. Let P be an extended program, I a two-valued interpretation. The GL–transformation $\frac{P}{I}$ is the program obtained from P by removing all rules containing a default literal *not A* such that $A \in I$, and by then removing all the remaining default literals from P. By definition, $\Gamma_P I$ is the least two-valued model of $\frac{P}{I}$.

To impose the coherence requirement we introduce:

Definition 2 Semi-normal version of a program. The semi-normal version of a program P is the program P_s obtained from P by adding to the (possibly empty) *Body* of each rule $L \leftarrow Body$ the default literal *not* $\neg L$, where $\neg L$ is the complement of L wrt. explicit negation.

Below we use $\Gamma(S)$ to denote $\Gamma_P(S)$, and $\Gamma_s(S)$ to denote $\Gamma_{P_s}(S)$.
It can be easily proved [3] that:

Theorem 3. *The operator $\Gamma\Gamma_s$ is monotonic, for arbitrary sets of literals.*

The above suggests introducing a definition of *WFSX$_p$* where the paraconsistent WFM (WFM_p) is given by the least fixpoint of $\Gamma\Gamma_s$:

Definition 4 Paraconsistent *WFSX*. Let P be an extended program whose least fixpoint of $\Gamma\Gamma_s$ is T. Then, the paraconsistent well-founded model of P is
$WFM_p(P) = T \cup not\ (\mathcal{H} - \Gamma_s T)$.

In this definition note how the use of Γ_s imposes coherence: if $\neg L$ is true, i.e. it belongs to T, then in $\Gamma_s T$, via semi-normality, all rules for L are removed and, consequently, $L \notin \Gamma_s T$, i.e. L is false by default. If condition $T \subseteq \Gamma_s T$ is satisfied, thereby guaranteeing that L and $\neg L$ do not belong both to the WFM_P, then $WFSX_p$ and $WFSX$ coincide.

Other paraconsistent semantics exist in the literature, probably the most representative ones being [7] and [27]. The former is an extension of definite logic programs with literals annotated with truth-values from Belnap's four valued logic. We are able to transform any annotated program to an extended logic program with isomorphic semantics. The latter is defined for extended logic programs but the coherence principle is not used. In fact, this semantics can easily be captured by viewing every explicit negated literal as a new literal, and it corresponds to a paraconsistent version of the semantics proposed by Przymuskinski in [24]. For lack of space we cannot explore these relations in more detail.

3 Three-valued Abductive Framework

After having presented a paraconsistent version of $WFSX$ (able therefore to detect contradiction in programs), we are now ready to foster the outlook, adopted in the introduction, that if a program is contradictory its revision is in order. By revision of a contradictory program P we mean a non-contradictory program P' obtained from P without introducing new predicates. One question immediately arises: what changes are allowed to P in order to obtain P'?

As argued in our previous work [21, 23], for the sake of simplicity and without loss of generality, the modification should be based solely on those literals that depend on no other literals, and by introducing or removing rules for such basic literals only. This is the stance we take here. The reader acquainted with abduction [10, 11, 12, 18] in logic programs will immediately notice the similarities between this condition on basic literals and conditions obeyed by abducibles.

In fact, for the purpose of defining revision, we will define a three-valued abductive framework using $WFSX_p$ as the underlying semantics. As usual, we will divide the logic program into a pair, which we call scenario: one element contains the subprogram that cannot be changed; the other contains the subprogram that can be changed in a predefined way, namely by adding rules of a simple fixed form for the abducible literals only. The latter part is such that it contains rules only for literals that depend on no other.

Until now we've considered contradiction as detected by $WFSX_p$. In general, however, there can be other reasons to consider a scenario unacceptable, due to violation of integrity knowledge about the intended program semantics. To encompass these situations, we now extend our language to deal with general integrity constraints with constraint checking defined according to theoremhood view [26, 19].

Definition 5 Integrity Constraints. An integrity constraint of an extended logic program P has the normal form $L_1 \vee L_2 \vee \ldots \vee L_n \Leftarrow L_{n+1} \wedge L_{n+2} \wedge \ldots \wedge L_{n+m}(n + m \geq 0)$, where each L_i is a literal belonging to the language of P. To an empty consequent (head) we associate the symbol $\mathbf{f}$ and to an empty antecedent (body) the symbol $\mathbf{t}$. An integrity theory is a set of integrity constraints, standing for their conjunction.

An integrity constraint is satisfied by an interpretation I if for every literal $L_{n+1}, \ldots, L_{n+m} \in I$ then at least one of the literals $L_1, \ldots, L_n$ is in I. Otherwise, the constraint is violated by I.

Notice that the literals appearing in the constraints can be objective or default ones. Violation of constraints of the form $\mathbf{f} \Leftarrow L, \neg L$ for each objective literal L detect paraconsistency.

It is only necessary to consider, as our basic building blocks, constraints of the form $L \Leftarrow \mathbf{t}$ and $\mathbf{f} \Leftarrow L$, as the table below shows.

Condition	Integrity Theory	Condition	Integrity Theory
$L = \mathbf{f}$	$not\ L \Leftarrow \mathbf{t}$	$L \neq \mathbf{f}$	$\mathbf{f} \Leftarrow not\ L$
$L = \mathbf{u}$	$(\mathbf{f} \Leftarrow L) \wedge (\mathbf{f} \Leftarrow not\ L)$	$L \neq \mathbf{u}$	$L \vee not\ L \Leftarrow \mathbf{t}$
$L = \mathbf{t}$	$L \Leftarrow \mathbf{t}$	$L \neq \mathbf{t}$	$\mathbf{f} \Leftarrow L$

Now, a program is contradictory iff its *WFSX* is contradictory, or violates some integrity constraint. In our previous work, we've always considered separately the two cases of removing contradiction by changing the truth value of revisable (or abducible) literals either into "undefined" [5, 21], or into true [23]. Note that, in the absence of integrity constraints, contradiction can always be removed by changing the truth value of abducibles into undefined only. However, in order to apply our contradiction removal approach to model based diagnosis [23], for instance, we also felt the need for stronger revisions, where the truth value of literals can be changed into true.

In the present work, as mentioned above, we will allow revision by introducing or removing rules for abducibles, such that we can change their truth values from any one value to any other value, in a mixed way. By relying on the above more general integrity constraints, a mixture of literal revisions to "undefined" and to the default complement value can be obtained.

The truth-varying basic beliefs are represented by a set of objective literals (the abducibles) whose truth-values are not determined beforehand. The fixed part of an abductive framework is an arbitrary extended logic program, which derives additional knowledge from the abducibles, once their value is known.

Definition 6 Three-valued Abductive Framework. An abductive framework is a triple $\langle P, A, I \rangle$. The first component, P, is an extended logic program, the second one a set of objective literals $A \subseteq \mathcal{H}(P)$ such that $L \in A$ iff $\neg L \in A$, and the third a set of integrity constraints. Set A contains all abducible literals. Additionally, there are no rules in P for any of the abducibles.

Given a three-valued abductive framework, a scenario for it is established by introducing, for each abducible, a rule defining its truth value.

Definition 7 Scenario. A scenario of an abductive framework $\langle P, A, I \rangle$ is a tuple $\langle P, P_{Var}, A, I \rangle$. For each literal $L \in A$ the variable part, P_{Var}, contains just one of the rules $L \leftarrow \mathbf{t}$ or $L \leftarrow \mathbf{u}$ or $L \leftarrow \mathbf{f}$. Moreover, if $L \leftarrow \mathbf{t} \in P_{Var}$ then $\neg L \leftarrow \mathbf{f} \in P_{Var}$ (and if $\neg L \leftarrow \mathbf{t} \in P_{Var}$ then $L \leftarrow \mathbf{f} \in P_{Var}$). No other rules belong to P_{Var}.

By definiton, the symbols $\mathbf{t}$ and *not* $\mathbf{f}$ belong to all models. Neither $\mathbf{u}$ nor *not* $\mathbf{u}$ belong to any model, where $\neg \mathbf{u} \equiv \mathbf{u}$.

An abductive explanation of a goal G can be expressed as satisfaction of an integrity constraint of the form $G \Leftarrow$. The variable component of a scenario determines an interpretation, and vice-versa. The coherence requirement on abducibles is imposed by the additional proviso in definition 7 and also guarantees non-contradiction in P_{Var}. By convention, when writing the variable part we omit all rules of the form $L \leftarrow \mathbf{f}$, since the effect is the same. To each scenario we associate two interpretations.

Definition 8. Let $\Omega = \langle P, A, I \rangle$ denote an abductive framework and σ a scenario $\langle P, P_{Var}, A, I \rangle$ of Ω. We define $\mathcal{M}(\sigma)$ by $WFSX_p(P \cup P_{Var})$, and $\mathcal{B}(\sigma)$ by $\mathcal{M}(\sigma) \cap \mathcal{H}(P_{Var})$.

Definition 9 Abductive solution. An abductive solution is a scenario σ of an abductive framework such that all integrity constraints in σ are satisfied by $\mathcal{M}(\sigma)$. Otherwise, the scenario is said to be contradictory.

The set of literals that we are willing to change in order to restore consistency is given a priori. These literals are called the revisables. These must have not rules for them in the program part. We enforce consistency with $\mathbf{f} \Leftarrow L, \neg L$ for all L:

Definition 10 Semantics of contradictory programs. Let logic program P be contradictory and R be a set of revisables. The semantics of P is given by the abductive solutions of the abductive framework:

$$\langle P, R, \{\mathbf{f} \Leftarrow L, \neg L \ : \ L \in \mathcal{H}(P)\} \rangle$$

A simple program transformation can be performed on the original program making every literal ersatz "revisable", even those with rules:

Definition 11. Let P be an extended logic program and P_{REV} the program obtained by applying the following transformation:

- For each literal $L \in \mathcal{H}(P)$ the rule $L \leftarrow true(L)$ is in P_{REV};
- Replace by $H \leftarrow B_1, \ldots, B_n, not \ false(id)$ any rule $H \leftarrow B_1, \ldots, B_n$ in P, where id is a new constant which uniquely identifies the new rule.

To define the semantics of a contradictory program, when the set of revisables is not given, we simply use definition 10 with P_{REV} and the revisables being all $true(\_)$ and $false(\_)$ literals. The same technique, restricted to the relevant rules, can be used to make revisable a literal which has rules.

4 Revision

First there exists an initial scenario. When it is contradictory abducibles must change so as to "reach" a new non-contradictory scenario, if possible. However this change is not arbitrary: the abducibles should change in a minimal way wrt the initial scenario. This raises the issue how to define a notion of "closeness" between two scenarios. Given a closeness relation we can then define (minimal) revisions of a scenario. We shall only consider revisions of a single scenario, thereof possibly obtaining several new scenarios[2].

Our notion of closeness is an extension, to the three-valued case, of Winslett's difference between pairs of models of the Possible Model Approach (PMA) [31]. Difference between models is defined as the facts upon which a pair of models disagree. In the PMA the models which have minimal difference, i.e. minimal change in the sense of set-inclusion, are preferred. Because we are in a three-valued setting, a refined notion of difference is necessary. What is distinctive in our approach is that if a literal is not "undefined" then we prefer to revise it to "undefined", rather than revising it to the opposite truth-value; if it is "undefined" then we revise it, with equal preferance, either to true or false. This is mainly motivated by the truth and knowledge orders among literals[3]: moving from *true* to *false* or vice-versa, *in either ordering*, is always via *undefined*.

Definition 12 Difference between interpretations. Let $I_1 = T_1 \cup not\ F_1$ and $I_2 = T_2 \cup not\ F_2$ be two interpretations. Then,

$$Diff(I_1, I_2) = (F_2 - F_1) \times \{\mathbf{f}\} \cup (F_1 - F_2) \times \{\mathbf{u}\} \cup (T_1 - T_2) \times \{\mathbf{u}\} \cup (T_2 - T_1) \times \{\mathbf{t}\}$$

Literals labeled with **t** (resp. **f**) are those that change their truth-value to **t** (resp. **f**), when going from I_1 to I_2. Literals labeled with **u** are those that change their original truth-value either to **u** or to their complementary truth-value. Some of these sets might not be disjoint. For more details see [3].

Definition 13 Closeness relation. Let M, A and B be interpretations wrt. a finite language. We say that A is closer to M than B iff $Diff(M, A) \subset Diff(M, B)$.

The closeness relation is defined between interpretations, not between scenarios. When scenarios are considered, one of the following two interpretations could be used: either the model determined by the scenario, $\mathcal{M}(\sigma)$, or else its associated beliefs, $\mathcal{B}(\sigma)$. We assume that what changes are the agent's basic beliefs, and that other changes of belief are just consequences of basic belief changes. So, when facing contradiction, the agent should minimally change its basic beliefs, not necessarily what it extracts from those beliefs (i.e. their consequences).

---

[2] In order to iterate revisions we should also consider revisions of sets of scenarios. That is left to future work.

[3] In the truth ordering $\mathbf{f} < \mathbf{u} < \mathbf{t}$, and in the knowledge one $\mathbf{u} < \mathbf{f}$ and $\mathbf{u} < \mathbf{t}$.

Definition 14 Revision of a scenario. Let σ be the initial scenario of an abductive framework Ω_P. A revision of σ is $\tau \in \Sigma_P$ (the set of all abductive solutions) such that: $\forall_{\delta \in \Sigma_P} \; Diff(\mathcal{B}(\sigma), \mathcal{B}(\delta)) \subseteq Diff(\mathcal{B}(\sigma), \mathcal{B}(\tau)) \Rightarrow \delta = \tau$

Notice, again, that revision is defined here wrt a single initial scenario.

5 Comparisons

In the first subsection below, we make the bridge with our contradiction removal methods, which are captured by simple abductive frameworks.

The rest of the section provides a reconstruction of some of the most important logic programming semantics, by strongly restricting our abduction and revision frameworks: namely we will forego the use of explicit and default negation in the programs, and abductive solutions are *always* two-valued. In fact, we use default negation in our transformed programs just in order to improve readability. These default literals can simply be thought of as new objective literals. In any case, default literals will have no rules for them; furthermore, they are always abducible.

Our approach has close relationships with the recent work on argumentation theoretic semantics [11, 8]. Thus, all that we do here can equivalently be done with the Generalized Stable Models approach. At the end of our comparisons, and as a side result, we prove the equivalence between *GSM* and our three-valued framework.

Notice that we started off with $WFSX_p$, a less expressive semantics (if $P \neq NP$), and that by adding abduction we obtain the same expressive power as an abductive framework built on top of a more expressive semantics (Stable Models). This is an advantage, because more efficient top-down proof procedures are defined for $WFSX$ [2].

Not using explicit negation in the abductive framework transformations needs some care. When we define our abductive frameworks, we will omit the condition "if L is an abducible then ¬L is also abducible." For the sake of simplicity, we assume we make all these explicitly negated literals abducibles. Because we are deliberately excluding the use of explicit negation, we enforce that in all abductive solutions $not \; \neg L$ must be true. This can be simply engendered by adding the integrity constraints $not \; \neg L \Leftarrow$ to the integrity theory, for all positive literals L in the language of the program.

5.1 Contradiction Removal Methods

Here we show how our three-valued (CRS) [21, 1] and two-valued contradiction [23] revision methods can be straightforwardly obtained from our abductive and revision frameworks [4].

---

[4] In [21, 1, 23] what we dubbed revisables had the conventional form "*not r*"; here we use their default complement "*r*" instead.

In *CRS* revisable literals are all assumed false by default at the start. Integrity constraints are restricted to denials of the form $f \Leftarrow L_1, \ldots, L_n$. In order to satisfy the integrity constraints, the revisable literals are allowed to change to the undefined value. Changes of revisables' truth-value to *true* are interdicted. The intended transformation is immediate, and the only restriction to enforce is that of discarding those abductive solutions which assign true to a revisable.

Theorem 15. *Let P be a logic program with revisables R and integrity constraints I. The abductive framework $\Omega_{CRS} = \langle P, R, I_{CRS} \rangle$, where $I_{CRS} = I \cup \{false \Leftarrow r : r \in R\}$. Then, σ_{CRS} is an abductive solution of Ω_{CRS} iff the pair $\langle \mathcal{M}(\sigma_{CRS}), \mathcal{B}(\sigma_{CRS}) \rangle$ is a non-contradictory submodel of P with I and revisables R.*

In our two-valued contradiction removal methods, the above restrictions to the integrity theory also apply, but now revisables are not allowed to change to undefined. If we need to revise an initially false assumption, then we can change its value only to true:

Theorem 16. *Let P be a logic program with revisables R and integrity constraints I in the sense of [23]. We define the abductive framework $\Omega_{2REV} = \langle P, R, I_{2REV} \rangle$, where $I_{2REV} = I \cup \{r \vee not\ r \Leftarrow t : r \in R\}$. Then, σ_{2REV} is a solution of Ω_{2REV} iff $\{r \mid r \in \mathcal{B}(\sigma_{2REV})\}$ is a revision of P wrt I and R. The abductive solutions obtained by revision of the initial scenario with variable part $\{\}$ are the minimal revisions of P.*

5.2 Two-valued semantics

The manner of capturing Stable Model semantics through abduction is well known from the literature [12, 17] and here we follow their approach. Also, the authors of Generalised Stable Models [18] present a transformation from an abductive framework with negation as failure in rules of programs into a pure abduction problem. In this subsection we simply recapture those transformations in our framework, showing that our setting is as powerful as GSM. The equivalence between our abductive framework and Kakas and Mancarella's one is discussed in a later next section.

Theorem 17. *Let P be a normal logic program. The abductive framework Ω_{SM} is $\langle P_{SM}, A_{SM}, I_{SM} \rangle$, where*

$$P_{SM} = \{H \leftarrow B_1, \ldots, B_n, C_1^*, \ldots, C_m^* \mid H \leftarrow B_1, \ldots, B_n, not\ C_1, \ldots, not\ C_m \in P\}$$
$$A_{SM} = \{L^* \mid L \in \mathcal{H}(P)\} \text{ and } I_{SM} = \{L \vee L^* \Leftarrow t, \quad not\ L \vee not\ L^* \Leftarrow t \mid L \in \mathcal{H}(P)\}$$

Then, σ_{SM} is a solution of Ω_{SM} iff $\{L \mid L^* \notin \mathcal{B}(\sigma_{SM})\}$ is a stable model of P.

The way of capturing *GSM* is by an extension of the above transformation as defined in [18]:

Theorem 18. *Let* $\langle P, A, I \rangle$ *be an abductive framework in the sense of [18]. The abductive framework* $\Omega_{GSM} = \langle P_{GSM}, A_{GSM}, I_{GSM} \rangle$ *is:*

$$P_{GSM} = P_{SM} \qquad\qquad A_{GSM} = A_{SM} \cup A \cup \{a^* \mid a \in A\}$$
$$I_{GSM} = \{L \vee L^* \Leftarrow t, not\ L \vee not\ L^* \Leftarrow t \mid L^* \in A_{GSM}\} \cup I^*$$

where I^* *is obtained from* I *by replacing every default literal* $not\ L$ *by* L^*. *Then,* σ_{GSM} *is a solution of* Ω_{SM} *iff* $M(\Delta) = \{L \mid L \notin A \wedge L^* \notin \mathcal{B}(\sigma_{SM})\}$, *where* $\Delta = \{a \in A \mid a \in \mathcal{B}(\sigma_{GSM})\}$, *is a GSM of* $\langle P, A, I \rangle$.

5.3 Three-valued semantics

Even with the strong restrictions to the abductive framework mentioned in the introduction of this section, we can still characterize how stationary (or, equivalently, partial stable) models semantics *(PSM)* [24] are obtained. Likewise, we also show how to obtain the *WFM* and Dung's preferred extensions [11], via revision.

What we intend to capture is the fixpoints of $T = \Gamma\Gamma(T)$ which obey the condition $T \subseteq \Gamma(T)$. Thus, we need a way of computing $\Gamma\Gamma(T)$ and $\Gamma(T)$. We achieve this by a program transformation and additional integrity constraints. Part of these integrity constraints enforce the inclusion condition above, and the others the fixpoint condition. We start with the program transformation:

Definition 19. Let P be a normal logic program. Program P_{T-TU} is obtained as follows. Consider a rule $H \leftarrow B_1, \ldots, B_n, not\ C_1, \ldots, not\ C_m$ of P. Then the following two rules belong to P_{T-TU}:

- $H^T \leftarrow B_1^T, \ldots, B_n^T, false(C_1), \ldots, false(C_m)$
- $H^{TU} \leftarrow B_1^{TU}, \ldots, B_n^{TU}, not\ true(C_1), \ldots, not\ true(C_m)$

The first kind of rule, labeled with superscript T, will be used to derive what is true in the program, i.e. it computes the outer Γ of $\Gamma\Gamma(T)$. The rules having TU superscripted literals determine what is true or undefined, i.e. it corresponds to the inner Γ of the alternating fixpoint definition of WFS.

As noticed before, the use of default literal "$not\ true(L)$" in the above rules can be substituted by, for instance, "$false\_or\_und(L)$." By removing all occurrences of default literals from the original program, cycles through negation by default disappear. The burden of default negation reasoning is taken on by abduction via integrity constraints. Now we introduce the integrity constraints used to guarantee that obtaining the *PSMs* of *WFS*:

Definition 20. To each literal L in the language of P we associate the following integrity theory I_L^{T-TU}:

$not\ L^{TU} \vee not\ false(L) \Leftarrow$ (1)	$not\ L^T \vee L^{TU} \Leftarrow$ (3)	$not\ L^T \vee true(L) \Leftarrow$ (4)
$L^{TU} \vee false(L) \Leftarrow$ (2)		$L^T \vee not\ true(L) \Leftarrow$ (5)

Integrity constraints (1) and (2) above link the computation of the inner Γ operator to the outer one. The abductive solutions satisfying (1) and (2) should have either one of two models: $\{L^{TU}, not\ false(L)\}$ or $\{not\ L^{TU}, false(L)\}$. Thus, if L is true or undefined iff it is not false.

Constraints (4) and (5) ensure the fixpoint condition. Any abductive solution conforming to these two ICs has either one of the two models: $\{L^T, true(L)\}$ or $\{not\ L^T, not\ true(L)\}$. Therefore L can be assumed true iff it is derived by the program. Finally, integrity constraint (3) enforces the condition $T \subseteq \Gamma(T)$. As a whole, the integrity constraints are satisfied by an abductive solution obeying one of the three following models:

$$\{not\ false(L), not\ true(L), not\ L^T, L^{TU}\}\ (M1)$$
$$\{false(L), not\ true(L), not\ L^T, not\ L^{TU}\}\ (M2)$$
$$\{not\ false(L), true(L), L^T, L^{TU}\}\qquad (M3)$$

Models M1, M2, and M3, correspond to the situation where literal L is, respectively, undefined, false, or true.

Now we can immediately define the abductive framework which captures all *PSMs* of an arbitrary normal logic program.

Theorem 21. *Let P be a normal logic program and Ω_{PSM} be the framework* $\left\langle P_{T-TU}, \{false(L), true(L)\,|\,L \in \mathcal{H}(P)\}, \bigcup_{L \in \mathcal{H}(P)} I_L^{T-TU} \right\rangle$. *Then σ_{PSM} is a solution of Ω_{PSM} iff $\{L\,|\,true(L) \in \mathcal{B}(\sigma_{PSM})\} \cup \{not\ L\,|\,false(L) \in \mathcal{B}(\sigma_{PSM})\}$ is a Partial Stable Model of P.*

Enforcing an additional constraint $false(L) \vee true(L)$ on each literal, we obtain again the stable models. This was to be expected, because stable models are two-valued partial stable models. The previous constraint simply discards model $(M1)$. The WFM is the F-least *PSM* of the program. This minimality condition can be captured by revision:

Theorem 22. *Let P be a normal logic program. Then, there is a unique revision of Ω_{PSM} wrt. the initial scenario σ with $\mathcal{B}(\sigma) = \{\}$. Furthermore, this revision determines the WFM of P as in theorem 21.*

In the other way around, Dung's preferred extensions are F-maximal *PSMs* of the program. These extensions can also be captured through revision:

Theorem 23. *Let P be a normal logic program. Then the revisions of Ω_{PSM} wrt. the initial scenario σ with $\mathcal{B}(\sigma) = \{false(L), true(L)\,|\,L \in \mathcal{H}(P)\}$ are in 1:1 correspondence with preferred extensions. Furthermore, they can be determined as in theorem 21.*

5.4   Semantics with Explicit Negation

Finally, we discuss how the *PSMs* of *WFSX* can be obtained. As a side effect we show how answer set semantics is also embedded. Afterwards, we go on to determine the paraconsistent WFM, and end this subsection by presenting the restricted abductive formulation of our three-valued framework. As a corollary, we prove that GSM has the same expressive power as our three-valued abductive framework.

Definition 24. Let P be an extended logic program. Program $P_{T-TU}^{\neg}$ is obtained as follows. Every explicitly negated literal $\neg L$ is replaced by a new atom $'\neg\_L'$ in the transformed program. Every rule of program P of the form $H \leftarrow B_1, \ldots, B_n, not\ C_1, \ldots, not\ C_m$ is replaced by the two rules:

- $H^T \leftarrow B_1^T, \ldots, B_n^T, false(C_1), \ldots, false(C_m)$
- $H^{TU} \leftarrow B_1^{TU}, \ldots, B_n^{TU}, not\ true(C_1), \ldots, not\ true(C_m), not\ true(N_H)$

Where N_H is L^T if $H = \neg\_L$, and $\neg\_H^T$ otherwise.

So we've replaced the inner Γ operator by the Γ_s one. For ease of presentation, we assume that the transformation between $\neg L$ in the language of the program and $\neg\_L$ in the language of $P_{T-TU}^{\neg}$ is implicitly applied in the remainder of the paper. For instance, in the theorem below, when we say "$\{L \in \mathcal{H}(P)\}$" we mean "$\{\neg\_L \mid \neg L \in \mathcal{H}(P)\} \cup \{L \mid L$ is a positive literal of $\mathcal{H}(P)\}$."

Theorem 25. *Let P be a normal logic program and Ω_{xpsm} be the framework $\left\langle P_{T-TU}^{\neg}, \{false(L), true(L) \mid L \in \mathcal{H}(P)\}, \bigcup_{L \in \mathcal{H}(P)} I_L^{T-TU} \right\rangle$. Then, σ_{xpsm} is a solution of Ω_{xpsm} iff $\{L \mid true(L) \in \mathcal{B}(\sigma_{xpsm})\} \cup \{not\ L \mid false(L) \in \mathcal{B}(\sigma_{xpsm})\}$ is a Partial Stable Model of P under WFSX.*

Answer Set Semantics [15] is obtained by adding the additional constraint $false(L) \vee true(L) \Leftarrow$ for every positive objective literal in $\mathcal{H}(P_{T-TU}^{\neg})$. As for WFS, the WFM of WFSX can be obtained through revision too:

Theorem 26. *Let P be an extended logic program. Then there is a unique revision of Ω_{XPSM} wrt. the initial scenario σ with $\mathcal{B}(\sigma) = \{\}$. Furthermore, this revision determines the WFM under WFSX of P, as per theorem 25.*

The paraconsistent WFM can also be obtained by revision of an abductive framework Ω_{WFMp}. This framework is gotten from Ω_{XPSM} by removing all third type integrity constraints from I_{XPSM}: these are the ones which ensure $T \subseteq \Gamma(T)$.

Theorem 27. *Let P be an extended logic program. Then there is a unique revision of Ω_{WFMp} wrt. the initial scenario σ with $\mathcal{B}(\sigma) = \{\}$. Furthermore, this revision determines the $WFMp$, as per theorem 25.*

Finally we introduce our last program transformation, describing our general framework abductive solutions in terms of the more restricted one.

Definition 28. Let $\langle P, A, I \rangle$ be a three-valued abductive framework as in definition 6. Let Ω_{ABD} be defined as follows:

$P_{ABD} = P_{T-TU}^{\neg} \cup \left\{ L^T \leftarrow true(L), L^{TU} \leftarrow true(L), L^{TU} \leftarrow und(L) \mid L \in A \right\}$

$A_{ABD} = \{false(L), true(L) \mid L \in \mathcal{H}(P)\} \cup \{und(L) \mid L \in A\}$

$I_{ABD} = \{not\ true(L) \vee not\ und(L) \Leftarrow,\ false(L) \Leftarrow true(\neg\_L) \mid L \in A\} \cup$
$\qquad I_{WFMp} \cup I_I$

where I_I is obtained from I by replacing every objective literal L in an integrity constraint by $true(L)$ and every default literal $not\ L$ by $false(L)$.

Our main theorem follows:

Theorem 29. *Let $\Omega = \langle P, A, I \rangle$ be a three-valued abductive framework. Then α is a revision of Ω_{ABD} wrt to the initial empty scenario iff σ, with $\mathcal{M}(\sigma) = \{L\,|\,true(L) \in \mathcal{B}(\alpha)\} \cup \{not\ L\,|\,false(L) \in \mathcal{B}(\alpha)\}$, is an abductive solution of Ω.*

The introduction of abducibles $und(\_)$ is only necessary to guarantee that all abductive solutions are incomparable. The minimality condition on the revisions has the purpose of finding the least fixpoints of $\Gamma\Gamma_s$, for each possible combination of values of abducibles, as per theorem 27.

Corollary 30. *The three-valued abductive framework has the same expressive power as GSM.*

Thus we find our abductive solutions without relying on a three-valued semantics. All is defined in terms of a two-valued abductive framework. Thus, if we translate our transformation into the *GSM* setting, the generalised stable models of the transformed abductive theory are in a 1:1 correspondence with our abductive solutions, when a minimality condition is enforced. This, in conjunction with the results of subsection 5.2, prove the equivalence of both abductive frameworks.

References

1. J. J. Alferes. *Semantics of Logic Programs with Explicit Negation.* PhD thesis, Universidade Nova de Lisboa, October 1993.
2. J. J. Alferes, C. V. Damásio, and L. M. Pereira. A top-down derivation procedure for programs with explicit negation. In M. Bruynooghe, editor, *ILPS'94*, pages 424–438. MIT Press, 1994.
3. J. J. Alferes, C. V. Damásio, and L. M. Pereira. A logic programming system for non-monotonic reasoning. *Special Issue of the JAR*, 1995. To appear.
4. J. J. Alferes and L. M. Pereira. On logic program semantics with two kinds of negation. In K. Apt, editor, *IJCSLP'92*, pages 574–588. MIT Press, 1992.
5. J. J. Alferes and L. M. Pereira. Contradiction: when avoidance equal removal. In R. Dyckhoff, editor, *4th Int. Ws. on Extensions of LP*, volume 798 of *LNAI*. Springer-Verlag, 1994.
6. C. Baral and M. Gelfond. Logic programming and knowledge representation. *J. Logic Programming*, 19/20:73–148, 1994.
7. H. A. Blair and V. S. Subrahmanian. Paraconsistent logic programming. *Theoretical Computer Science*, 68:135–154, 1989.
8. A. Bondarenko, F. Toni, and R. Kowalski. An assumption–based framework for nonmonotonic reasoning. In L. M. Pereira and A. Nerode, editors, *2nd Int. Ws. on LP & NMR*, pages 171–189. MIT Press, 1993.
9. K. Clark. Negation as failure. In H. Gallaire and J. Minker, editors, *Logic and Data Bases*, pages 293–322. Plenum Press, 1978.

10. M. Denecker and D. D. Schreye. SLDNFA: an abductive procedure for normal abductive programs. In K. Apt, editor, *IJCSLP'92*, pages 686–700. MIT Press.

11. P. M. Dung. Negation as hypotheses: An abductive framework for logic programming. In K. Furukawa, editor, *8th Int. Conf. on LP*, pages 3–17. MIT Press, 1991.

12. K. Eshghi and R. Kowalski. Abduction compared with negation by failure. In *6th Int. Conf. on LP*. MIT Press, 1989.

13. A. V. Gelder, K. A. Ross, and J. S. Schlipf. The well-founded semantics for general logic programs. *Journal of the ACM*, 38(3):620–650, 1991.

14. M. Gelfond and V. Lifschitz. The stable model semantics for logic programming. In R. Kowalski and K. A. Bowen, editors, *ICLP'88*, pages 1070–1080. MIT Press.

15. M. Gelfond and V. Lifschitz. Logic programs with classical negation. In Warren and Szeredi, editors, *7th Int. Conf. on LP*, pages 579–597. MIT Press, 1990.

16. M. Gelfond and V. Lifschitz. Representing actions in extended logic programs. In K. Apt, editor, *IJCSLP'92*, pages 559–573. MIT Press, 1992.

17. A. Kakas, R. Kowalski, and F. Toni. Abductive logic programming. *Journal of Logic and Computation*, 2:719–770, 1993.

18. A. C. Kakas and P. Mancarella. Generalized stable models: A semantics for abduction. In *Proc. ECAI'90*, pages 401–405, 1990.

19. J. Lloyd and R. Topor. A basis for deductive database systems. *Journal of Logic Programming*, 2:93–109, 1985.

20. L. M. Pereira and J. J. Alferes. Well founded semantics for logic programs with explicit negation. In B. Neumann, editor, *ECAI'92*, pages 102–106. Wiley, 1992.

21. L. M. Pereira, J. J. Alferes, and J. N. Aparício. Contradiction Removal within Well Founded Semantics. In A. Nerode, W. Marek, and V. S. Subrahmanian, editors, *LP & NMR*, pages 105–119. MIT Press, 1991.

22. L. M. Pereira, J. N. Aparício, and J. J. Alferes. Non–monotonic reasoning with logic programming. *Journal of Logic Programming. Special issue on Nonmonotonic reasoning*, 17(2, 3 & 4):227–263, 1993.

23. L. M. Pereira, C. Damásio, and J. J. Alferes. Diagnosis and debugging as contradiction removal. In L. M. Pereira and A. Nerode, editors, *2nd Int. Ws. on LP & NMR*, pages 316–330. MIT Press, 1993.

24. T. Przymusinski. Extended stable semantics for normal and disjunctive programs. In Warren and Szeredi, editors, *ICLP'90*, pages 459–477. MIT Press, 1990.

25. R. Reiter. On closed–world data bases. In H. Gallaire and J. Minker, editors, *Logic and DataBases*, pages 55–76. Plenum Press, 1978.

26. R. Reiter. Towards a logical reconstruction of relational database theory. In M. Brodie and J. Mylopoulos, editors, *On Conceptual Modelling*, pages 191–233. Springer–Verlag, 1984.

27. C. Sakama. Extended well–founded semantics for paraconsistent logic programs. In *Fifth Generation Computer Systems*, pages 592–599. ICOT, 1992.

28. F. Teusink. A proof procedure for extended logic programs. In *Proc. ILPS'93*. MIT Press, 1993.

29. G. Wagner. Logic programming with strong negation and innexact predicates. *J. of Logic and Computation*, 1(6):835–861, 1991.

30. G. Wagner. Neutralization and preeemption in extended logic programs. Technical report, Freien Universitat Berlin, 1993.

31. M. Winslett. Reasoning about action using a possible model approach. In *7th AAAI*, pages 89–93, 1988.

On Logical Constraints in Logic Programming

V. Wiktor Marek[1] and Anil Nerode[2] and Jeffrey B. Remmel[3]

[1] Department of Computer Science, University of Kentucky, Lexington, KY 40506, Research supported by NSF grant IRI-9400568. E-mail: *marek@cs.engr.uky.edu*
[2] Mathematical Sciences Institute, Cornell University, Ithaca, NY 14853, Research supported by US ARO contract DAAL03-91-C-0027. E-mail: *anil@math.cornell.edu*
[3] Department of Mathematics, University of California, La Jolla, CA 92093. Research partially supported by NSF grant DMS-9306427. E-mail: *remmel@kleene.ucsd.edu*

Abstract. We introduce a new form of logic programming with constraints. The constraints that we consider are not restricted to statements on real numbers as in CLP($\mathbf{R}$), see Jaffar and Lassez [10]. Instead our constraints are arbitrary *global* constraints. The basic idea is that the applicability of a given rule is not predicated on the fact that individual variables satisfy certain constraints, but rather on the fact that the least model of the set rules that are ultimately applicable satisfy the constraint of the rule. Thus the role of clauses is slightly different than in the usual Logic Programming with constraints. In fact, the paradigm we present is closely related to stable model semantics of general logic programming, Gelfond and Lifschitz [9]. We define the notion of a constraint model of a constraint logic program and show that stable models of logic programs as well as the supported models of logic programs are just special cases of constraint models of constraint logic programs. In the general definition of a constraint logic program, the constraint of a clause is not restricted to be of a certain form or even to be expressible in the underlying language of the logic program. This feature is useful for certain applications in hybrid control systems and database applications that we have in mind. In this paper, however, we focus on the properties of constraint programs and constraint models in the simplest case where the constraints are expressible in the language of underlying program.

1 Introduction

Constraint Logic Programming received a good deal of attention since its introduction in the seminal paper of Jaffar and Lassez in 1986 [10]. The crucial observation of Jaffar and Lassez is that not all atoms in the body of a logic program play the same role in the evaluation process. They noted that it is actually beneficial to separate some goals, i.e. goals of the form $f(X_1, \ldots, X_n) = a$ and goals of the form $f(X_1, \ldots, X_m) \leq a$, from the other goals of the clauses and evaluate or solve such goals using specialized resources of the underlying computer system. Thus a part of the goals act as constraints in the space of possible solutions. One can interpret this phenomenon with a slightly shifted emphasis.

Namely we can think of the constraints as conditions that need to be satisfied by (some) parameters before we start to evaluate the remaining goals in the body of the clause. Thus the constraint controls the applicability of the clause, that is, if it fails, then the rule cannot fire. It is this point of view that we shall generalize in this paper. Our idea is that constraints can be quite arbitrary and that their role is to control the applicability of the rule. Nevertheless we shall see that in certain special cases our generalization still fits quite naturally with the idea of solving the constraints via "specialized equipment".

The constraints in constraint logic programming are evaluated in some fixed domain. This is generally some fixed relational system. In the specific schemes proposed for constraint logic programming, it can be the set of reals with some specific operations and/or relations, as in the original $CLP(R)$ or a Boolean Algebra as in *CHIP* see Dincbas et al. [8] or the set of finite or infinite trees as in Colmerauer [7]. Here we will not make that type of assumption. That is, the domain where the constraints are evaluated is **not** necessarily fixed upfront. This decision makes the resulting theory much more flexible. In fact all the previously mentioned schemes can be embedded in our scheme. It should be noted that already Jaffar and Lassez showed that the constraint logic programming scheme $CLP(R)$ can be embedded in logic programming. The gain is not in the conceptual level but in the effectiveness of processing. By contrast, our notions of constraint programs and constraint models is a genuine extension of logic programming and logic programming with negation as failure. We will show that appropriate translations allow us to express both the supported semantics of logic programs, Clark [6] and the stable semantics of logic programs (therefore also the perfect semantics, Blair et al. [3] as well) as the particular cases of our scheme. Nevertheless the relaxation of the conditions on the constraints still leads to a coherent and clean theory of what we call below constraint models of constraint programs.

The novel feature of our theory is that the constraints are supposed to be true in the model consisting of the atoms computed in the process. That is, we allow for an *a posteriori* justification of the rules applied in the process of computation. At first glance a different phenomenon occurs in constraint logic programming, namely the constraints are applied *a priori*. However we shall see below that both approaches essentially coincide in the case of upward preserved constraints. Upward preserved constraints are constraints which have the property that they are preserved as we go from a smaller to a bigger model, see section 4. This is precisely what happens in constraint logic programming. Once a constraint is computed, it is maintained. Indeed, in $CLP(R)$, once we find the multidimensional region containing the solutions of the constraint, the subsequent steps can only refine this description, the solutions must come from that region. We will see, however, that even in the case of upward preserved constraints, we get a scheme more powerful than ordinary logic programming. We shall show the least constraint model always exists and can be computed via the usual construction of the least fixed point of a monotone operator but that the closure ordinal for the least constraint model can be strictly bigger than ω. That is, the operators corresponding to such programs are monotone but are not always compact. Thus there will be the least and largest fixpoints for such operators. The least fixpoint will be the least constraint model. The largest fixpoint will be the largest constraint model but of a transformed program. The fact that both the upward and downward iterations of the operator can be transfinite make our class of programs less irregular in a certain sense than the usual operators for Horn logic programming. Recall in Horn logic programming, the least fixpoint is reached in at most ω steps whereas the greatest fixed point can take up to ω_1^{CK} iterations, see Blair [5].

The constraints we allow can be very diverse and complex formulas. They may be formulas in the underlying language of the program or they may be formulas in second order or even infinitary logic. An example of this type of constraint is the parity example where constraint is a statement about the parity of the (finite) putative model of the program. This type of the constraint is a formula of infinitary logic if the Herbrand base is infinite. What is important is that there is a method to check if the constraint holds relative to a possible model. Thus we will assume that we have a *satisfaction relation* between subsets of Herbrand base and formulas that can be used in constraints. Of course, $CLP(R)$ and similar schemes make the same type of assumptions.

Our motivation for allowing more general constraints originally came from certain applications in control theory of real-time systems. The basic idea is that we sample the plant state at discrete intervals of time, $\Delta, 2\Delta, \ldots$. Based on the plant measurements, a logic program computes the control to be used by the plant for the next Δ seconds so that the plant trajectory will meet certain required specifications. One possible way for such logic programs to operate is that the set of rules with which we compute at any give time $n\Delta$ is a function of the observations of the state of the plant at time $n\Delta$. In this fashion, we can view the plant state at time $n\Delta$ as determining which constraints of the rules of the logic program are satisfied and hence which rules can fire at time $n\Delta$. In such a situation, we cannot expect that our constraints are upward preserved or even that our constraints should necessarily be in the same language as the underlying language of the program. Thus it seems necessary to step out of the current constraint logic programming paradigm. Another application of the same type is to have a logic program controlling inventory via a deductive database. In this case, we may want to change the rules of the deductive database as a function of outside demand. Once again one could view the satisfaction of the constraints as depending on the current state of the inventory and the set of orders that come in at a given time. In this way one can vary the set of applicable rules as function of the current inventory and demand.

The paper is organized as follows. We investigate the basic properties of constraint programs in Section 2. Propositional programs with propositional constraints are studied in Section 3. It turns out that both stable and supported models of programs are special cases of constraint models of such constraint programs. Moreover we show that the existence problem for constraint models of constraint logic programs with propositional constraints is NP-complete. In Section 4 we discuss programs with whose constraints satisfy certain preservation properties. For example, we show that programs with upward-preserved constraints always have a least constraint model, although the closure ordinal of the construction is no longer ω. Section 5 contains a proof theoretic characterization of constraint models. In Section 6 we show how Default Logic can be simulated by programs with constraints. Section 7 describes a foward chaining construction of constraint models. This technique is complete for both upward-preserved and downward-preserved constraints. We show that the alternating fixpoint technique for constructing a three-valued model can be used for constraint programs with downward-preserved constraints.

2 Preliminaries

First, we shall introduce our syntax. The clauses we consider are of the form:

$$C = p \leftarrow q_1, \ldots, q_m : \Phi$$

Here $p, q_1, \ldots, q_m$ are (ground) atoms. Φ is a formula of some formal language. In the most general case we do not put restrictions on what Φ will be. The (Horn) clause $p \leftarrow q_1, \ldots, q_m$ will be denoted by H_C. The formula Φ is called the constraint of C and is denoted by Φ_C.

The idea behind this scheme is the following: Given a subset of Herbrand base, M, we consider the formulas Φ_C for clauses C of the program. Some of these formulas are satisfied by the structure determined by M. This satisfaction is checked as follows. In the case of predicate logic, M determines a relational system $\mathcal{M}_M$ which is used for testing if Φ is true. Notice that, in principle, Φ can be any sort of formula. For instance, Φ may be a formula of second order logic or of some infinitary logic. In the case of propositional logic, M determines a valuation that can be used to test Φ is true. Then we let A_M be the set of all clauses C such that $\mathcal{M}_M \models \Phi_C$. If we drop the constraints from the clauses in A_M, we get a set of Horn clauses H_C, i.e. a Horn program, which we denote by P_M. Then P_M has a least Herbrand model N_M and we say that M is a constraint model for P if $N_M = M$.

Although the use of formulas from outside of propositional logic or predicate logic may not seem very natural, it is easy to construct natural examples where the constraints naturally go beyond the expressibility in these logics. Indeed there has been significant work in deductive database theory on extensions of Datalog where one extends the underlying language by adding constructs like a transitive closure operator, Aho and Ullman [1]. Here is a simple example involving parity constraints.

Example 1. We consider a propositional program P with constraints on the parity of the target model. In a infinite propositional language, the formula expressing the fact that the number of the atoms in the model is even (or odd) is infinitary formula, i.e. it will be an infinte disjunction of infinite conjuctions. Let P be the following program with logical constraints:

$$p \leftarrow \; : \top \qquad\qquad q \leftarrow p \; : \text{"parity odd"}$$
$$s \leftarrow q \; : \top \qquad\qquad r \leftarrow p \; : \text{"parity even"}$$

Here the formula "parity odd" is an infinitary formula saying that the number of atoms true in the model is odd. Likewise "parity even" is the infinitary formula saying that the number of atoms true in the model is even. In a model with *finite* number of true atoms, one of these formulas will be true. Of course, in the model with infinite number of true atoms both of these formulas will be false.

It is easy to check that P has two constraint models. The first one $\{p, q, s\}$ has odd parity, and the other one $\{p, r\}$ has even parity.

Let us add now to this program P an additional clause: $C = w \leftarrow r \; : \top$
Then the program $P \cup \{C\}$ has just one constraint model, $\{p, q, s\}$. The reason the other model disappeared is that the presence of an (unconditional) clause C allows us to derive w once r has been derived. But the derivation of w changes the parity of the constructed set of atoms and hence invalidates the derivation of r which in turn invalidates the derivation of w as well.

3 Propositional programs with constraints

In this section we shall look at the simplest case where the underlying logic is propositional logic and the constraints Φ_C are also formulas of the propositional logic.

First of all notice that if all constraints are equal to $\top$, then our construction of a constraint model reduces to the the least Herbrand model as in ordinary Horn logic programming. Thus in that case, the constraint model coincides with the usual intended model of the program, namely, the least model.

Next, consider the case when the formulas Φ are of the form, $\neg r_1 \wedge \ldots \wedge \neg r_n$, that is, when all the constraints are goals. It should be noted here that this type of constraint is frequently used in Artificial Intelligence. In this case constraint models are just stable models of an associated general logic program. That is, given a constraint clause with the constraint in the form of a goal

$$p \leftarrow q_1, \ldots, q_m : \neg r_1 \wedge \ldots \wedge \neg r_n,$$

assign to it a general clause

$$C' = p \leftarrow q_1, \ldots, q_m, \neg r_1, \ldots, \neg r_n$$

Then set $P' = \{C' : C \in P\}$. We then have the following proposition.

Proposition 1. *If P is a constraint program whose clauses have constraints in the form of a goal, then for every $M \subseteq H$, M is a constraint model of P if and only if M is a stable model of P'.*

Next, notice that the clauses with equivalent constraints have the same effect. Specifically we have the following proposition

Proposition 2. *If $\Phi \equiv \Psi$ and P is a constraint program such that for some clause C in P, $\Phi_C = \Phi$ and $\bar{P}$ a new program which results from P by substituting Ψ for Φ in every place it occurs, then P and $\bar{P}$ have precisely the same constraint models.*

Next, consider the case when a formula $\Phi \equiv \Psi_1 \vee \ldots \vee \Psi_k$. Since changing Φ to $\Psi_1 \vee \ldots \vee \Psi_k$ does not change the class of constraint models, we can assume that $\Phi = \Psi_1 \vee \ldots \vee \Psi_k$. Given a clause $C = p \leftarrow q_1, \ldots, q_m : \Phi$, let

$$C'' = \{p \leftarrow q_1, \ldots, q_m : \Psi_1, \ldots, p \leftarrow q_1, \ldots, q_m : \Psi_k\}$$

Given a program P, define $P'' = \bigcup_{C \in p} C''$.

Proposition 3. *Let $M \subseteq H$. Then M is a constraint model of P if and only if M is a constraint model of P''.*

Now call Φ purely negative if Φ is of the form $\Psi_1 \vee \ldots \vee \Psi_k$ and each Ψ_j is a goal. Propositions 1 and 3 imply the following

Proposition 4. *If each constraint in P is purely negative, then M is a constraint model of P if and only if M is a stable model of $(P'')'$.*

Since stable models of a general logic program form an antichain (i.e. are inclusion incompatible), the same holds for programs with purely negative constraints. The inclusion-incompatibility of constraint models does not hold for arbitrary constraints. To this end consider this example:

Example 2. Let $P = \{p \leftarrow: p \wedge q, q \leftarrow: p \wedge q\}$. Then it is easy to see that both $\emptyset$ and $\{p, q\}$ are constraint models of P.

An analysis of example 2 lead us to the realization that the supported models of general logic programs can easily be described by means of a suitable transformation into programs with constraints. To this end let

$$C = p \leftarrow q_1, \ldots, q_m, \neg r_1, \ldots, \neg r_n$$

be a general logic program clause and assign to C the following constraint clause

$$C''' = p \leftarrow: q_1 \wedge \ldots \wedge q_m \wedge \neg r_1 \wedge \ldots \wedge \neg r_n$$

Then set $P'''' = \{C'''' : C \in ground(P)\}$. We then have the following proposition.

Proposition 5. *Let P be a general logic program. Then M is a supported model of P if and only if M is a constraint model of P'''.*

Next we consider another very natural case, namely when all constraints are purely positive. Here a formula is said to be purely positive if it is built from propositional letters using only conjunctions and disjunctions. In this case constraint programs reduce (at a cost) to usual Horn programs. Here is how is it done. First we can assume that our constraints are in disjunctive normal form (Proposition 2) and then we can distribute constraints in disjunctions (this fact has been used in the proof of Proposition 3). A this point all our constraints will be conjunctions of atoms which can put into the bodies of their respective clauses. This is precisely how Jaffar and Lassez reduce $CLP(R)$ to the usual logic programming. We leave to the reader to show that the constraint model of the original program coincides with the least model of the transformed program.

The existence problem for constraint models of constraint programs with propostional constraints are, generally, on the first level of the polynomial hierarchy. Specifically, we can use Proposition 4 to prove the following:

Proposition 6. *The problem of existence of constraint model of a constraint program with propositional constraints is NP-complete.*

4 Arbitrary constraints

In this section, we study constraints programs with constraints possessing some preservation properties. The first preservation property we consider is being preserved upwards. Let H denote the Herbrand bases of the constraint program being considered.

Definition 7. A formula Φ is preserved upwards if whenever $M \subseteq N \subseteq H$ and $M \models \Phi$, then $N \models \Phi$.

The nature of the formula Φ is immaterial here, it can be a propositional formula, a formula of the predicate calculus or even a formula of higher-order predicate calculus. What is important is that we have a satisfaction relation $\models$ which is used to check when a model satisfies a constraint. An example of an upward preserved constraint is a formula Φ of propositional calculus built out of propositional variables by means of conjunction and disjunction only. Similarly in predicate calculus, formulas built by use of existential quantifiers in addition to conjunction and disjunction are also preserved upwards. The same happens in case of higher-order existential quantifiers.

Since constraint models of programs with purely negative constraints are stable models of an appropriate translation, it follows easily that some constraint programs have no constraint models. However for constraint programs in which all constraints are preserved upwards, there always exists a least constraint model.

Theorem 8. *Let P be a constraint program with upward preserved constraints. Then P has the least constraint model.*

As noticed above, when the constraints are propositional positive formulas, they certainly are preserved upwards. This means that for constraint programs with positive propositional constraints there always exists the least constraint program. This last result can be obtained independently via the simpler proof outlined at the end of section 3. Moreover, it can be proven that in that case the least model can be reached by iterating operator S_P described below for ω steps. In addition, if the program P is finite, the model M can be constructed in polynomial time. However in the case when the constraints are formulas of predicate calculus and are upward preserved, it may take more than ω steps to construct the least model.

Example 3. Let P be the following program:

$$p(0) \leftarrow: \top \qquad p((s(s(X)))) \leftarrow p(s(X)) : p(X)$$
$$p((s(0))) \leftarrow: \top \qquad q(0) \leftarrow: \forall_X p(X)$$

Here the construction of the least constraint model takes precisely $\omega + 1$ steps. At step 0 we get $p(0)$ and $p(s(0))$ and at all subsequent steps we get a single new atom. In ω steps, we construct all the atoms $p^n(0)$ for $n \in \omega$ and since the Herbrand universe (as opposed to Herbrand base) is ω, the sentence $\forall_X p(X)$ becomes true. Thus the last rule is activated and in the ωth step we get $q(0)$ which completes the construction of the least constraint model.

We notice that constraint programs with longer closure ordinals can easily be constructed using the same idea. In fact, we can prove that every constructive ordinal is the closure ordinal of a finite predicate logic constraint program.

It is natural to look at the "one step" provability operator associated with a constraint program. Here is the definition of the operator. The variable I ranges over the subsets of the Herbrand base. $\models$ denotes the satisfaction relation between the formulas used for constraints and subsets of Herbrand base.

$S_P(I) = \{p : \text{for some } q_1, \ldots, q_n \in I \text{ and } I \models \Phi, \ p \leftarrow q_1, \ldots, q_n : \Phi \text{ belongs to } ground(P)\}$

Proposition 9. *If all the constraints in P are preserved upwards, then S_P is a monotone operator. Hence S_P possesses least and largest fixpoints. Moreover, the least fixed point of S_P is the least constraint model of P.*

The largest fixpoint of P is also a constraint model but of a transformed program. This is the case because we did not introduce the notion of a supported model and so we need to make a program transformation first.

Specifically, given a clause with a logical constraint, $C = p \leftarrow q_1, \ldots, q_n : \Phi$, let C'^{IV} be the clause

$$C^{IV} = p \leftarrow : q_1 \wedge \ldots \wedge q_n \wedge \Phi$$

We then have the following fact:

Proposition 10. *Assume that all the constraints in clauses of P are preserved upwards. Then fixpoints of operator S_P are precisely constraint models of P^{IV}. Thus the largest fixpoint of S_P is the largest constraint model of P^{IV}.*

Next we consider constraint programs where the constraints are preserved downwards. Formally, Φ is preserved downwards if $M_1 \subseteq M_2$ and $M_2 \models \Phi$ implies $M_1 \models \Phi$. Clearly, purely negative formulas of propositional language are preserved downwards but in other languages there are many more such formulas. For instance any negation of a purely positive formula is preserved downwards. We have the following proposition.

Proposition 11. *If all the constraints in the constraint program P are preserved downwards then the constraint models of P are inclusion-incompatible.*

5 Proof theory for constraint programs

Although Logic Programming is often associated with top-down evaluation of queries (i.e. backward chaining), we can also look at Horn logic programming in a forward chaining fashion (i.e. with bottom-up evaluation). In this paradigm, we treat the clauses of the program as rules of proof which fire when their premises are proven. This is how the layers of the iterations of the standard operator T_P are produced.

In the case of constraint programs, a similar construction is possible. However for constraint programs, we need to take into account the fact that the application of a rule requires that its constraint be satisfied.

Assume now that all the constraints of a constraint program P come from some formal language. The nature of this language is immaterial as long as we have a satisfaction relation between the subsets of the Herbrand base and formulas of that language.

We define the notion of M-proof of an atom p from program P (here M is a subset of Herbrand base) inductively.

Definition 12. 1. If $p \leftarrow: \Phi$ belongs to P and $M \models \Phi$ then $\langle p \rangle$ is an M-proof of length 1.

2. If $p \leftarrow q_1, \ldots, q_n : \Phi$ belongs to P and $S_1, \ldots, S_n$ are M-proofs, respectively, of $q_1, \ldots, q_n$ of length, respectively, $m_1, \ldots, m_n$ and $M \models \Phi$ then $S_1 \frown \ldots \frown S_n \frown \langle p \rangle$ is an M-proof of p of length $m_1 + \ldots + m_n + 1$.

3. p is M-provable from P if p has an M-proof of some length from P.

With this concept we have the following characterization theorem

Theorem 13. *M is a constraint model of P if and only if M coincides with the set of atoms possessing an M-proof from P.*

6 Default Logic and programs with logical constraints

Our constraint programs provide us with some useful insights into Reiter's Default Logic [13] and on some possible extensions of Reiter's formalism.

Recall that a default rules is a rule of the form:

$$ d = \frac{\alpha : M\beta_1, \ldots, M\beta_k}{\gamma} \tag{1} $$

A default theory is a pair $\langle D, W \rangle$ where D is a set of default rules and W is a set of formulas.

We shall see now how constraint programs can be used to get a nice rendering of default logic. Our Herbrand base (the set of atoms) is the set of all formulas of the propositional language $\mathcal{L}$. The constraints, however, are formulas of the form

$$ \neg\delta_1 \notin S \& \ldots \& \neg\delta_k \notin S \tag{2} $$

We need to define when a set of atoms, say I, satisfies a constraint of the form (2). It is defined in the most natural way; namely $I \models \neg\delta_1 \notin S \& \ldots \& \neg\delta_k \notin S$ if none of $\neg\delta_i$, $1 \leq i \leq k$ belongs to I,

We translate the rule d as in (1) into the following constraint program clause:

$$ t(d) = \gamma \leftarrow \alpha : \neg\beta_1 \notin S \& \ldots \& \neg\beta_k \notin S $$

Two more sets of rules must be included in addition to the translations of the default rules in D. First we need to translate the formulas of W. This is done by adding clauses of the form

$$t(\gamma) = \gamma \leftarrow : \top$$

for each $\gamma \in W$. Moreover we need to add the processing rules of logic. This is done by a (uniform) construction of an auxiliary program P_{logic}.

$\alpha \leftarrow : \top$ for all tautologies α, and
$\beta \leftarrow \alpha \supset \beta, \alpha : \top$ for all pairs of formulas α and β.

Now form a translation of the default theory $\langle D, W \rangle$ as follows:

$$t(D, W) = \{t(d) : d \in D\} \cup \{t(\gamma) : \gamma \in W\} \cup P_{\text{logic}}$$

We then have the following result.

Theorem 14. *Let $\langle D, W \rangle$ be a default theory, and let $t(D, W)$ be its translation as a constraint program as defined above. Then I is a constraint model of $t(D, W)$ if and only if I is a default extension of $\langle D, W \rangle$.*

We notice that once this result is proved, it becomes clear that Reiter's Default Logic is just one of many other techniques for constructions of constraint-controlled rules of proof in propositional logic. We shall not pursue the matter further here, although it should be clear that the results on upward preserved constraints are also applicable for default theories with such constraints.

7 Programs with Downward-Preserving Constraints

In this section we shall outline a foward chaining type construction to produce constraint models for the case of constraint programs with downward-preserved constraints. This construction is a modification of the forward chaining construction of the stable models of a logic program, see [12]. The input to our construction is a well ordering of constraint rules with nontrivial constraints of a constraint program P. The output of our construction will either be a constraint model of P or a subprogram P' of P and constraint model of P'. We shall show that as long as the constraints are purely negative, any constraint model of P will be the output of our foward chaining construction for a suitable well ordering of the rules of P. However this well ordering may be forced to be of length greater than ω. Therefore we will have to require that the constraints used in the construction are preserved under limits of increasing sequences. This requires that we introduce the following definition.

Definition 15. We say a formula Φ is limit-preserving if the following holds: Whenever $\langle M_\xi \rangle_{\xi < \alpha}$ is increasing family of subsets of Herbrand base (i.e for $\xi_1 < \xi_2 < \alpha$, $M_{\xi_1} \subseteq M_{\xi_2}$) and for all $\xi < \alpha$, $M_\xi \models \Phi$, then $\bigcup_{\xi < \alpha} M_\xi \models \Phi$.

First-order formulas limit-preserving formulas are Π_2 formulas, that is formulas which in their prenex form have one at most one alternation of quantifiers with universal quantifier as the first one.

Limit-preserving formulas are closed under conjunctions and disjunctions (finite or infinitary). It should be clear that atomic sentences are limit preserving.

We shall describe now our foward chaining construction which given a constraint program will always produce a constraint model of a subprogram of the original constraint program even in the case when the original program has no constraint models. For constraints which are disjunctions (even possibly infinite disjunctions) of negative literals, we prove a completeness result as well. Specifically, in such cases, we will show that every constraint model can be obtained by means of our construction.

Let P be a constraint program. Let $\preceq$ be a well-ordering of ground version of P. We also assume that all the constraints in clauses of P are limit-preserving. By $H(P)$ we mean the part of program P consisting of these clauses C for which Φ_C is a tautology. That is $H(P)$ is the Horn part of P. Thus $T_{H(P)}$ is an operator associated with the Horn part of P, van Emden and Kowalski [14]. The least fixpoint of this operator is the least model of $H(P)$. We define an increasing family of subsets of Herbrand base of P, $\langle A_\xi \rangle$ and a sequence of clauses $\langle C_\xi \rangle$ as follows:

Definition 16. .
(I) $A_0 = T_{H(P)} \uparrow \omega(\emptyset)$ and $C_0 = \top \leftarrow: \top$.

(II) If α is limit ordinal and A_ξ are defined for ordinals $\xi < \alpha$, then $A_\alpha = \bigcup_{\xi < \alpha} A_\xi$ and $C_\alpha = \top \leftarrow: \top$.

(III) If α is a non-limit ordinal, say $\alpha = \beta + 1$, and A_β already defined, then consider clauses C with the following properties:
(1) $head(C) \notin A_\beta$, (2) $body(C) \subseteq A_\beta$, (3) $A_\beta \models \Phi_C$,
(4) $T_{H(P)} \uparrow \omega(A_\beta \cup \{head(C)\}) \models \Phi_\xi$ for all $\xi < \alpha$,
(5) $T_{H(P)} \uparrow \omega(A_\beta \cup \{head(C)\}) \models \Phi_C$

If there is no clause C satisfying all the above conditions then C_α is not defined and we set $A_\alpha = \bigcup_{\xi < \alpha} A_\xi$.

Otherwise C_α is the $\prec$-first clause satisfying the above conditions and set

$$A_\alpha = T_{H(P)} \uparrow \omega(A_\beta \cup \{head(C)\})$$

Finally we set $A = \bigcup A_\alpha$. A is the output of our construction and is called the set of *accepted elements*. The least ordinal α such that C_α is not defined is called the *length of the construction*. This ordinal number α depends on the ordering $\prec$ and we denote it $\alpha_\prec$.

It is easy to see that $A = \bigcup_{\xi < \alpha_\prec} A_\xi$.

We say that a clause C is called contradictory if $head(C) \notin A$, $body(C) \subseteq A$, $A \models \Phi_C$ but $T_{H(P)} \uparrow \omega(A \cup \{head(C)\})$ does not satisfy Φ_{C_ξ} for some $\xi < \alpha_\prec$ or $T_{H(P)} \uparrow \omega(A \cup \{head(C)\})$ does not satisfy Φ_C.

The contradictory clauses are clauses which meet all the preconditions for application, but after it has been fired either some of the constraints which were previously applied are no longer valid or the constraint of C (which was valid in A) is no longer valid.

Whether a clause is contradictory depends on $\prec$. A clause C can be non-contradictory for various reasons. For instance, the head of C may already be in A in which case, there is no need to look at the body. Similarly the body of C may be not satisfied in A or, finally, the constraint of C may be false in A. Finally, a clause which has been applied during the construction of A is also non-contradictory.

Let us denote by $I^\prec$ the set of clauses that are contradictory. We can then prove the following.

Theorem 17. *1. Let $\prec$ be a well-ordering of P. Then $A^\prec$ is a constraint model of $P \setminus I^\prec$.*

  2. If all the constraints in P are conjunctions of negative literals (finite or infinite) or disjunctions of conjunctions of negative literals, then for every constraint model M of P there exists a well-ordering $\prec$ of P such that $A^\prec = M$.

A version of the well-founded semantics is also available for constraint programs with downward preserved constraints. The operator S_P introduced in Section 4 for positive programs can be defined for arbitrary constraint programs. For constraint programs with downward-preserved constraints, the operator S_P is no longer monotone. However, it is antimonotone.

Proposition 18. *When all the constraints in P are downward-preserving, S_P is antimonotone operator.*

Therefore, as noticed in Baral and Subrahmanian [4], the operator S_P^2 is monotone. Thus S_P^2 possesses both least and largest fixpoints. Moreover, these fixpoints, say A and B, are alternating, i.e. $S_P(A) = B$ and $S_P(B) = A$. Thus, although we cannot be sure that that for constraint programs with downward preserved constraints, there is a constraint model (as was the case of programs with upward-preserving constraints), we have an alternating pair of fixed points of S_P^2. Notice that $S_P(A)$ consists of these atoms that can be computed when the constraints are checked with respect to A and hence is the analogue of the usual T_P operator of logic programming.

References

1. A.V. Aho and J.D. Ullman. Universality of Data Retrieval Languages. In: *ACM Symposium on Principles of Programming Languages*, pages 110–120, 1979.
2. K. Apt. Logic programming. In: J. van Leeuven, editor, *Handbook of Theoretical Computer Science*, pages 493–574. MIT Press, Cambridge, MA, 1990.
3. K.R. Apt, H.A. Blair, and A. Walker. Towards a Theory of Declarative Knowledge. In: J. Minker, editor, *Foundations of Deductive Databases and Logic Programming*, pages 89–142, Los Altos, CA, 1987. Morgan Kaufmann.
4. C. Baral and V.S. Subrahmanian. Stable and Extension Class Theory for Logic Programs and Default Theories. *Journal of Automated Reasoning* 8, pages 345–366, 1992.
5. H.A. Blair. The recursion-theoretic complexity of predicate logic as programming language. *Information and Control* 54, pages 25–47, 1982.
6. K.L. Clark. Negation as failure. In: H. Gallaire and J. Minker, editors, *Logic and data bases*, pages 293–322. Plenum Press, 1978.
7. A. Colmerauer. PROLOG III Reference and Users Manual, PrologIA, Marseilles 1990.
8. M. Dincbas and H. Simonis and P. Van Hententryck and A. Aggoun. The Constraint Logic Programming Language CHIP. In: *Proceedings of the 2nd International Conference on Fifth Generation Computing Systems*, pages 249–264. 1988.
9. M. Gelfond and V. Lifschitz. The Stable Semantics for Logic Programs. In: *Proceedings of the 5th International Symposium on Logic Programming*, pages 1070–1080, Cambridge, MA., MIT Press, 1988.
10. J. Jaffar and J.-L. Lassez. Constraint Logic Programming. In: *Proceedings of the 14th ACM Symposium on Principles of Programming Languages*, pages 111–119, Münich, 1987.
11. J. Jaffar and M. Maher. Constraint Logic Programming: A Survey. *Journal of Logic Programming* 19-20, pages 503–581, 1994.
12. V. Marek, A. Nerode, and J.B. Remmel. Rule Systems, well orderings, and foward chaining. Cornell University, MSI Technical Report, 1994.
13. R. Reiter. A logic for default reasoning. *Artificial Intelligence* 13, pages 81–132, 1980.
14. M.H. van Emden and R.A. Kowalski. The Semantics of Predicate Logic as a Programming Language. *Journal of the ACM* 23 pages 733–742, 1976.

An Operator for Composing Deductive Data Bases with Theories of Constraints*

D. Aquilino[1] and P. Asirelli[2] and C. Renso F. Turini[3]

[1] Intecs Sistemi SpA, Via Gereschi 32, 56125 Pisa, Italy,
aquilino@pisa.intecs.it
[2] Istituto di Elaborazione dell'Informazione-CNR,Via S.Maria 46, 56126 Pisa, Italy,
asirelli@iei.pi.cnr.it
[3] Dipartimento di Informatica,Università di Pisa, Corso Italia 40, 56125 Pisa, Italy,
{renso,turini}@di.unipi.it

Abstract. An operation for restricting deductive databases represented as logic programs is introduced. The restrictions are represented in a separate deductive database. The operation is given an abstract semantics in terms of the immediate consequence operator. A transformational implementation is given and its correctness is proved with respect to the abstract semantics.

1 Introduction

Deductive Databases can potentially solve many problems in the field of data and knowledge management. However, in order to make the approach really viable, it is necessary to define an environment at least as rich as the one which has been developed for other kinds of database management systems.

At present, in the database area, a lot of attention is devoted to studying the possibility of combining different databases or, in any case, databases which have been developed within other projects, and that may be resident at different sites. The goal generally can be seen from the point of view of building applications by means of cooperative, interoperable and/or distributed databases. Our claim is that one approach to the above problems is the definition of an environment that provides a set of basic operators, with a clear formal semantics, for the combination of different databases.

One interesting point is the ability of defining a particular view of the database as a refinement of its relations.

The notion of view we have in mind should allow one, among other possible operations, to impose dynamically a set of restrictions on a database in order to filter elements which do not satisfy certain properties out of the relations of the database. More precisely, we propose an operation that, given a deductive database and a set of restrictions, computes a new deductive database, the relations of which are the original ones properly filtered according to the restrictions.

---

* Work partially supported by "The Exploratory activity EC-US" Nr. 033 and by ESPRIT project Compulog II, #6810

It is worth noting now that the restrictions are represented as a deductive database itself, i.e. as a collection of deductive rules which establish conditions about the belonging of a tuple to a relation. In other words, the database of restrictions can be seen as an incremental refinement of the original database.

Given that we represent deductive databases as logic programs, our aim is the definition of an operation that, given two logic programs - the original database and the one containing the restrictions - computes a new logic program that behaves as the restricted database.

The following example clarifies the expected behavior of the operation. We have a theory graph which defines two extensional relations, node and edge, that represent a graph, and two intensional relations, path and bidirectional_edge, with the straightforward meaning.

Theory Graph

```
node(a)              edge(d,c)
node(b)              edge(b,a)
node(c)              edge(a,b)
node(d)              edge(b,d)
                     edge(d,b)

path(X,Y) ← edge(X,Y)
path(X,Y) ← edge(Z,Y),path(X,Z)

bidirectional_edge(X,Y) ← edge(X,Y), edge(Y,X)
```

We can now *constrain* the above database by means of rules which impose further restrictions on the relations. For example the rule

node(X) ← path(a,X)

establishes that we accept only nodes that are reachable from the node a.

The following rule establishes that in the constrained graph each node has degree al least two

node(X) ← bidirectional_edge(X,Y), bidirectional_edge(X,Z), $Y \neq Z$

In the following, we show the results of some queries on the theory graph w.r.t. the constraints.

The query ← $node(b)$ satisfies both constraints, while the query ← $node(d)$ satisfies the first constraint but not the second one.

The approach presented in this paper stems from two separate lines of research:

- the general study of operators on logic programs [7]
- the transformational approach to constraints handling in deductive databases [4]

The study of operators on logic programs consists in the abstract definition of a suite of operations, in the study of their formal properties, in the design of several implementation strategies, and in the application to several problems in the field of knowledge representation and software engineering.

The transformational approach to integrity constraints checking was defined in [4] for databases represented by positive logic programs. There, the theory of constraints was not a logic program itself but a conjunction of formulas denoting the "if then" part of relations. The assumption to constraints satisfaction was that the constraints should have been logical consequences of the database, i.e. the formulas ought to be true in the minimal model of the database. According to this view, the considered approach defined a new logic program, "modified" by the constraints formulas, so that the minimal model of the resulting database was a model of the theory of constraints as well. The approach of [4] was afterwards extended by [8] to deal with stratified logic program, in view of programming by exception as an approach to default reasoning.

The paper is organized as follows. In Section 2, we describe a complex application of the incremental restriction of deductive databases in the field of building software engineering tools. In Section 3 we present the operation of restriction via its abstract semantics. The abstract semantics is given compositionally in terms of the consequence operators associated to the deductive data base and the database containing the restriction rules. Section 4 discusses an implementation of the operator based on a transformation of the database and the restricting database into a new *constrained* database. We establish also the correctness of the implementation with respect to the abstract semantics. Section 5 presents the comparison of our approach with other approaches in the literature and discusses future work.

2 A Motivating Example

The use of constraints as a restriction in deductive databases [4] has been experimented in Gedblog which is a multi-theory deductive database management system [3]. Gedblog supports the consistent design and prototyping of graphic applications through an incremental development and/or by combining pre-defined theories and constraints. Knowledge is expressed as a deductive database spread into multiple extended logic theories. Each theory entails a piece of application data model and behavior by means of

- facts and rules to express the "general" knowledge;
- integrity constraints formulas to express either "exceptions" to the general knowledge, or general "requirements" of the application being developed;
- transactions to express atomic update operations that can be performed on the knowledge-base.

A very interesting and practical experiment in using Gedblog has been carried out by realising a graphical editor for Oikos process structure [2]. Oikos is an environment that provides a set of functionalities for the easy construction of process-centered software development environments. In the following a brief description of Oikos is provided. This description will highlight the characteristics that have a direct impact on the realization of the Oikos editor.

In Oikos [1] a software process model is a set of hierarchical entities. Each entity is an instance of one of the Oikos classes and represent a modelling concept. Classes are: Process, Environment, Desk, Cluster, Role and Coordinator.

An entity can be either structured (i.e. formed by other entities) or simple (i.e. a leaf in the model structure). Oikos defines a top-down method to construct process models using two descriptions of the entities: abstract and concrete. The abstract entities are introduced first and then refined in the concrete ones. The method establishes some constraints about the entities and their use as subentities during the model refinement.

The Oikos Process Editor consists of several databases. There are two kinds of theories:

- the ones representing the Oikos model;
- the ones imposing constraints on the modelling process.

The theory Oikos-instance contains a specific instance of an Oikos process and the theory Oikos-model describes the types of the entities of the Oikos model and some relations between them (consistency, refinement...). In the following, we will refer to the union of this two theories as the Oikos model. The theory Oikos-constraints contains all the constraints on the Oikos processes.

The following theory defines the schema of a process model, where slc stands for Software Life Cycle. The slc process is structured into a management activity (manager), a verification environment (v_v) and an automatic control (assistant).

Oikos-instance

```
compound(slc,process)
management(slc,manager,env)
angel(v_v, ang_env)
part_of(slc,process,v_v,ang_env)
part_of(slc,process,assistant,coord)
```

Oikos-model

The following clauses define which kinds are concrete, which ones are compound, which are the abstract types and which are the management ones:

```
concrete_kind(process).        compound_kind(process).
concrete_kind(env).            compound_kind(env).
concrete_kind(desk).           compound_kind(desk).
concrete_kind(cluster).        compound_kind(cluster).
concrete_kind(role).
concrete_kind(coord).
```

```
abstract_kind(ang_process).      management_kind(process).
abstract_kind(ang_env).          management_kind(env).
abstract_kind(ang_desk).         management_kind(desk).
abstract_kind(ang_cluster).
abstract_kind(ang_role).
```

The following rule states that each compound entity is necessarily concrete:

```
concrete(Name,Kind) <-- compound(Name,Kind)
```

The definition of the `consistent` predicate states that the entity given as the second argument can be used inside the entity which appears as the first argument. The `is_refinement` predicate states that the second argument is a refinement of the first one.

```
consistent(process,ang_process).    is_refinement(ang_process,process).
consistent(process,ang_env).        is_refinement(ang_env,env).
consistent(process,ang_role).
```

We can now discuss a number of constraints on the database that are given as rules restricting the relations of the database.

Oikos-constraints

An Oikos compound entity has compound kind and at least a coordinator:

```
1)    compound(Name,Kind) <-  compound_kind(Kind),
                              part_of(Name,Kind,Coord,coord).
```

An Oikos angelic entity has abstract kind and is part of a concrete one:

```
2)    angel(Name,Kind) <- part_of(Name1,Kind1,Name,Kind),
                          compound_kind(Kind1),
                          abstract_kind(Kind).
```

An Oikos entity which has parts is compound and the relationship between kinds is consistent (predicate 'consistent'):

```
3)    part_of(Name1,Kind1,Name2,Kind2) <- compound(Name1,Kind1),
                                          consistent(Kind1,Kind2).
```

Name2 may be a refinement of Name1 if Name1 is angelic, Name2 is concrete and the kind of Name2 is a refinement of the kind of Name1 (predicate 'is_refinement'):

```
4)    refinement(Name1,Name2) <- angel(Name1,Kind1),
                                  concrete(Name2,Kind2),
                                  is_refinement(Kind1,kind2).
```

Each process has a management entity:

```
5)    compound(Name,process) <- management(Name,Name1,Kind),
                                management_kind(Kind).
```

Here we consider some examples of successful and failing queries. Suppose to add a set of new instances to the theory Oikos-instance. Let Oikos-new-inst be the following:

Oikos-new-inst

```
    part_of(desk1, desk, slc, process).
    compound(role1,role).
    refinement(slc,desk1).
    angel(coord1,coord).
```

The new instance of the Oikos process is the theory Oikos-instance ∪ Oikos-new-inst. Notice that the query ← *part_of(desk1, desk, slc, process)* succeeds if we evaluate it in the unconstrained Oikos process (Oikos-instance ∪ Oikos-new-inst ∪ Oikos-model), while it fails when we evaluate it in the constrained theory:

(Oikos-instance ∪ Oikos-new-inst ∪ Oikos-model) $/_{IC}$ Oikos-constraints

where $A/_{IC}B$ denotes the operation of restricting A by means of B.

In fact constraint number 3 constrains the facts of the part_of relation. In our case, desk is a compound kind (see Oikos-model), but desk in not consistent with process, that is, we cannot have a process inside a desk.

An analogous case is the query ← *compound(role1,role)*. The constraint 1 is not satisfied, because role does not have a compound_kind definition in Oikos-model.

The query ← *refinement(slc,desk1)* does not satisfy the constraint 4 because *slc* is not an angel.

Finally, the query ← *angel(coord1,coord)* fails because of the failure of the constraint 2 which states that coord1 must be a part of a compound_kind and it must also be an abstract_kind.

3   Abstract definition

We are now in the position of giving an abstract definition of the operation $/_{IC}$ that restrict a deductive database by means of a database of constraints. The abstract definition is given via the notion of *immediate consequence operator*. The immediate consequence operator T_P associated to a logic program P is a function that points out which consequences are deducted from a given interpretation by the program P in a single inference step. The abstract definition of an operation on logic programs can then be given in a compositional way by defining the immediate consequence operator of the resulting logic program as a composition of the immediate consequence operators of the argument programs.

We recall now from [7] the basic definition of immediate consequence operator and the definition of two basic operations on logic programs: union and intersection.

For a logic program P, the immediate consequence operator $T(P)$ is a continuous mapping over Herbrand interpretations defined as follows [10]. For any Herbrand interpretation I:

$$A \in T(P)(I) \iff (\exists \bar{B} : A \leftarrow \bar{B} \in ground(P) \wedge \bar{B} \subseteq I)$$

where $\bar{B}$ is a (possibly empty) conjunction of atoms.

The semantics of program operations is given in a compositional way by extending the definition of T with respect to the first argument. We assume that the language in which programs are written is fixed. Namely, the Herbrand base we refer to is determined by a set of function and predicate symbols that include all function and predicate symbols used in the programs being considered.

For any Herbrand interpretation I:

$$T(P \cup Q)(I) = T(P)(I) \cup T(Q)(I)$$
$$T(P \cap Q)(I) = T(P)(I) \cap T(Q)(I)$$

The above definition generalises the notion of immediate consequence operator from programs to compositions of programs. The operations of union and intersection of programs directly relate to their set-theoretic equivalent. The set of immediate consequences of the union (resp. intersection) of two programs is the set-theoretic union (resp. intersection) of the sets of immediate consequences of the separate programs.

Let us now go back to the new operator we want to define. Informally, a program P constrained by another program Q is obtained by the union of two parts. One is the intersection of the two theories, that forces the theories to agree during the deduction. But the intersection alone is not enough, because some clauses would be missing in the result. In particular, we miss all the clauses which are defined in P and not constrained by Q. Those are of two kinds: the ones which do not have a definition in Q, and those which have a definition in Q that constrains only a subset of atoms potentially derivable in P. In the last case the clause defining a given predicate r in P is not fully instatiated while the definition of r in Q has at least a ground argument. In order to include both the first and the second kinds of clauses in the resulting program we define a new operator $/_{HIC}$. The definitions of these new operators are given as follows.

Definition 1. $T(P /_{IC} Q) = \lambda I.\ T(P /_{HIC} Q)(I) \cup T(P \cap Q)(I)$

Definition 2. $T(P /_{HIC} Q) = \lambda I.\ T(P_1)(I) \cup (T(P_2)(I) \setminus T(Q)(\mathcal{B}_P))$

where

P_1 contains the predicate definition of P which are not defined in Q

P_2 contains the predicate definition of P which are also defined in Q, such that $P = P_1 \cup P_2$ and $\mathcal{B}_P$ is the Herbrand Base associated to P

We are mainly interested in a transformational definition of the $/_{IC}$ operator. However, although the above definition, based on set-theoretic difference, is easy to understand, it is not easy to implement in a transformational way. Hence, we need to find another abstract definition of the operator, equivalent to the previous one, easier to implement. Of course, we want to show that the two abstract definitions coincide. The idea is to exploit the observation that the operation of set-theoretic difference can be turned into a combination of intersection and complement. In our case, the complement of a program is not defined, so we need first to give an abstract semantics and a transformational definition for it.

Definition 3. $T(P /_{HIC} Q) = \lambda I.T(P_1)(I) \cup T(P_2 \cap (Q,P)^c)(I)$

where the semantics of $(Q,P)^c$ is defined as follows

Definition 4. $T((Q,P)^c) = \lambda I.\ \mathcal{B}_{P_\pi} \setminus T(Q)(\mathcal{B}_P)$

We denote with $\mathcal{B}_{P_\pi}$ the subset of the Herbrand base of P obtained by selecting those predicate names which have a definition in Q

Theorem 5. *Let P, Q be programs, P_1 and P_2 defined as above, A be an atom. Then*
$$A \in T(P_1)(I) \cup T(P_2 \cap (Q,P)^c)(I) \iff A \in T(P_1)(I) \cup (T(P_2)(I) \setminus T(Q)(\mathcal{B}_P))$$

Proof

$A \in (T(P_1)(I) \cup T(P_2 \cap (Q,P)^c)(I))$

$\iff$ *definition of* P_2

$A \in (T(P_1)(I) \cup T(P \cap (Q,P)^c)(I))$

$\iff$ *definition of* $\cup$ *and* $\cap$

$A \in (T(P_1)(I) \vee (\ A \in T(P)(I) \wedge A \in T((Q,P)^c)(I)))$

$\iff$ *definition of* $T((Q,P)^c)(I)$

$A \in T(P_1)(I) \vee (\ A \in T(P)(I) \wedge A \in (\mathcal{B}_\pi \setminus T(Q)(\mathcal{B}_P)))$

$\iff$ *definition of set difference*

$A \in T(P_1)(I) \vee (\ A \in T(P)(I) \wedge A \in \mathcal{B}_{P_\pi} \wedge A \notin T(Q)(\mathcal{B}_P))$

$\iff$ *definition of* $\cap$

$A \in T(P_1)(I) \vee (A \in T(P)(I) \cap \mathcal{B}_{P_\pi} \wedge A \notin T(Q)(\mathcal{B}_P))$

$\iff$ *definition of* P_2 *and* $\mathcal{B}_{P_\pi}$

$A \in T(P_1)(I) \vee (A \in T(P_2)(I) \wedge A \notin T(Q)(\mathcal{B}_P))$

$\Longleftrightarrow$ *definition of set difference*

$A \in T(P_1)(I) \vee (A \in T(P_2)(I) \setminus T(Q)(\mathcal{B}_P))$

$\Longleftrightarrow$ *definition of* $\cup$

$A \in (T(P_1)(I) \cup (T(P_2)(I) \setminus T(Q)(\mathcal{B}_P)))$ $\qquad\qquad\qquad$ □

4 Transformational Definition

We address now the problem of providing an implementation for our abstract operator. There are several possibilities for implementing operations on logic programs and they can be classified in two broad categories: an interpretation oriented approach, and a compilation oriented approach. As discussed in [7] the first one can rely upon either metaprogramming techniques or the design of specific abstract machines, whilst the second approach consists in designing proper transformations which map logic programs into logic programs. In this paper we give a transformation oriented implementation of the operator $/_{IC}$. We first recall from [7] the transformational implementation of $\cup$ and $\cap$. We overload the symbols $\cup$ and $\cap$ and we use them also for denoting the transformations.

Given two programs $\mathcal{P}$ and Q, the program corresponding to the union $\mathcal{P} \cup Q$ is just the set theoretic union of the clauses of the two programs:

$$\mathcal{P} \cup Q = \{A \leftarrow B \mid (A \leftarrow B \in \mathcal{P}) \vee (A \leftarrow B \in Q)\}.$$

On the other hand, the implementation of the $\cap$ operation exploits the basic unification mechanism of logic programming. Given two programs $\mathcal{P}$ and Q, a clause $A \leftarrow B$ belongs to $\mathcal{P} \cap Q$ if there is a clause $A' \leftarrow B'$ in $\mathcal{P}$ and there is a clause $A" \leftarrow B"$ in Q such that their heads unify, that is $\exists \vartheta : \vartheta = mgu(A', A")$, and $A = (A')\vartheta$, and B is obtained from $(B', B")\vartheta$ by removing duplicated atoms.

$$\mathcal{P} \cap Q = \{A \leftarrow B \mid (A' \leftarrow B' \in \mathcal{P}) \wedge (A" \leftarrow B" \in Q) \wedge$$
$$\vartheta = mgu(A', A") \wedge (A = A'\vartheta) \wedge (B = B'\vartheta \cup B"\vartheta)\}.$$

For example, consider the programs

$\mathcal{P}$	Q	$\mathcal{R}$
$r(x, y) \leftarrow s(x)$	$t(x) \leftarrow$	$r(x, y) \leftarrow s(x), t(y)$
$s(f(x)) \leftarrow$		$s(x) \leftarrow$

The program corresponding to the program expression $(\mathcal{P} \cup Q) \cap \mathcal{R}$ is obtained as follows:

$\mathcal{P} \cup Q$	$(\mathcal{P} \cup Q) \cap \mathcal{R}$
$r(x, y) \leftarrow s(x)$	$r(x, y) \leftarrow s(x), t(y)$
$s(f(x)) \leftarrow$	$s(f(x)) \leftarrow$
$t(x) \leftarrow$	

We are now in the position of giving the definition of the constraining operator.

Definition 6. $P /_{IC} Q = P_1 \cup (P_2 \cap Q^c) \cup (P \cap Q)$

where
P_1 and P_2 are defined as above and Q^c is defined as follows.

Definition 7. For each predicate p in Q let C_p be the set of clauses defining it. Then we define a correspondent set of unit clauses C_p' in Q^c in the following way:

- if C_p is a ground unit clause
  $p(t_1, \ldots, t_n) \leftarrow$
  then C_p' is
  $p(X_1, \ldots, X_n) \leftarrow X_1 \neq t_1$
  $\vdots$
  $p(X_1, \ldots, X_n) \leftarrow X_n \neq t_n$

- if $C_p = p(X_1, \ldots, X_n)$ is a unit clause where X_i is a variable and $X_i \neq X_j$ $\forall$ i,j $\in [1, \ldots n]$ then C_p' is the empty set
- if $C_p = p(X_1, \ldots, X_n)$ is a unit clause where X_i is a variable $\forall$ i $\in [1, \ldots n]$ and $X_i = X_j$ for some i,j , then C_p' is
  $p(X_1, \ldots, X_n) \leftarrow X_i \neq X_j$
- if $C_p = p(t_1, \ldots, X_n)$ is a unit clause containing both ground and non ground arguments, then C'_p is obtained by applying the rules defined above for each argument, according to its nature.
- if $C_p = p(t_1, \ldots, t_n) \leftarrow q(s_1, \ldots, s_m), \ldots, r(v_1, \ldots, v_k)$ is a clause of Q, not necessary ground, then C_p' is obtained by "forgetting" the body $q(s_1, \ldots, s_m), \ldots, r(v_1, \ldots, v_k)$ and applying the method to the remaining unit clause
- if C_p is a set of clauses c1, c2, $\ldots$, cm defining the predicate p, then C_p' is obtained in the following way:
  $C_p' = c1^c \cap c2^c, \cap \ldots, \cap cm^c$

Then Q^c is the set of all the clauses in C_p' for each predicate p defined in Q

From now on we will use the *Ground* predicate with two arguments: the first is the Herbrand base with respect to which we instantiate the program given as the second argument which is a program. So, $\text{Ground}(\mathcal{B}_P)(Q^c)$ denotes a set of ground unit clauses which derive all the facts $(\in \mathcal{B}_P)$ that cannot unify with the heads of the clauses in Q

Observation *Consider the program Q' obtained from Q taking only the heads of the clauses of Q. If we transform Q' into a program in Constraint Logic Programming (with '=' constraints), then Q^c is obtained by replacing each '=' symbol occurring in the body, with the $\neq$ symbol.*

Now, we want to show that a ground atom A is an immediate consequence of the transformed program Q^c with respect to an interpretation I if and only if A belongs to $T((Q,P)^c)(I)$. In order to give the proof, we need a preliminary result

Lemma 8. $A \leftarrow \; \in Ground(\mathcal{B}_P)(Q^c) \Longleftrightarrow A \notin T(Q)(\mathcal{B}_P) \wedge A \in \mathcal{B}_{P\pi}$

Proof

By contradiction

Suppose A is $p(t_1, \ldots, t_n) \in T(Q)(\mathcal{B}_P)$. This means that it is a ground instance of a head of a clause in Q. The head of the clause in Q can have one of three possible forms:

1) $p(t_1, \ldots, t_n) \leftarrow$
2) $p(X_1, \ldots, X_n) \leftarrow$
3) $p(X_1, \ldots, t_n) \leftarrow$

Then, in Q^c there must be the complement of it. In case 1) in Q^c there must be:

$p(X_1, \ldots, X_n) \leftarrow X_1 \neq t_1, \ldots, X_n \neq t_n, \ldots$

but this is a contradiction because $A \leftarrow \; \in Ground(\mathcal{B}_P)(Q^c)$ by hypothesis.

In case 2) either the variables are all different or some of them are the same. In the former case no new definitions of p are introduced.

In the latter case, if $X_i = X_j$, then in Q^c there will be

$p(X_1, \ldots, X_n) \leftarrow X_i \neq X_j$

that contradicts the hypothesis.

Finally, in case 3), in Q^c there will be:

$p(X_1, \ldots, X_n) \leftarrow X_n \neq t_n,$

and this is a contradiction because $A \leftarrow \; \in Ground(\mathcal{B}_P)(Q^c)$ by hypothesis. $\quad\square$

The following theorem establishes the correctness of the *complement* operation with respect to the natural interpretation in terms of set-difference with respect to the universe.

Theorem 9. *Given a program Q, let R be the program obtained from the transformation Q^c. Then $T(R)(I) = \mathcal{B}_{P\pi} \setminus T(Q)(\mathcal{B}_P)$*

Proof

Let A an atom

$A \in T(R)(I)$

$\Longleftrightarrow$ *definition of* T

$\exists \, B \mid A \leftarrow B \in Ground(\mathcal{B}_R)(R) \wedge B \subseteq I$

$\Longleftrightarrow$ *R contains only unit clauses*

$A \leftarrow \; \in Ground(\mathcal{B}_R)(R)$

$\Longleftrightarrow$ *lemma 1*

$A \notin T(Q)(\mathcal{B}_P) \wedge A \in \mathcal{B}_{P\pi}$

$\Longleftrightarrow$ *definition of* $\setminus$

$A \in (\mathcal{B}_{P\pi} \setminus T(Q)(\mathcal{B}_P))$ $\qquad\qquad\qquad\qquad\qquad\qquad\qquad\qquad\square$

We are, at last, in the position of proving the correctness of the transformation oriented implementation of $/_{IC}$ with respect to its abstract semantics.

Theorem 10. *Let P, Q be programs and P_1, P_2 defined as in the previous section. Then*

$$P /_{IC} Q = P_1 \cup (P_2 \cap Q^c) \cup (P \cap Q) \text{ is correct w.r.t. } T(P /_{IC} Q)$$

Proof

The correctness of union and intersection was proved in [6]. The correctness of the ^{c} operator is established by theorem 2.   □

The following example clarifies the transformational approach

Example 1. Let us consider the programs

$$
\begin{array}{ll}
P & Q \\
A(a,a) \leftarrow & A(a,y) \leftarrow C(a,y) \\
A(a,b) \leftarrow & \\
B(b,b) \leftarrow & \\
B(c,c) \leftarrow & \\
C(b,a) \leftarrow & \\
A(x,y) \leftarrow B(x,y) &
\end{array}
$$

We split P into P_1 and P_2 as stated in section 3

$$
\begin{array}{ll}
P_1 & P_2 \\
B(b,b) \leftarrow & A(a,a) \leftarrow \\
B(c,c) \leftarrow & A(a,b) \leftarrow \\
C(b,a) \leftarrow & A(x,y) \leftarrow B(x,y)
\end{array}
$$

The complement of Q, built according to the definition 6, is

$$
\begin{array}{l}
Q^c \\
A(x,y) \leftarrow x \neq a
\end{array}
$$

The first step is to compute the theory obtained from the intersection of P_2 and Q^c. Namely, we want to obtain those rules the head of which has a predicate defined in Q, but are not constrained by Q. This is the case in which Q constraints a specific instantiation of rules of P. The second step is to build $P \cap Q$, that is, all the clauses of P which agree with the constraints. So,

$$
\begin{array}{ll}
P_2 \cap Q^c & P \cap Q \\
A(x,y) \leftarrow x \neq a, B(x,y) & A(a,a) \leftarrow C(a,a) \\
 & A(a,b) \leftarrow C(a,b) \\
 & A(a,y) \leftarrow B(a,y), C(a,y)
\end{array}
$$

Thus, the transformed program which satisfies the constraints is

$$P_1 \cup (P \cap Q^c) \cup (P \cap Q)$$
$$A(x,y) \leftarrow x \neq a, B(x,y)$$
$$A(a,a) \leftarrow C(a,a)$$
$$A(a,b) \leftarrow C(a,b)$$
$$A(a,y) \leftarrow B(a,y), C(a,y)$$
$$B(b,b) \leftarrow$$
$$B(c,c) \leftarrow$$
$$C(b,a) \leftarrow$$

5 Conclusions

We have presented an operation for the definition of restricting views of deductive databases, represented as logic programs. The approach allows the definition and use of databases as theories of constraints for another existing database. The main advantage of the approach is that the database and the theories of constraints can be defined separately by different users and at different stages of the development of an application. The example of Section 2 illustrates exactly this point: deductive databases supporting software engineering tools are incrementally restricted by means of theories of constraints.

The operation of restriction has been given an abstract semantics. The semantics has been given compositionally, in terms of the immediate consequence operators associated to the deductive database to restrict and the databases containing the restriction rules. An implementation of the operation has been presented. Such an implementation is based on a transformation of the database and the restricting database into a new constrained database. The correctness of the implementation with respect to the abstract semantics has been proved.

The example of Section 2 is part of a complex application developed in [2]. The transformed program resulting from the implementation of operation $/_{IC}$ is also obtainable by the so called *Modified Program Approach* of [4] supported by GEDBLOG [3]. The novelty is that the transformation is now associated to an operation on logic program with a clear formal semantics. We consider this as a step towards the design of a semantically well founded environment for prototyping and developing deductive databases. The development can be carried on by using the operation of these paper and other ones to put together theories developed either elsewhere, or by different users, or resident in different sites, i.e. a formal basis to study applications obtained by means of cooperative, interoperable and/or distributed databases.

A transformational approach for constraints and default reasoning is described in [8]. Although the transformation is similar to the one proposed here and in GEDBLOG, no formalization in terms of operation on logic programs is given. On the other hand in [8] also normal stratified logic programs are dealt with. We are currently addressing the problem of handling normal programs within a more general framework. In fact, we are developing conditions for the deductive database and the theory of constraints that guarantee the conservation of given properties through the application of the operator. In particular, we have preliminary results on conditions which mantain stratification properties, and, consequently, allow us to extend the use of our operator to stratified normal programs. Roughly speaking, if the deductive database is stratified and the theory of constraints is stratified with respect to it, then the resulting database is stratified, too. From an abstract viewpoint, the definition of the semantics of the operator when applied to normal programs is given by referring to Fitting's immediate consequence operator [9]. From an implementational viewpoint we are considering both a transformational approach based on intensional negation [5] and an interpretation oriented approach.

References

1. V. Ambriola and C. Montangero. Oikos: Constructing process-centered SDEs, Software Process Modelling and Technology. In A. Finkelstein, J. Kramer, and B. Nuseibeh, editors, *Research Study Press*. J. Wiley and sons, 1994.
2. D. Apuzzo, D. Aquilino, and P. Asirelli. A Declarative Approach to the Design and Realization of Graphic Interfaces. Technical Report B4-39, October 1994, IEI-CNR Internal Report, 1994. submitted for publication.
3. P. Asirelli, D. Di Grande, P. Inverardi, and F. Nicodemi. Graphics by a logic database management system. *Journal of the Visual languages and Computing*, 1994. to appear.
4. P. Asirelli, M. De Santis, and M. Martelli. Integrity Constraints in Logic Databases. *Journal of Logic Programming*, 3:221,232, 1985.
5. R. Barbuti, P. Mancarella, D. Pedreschi, and F. Turini. A trasformational approach to negation in logic programming. *Journal of Logic Programming*, 2:201–228, 1990.
6. A. Brogi. *Program Construction in Computational Logic*. PhD thesis, University of Pisa, March 1993.
7. A. Brogi, A. Chiarelli, P. Mancarella, V. Mazzotta, D. Pedreschi, C. Renso, and F. Turini. Implementations of program composition operations. In M. Hermenegildo and J. Penjam, editors, *Sixth International Symposium on Programming Languages Implementation and Logic programming*, pages 292–307. Springer-Verlag, 1994.
8. R.A. Kowalski and F. Sadri. Logic programs with exceptions. In DHD Warren and P. Szeredi, editors, *7th International Conference on Logic Programming, Proceedings*, pages 598–613. The MIT Press, Cambridge, Mass., 1990.
9. Fitting M. A kripke-kleene semantics for general logic programs. *Journal of Logic Programming*, 2:295–312, 1985.
10. van Emden M. H. and R. A. Kowalski. The semantics of predicate logic as a programming language. *Journal of the ACM*, 23(4):733–742, 1976.

Update Rules in Datalog Programs [*]

M. Halfeld Ferrari Alves[2], D. Laurent[1,2], N. Spyratos[2]

[1] Université d'Orléans, LIFO, F-45067 Orléans Cedex 2, France
[2] Université de Paris-Sud, LRI, U.R.A. 410 du CNRS, Bât. 490
F-91405 Orsay Cedex, France

Abstract. We consider Datalog^{neg} databases containing two kinds of rules: update rules and query rules. We regard update rules as constraints, *all* consequences of which must hold in the database until a new update. We introduce a semantics framework for database updates and query answering based on the well-founded semantics. In this framework, updating over intensional predicates is deterministic.

1 Introduction

In this paper, we present formal semantics for Datalog databases containing two kinds of rules: *update rules* and *query rules.* Update rules are activated during an update request and query rules are activated during a query request. Thus, the database comprises two inference mechanisms, one for updates and one for queries. We assume that after every update, *all* consequences of update rules must hold in the database until a new update. In the remaining of this section, we illustrate by examples how these two inference mechanisms work and how they interact with each other.

Update Rules - In our context, an update means the insertion or deletion of a single fact in the database, and an update rule describes a possible side effect of an update. Therefore, the update rules that we consider are rules of the form $L_0 \leftarrow L_1$ where L_0 and L_1 are literals. The literal L_1 in the body represents the fact being inserted or deleted while the literal L_0 in the head represents a side effect. We denote an update as $l \leftarrow$, where l is a ground literal, *i.e.,* we denote updates as update rules without body. Moreover, if $l = A$, where A is a ground atom, then we write $A \leftarrow$ and we read *insert A*, and if $l = \neg A$ then we write $\neg A \leftarrow$ and we read *delete A*. We illustrate the intuitive meaning of updates and update rules in the following example.

Running Example 1. Consider the predicate symbols *cs*, *math* and *sci* (meaning computer scientist, mathematician and scientist, respectively) together with the predicate symbols *crazy*, *happy* and *sad* (with their obvious meaning). We shall use the following updates and update rules as our running example.
Updates:

$U_1 : sci(Bob) \leftarrow$ $\qquad\qquad\qquad\qquad$ $U_2 : math(Bob) \leftarrow$

---

[*] Work partially supported by the French National Project GDR-PRC *BD3* and the Brazilian Government (CNPq and Universidade Federal do Paraná - Brazil)

Update rules:

$$UR_1: \neg\, crazy(X) \leftarrow\ cs(X) \qquad\qquad UR_2: \qquad \neg\, cs(X) \leftarrow\ crazy(X)$$
$$UR_3: \qquad sad(X) \leftarrow\ \neg\, crazy(X) \qquad UR_4: \neg\, happy(X) \leftarrow\ sad(X)$$

Update U_1 means *insert sci(Bob)* and update U_2 means *insert math(Bob)*. On the other hand, rule UR_1 means that if $cs(a)$ becomes true then $crazy(a)$ must become false; UR_3 means that if $crazy(a)$ becomes false then $sad(a)$ must become true; and so on. We note that the body of two or more rules can be the same literal L. $\quad\square$

An important aspect of update rules is that they act as constraints. For instance, rules UR_1 and UR_2 together express the constraint that nobody can be *computer scientist* and *crazy* at the same time.

Insertions and deletions are implemented by storing positive or negative literals in the database. But storing both positive and negative literals can generate inconsistencies. The system is going to perform actions in order to restore consistency, as shown by the following example.

Running Example 2. The updates U_1 and U_2 are implemented by storing the positive literals $sci(Bob)$ and $math(Bob)$ in the database. Denoting by $\mathcal{L}$ the set of stored ground literals, we have: $\mathcal{L} = \{sci(Bob),\ math(Bob)\}$.

If we insert now $crazy(Bob)$ then rule UR_2 is activated and derives $\neg\, cs(Bob)$. Since the set $\mathcal{L}$ does not contain $cs(Bob)$, there is no inconsistency and the insertion is performed by simply storing $crazy(Bob)$ in the set $\mathcal{L}$. Note that we do not store side effects (such as $\neg\, cs(Bob)$) in $\mathcal{L}$, but we *do* take them into account as will be explained shortly. Thus, after the insertion of $crazy(Bob)$, we have: $\mathcal{L} = \{sci(Bob),\ math(Bob),\ crazy(Bob)\}$. Suppose next that we insert $cs(Bob)$. Rule UR_1 is activated and derives $\neg\, crazy(Bob)$, thus creating an inconsistency with $\mathcal{L}$. In order to restore consistency we remove $crazy(Bob)$ from $\mathcal{L}$ and the updated database becomes: $\mathcal{L} = \{sci(Bob),\ math(Bob),\ cs(Bob)\}$. $\quad\square$

Two remarks are in order at this point. First, to maintain consistency, it may be necessary to remove previously stored ground literals. Thus, for example, the insertion of $cs(Bob)$ above required the removal of $crazy(Bob)$. In doing so, our model privileges "new knowledge" over "old knowledge". Second, in our model, database updating comprises two distinct levels, the *user level* and the *system level*. At user level, the user disposes of two operations, *insert* and *delete*. At system level, the system disposes again of two operations, *store* and *remove*. An insert or delete request at user level is translated by the system into a sequence of store/remove operations whose purpose is twofold: to store the positive or negative literal as requested by the user update, and to maintain database consistency. Thus, for example, *insert cs(Bob)* was translated by the system into the sequence: *remove crazy(Bob); store cs(Bob)*.

In the previous example, we saw that the update rules were activated during an insertion or deletion in order to compute the new set $\mathcal{L}$, based on the old $\mathcal{L}$ and on the ground literal being inserted or deleted. Now, given that the literals of $\mathcal{L}$ hold in the database, the update rules imply that some other literals must also hold. Let us illustrate this point by an example.

Running Example 3. We recall that the set of stored ground literals after the insertion of $cs(Bob)$ is $\mathcal{L} = \{sci(Bob),\ math(Bob),\ cs(Bob)\}$. As these literals hold in the database, applying the update rules, we find that the literals $\neg crazy(Bob)$, $sad(Bob)$ and $\neg happy(Bob)$ must also hold. Let ξ_L denote the set of all ground literals that hold in the database. This set consists of the literals in $\mathcal{L}$ and their side effects, *i.e.*, all literals that can be obtained from $\mathcal{L}$ by repeated application of the update rules. It follows that, after the insertion of $cs(Bob)$, we have:

$$\xi_{\mathcal{L}} = \{sci(Bob), \, math(Bob), \, cs(Bob), \, \neg crazy(Bob), \, sad(Bob), \, \neg happy(Bob)\}. \qquad \square$$

So, when the database is updated, the update rules are activated to compute:

1. the new set $\mathcal{L}$, based on the old set $\mathcal{L}$ (*i.e.*, the one before the update) and the ground literal being inserted or deleted,
2. the new set $\xi_{\mathcal{L}}$ of all literals that must hold in the database until the next update.

These computations reflect our philosophy of regarding update rules as constraints, *all* consequences of which must hold in the database. The *only* event that can change the truth values of the literals in $\xi_{\mathcal{L}}$ is a new update. However, until that new update happens, the literals of $\xi_{\mathcal{L}}$ continue to hold in the database and form the basis for query answering, as we shall now explain.

Query Rules - Our query rules are standard Datalog^{neg} rules. However, in order to perform query processing we must first settle two matters.

(1) Our update rules allow explicit derivation of negative information, so we can consider that the falsity of a literal corresponds to classical negation. On the other hand, to perform query processing, we use well-founded semantics [5], with one important difference. Namely, in the usual computation of well-founded semantics, the initial interpretation is empty, whereas, in our model, the initial interpretation is $\xi_{\mathcal{L}}$. As a consequence, two kinds of negation are present: *classical negation*, denoted by $\neg$, and *negation of well-founded semantics*, denoted by *not*. Now, classical negation does not coincide with negation in the well-founded semantics because negation in well-founded semantics is not produced explicitly by rules. So, if we want to use negative literals of $\xi_{\mathcal{L}}$ with query rules, we must first say how classical negation is related to negation of well-founded semantics. In this respect we make the following assumption (in the spirit of [1, 6]):

Assumption: *for every atom A, $\neg A$ is transformed into not A.*

(2) As the literals of $\xi_{\mathcal{L}}$ hold in the database (and must hold until a new update), the ground literals derived by query rules must not be in contradiction with those of $\xi_{\mathcal{L}}$. We cope with this problem by conditioning each derivation step of a ground literal l through query rules as follows:

- if l is in $\xi_{\mathcal{L}}$ then l holds
    else if $\neg l$ is not in $\xi_{\mathcal{L}}$ then
        if there is a query rule $l \leftarrow L_1, \ldots, L_n$ and $L_1, \ldots, L_n$ hold then l holds.

In a sense, the literals of $\xi_{\mathcal{L}}$ act as exceptions to query rule derivations. We feel therefore justified in calling $\xi_{\mathcal{L}}$ the *exception set* of the database.

Running Example 4. Suppose that we have the following query rules (along with the update rules UR_1, UR_2, UR_3 and UR_4 seen earlier):

$QR_1 : crazy(X) \leftarrow sci(X),\ not\ cs(X)$
$QR_2 : \quad cs(X) \leftarrow math(X),\ not\ crazy(X)$
$QR_3 : happy(X) \leftarrow math(X),\ cs(X)$

As $math(Bob)$ and $cs(Bob)$ are in $\xi_\mathcal{L}$, QR_3 applies and derives $happy(Bob)$. However, as $\neg happy(Bob)$ is in $\xi_\mathcal{L}$ this derivation is not valid, *i.e.*,$\neg happy(Bob)$ remains true. $\square$

As we assume the transformation $\neg A \rightarrow not\ A$, we feel justified in using, henceforth, *only* one negation symbol, namely *not*.

In the following Section 2 we recall briefly some facts concerning Datalog^{neg} databases under well-founded semantics and, in Section 3, we introduce *literal rules* as a tool for modeling update rules. In Section 4, we present formal semantics for databases containing both update rules and query rules, and in Section 5 we present formal semantics for database updating. In Section 6 we describe briefly the principal ideas of [6] and [14] concerning extended logic programs; and, in Section 7, we show how our approach is related to extended logic programs. Finally, in Section 8, we discuss alternatives opened by our approach. Due to limitation of space, proofs are omitted; they can be found in the full paper [12].

2 Datalog^{neg} Databases

A Datalog^{neg} database is a finite function-free set of facts and rules over an alphabet **A** consisting of constants, variables and predicates [4, 19]. There is an infinite set CONST of constants usually denoted by lower case characters such as a, b, etc and an infinite set VAR of variables usually denoted by lower case characters such as x, y, etc. There is a *finite* set PRED of predicates usually denoted by lower case characters such as p, q, etc; each predicate is associated with a positive integer called its *arity*. We assume that the sets CONST, VAR and PRED are mutually disjoint. We note that, since we consider function-free languages, the only possible terms are constants or variables.

As a notational convenience, if p is an n-ary predicate, and if $t_1, t_2, \ldots, t_n$ are terms, then the formula $p(t_1, t_2, \ldots, t_n)$ is denoted by $p(\tilde{t})$. A formula of the form $p(\tilde{t})$ is referred to as *positive literal* and a formula of the form $not\ p(\tilde{t})$ is referred to as *negative literal*. A literal is said to be *ground* if it contains no variable. A positive ground literal is also called a *fact*.

A *rule* is a formula of the form $p(\tilde{t}) \leftarrow L_1, L_2, \ldots, L_k$ where, for every $i = 1, 2, \ldots, k$, L_i is a literal. The positive literal $p(\tilde{t})$ is called the *head* of the rule and the set of literals $L_1, L_2, \ldots, L_k$ is called the *body* of the rule.

We now recall the definition of well-founded semantics of Datalog^{neg} [5, 2]. Well-founded semantics are based on the notion of *partial* interpretation defined as follows. Let HB be the Herbrand base of a database alphabet **A**. For every subset S of HB, let $not.S$ denote the set $\{not\ f \mid f \in S\}$; in particular, $not.HB$ denotes the set $\{not\ f \mid f \in HB\}$. A subset C of $HB \cup not.HB$ is called consistent if C does not contain a fact f and its negation $not\ f$. That is, C is *consistent* if there are disjoint subsets $pos(C)$ and $neg(C)$ of HB such that $C = pos(C) \cup not.neg(C)$. A *partial interpretation* of **A** is just a consistent subset I of $HB \cup not.HB$.

A Datalog^{neg} database is a set of facts and rules denoted as $DB = (F, R)$, where F stands for the set of facts and R for the set of rules. We denote by *inst_DB* the set of all facts in F together with all instantiations of the rules in R. Given a partial interpretation I and a Datalog^{neg} database DB, consider the following operators.

- Define the *immediate consequence* operator $T^{\in}$ by:
  $$T^{\in}_{DB}(I) = \{head(r) \mid r \in inst\_DB \ \wedge \ \forall L \in body(r), L \in I\}.$$
- Define a set of facts U to be unfounded with respect to I, if for all f in U and for all instantiated rules r in *inst_DB* the following holds:
  $$(head(r) = f) \ \Rightarrow \ \exists L \in body(r), \ [not \ L \in I \vee L \in U].$$
  Define the *unfounded-set operator* U_{DB} by: $U_{DB}(I)$ is the greatest set of unfounded facts with respect to DB and I.

It can be shown [5] that the operator $T^{\in}_{DB} \cup not.U_{DB}$ has a least fixpoint which is a partial interpretation of **A**. It is precisely this fixpoint that is referred to as the *well-founded model* of DB.

3 Literal Rules

We call *literal rule* a rule whose head and body contain exactly one literal. There are two differences between a literal rule and a usual rule. First, the head of a literal rule can be a positive or a negative literal, while the head of a usual rule is always a positive literal. Second, the body of a literal rule contains exactly one literal while, the body of a usual rule may contain more than one literal.

Definition 1. - Literal Rule. A literal rule, or l-rule for short, is a rule of the form $L_1 \leftarrow L_2$, where L_1 and L_2 are literals. □

Given a set of l-rules LR, we denote by *inst_LR* the set of all instantiations of the rules of LR (with respect to the underlying alphabet **A**). We associate LR with an operator $^{LR}\xi$, called *exception operator*, defined as follows: for every set of ground literals I, $^{LR}\xi(I) = I \cup \{head(r) \mid r \in inst\_LR \wedge body(r) \in I\}$.

Clearly, $^{LR}\xi$ is a monotonic operator over the power-set of $HB \cup not.HB$. In the remaining of this paper, when a specific set of literal rules is understood, we shall write simply ξ instead of $^{LR}\xi$. Now, as ξ is monotonic, the sequence defined by: $\xi^0(I) = I$ and $\xi^k(I) = \xi(\xi^{k-1}(I))$, for every integer $k > 0$, has a limit. This limit, referred to as the *least fixpoint of ξ with respect to I*, is denoted by $lfp(\xi(I))$ or by ξ_I.

Definition 2. - Consistent Set of L-Rules. A set LR of l-rules is consistent if for every ground literal L the set $\xi_{\{L\}}$ is a partial interpretation. □

NOTE: In the remaining of the paper, when no confusion is possible, a singleton $\{L\}$ is simply denoted by its single element L. Thus, for example, we shall write ξ_L instead of $\xi_{\{L\}}$.

Now, a set LR of l-rules can be associated with a directed graph $G(LR)$ whose nodes and links are defined as follows:

- *Nodes:* For every literal L of LR define

$$node(L) = \begin{cases} p, & \text{if } L = p(\tilde{t}) \\ not\ p, & \text{if } L = not\ p(\tilde{t}) \end{cases}$$

Then, for every l-rule $L_1 \leftarrow L_2$ in LR, $node(L_1)$ and $node(L_2)$ are nodes of $G(\text{LR})$.
- *Links:* For every l-rule $L_1 \leftarrow L_2$ in LR, there is a link from $node(L_2)$ to $node(L_1)$.

The following proposition gives a sufficient condition for testing the consistency of LR.

Proposition 3. *Let* LR *be a set of l-rules and let* $G(\text{LR})$ *be its associated graph. If* LR *is inconsistent then one of the following holds:*

1. *there is a predicate symbol p such that $G(\text{LR})$ contains a path from p to $(not\ p)$ or a path from $(not\ p)$ to p.*
2. *there are predicate symbols p and q such that $G(\text{LR})$ contains paths from q to p and from q to $(not\ p)$.* □

We note, however, that the converse of the above proposition does not hold. Indeed, if LR contains the single instantiated l-rule $not\ p(b) \leftarrow p(a)$, then $G(\text{LR})$ contains a link from p to $not\ p$ but LR is consistent (in the sense of Definition 2).

Lemma 4. *Let* LR *be a set of l-rules, and let* ξ *be the associated exception operator. For every partial interpretation I, we have: (1) $I \subseteq \xi_I$, and (2) $\xi_I = \bigcup_{L \in I} \xi_L$.* □

Let LR be a set of update rules. Since the head and the body of every l-rule in LR contains a single literal, we can associate LR with an "inverse" operator ϑ as follows: For every set of ground literals I, define:

$$\vartheta(I) = I \cup \{body(r) \mid r \in inst\_\text{LR} \wedge head(r) \in I\},$$

It is easy to see that, for all literals L and L', we have: $L \in \xi(L') \Leftrightarrow L' \in \vartheta(L)$.

Clearly, ϑ enjoys the same properties as ξ (see [9]). As a consequence, ϑ is monotonic, and we denote by ϑ_I the least fixpoint of ϑ with respect to a set of ground literals I. Moreover, ϑ satisfies points (1) and (2) of Lemma 4.

4 Database Semantics

In this section, we present formal semantics for databases containing both updates rules and query rules.

Definition 5. - Database. A database is a triple $\Delta = (\mathcal{L}, UR, QR)$ where:
- $\mathcal{L}$ is a set of ground literals (meant to be the set of literals stored in the database).
- UR is a consistent set of l-rules (meant to be the update rules).
- QR is a set of usual Datalog^{neg} rules (meant to be the query rules). □

Given a database Δ, let ξ be the exception operator associated with the set UR. The least fixpoint of ξ with respect to $\mathcal{L}$, denoted $\xi_{\mathcal{L}}$, will be referred to as the *exception set* of Δ, and the literals of $\xi_{\mathcal{L}}$ as the *exceptions* of Δ. It follows from Lemma 4 that: (1) $\mathcal{L} \subseteq \xi_{\mathcal{L}}$, and (2) $\xi_{\mathcal{L}} = \left(\bigcup_{l \in \mathcal{L}} \xi_l\right)$.

The following definition says that a database Δ is consistent if the literals of $\mathcal{L}$ do not lead to inconsistencies between exceptions.

Definition 6. - Consistent Database. A database Δ is said to be consistent if $\xi_{\mathcal{L}}$ is a partial interpretation. □

As $\mathcal{L} \subseteq \xi_{\mathcal{L}}$, the above definition implies that: if Δ is a consistent database then $\mathcal{L}$ is a partial interpretation. This simply means that, in a consistent database, $\mathcal{L}$ cannot contain a fact f and its negation.

Running Example 5. Let us call $\Delta = (\mathcal{L}, UR, QR)$, the database of our running example after the insertion of $math(Bob)$ and $cs(Bob)$ (with $\neg$ replaced by not). Therefore, we have:

- $\mathcal{L} = \{math(Bob), cs(Bob)\}$,
- UR:
  $\quad UR_1$: $not\, crazy(X) \leftarrow cs(X)$ $\qquad\qquad UR_2$: $\quad not\, cs(X) \leftarrow crazy(X)$
  $\quad UR_3$: $\qquad sad(X) \leftarrow not\, crazy(X)$ $\qquad UR_4$: $not\, happy(X) \leftarrow sad(X)$
- QR:
  $\quad QR_1$: $crazy(X) \leftarrow sci(X),\ not\, cs(X)$
  $\quad QR_2$: $\quad cs(X) \leftarrow math(X),\ not\, crazy(X)$
  $\quad QR_3$: $happy(X) \leftarrow math(X),\ cs(X)$

The operator ξ works over the update rules, beginning with the partial interpretation $\mathcal{L}$. Thus, $\xi_{\mathcal{L}}$ is $\{math(Bob), sad(Bob), cs(Bob), not\, crazy(Bob), not\, happy(Bob)\}$. □

We now explain how the database semantics is computed using the set $\xi_{\mathcal{L}}$ and the operators of well-founded semantics. First, following [10], we define two operators, T^* and U^*, as follows: for every partial interpretation I, define

$$T^*(I) = (T^{\in}(I) \setminus neg(\xi_{\mathcal{L}})) \cup pos(\xi_{\mathcal{L}}) \text{ and } U^*(I) = (U(I) \setminus pos(\xi_{\mathcal{L}})) \cup neg(\xi_{\mathcal{L}}).$$

Here $T^{\in}$ and U are the well-founded operators associated with the positive literals in $\xi_{\mathcal{L}}$ and with the query rules of QR. It can be shown [11] that the operator $T^* \cup not.U^*$ is monotonic. It follows that, for every integer $k > 0$, the sequence defined by:

$$SEM^0 = \xi_{\mathcal{L}} \text{ and } SEM^k = T^*(SEM^{k-1}) \cup not.U^*(SEM^{k-1}),$$

has a limit. This limit, is referred to as the *semantics* of Δ, and is denoted by $SEM(\Delta)$. Moreover, we have the following theorem.

Theorem 7. *For every consistent database* $\Delta = (\mathcal{L}, UR, QR)$, $SEM(\Delta)$ *is a partial interpretation such that* $\mathcal{L} \subseteq \xi_{\mathcal{L}} \subseteq SEM(\Delta)$. □

The above theorem shows that a consistent database Δ always has consistent semantics. Moreover, the underlying Herbrand base HB_Δ is partitioned into the following three sets: the set $pos(SEM(\Delta))$, the set $neg(SEM(\Delta))$ and the set $HB_\Delta \setminus (pos(SEM(\Delta)) \cup neg(SEM(\Delta)))$. The ground literals of these sets are called respectively *true*, *false* and *unknown* (in Δ). The presence of unknown facts shows that our approach is *not* based on the so-called *Closed World Assumption* of [17].

Running Example 6. Let us compute the semantics $SEM(\Delta)$ of the database Δ of our running example.
In *step 0* we have: $SEM^0 = \xi_{\mathcal{L}}$, while in *step 1* we have:

$T^*(SEM^0) = (T^\in(SEM^0) \setminus neg(\xi_{\mathcal{L}})) \cup pos(\xi_{\mathcal{L}}) = \{math(Bob), sad(Bob), cs(Bob)\}$.
and, computing $U(SEM^0)$ following the method of [3], we obtain:
$U^*(SEM^0) = (U(SEM^0) \setminus pos(\xi_{\mathcal{L}})) \cup neg(\xi_{\mathcal{L}}) = U(SEM^0) \cup \{happy(Bob)\}$.
Thus: $SEM^1 = \{math(Bob), sad(Bob), cs(Bob),$
$\qquad\qquad not\ sci(Bob), not\ crazy(Bob), not\ happy(Bob)\}$.

As $SEM^2 = SEM^1$, we have $SEM(\Delta) = SEM^1$. We note that, in the above computation, $happy(Bob)$ was derived as a true fact by the query rule QR_3 ($happy(Bob) \in T^\in(SEM^0)$), but also as a false exception ($happy(Bob) \in neg(\xi_{\mathcal{L}})$). As we explained earlier, priority is given to derivations driven by update rules, so $not\ happy(Bob)$ is included in the semantics and $happy(Bob)$ is ignored. In other words, $not\ happy(Bob)$ acts as an exception to QR_3. □

5 Update Semantics

We show now that our update processing transforms, in a *deterministic* way, a consistent database into a new database which is again consistent, and in which the requested update has been effectively performed.

In order to define the way in which the set $\mathcal{L}$ is modified during updating, we extend the notation $not.I$ to partial interpretations as follows: given a partial interpretation $I = pos(I) \cup not.neg(I)$, define $not.I = neg(I) \cup not.pos(I)$.

As we have explained in the introduction, updating a database means inserting or deleting a fact. In both cases, the update requires to store a ground literal l in $\mathcal{L}$, while leaving the database consistent. However, storing l in $\mathcal{L}$ may create inconsistencies. To restore consistency, we remove from $\mathcal{L}$ all literals which generate exceptions contradicting the exceptions generated by l. The following definition summarizes our discussion.

Definition 8. - Database Update. Let $\Delta = (\mathcal{L}, UR, QR)$ be a database and let l be a literal. Define the update of Δ by l, denoted $upd(l, \Delta)$, to be a database $(\mathcal{L}', UR, QR)$ defined by: $\mathcal{L}' = (\mathcal{L} \setminus lfp(\vartheta(not.\xi_l))) \cup \{l\}$. □

Running Example 7. First we consider the deletion of $sad(Bob)$ from Δ. Recall that in our running example $\mathcal{L} = \{math(Bob), cs(Bob)\}$.
For $l = not\ sad(Bob)$, we obtain $\xi_l = \{not\ sad(Bob)\}$. Clearly, $not.\xi_l = \{sad(Bob)\}$ and we find that $lfp(\vartheta(not.\xi_l)) = \{sad(Bob), cs(Bob), not\ crazy(Bob)\}$. Thus the database $\Delta' = upd(not\ sad(Bob), \Delta)$ is defined by $\mathcal{L}' = (\mathcal{L} \setminus \{sad(Bob), cs(Bob), not\ crazy(Bob)\}) \cup \{not\ sad(Bob)\} = \{math(Bob), not\ sad(Bob)\}$.
Now, consider the insertion of $crazy(Bob)$ in Δ'. We have $l = crazy(Bob)$, $\xi_l = \{crazy(Bob), not\ cs(Bob)\}$ and $lfp(\vartheta(not.\xi_l)) = \{crazy(Bob), cs(Bob)\}$. Therefore, nothing has to be removed from $\mathcal{L}'$ and the database $\Delta'' = upd(crazy(Bob), \Delta')$ is defined by $\mathcal{L}'' = \{math(Bob), crazy(Bob), not\ sad(Bob)\}$. □

Proposition 9. *Let Δ be a database and let l be a literal. If Δ is consistent then $\Delta' = upd(l, \Delta)$ is consistent and $l \in SEM(\Delta')$.* □

What this proposition implies is the following: if we start with the empty database (which is consistent), then our method always maintains the database consistent during updating.

6 Related Works

In all semantics of logic programming that uses the Closed World Assumption [17], an answer to a ground query is either *yes* or *no*. Therefore, in the Closed World Assumption (two-valued logic), if a ground literal cannot be derived as true, then it is assumed false. Based on this assumption, a *general logic program* [15] is defined to be a set of rules of the form: $A_0 \leftarrow A_1, \ldots, A_m$, *not* $A_{m+1}, \ldots$, *not* A_n, where $n \geq m \geq 0$ and A_i is an atom.

The notion of *extended logic program* [6, 14] was introduced in the context of three-valued logic. In an extended logic program, the answer to a ground query can be: *yes* (ground literal derived as true), *no* (ground literal derived as false) or *unknown* (ground literal derived neither as true nor as false). To represent these three values, two types of negation are used: *not*, as negation by failure, meaning that no information is available (*i.e.*, the value is unknown); and ¬, as classical negation, meaning that a negative information can be inferred from the rules (*i.e.*, explicitly). An extended logic program can be defined as a set of rules of the form $L_0 \leftarrow L_1, \ldots, L_m$, *not* $L_{m+1}, \ldots$, *not* L_n, where $n \geq m \geq 0$ and where L_i is a literal (a literal is a formula of the form A or $\neg A$ where A is an atom).

In the context of extended logic programs, a relationship between our work and the approaches proposed in [6] and [14] can be established. Before doing that, let us describe briefly the principal ideas of [6] and [14].

Extended Logic Programs - The semantics of extended logic programs [6] is based on the notion of stable models [7], and consists in defining when a set S of ground literals is an *answer set* of a program. Since a rule with variables is considered as a shorthand for a set of ground instances, it is sufficient to define answer sets for extended logic programs without variables. The definition is given in two steps that constitute a procedure to verify if an answer set proposed by a rational agent is valid.

– STEP 1: Suppose that Π is an extended logic program without variables and without *not*, and that *Lit* is the set of ground literals in the language of Π. The *answer set* of Π, denoted by $\alpha(\Pi)$, is the smallest subset S of *Lit* such that:
  1. for any rule $L_0 \leftarrow L_1 \ldots L_m$ in Π, if $L_1 \ldots L_m \in S$ then $L_0 \in S$ and
  2. if S contains a pair of complementary literals, then $S = Lit$ (and Π is contradictory).

– STEP 2: Suppose that Π is an extended logic program without variables and *Lit* is the set of ground literals in the language of Π. For any set $S \subset Lit$, let Π^S be the extended logic program obtained from Π by deleting:
  1. each rule that has a formula *not* L in its body with $L \in S$ and
  2. all formulas of the form *not* L in the body of the remaining rules.

Clearly, Π^S does not contain *not*, so its answer set can be defined using STEP 1. If $\alpha(\Pi^S)$, the answer set of Π^S (found by STEP 1), is equal to S then S is the answer set of Π. Therefore, the answer sets of Π are characterized by the equation $S = \alpha(\Pi^S)$.

Example 1. Consider the following extended logic program Π:

(1) $crazy(X) \leftarrow sci(X),\ not\ \neg crazy(X)$
(2) $cs(X) \leftarrow math(X),\ not\ \neg cs(X)$
(3) $\neg crazy(X) \leftarrow cs(X)$
(4) $\neg cs(X) \leftarrow crazy(X)$
(5) $sci(Bob)$
(6) $math(Bob)$

The following two sets are answer sets for this program: $S_1 = \{sci(Bob), math(Bob),$ $crazy(Bob), \neg cs(Bob)\}$ and $S_2 = \{sci(Bob), math(Bob), \neg crazy(Bob), cs(Bob)\}$.   □

In [6] the authors also show how an extended logic program can be transformed into a general logic program by renaming predicates (see also [18]). This is done by (a) transforming every literal of the form $L = \neg\ p(\tilde{t})$ into the literal $L^+ = nonp(\tilde{t})$, and (b) transforming the extended logic program Π into the general logic program Π^+, obtained from Π by replacing every rule of the form: $L_0 \leftarrow L_1, \ldots, L_m,\ not\ L_{m+1}, \ldots,\ not\ L_n$ by the rule $L_0^+ \leftarrow L_1^+, \ldots, L_m^+,\ not\ L_{m+1}^+, \ldots,$ $not\ L_n^+$. It has been shown [6] that a consistent set S is an answer set of Π if and only if S^+ is an answer set of Π^+.

Exception Rules - A modification of the answer set semantics defined in [6] is proposed in [14]. Two types of rules are used: *general rules* and *exception rules*. Both rules follow the syntax proposed by extended logic programs.

Example 2. Consider the following program Π:

GENERAL RULES

(1) $crazy(X) \leftarrow\ sci(X)$ (2) $cs(X) \leftarrow\ math(X)$
(3) $sci(Bob)$ (4) $math(Bob)$

EXCEPTION RULES

(1) $\neg crazy(X) \leftarrow cs(X)$ (2) $\neg cs(X) \leftarrow crazy(X)$ □

The difference between these two types of rules is that general rules have always positive heads while exception rules have always negative heads. Moreover, exception rules have higher priority than general rules, *i.e.*, negative information is preferred over positive information.

The modification with respect to [6] is done by adding a third provision in STEP 2, thus redefining Π^S. The goal of this new provision is to delete any rule in Π having conclusion L with $\neg L \in S$. Therefore, STEP 2 for [14] is:

  – NEW STEP 2: For any set $S \subset Lit$, let $^S\Pi$ be the extended logic program obtained from Π by appling STEP 2, as defined above, and by deleting every rule having a positive conclusion L, with $\neg L \in S$.

As before, we can use STEP 1 to find the answer set of $^S\Pi$. Therefore, for any variable-free program Π, S is defined as an *e-answer set* if and only if $S = \alpha(^S\Pi)$.

7   Comparison to our Approach

The approach of [14], where two types of rules are used, general rules and exception rules, is closely related to our approach as we also use two types of rules: query rules and update rules. However, the objective of [6, 14] is to study essentially queries, whereas our objective is to study database queries *and* updates.

Regarding the interaction between the two types of rules there is an important difference. Namely, the derivations by the two types of rules are interleaved in [14] whereas they are completely separated in our approach. In [14], as Example 2 illustrates, $sci(Bob)$ is used in the rule $crazy(Bob) \leftarrow sci(Bob)$ in order to derive $crazy(Bob)$. Afterwards, an interaction with exception rules takes place and $crazy(Bob)$ is used by the rule $\neg cs(Bob) \leftarrow crazy(Bob)$ meaning that, if *Bob* is crazy he cannot be computer scientist. In other words, the facts derived by general rules are used as "input" to derivations by exception rules.

In our approach, update rule derivations are completely independent from query rule derivations. For instance, if $crazy(Bob) \leftarrow sci(Bob)$ is a query rule, its consequence ($crazy(Bob)$) is not used by an update rule to generate an exception. Update rules are activated *only* over the literals stored in the database (*i.e*, the literals of $\mathcal{L}$) to compute $\xi_{\mathcal{L}}$. The literals of $\xi_{\mathcal{L}}$ are used as *input* to the derivations by query rules and a derivation by a query rule is accepted only if it is not in contradiction with a literal in $\xi_{\mathcal{L}}$. However, $\xi_{\mathcal{L}}$ does not change during query-driven derivations: $\xi_{\mathcal{L}}$ is recalculated *only* when an update is performed.

Example 3. Consider $\Delta = (\mathcal{L}, UR, QR)$ where $\mathcal{L} = \{sci(Bob), math(Bob)\}$ and:

UPDATE RULES
(1) $\neg crazy(X) \leftarrow cs(X)$
(2) $\quad \neg cs(X) \leftarrow crazy(X)$

QUERY RULES
(1) $crazy(X) \leftarrow sci(X), not\ cs(X)$
(2) $\quad cs(X) \leftarrow math(X), not\ crazy(X)$

Following our approach, the semantics of the database is given by $SEM(\Delta) = \{sci(Bob), math(Bob)\}$, which is the intersection of the two e-answer sets of Example 1.

Consider now the insertion of $crazy(Bob)$ in the program of examples 1 and 2. We note that if $crazy(Bob)$ is added to the general rules of Example 2, there will be no change in the semantics of that program. That is, both e-answer sets, $S_1 = \{sci(Bob), math(Bob), crazy(Bob), \neg cs(Bob)\}$ and $S_2 = \{sci(Bob), math(Bob), \neg crazy(Bob), cs(Bob)\}$, will still be correct. However, if we consider the insertion of $crazy(Bob)$ in the extended logic program of Example 1, only the answer set S_1 will be found.

In our approach, the goal is to privilege the inserted literals. Therefore, $sci(Bob)$, $math(Bob)$ and $crazy(Bob)$ will be true and $cs(Bob)$ will be false because: (*i*) $sci(Bob)$, $math(Bob)$ and $crazy(Bob)$ are inserted literals, and (*ii*) $crazy(Bob)$ and $cs(Bob)$ cannot be true at the same time. $\square$

It is important to note that, in a certain sense, our approach is an extension of the one proposed in [14], since exceptions in our approach can have true or false heads. Therefore, in our approach, there is no priority between a negative and a positive fact. Instead, there is priority between derivations: update rule derivations have priority over query rule derivations. Another important difference is that, whereas in [14] a fact can activate just one exception rule, in our approach, we perform fixpoint computation in order to find all facts derived by update rules. This is in keeping with our philosophy of considering update rules as constraints, *all* consequences of which must hold in the database, until a new update.

On the other hand, update rules in our approach are special rules having exactly one literal in the body and one in the head (we have called these rules literal rules). Therefore, they are less general than those proposed in [14]. The reason why we consider only literal rules is because we are interested only in modeling the elementary update operations (the insertion or deletion of a *single* fact) and their possible side effects. If one is interested only in the transitions from one database state to another, then literal rules seem to be quite adequate. However, if one is interested in stating "constraints" that any given database state must satisfy then, rules with more that one literal in the body are required. In the presence of such rules, updating becomes nondeterministic. Indeed, suppose that $q \leftarrow p, t$ is a constraint and that the database contains p and t. If we want to delete q then there are three possibilities: (i) delete p, (ii) delete t or (iii) delete p and t. Our objective in this paper is not how to choose one of these possibilities but, assuming that a choice has been made, how to perform the required update.

8 Alternative Approaches

We end this paper with a few remarks on the alternatives offered by our approach. First, let us recall that our approach relies on two basic concepts:

1. Update rules of the form $L_0 \leftarrow L_1$, where L_0 and L_1 are literals.
2. Query rules as in general logic programs.

In order to implement these concepts, the choices made in this paper were to use (in the framework of a three-valued logic):

1. Fixpoint semantics with classical negation for update rules.
2. Well-founded semantics for query rules.
3. Transform classical negation into negation of well-founded semantics (in the spirit of [1, 6]).

However, other choices are possible. For example, another choice would be:

(1) To use query rules not as in general logic programs but as in extended logic programs. In this case, rule QR_1 of our running example would have to be rewritten as:

$$crazy(X) \leftarrow sci(X),\ not\ cs(X),\ not\ \neg crazy(X),$$

and interpreted as follows: *if someone is a scientist and there is no evidence that he is a computer scientist and there is no evidence that he is not crazy then he is crazy.* The intended semantics of a query rule is that a fact can be derived only if there is no exception to this fact.

(2) Instead of using well-founded semantics and transforming one type of negation to another, to use the method proposed in [6]. In this case, given a consistent database $\Delta = (\mathcal{L}, UR, QR)$, we would associate Δ with an extended logic program $\Pi(\Delta)$ as follows:

- $\xi_{\mathcal{L}}$ is first computed (with classical negation) and its literals are put in $\Pi(\Delta)$.
- If $p(\tilde{t}) \leftarrow L_1, L_2, \ldots, L_k$ is in QR then the rule $p(\tilde{t}) \leftarrow L_1, L_2, \ldots, L_k,$ $not\ \neg p(\tilde{t})$ is in $\Pi(\Delta)$.

Running Example 8. Transforming the database Δ of Running Example 5, we have the following extended logic program $\Pi(\Delta)$.

1) $crazy(X) \leftarrow sci(X), \; not \; cs(X), \; not \; \neg crazy(X)$
2) $cs(X) \leftarrow math(X), \; not \; crazy(X), \; not \; \neg cs(X)$
3) $happy(X) \leftarrow math(X), \; cs(X), \; not \; \neg happy(X)$
4) $cs(Bob)$
5) $math(Bob)$
6) $sad(Bob)$
7) $\neg crazy(Bob)$
8) $\neg happy(Bob)$

To find the semantics of $\Pi(\Delta)$ we can use the method of [6]. We then find the answer set: $S = \{math(Bob), sad(Bob), cs(Bob), \neg \; crazy(Bob), \neg \; happy(Bob)\}$.   □

Since it is possible to translate Δ into $\Pi(\Delta)$, the results of [6] apply here, *i.e.*, the database model can be defined in terms of non-monotonic formalisms [8, 13, 16]. Indeed, it is easy to see that query rules work as default rules with respect to update rules.

Moreover, applying the transformation used in [6, 18] that renames classically negated predicates, QR_1 would be rewritten as

$$crazy(X) \leftarrow sci(X), \; not \; cs(X), \; not \; noncrazy(X).$$

Recall that using this transformation, we can obtain $\Pi^+(\Delta)$ from $\Pi(\Delta)$. In [12], we show that a consistent database Δ can be associated with a Datalog database Δ^+ such that $SEM(\Delta)$ is contained in the well-founded model of Δ^+. Moreover, we note that Δ^+ corresponds to a logic program identical to $\Pi^+(\Delta)$. A comparison between $SEM(\Delta)$ and the answer sets of $\Pi(\Delta)$ is provided in [12].

9 Concluding Remarks

We have seen a model for Datalog^{neg} databases which contain query rules as well as update rules. The essential difference between these two types of rules lies in the fact that update rules have higher priority than query rules. Ground literals generated by update rules act as exceptions to derivations by query rules.

An important contribution of our approach is that it provides a *deterministic* way to insert or delete any fact in a database. In this respect, update rules work as constraints which trigger the *side effects* that are necessary in order to maintain automatically the database consistent.

Two different types of negation appear in our approach. Thus, to compute a semantics for our database, we were led to transform one type of negation into the other. We have shown, however, that another choice is possible, namely, to use the answer set semantics proposed in [6], where extended logic programs are considered.

We can also transform our database Δ into an extended logic program $\Pi(\Delta)$ or into a Datalog program $\Pi^+(\Delta)$. Regarding the translation of Δ into $\Pi(\Delta)$, we note that our database model can be defined in terms of non-monotonic formalisms, since the results of [6] apply here.

References

1. J. J. Alferes and L. M. Pereira. On logic of program semantics with two kinds of negation. In *Proceedings of the Joint International Conference and Symposium of Logic Programming*, 1992.
2. N. Bidoit. Negation in rule-based database languages: a survey. *Theoretical Computer Science*, 78(1), 1991.
3. N. Bidoit and Ch. Froidevaux. Negation by default and unstratifiable logic programs. *Theoretical Computer Science*, 78(1), 1991.
4. S. Ceri, G.Gottlob, and L.Tanca. *Logic Programming and Databases*. Springer-Verlag, 1990.
5. A. Van Gelder, K.A. Ross, and J.S. Schlipf. The well-founded semantics for general logic programs. *Journal of the ACM*, 38(3), 1991.
6. M. Gelfond and V. Lifschitz. Logic programming with classical negation. In *Proceedings of the Seventh International Conference of Logic Programming*, 1990.
7. M. Gelfond and V. Lifschitz. The stable model semantics for logic programming. In *Proceedings of the Fifth International Conference and Symposium of Logic Programming*, 1990.
8. M. Gelfond, H. Przymusinka, and T. Przymusinki. On the relationship between circumscription and negation as failure. *Artificial Inteligence*, 38(1), 1989.
9. M. Halfeld Ferrari Alves, D. Laurent, and N. Spyratos. Passive and active rules in deductive databases. Technical Report LRI, Université de Paris-Sud, 1994.
10. M. Halfeld Ferrari Alves, D. Laurent, and N. Spyratos. Passive and active rules in deductive databases. In *Proceedings of the International Symposium on Mathematical Foundations of Computer Science*, number 841 in LNCS - Lecture Notes in Computer Science. Springer-Verlag, 1994.
11. M. Halfeld Ferrari Alves, D. Laurent, and N. Spyratos. Update driven rules in Datalog^{neg} databases. Technical Report 94-06, LIFO, Université d'Orléans, 1994.
12. M. Halfeld Ferrari Alves, D. Laurent, and N. Spyratos. Update rules in Datalog programs. Technical Report LRI, Université de Paris-Sud, 1995.
13. K. Konolige. On the relation between default and autoepistemisc logic. *Artificial Inteligence*, 35, 1989.
14. R. A. Kowalski and F. Sadri. Logic programs with exceptions. In *Proceedings of the Seventh International Conference of Logic Programming*, 1990.
15. J. W. Lloyd. *Foundations of Logic Programming*. Springer-Verlag, second extended edition, 1987.
16. T. C. Przymusinki. On the relationship between logic programming and non-montonic reasoning. In *The seventh National Conference of Artificial Intelligence*, Saint Paul, Minnesota, 1988.
17. R. Reiter. On closed word databases. In H. Gallaire and J. Minker, editors, *Logic and data bases*. Plenum Press, New York, 1978.
18. J. C. Shepherdson. Negation in logic programming. In J. Minker, editor, *Foundations of Deductive Databases and Logic Programming*. Morgan Kaufmann Publishers, 1988.
19. J.D. Ullman. *Principles of Databases and Knowledge Base Systems*, volume I and II. Computer Science Press, 1989.

Characterizations of the Stable Semantics by Partial Evaluation

Stefan Brass[1] and Jürgen Dix[2]

[1] University of Hannover, Inst. f. Informatik, Lange Laube 22, D-30159 Hannover, Germany, sb@informatik.uni-hannover.de
[2] University of Koblenz, Dept. of Computer Science, Rheinau 1, D-56075 Koblenz, Germany, dix@informatik.uni-koblenz.de

Abstract. There are three most prominent semantics defined for certain subclasses of disjunctive logic programs: GCWA (for positive programs), PERFECT (for stratified programs) and STABLE (defined for the whole class of all disjunctive programs). While there are various competitors based on 3-valued models, notably WFS and its disjunctive counterparts, there are no other semantics consisting of 2-valued models. We argue that the reason for this is the *Partial Evaluation*-property (also called *Unfolding* or *Partial Deduction*) wellknown from Logic Programming. In fact, we prove characterizations of these semantics and show that if a semantics SEM satisfies *Partial Evaluation* and *Elimination of Tautologies* then *(1) SEM is based on 2-valued minimal models for positive programs*, and *(2) if SEM satisfies in addition* Elimination of Contradictions, *it is based on stable models*. We also show that if we require *Isomorphy* and *Relevance* then STABLE is *completely* determined on the class of all stratified disjunctive logic programs. The underlying notion of a semantics is very general and our abstract properties state that certain *syntactical transformations* on programs are equivalence preserving.

1 Introduction

The generalized closed world assumption GCWA for positive disjunctive programs (introduced in [Min82]), the perfect semantics PERFECT for stratified programs (introduced in [Prz88]) and the stable semantics STABLE for the class of all disjunctive programs (introduced in [GL91, Prz91]) are the most prominent semantics based on two-valued models. Why are there no other such semantics? We answer this question by introducing a framework that enables us to prove characterizations of these semantics and thus to detect the real principles behind them. The starting point is the observation that all these semantics satisfy certain abstract conditions — the most important one being the *Partial Evaluation*-property known from Logic Programming. This has been shown in [BD94b] and, independently at about the same time, in [SS94]. Our aim in this paper is to prove the converse: any semantics satisfying *Partial Evaluation* and some additional properties is already uniquely determined.

We distinguish between *partial* and *complete* characterizations of a semantics SEM. While the first notion states that any semantics (satisfying certain properties) is already *contained* in SEM, the latter notion states that any such semantics in fact *coincides* with SEM.

Our approach is based on purely abstract properties of a semantics SEM. These conditions come in the form of syntactical transformations $\Phi \mapsto \Phi'$ on instantiated logic programs. For a particular syntactical transformation, the corresponding condition states that the transformation is equivalence preserving, i.e. it does not change the underlying semantics: SEM(Φ)=SEM(Φ'). The underlying notion of a semantics is very general. We do not even require that SEM is based on 2-valued models, although for some characterizations we have to add a property (*Elimination of Contradictions*) that implies this.

Our abstract properties can be illustrated by introducing some additional semantics weaker than GCWA and STABLE. This means that our framework allows us to distinguish very carefully between various possible semantics. We generalize the notion of *supported* model from the non-disjunctive to the disjunctive context and get two different notions: the *weakly supported* (Weak-SUPP) and the *supported* (SUPP) models. Weakly supported models are obtained from our underlying notion of a semantics by requiring *Elimination of Contradictions* and *GPPE*, the *G*eneralized *P*rinciple of *P*artial *E*valuation (see Lemma 18). Note that weakly supported and supported models collapse for normal programs to the well-known notion of *supported model* defined in [ABW88]. Minimal models for positive disjunctive programs are obtained from *GPPE* and *Elimination of Tautologies* (see Theorem 15). Stable models for all disjunctive programs are obtained by still adding *Elimination of Contradictions* (see Corollary 21).

To get the whole set of *all* minimal models for positive programs (i.e. characterizing GCWA completely) or to get the whole set of *all* stable models for stratified disjunctive programs (i. e. characterizing PERFECT completely) we only have to assume two additional properties: *Relevance* and *Isomorphy* (see Theorem 16 and Theorem 24).

Abstract properties of logic programming semantics have been already investigated in [Dix92] for normal (i.e. non-disjunctive), and in [DM94a] for disjunctive programs. Here we build on and use some results of the recent [BD95b]. In the companion paper [BD95a] we investigate the normal form of a program and show how it can be used to compute various semantics.

The paper is organized as follows. Section 2 introduces the semantics Weak-SUPP, SUPP and reviews GCWA, PERFECT and STABLE. We also introduce our abstract properties as certain equivalence preserving transformations. In Section 3 we investigate which of our properties are satisfied for these semantics. Finally, in Section 4, we give characterizations of these semantics: any semantics satisfying certain abstract properties is characterized by these. Our main results are Theorems 15, 16, 21, and 24. We conclude with Section 5.

2　Semantics and Transformations

In this section we first introduce our setting of a what we call a semantics (Subsection 2.1). We then present our syntactical transformations (Subsection 2.2), and finally we define four properties to ensure a "good" behaviour of a semantics (Subsection 2.3).

2.1 Semantics

We consider disjunctive programs over some fixed infinite signature Σ.

Definition 1 Program Φ.
A program Φ is a finite set of ground rules of the form $A_1 \vee \cdots \vee A_k \leftarrow B_1 \wedge \cdots \wedge B_m \wedge \neg B_1' \wedge \cdots \wedge \neg B_n'$, where the $A_i / B_i / B_i'$ are Σ-atoms, $k \geq 1$, $m, n \geq 0$.

We identify such a rule with the triple consisting of the *set* of atoms $\mathcal{A} := \{A_1, \ldots, A_k\}$, the *multi-set*[1] of positive atoms $\mathcal{B} := \{B_1, \ldots, B_m\}$ and the set of negative ground atoms $\{\neg B_1', \ldots, \neg B_n'\}$ which we denote by $\neg \mathcal{C}$ (thus, $\mathcal{C}$ is an abbreviation for the disjunctively connected set $\{B_1', \ldots, B_n'\}$).

Definition 2 Three-Valued Model I.
A three-valued Herbrand interpretation I (or short: an interpretation) is a mapping which assigns to every ground atom A a number $I[A] \in \{-1, 0, 1\}$. We identify -1 with false (**f**), 0 with undefined (**u**), and 1 with true (**t**).

An interpretation I is a model of a logic program Φ iff for every $A_1 \vee \cdots \vee A_k \leftarrow B_1 \wedge \cdots \wedge B_m \wedge \neg B_1' \wedge \cdots \wedge \neg B_n'$ in Φ the following holds:

$$\max\{I[A_1], \ldots, I[A_k]\} \geq \min\{I[B_1], \ldots, I[B_m], -I[B_1'], \ldots, -I[B_n']\}.$$

We also use True(I), False(I) and Undef(I) to represent the *true*, resp. *false*, resp. *undefined* ground atoms of I. We restrict to the atoms actually occurring in the underlying program, so that all these sets are finite.

Definition 3 Semantics SEM.
A semantics SEM is a mapping from a class of logic programs Φ over Σ into the set of three-valued Herbrand models

$$\text{SEM}(\Phi) \subseteq \text{MOD}_{3-val}^{Herbrand}(\Phi).$$

We assume that if a ground atom $X \in \Sigma$ is not contained in Φ, then SEM(Φ) $\models \neg X$. We also assume that a literal may occur more than once in the body of a rule (or an atom may occur more than once in the head) without changing the semantics. We call a semantics *trivial* iff SEM(Φ) $= \emptyset$ *for all Φ*.

Note that not all semantics are defined on the whole class of *all* disjunctive programs. Our results hold for all classes of programs that are closed under the transformations to be introduced below and that contain with any program Φ also $\Phi \cup \{X \leftarrow \text{body}\}$ where X is a new atom. In particular, they hold for the classes of *positive disjunctive, general disjunctive, positive non-disjunctive, stratified non-disjunctive* and *general non-disjunctive* programs.

We could also have defined a semantics as a mapping into a set of two-valued models. In fact, our *Elimination of Contradictions* condition just implies this.

---

[1] The need to allow for duplicate occurrences in $\mathcal{B}$ is a technical one. It is only needed in the proof of Lemma 18.

But this assumption is not needed for our characterization theorems for GCWA (Theorems 15 and 16). And, obviously, the weaker the underlying notion of a semantics, the stronger the characterization theorems.

The semantics we are interested in are indeed based on two-valued models:

Definition 4 (Weakly) Supported, Stable.
A two-valued model I of a (disjunctive) logic program Φ is a **a)** *supported*, resp. **b)** *weakly supported*, resp. **c)** *stable* model of a disjunctive logic program Φ iff

a) for every ground atom A with $I \models A$ there is a ground instance $\mathcal{A} \leftarrow \mathcal{B} \wedge \neg \mathcal{C}$ of a rule in Φ with $A \in \mathcal{A}$ and (1.) $I \models \mathcal{B} \wedge \neg \mathcal{C}$, and (2.) $I \not\models \mathcal{A} - \{A\}$.
b) in a) only (1.) is required.
c) I is a minimal model of the positive disjunctive program Φ/I.

Here Φ/I is the GL-Transform of Φ wrt I. It is obtained from Φ by eliminating all occurrences of *negative* literals according to I: if $I \models \neg A$ then drop $\neg A$ from the clause, if $I \models A$ then drop the whole clause containing $\neg A$.

We use SUPP(Φ), Weak-SUPP(Φ), STABLE(Φ) for the respective set of intended models. Another famous semantics is GCWA: it is only defined for *positive* disjunctive programs and given by the set of all minimal (two-valued) models Min-MOD$_{2-val}^{Herbrand}(\Phi)$. While STABLE extends GCWA (in the sense that both coincide for positive disjunctive programs) this is neither true for Weak-SUPP nor for SUPP. It is well-known that PERFECT (introduced by Przymusinski in [Prz88]) coincides for stratified programs with STABLE (STABLE for disjunctive programs has been introduced independently by Gelfond and Lifschitz in [GL91] and by Przymusinski in [Prz91]) and we will think therefore of PERFECT as the restriction of STABLE to this class of programs.

2.2 Transformations

To illustrate very clearly the differences of logic programming semantics we base our discussion on abstract properties. All of them require that certain elementary transformations do not change the semantics of a given logic program, i.e. are SEM-equivalence transformations:

Definition 5 (SEM-Equivalence) Transformation.
A transformation $\mapsto$ is an arbitrary binary relation on the class of all programs. We call it a *SEM-equivalence transformation* iff SEM(Φ) = SEM(Φ') for all Φ, Φ' with $\Phi \mapsto \Phi'$.

Definition 6 Elimination of Rules.
A semantics SEM allows **a)** *Elimination of Tautologies*, resp. **b)** *Elimination of Contradictions* iff the following transformations on instantiated logic programs are SEM-equivalence transformation:

a) *Delete a rule $\mathcal{A} \leftarrow \mathcal{B} \wedge \neg \mathcal{C}$ with $\mathcal{A} \cap \mathcal{B} \neq \emptyset$.*
b) *Delete a rule $\mathcal{A} \leftarrow \mathcal{B} \wedge \neg \mathcal{C}$ with $\mathcal{C} \cap \mathcal{B} \neq \emptyset$.*

These transformations allow us to *eliminate* certain rules. They are special cases of the D-reduction introduced in [DM94a].

The most important transformation, however, is partial evaluation in the sense of the "unfolding" operation (see [DM94b, BD94b]):

Definition 7 GPPE.
A semantics SEM satisfies *GPPE* iff the following transformations on instantiated logic programs are SEM-equivalence transformations: *Replace a rule* $\mathcal{A} \leftarrow \mathcal{B} \wedge \neg \mathcal{C}$ *where* $\mathcal{B} := \{B_1, \ldots, B_l, B, B_{l+1}, \ldots, B_m\}$ *contains at least one (distinguished) occurrence of an atom* B *by the* n *rules*

$$\mathcal{A} \cup (\mathcal{A}_i - \{B\}) \leftarrow \{B_1, \ldots B_l, B_{l+1}, \ldots, B_m\} \wedge \mathcal{B}_i \wedge \neg(\mathcal{C} \cup \mathcal{C}_i),$$

where $\mathcal{A}_i \leftarrow \mathcal{B}_i \wedge \neg \mathcal{C}_i$ $(i = 1, \ldots, n)$ are all rules with $B \in \mathcal{A}_i$.

2.3 Consistency, Independence, Relevance and Isomorphy

Besides allowing the above transformations, we would require from a good semantics that it satisfies also the following four natural transformations. Three of them point to a weakness of the stable semantics discussed elsewhere ([DM94b]):

Definition 8 Consistency and Independence.
a) A semantics SEM satisfies *Consistency*, iff SEM(Φ) $\neq \emptyset$ for all programs Φ.
b) A semantics SEM satisfies *Independence*, iff

$$\text{SEM}(\Phi) \models \psi \iff \text{SEM}(\Phi \cup \Phi') \models \psi,$$

provided that the predicates occurring in Φ and Φ' are disjoint, and ψ contains only predicates from Φ.

The requirement that a semantics be *consistent* is immediate, although one could object that not even classical logic satisfies it. But this is no convincing argument because in the case of logic programs we have a very restricted language and it should not be possible to explicitly express inconsistency (note that no negative literals can appear in the head of program clauses).

Independence goes a small step further and formalizes the idea that if a program Φ can be split into two disjoint parts that have nothing to do with each other (i. e. $\Phi = \Phi' \cup \Phi''$ and no predicate occurs in both parts), then the meaning of a predicate P with respect to the whole program Φ coincides with P's meaning with respect to the part it belongs. *Independence* expresses a kind of *consistency-persistence*: it holds in classical logic, provided that Φ and Φ' are consistent. There, it corresponds to the well-known notion of *conservative extension*.

Independence is implied by the next condition, which has been introduced in [Dix92] (see also [DM94b]). This condition is very natural because it is the underlying principle of all Top-Down Query Evaluation methods: clauses that contain only predicates that have nothing to do with a given literal A, should not affect A's truth-value. More precisely: the truth-value of a literal should only depend on the *call-graph* below it. It is well known that any program Φ induces a notion of *dependency* between its atoms. We say that A *depends immediately* on B iff B appears in the body of a clause in Φ, such that A appears in its head or if A and B both appear in the head of a clause. The binary relation *depends on* is the reflexive and transitive closure of *depends immediately on*. The *dependencies of* and the *rules relevant for an atom* X are now defined by:

- *dependencies_of*$(X) := \{A | X$ depends on $A\}$,
- *rel_rul*(Φ, X) is the set of *relevant rules* of Φ with respect to X, i.e. the set of rules that contain an $A \in$ *dependencies_of*(X) in their head.

Analogously, the two sets are defined for an arbitrary set M consisting of atoms.

Our condition formalizes that if we are given a program Φ but are only interested in (determining the truth-values of) atoms belonging to a certain set M, then it is completely sufficient to look at the subset of Φ consisting of the rules relevant for M. Since this set Φ' usually is a proper subset of Φ formulated in a smaller language, the elements of SEM(Φ') and SEM(Φ) are in general incomparable. Therefore we need the notion "$\mathcal{RED}_{\Phi'}(I)$" of a "*reduct* of an interpretation I in SEM(Φ) to a model of Φ'":

$$\mathcal{RED}_{\Phi'}(I)[A] := \begin{cases} I[A] & \text{if } A \text{ occurs in } \Phi', \\ \mathbf{f} & \text{otherwise.} \end{cases}$$

Definition 9 Relevance.
Let Φ be a program and M be a set of atoms occurring in Φ. A semantics SEM satisfies *Relevance* iff

1. SEM$(rel\_rul(\Phi, M)) = \mathcal{RED}_{rel\_rul(\Phi, M)}(\text{SEM}(\Phi))$, i.e. SEM$(rel\_rul(\Phi, M))$ consists exactly of the reducts of SEM(Φ) to $rel\_rul(\Phi, M)$, and
2. If $I \in$ SEM$(\Phi \cup \{X_{new} \vee A\})$ (where X_{new} is a new atom not occurring in Φ) and $I[\neg A] = \mathbf{t}$, then $\mathcal{RED}_\Phi(I) \in$ SEM(Φ).

A very special case of *Relevance* that we need later is the following. Let A occur in Φ and X_{new} be a new atom not occurring in Φ. Then

$$\mathcal{RED}_\Phi(\text{SEM}(\Phi \cup \{X_{new} \leftarrow \neg A\})) = \text{SEM}(\Phi).$$

Again this is a technical property needed to prove "(1)$\Longrightarrow$ (2)" in Theorem 20.

Let us note that all our transformations are still very weak. As an example, it is possible to construct a semantics SEM (satisfying all our properties) which selects only one minimal model from the program consisting of "$A \vee B$". This is strange because the program is completely symmetric in A and B and therefore if $\{A\} \in$ SEM$(\{A \vee B\})$ then also $\{B\} \in$ SEM$(\{A \vee B\})$ should hold. It turns out later (Theorems 16 and 24) that the following property is indeed sufficient to exclude such anomalous behaviour:

Definition 10 Isomorphy.
A semantics SEM satisfies *Isomorphy*, iff

$$\text{SEM}(\mathcal{I}(\Phi)) = \mathcal{I}(\text{SEM}(\Phi))$$

for all programs Φ and isomorphisms $\mathcal{I}$ on the set of all Σ-ground atoms.

This condition ensures that a semantics is invariant under a renaming (namely an isomorphism $\mathcal{I}$) of the underlying signature. The programs $\mathcal{I}(\Phi)$ and Φ are syntactically different, but considered to represent *equivalent* programs, in the sense that their semantics coincide via $\mathcal{I}$.

It is easy to see that the first three properties are of increasing strength:

Lemma 11 Relevance $\Longrightarrow$ Independence $\Longrightarrow$ Consistency.
a) *If SEM satisfies Relevance then it also satisfies Independence.*
b) *Let SEM be a non-trivial semantics satisfying* Isomorphy.
If SEM satisfies Independence then it also satisfies Consistency.

3 Properties of the Semantics

The next theorem illustrates how most of the semantics behave according to our conditions. The proofs of most of these results are sometimes technical (and nontrivial): we refer the reader to the forthcoming full version.

Theorem 12 Properties of various Semantics.
The following table summarizes the properties of various semantics[2].

Properties of Logic-Programming Semantics								
Semantics	Domain	El. Taut.	El. Contr.	GPPE	Cons.	Indep.	Relev.	Isom.
M_Φ^{supp}	strat. norm.	•	•	•	•	•	•	•
comp	norm.	—	•	•	•	•	•	•
WFS	norm.	•	----	•	•	•	•	•
GCWA	pos. disj.	•	•	•	•	•	•	•
PERFECT	strat. disj.	•	•	(•)	•	•	•	•
Weak-SUPP	disj.	—	•	----	---	---	----	•
SUPP	disj.	—	•	•	—	----	----	•
STABLE	disj.	•	•	•	—	---	---	•
STATIC	disj.	•	----	•	•	•	•	•
D-WFS	disj.	•	----	•	•	•	•	•

While GCWA satisfies all the properties introduced in the last section, STABLE satisfies all except for *Independence* and *Relevance* (see [BD94b]). Let us note that from all our transformations only GPPE is not closed for a particular class of programs: GPPE might transform a stratified disjunctive program into a non-stratified one. Nevertheless we can say, that PERFECT satisfies GPPE, because STABLE does (for general disjunctive programs) and PERFECT coincides with the restriction of STABLE to stratified disjunctive programs. While both Weak-SUPP and SUPP fail to satisfy elimination of tautologies, at least SUPP satisfies GPPE (Weak-SUPP does not). Neither Weak-SUPP, SUPP nor STABLE satisfy *Independence*.

As already noted in the Introduction, GPPE for STABLE and GCWA has also been established independently by Sakama and Seki in [SS94].

---

[2] M_Φ^{supp} is the perfect Herbrand model introduced in [ABW88].

The supported model semantics does not allow the elimination of tautologies. For instance, $I := \{A\}$ is a supported model of $\Phi := \{A \leftarrow A\}$, but it is not a supported model of $\Phi' := \emptyset$. However, a supported model of Φ' is also a supported model of Φ. This example also applies, mutatis mutandis, to Weak-SUPP. In fact, Weak-SUPP also fails GPPE:

Example 1. Weak-SUPP does not satisfy GPPE, nor even *Elimination of non-minimal Rules.* Consider the following logic program and its partial evaluation:

$A \leftarrow B.$ $A \vee C.$

$B \vee C.$ $B \vee C.$

Now consider $I := \{B, C\}$. This is a weakly supported model of the resulting program, but it does not satisfy the first rule in the original program.

4 Characterizations of the Semantics

In this section we give characterizations of the semantics Weak-SUPP, SUPP, GCWA, PERFECT and STABLE in terms of our abstract properties. We begin with a useful lemma (DUNG/KANCHANSUT in [DK89], BRY in [Bry90], and HU/YUAN in [HY91] also considered rules with only negative literals):

Lemma 13 Normal Form.
Let SEM be a semantics satisfying GPPE *and* Elimination of Tautologies.
Then any program Φ is SEM-equivalent to a program Φ' where all clauses have the form $A \leftarrow \neg C$, i.e. there do not occur any positive atoms in the bodies.

Moreover, if Φ is a positive program, then Φ' is a set of positive disjunctions (containing no body literals at all).

Lemma 14. *An interpretation I is a supported model of a program Φ without positive body literals iff it is a stable model of Φ.*

In fact, for all programs a stable model is supported, and the converse is true for all programs without positive body literals.

Theorem 15 Partial Characterization of GCWA.
Let SEM be a semantics satisfying GPPE *and* Elimination of Tautologies.
Then: $SEM(\Phi) \subseteq Min\text{-}MOD_{2-val}^{Herbrand}(\Phi)$ for any positive disjunctive program Φ.

I.e. any such semantics is already based on 2-valued minimal models. In particular, GCWA is the weakest semantics with these properties.

Proof. We give an indirect but nevertheless constructive proof. Let $I \in SEM(\Phi)$ with $I \notin Min\text{-}MOD_{2-val}^{Herbrand}(\Phi)$. We have to derive a contradiction. Due to Lemma 13 we can assume w. l. o. g. that Φ only consists of positive disjunctions. Let $\perp$ be a new atom. We consider the program

$(*)$ $\Phi' = \Phi \cup \{\perp \leftarrow True(I), Undef(I)\}$.

Now we only have to show

$(**)$ Φ' is SEM-equivalent to Φ (i.e. $SEM(\Phi') = SEM(\Phi)$), and we arrive at a contradiction because $I[\bot] = \mathbf{f}$ ($I \in \text{SEM}(\Phi)$ and Φ does not contain $\bot$) and therefore I can not be a 3-valued model of $\bot \leftarrow \text{True}(I)$, $\text{Undef}(I)$.

Indeed, we will show (**) by applying GPPE and *Elimination of Tautologies* to (*): this allows us to eliminate the whole new rule in (*).

Case 1: $\text{Undef}(I) \neq \emptyset$.

Let $A \in \text{Undef}(I)$. We apply GPPE in (*) to replace A, i.e. we get for any $A \vee \mathcal{A} \in \Phi$ a new rule R_A. But since $I[A] = \mathbf{u}$, and I is a model of $A \vee \mathcal{A}$, $\mathcal{A}$ contains another atom which evaluates to $\mathbf{t}$ in I and therefore R_A can be eliminated using *Elimination of Tautologies*.

Case 2: $\text{Undef}(I) = \emptyset$.

This means that $I \in \text{MOD}_{2-val}^{Herbrand}(\Phi)$ but I is not a minimal model (our general assumption). There is therefore a $I' \models \Phi$ with $I' \prec I$. Thus there exists an atom Y with $I'[Y] = \mathbf{f} \neq \mathbf{t} = I[Y]$. We apply GPPE to (*) and replace Y. For any $Y \vee \mathcal{A} \in \Phi$: $I'[\mathcal{A}] = \mathbf{t}$ (since $I' \models \Phi$) and therefore also $I[\mathcal{A}] = \mathbf{t}$. So any of the disjunctions $Y \vee \mathcal{A}$ contains an atom true in I and therefore all rules can be eliminated by *Elimination of Tautologies*. □

The last theorem only tells us that any semantics satisfying our two conditions selects minimal models. It still leaves open the possibility to select a proper subset of them. To get the whole set of all minimal models, we have to add *Isomorphy* and *Relevance*:

Theorem 16 Complete Characterization of GCWA.
Any non-trivial semantics satisfying GPPE, Elimination of Tautologies, Isomorphy *and* Relevance *coincides with GCWA on positive disjunctive programs.*

Proof. Let SEM be a semantics satisfying these conditions. By Lemma 13, it suffices to consider only programs Φ without body literals. The preceding theorem has shown that $\text{SEM}(\Phi) \subseteq \text{Min-MOD}_{2-val}(\Phi)$. Now we have to show the converse. So let $I \in \text{Min-MOD}_{2-val}(\Phi)$.

In order to show that $I \in \text{SEM}(\Phi)$, we first transform Φ into a program Φ' by replacing false head literals by true body literals. More formally, we introduce a new atom X_A for every atom A which is false in I. Then let

$$\Phi_0' := \{A \vee X_A \mid A \text{ is false in } I\}.$$

Now let Φ_1' contain for every disjunction $\mathcal{A} \leftarrow \in \Phi$ the rule

$$\{A \in \mathcal{A} \mid I \models A\} \leftarrow \{X_A \mid A \in \mathcal{A}, I \not\models A\}.$$

For instance, if p is true and q is false, we transform $p \vee q \leftarrow$ into $p \leftarrow X_q$. Let finally $\Phi' := \Phi_0' \cup \Phi_1'$.

Now consider the model I_0' of Φ_0' which makes all X_A true and all A false. It obviously is a minimal model of Φ_0'. We will show that $I_0' \in \text{SEM}(\Phi_0')$. By Lemma 11, SEM is consistent, so there must be an $\hat{I}_0 \in \text{SEM}(\Phi_0')$. By the preceding theorem, $\hat{I}_0$ is a minimal model. But because of the simple structure of Φ_0', there is an isomorphism which transforms $\hat{I}_0$ into I_0'. So our isomorphy condition yields $I_0' \in \text{SEM}(\Phi_0')$.

By Relevance, there must be an $I' \in \text{SEM}(\Phi')$ such that $\mathcal{RED}_{\Phi_0'}(I') = I_0'$. We will show that I' must make all atoms true which are true in I (and since it extends I_0' it must also make all atoms false, which are false in I). So let A be an atom with e $I \models A$. Because of the minimality of I, Φ must contain a disjunction in which A is the only true atom (otherwise we could make A false and still have a model). But this means that Φ' contains a rule

$$A \leftarrow X_{A_1} \wedge \cdots \wedge X_{A_n},$$

where all the body atoms are true. Therefore, $I' \models A$.

So $I' \in \text{SEM}(\Phi')$ is an extension of I. We can now finally apply the second part of the Relevance condition to get $I = \mathcal{RED}_\Phi(I') \in \text{SEM}(\Phi)$. $\qquad\square$

It is worth noting that we do not assume in the last theorem that SEM is based on two-valued models. We get this automatically from our conditions, although this does not follow for arbitrary programs (not even for non-disjunctive programs, where WFS is a counterexample as can be seen from Theorem 12).

For the next theorems we need the *Elimination of Contradictions*. This condition implies that a semantics is based on 2-valued models:

Lemma 17.
Let SEM satisfy Elimination of Contradictions.
Then: $SEM(\Phi) \subseteq MOD_{2-val}^{Herbrand}(\Phi)$.

Lemma 18 Partial Characterization of Weak-SUPP.
Let SEM be a semantics satisfying GPPE *and* Elimination of Contradictions.
Then: $SEM(\Phi) \subseteq Weak\text{-}SUPP(\Phi)$.

Obviously, Weak-SUPP is not the weakest semantics satisfying these principles, because GPPE does not hold. For non-disjunctive programs, Weak-SUPP and SUPP collapse and SUPP satisfies GPPE:

Corollary 19 SUPP for non-disjunctive programs.
Let SEM be a semantics for non-disjunctive programs satisfying GPPE *and* Elimination of Contradictions. *Then:* $SEM(\Phi) \subseteq SUPP(\Phi)$.

In particular, SUPP is the weakest semantics for non-disjunctive programs with these properties.

Theorem 20 First Partial Characterization of STABLE.
Let SEM be a semantics satisfying GPPE *and* Elimination of Tautologies.
The following three conditions are equivalent[3]:

(1) $SEM(\Phi) \subseteq Weak\text{-}SUPP(\Phi)$ *for all* Φ.
(2) $SEM(\Phi) \subseteq SUPP(\Phi)$ *for all* Φ.
(3) $SEM(\Phi) \subseteq STABLE(\Phi)$ *for all* Φ.

---

[3] In *(1) implies (2)* we need the instance of *Relevance* introduced after Definition 9.

(2) of Theorem 20 shows that STABLE is the weakest semantics selecting only supported models and satisfying *GPPE* and *Elimination of Tautologies*. Since SUPP itself satisfies *GPPE*, this shows that the *Elimination of Tautologies* is, in a strict sense, the only difference between STABLE and SUPP.

(1) of Theorem 20 shows that STABLE is the weakest semantics selecting only weakly supported models and satisfying *GPPE*, *Elimination of Tautologies*. Note: the assumption that an atom has only a weak support is indeed a very weak assumption.

Corollary 21 Second Partial Characterization of STABLE.
Let SEM be a semantics satisfying GPPE, Elimination of Tautologies, *and* Elimination of Contradictions. *Then:* $SEM(\Phi) \subseteq STABLE(\Phi)$.

By Corollary 21, STABLE is the weakest semantics satisfying GPPE, *Elimination of Tautologies* and *Elimination of Contradictions*. In particular, if Φ is stratified, then SEM(Φ) consists of perfect models. For non-disjunctive programs we immediately get

Corollary 22 M_Φ^{supp} for non-disjunctive programs.
Let SEM be a semantics for non-disjunctive programs satisfying GPPE, *Elimination of Tautologies and* Elimination of Contradictions.
Then: $SEM(\Phi) \subseteq \{M_\Phi^{supp}\}$ *for all stratified programs* Φ.
In particular, if $SEM(\Phi) \neq \emptyset$ *then* $SEM(\Phi) = \{M_\Phi^{supp}\}$.

Our next result is an impossibility result. It is well known that STABLE is not always consistent, i.e. it is possible that STABLE(Φ) = $\emptyset$. But even Weak-SUPP is already inconsistent for programs of the form $X \leftarrow \neg X$. Since we have proven SEM(Φ) $\subseteq$ Weak-SUPP(Φ), this also applies to any semantics SEM with the above properties. Thus we have

Corollary 23 Impossibility Result.
Let SEM be a semantics defined for all disjunctive logic programs satisfying GPPE, Elimination of Contradictions *and* Independence.
Then SEM is trivial, i. e. $SEM(\Phi) = \emptyset$ *for all programs* Φ.

The argument is simple. Choose $\Phi' := \{A \leftarrow \neg A\}$ with a new predicate A. Then by Lemma 18 SEM($\Phi \cup \Phi'$) $\subseteq$ Weak-SUPP($\Phi \cup \Phi'$) = $\emptyset$. So any formula ψ would follow from SEM($\Phi \cup \Phi'$), and therefore also from SEM(Φ), for any logic program Φ.

The failure of *Independence* is a weakness of both Weak-SUPP and the stable semantics. We already introduced in Section 2.2 a principle strongly related to this: *Relevance* (see also [DM94b, Dix95a]). Using our framework it is possible to show

Theorem 24 Complete Characterization of PERFECT.
Any non-trivial semantics SEM satisfying SEM $\subseteq$ *Weak-SUPP*, GPPE, Elimination of Tautologies, Isomorphy *and* Relevance *already coincides with PERFECT on stratified disjunctive programs.*

The question arises whether there exist semantics satisfying our properties on classes of programs that significantly extend the class of stratified (or locally stratified) programs. The general feeling is no, but the proof of such a statement is not trivial, because a given program might be non-stratified, but easily equivalent (using some of our transformations) to a stratified one. For example although the program Φ_{ns}: "$A \leftarrow$, $A \leftarrow \neg A$" is non-stratified, it is certainly equivalent to "$A \leftarrow$" for any reasonable semantics and thus its intended model should be $\{A\}$.

We can eliminate such *non-genuine* cycles through negation by using our work in [BD95b] where we have associated to any program Φ a certain normal form $\hat{\Phi}$, the *residual program*. $\hat{\Phi}$ is obtained from Φ by using our transformations and some very weak reductions (if there is a clause $A \leftarrow$ then any occurrence of A (resp. $\neg A$) can be replaced by true (resp. false)). For the above program Φ_{ns} we get $\hat{\Phi}_{ns} = A \leftarrow$.

We can also formally define the class of all programs that possess a stable model. This class extends the class of stratified programs but STABLE is not relevant on this class (see [DM94b]). It is easy to see that STABLE on this class satisfies *Independence* — thus *Relevance* is strictly stronger. Recently, however, Li-Yan Yuan noted that the stable semantics on the smaller class of all programs without an *odd* number of negative edges through negation satisfies *Relevance*.

We believe that this class can not only be extended using our idea of deleting *non-genuine* cycles through negation, but it also represents the maximal such class. More formally, we have the following

Conjecture 25 No Semantics for genuine Non-Stratified Programs.
Let SEM be a non-trivial semantics satisfying GPPE, Elimination of Tautologies, Elimination of Contradictions *and* Relevance.
Then SEM is only defined on the class of all programs Φ such that $\hat{\Phi}$ contains no cycles with an odd number of negative edges. There is no semantics beyond this class.

Note that if we cancel the *Elimination of Contradictions* then there are semantics defined on the whole class of programs, e.g. the static semantics of Przymusinski ([Prz94]) or the D-WFS introduced by the authors ([BD94b, BD95b]).

We believe our conjecture to be true both for disjunctive and non-disjunctive programs. In the latter case already the wellfounded semantics WFS satisfies all properties except *Elimination of Contradictions*.

We think that the last two results and our conjecture show us that if we leave the class of stratified disjunctive programs then a semantics should be based on three-valued models, i.e. *Elimination of Contradictions* should be given up.

5 Conclusions

In this paper, we have shown that partial evaluation is an interesting property. It not only holds for various semantics but it also characterizes these semantics together with some other weak transformation conditions. Let us note, that GPPE also holds for Przymusinski's *static* and *stationary* semantics.

GPPE is a powerful principle. Together with *Elimination of Tautologies* it enables us to define a normal form of a program (Lemma 13). Both properties are sufficient to ensure that a semantics only selects minimal two-valued models for positive disjunctive programs (Theorem 15). Together with *Elimination of Contradictions* (resp. the assumption that a semantics is based on weakly supported models), we can partially characterize the disjunctive stable semantics (Corollary 21 resp. Theorem 20).

We were also able to characterize GCWA (on positive disjunctive programs) and STABLE on stratified disjunctive programs completely, by simply adding *Isomorphy* and *Relevance* (Theorems 16 and 24).

It is interesting that for programs in normal form supported models *coincide* with stable models. This fact can be used to compute these semantics (see [BD95a] for more details).

Our impossibility result tells us that a reasonable semantics for the class of all disjunctive programs should not satisfy *Elimination of Contradictions*, i.e. should be based on three-valued models.

Finally, our conjecture formally states that there are no semantics (besides the trivial one) having our properties on non-trivial extensions of the class of all stratified programs.

Acknowledgements

We are indebted to Li-Yan Yuan, F. Miguel Dionisio and to two anonymous referees for helpful comments on a draft of this paper.

References

[ABW88] K. Apt, H. Blair, and A. Walker. Towards a theory of declarative knowledge. In Jack Minker, editor, *Foundations of Deductive Databases*, chapter 2, pages 89–148. Morgan Kaufmann, 1988.

[BD94a] Stefan Brass and Jürgen Dix. A Characterization of the Stable Semantics by Partial Evaluation. In *Proc. of the 10th Workshop on Logic Programming, Zuerich, October 1994*, 1994.

[BD94b] Stefan Brass and Jürgen Dix. A disjunctive semantics based on unfolding and bottom-up evaluation. In Bernd Wolfinger, editor, *Innovationen bei Rechen- und Kommunikationssystemen, (IFIP-Congress, Workshop FG2: Disjunctive Logic Programming and Disjunctive Databases)*, pages 83–91. Springer-Verlag, 1994.

[BD95a] Stefan Brass and Jürgen Dix. A General Approach to Bottom-Up Computation of Disjunctive Semantics. In J. Dix, L. Pereira, and T. Przymusinski, editors, *Nonmonotonic Extensions of Logic Programming*, pages 127–155, Springer LNCS, to appear, 1995.

[BD95b] Stefan Brass and Jürgen Dix. Disjunctive Semantics based upon Partial and Bottom-Up Evaluation. In Leon Sterling, editor, *Proceedings of the 12th Int. Conf. on Logic Programming, Tokyo*. MIT, June 1995.

[Bry90] François Bry. Negation in logic programming: A formalization in constructive logic. In Dimitris Karagiannis, editor, *Information Systems and Artificial Intelligence: Integration Aspects*, pages 30–46. Springer, 1990.

[Dix92] Jürgen Dix. A Framework for Representing and Characterizing Semantics of Logic Programs. In B. Nebel, C. Rich, and W. Swartout, editors, *Principles of Knowledge Representation and Reasoning: Proceedings of the Third International Conference (KR '92)*, pages 591–602. San Mateo, CA, Morgan Kaufmann, 1992.

[Dix95a] Jürgen Dix. A Classification-Theory of Semantics of Normal Logic Programs: II. Weak Properties. *Fundamenta Informaticae*, XXII(3), pages 257–288, 1995.

[DM94a] Jürgen Dix and Martin Müller. An Axiomatic Framework for Representing and Characterizing Semantics of Disjunctive Logic Programs. In Pascal Van Hentenryck, editor, *Proceedings of the 11th Int. Conf. on Logic Programming, S. Margherita Ligure*, pages 303–322. MIT, June 1994.

[DM94b] Jürgen Dix and Martin Müller. Partial Evaluation and Relevance for Approximations of the Stable Semantics. In Z.W. Ras and M. Zemankova, editors, *Proceedings of the 8th Int. Symp. on Methodologies for Intelligent Systems, Charlotte, NC, 1994*, pages 511–520. Springer, Lecture Notes in Artificial Intelligence 869, 1994.

[DK89] P. M. Dung and K. Kanchansut. A fixpoint approach to declarative semantics of logic programs. In E.L. Lusk and R.A. Overbeek, editors, *Proceedings of North American Conference Cleveland, Ohio, USA*. MIT, October 1989.

[GL91] Michael Gelfond and Vladimir Lifschitz. Classical Negation in Logic Programs and Disjunctive Databases. *New Generation Computing*, 9:365–387, 1991.

[HY91] Yong Hu and Li Yan Yuan. Extended Well-Founded Model Semantics for General Logic Programs. In Koichi Furukawa, editor, *Proceedings of the 8th Int. Conf. on Logic Programming, Paris*, pages 412–425. MIT, June 1991.

[Min82] Jack Minker. On indefinite databases and the closed world assumption. In *Proceedings of the 6th Conference on Automated Deduction, New York*, pages 292–308. Springer, 1982.

[Prz88] Teodor Przymusinski. On the declarative semantics of deductive databases and logic programs. In Jack Minker, editor, *Foundations of Deductive Databases*, chapter 5, pages 193–216. Morgan Kaufmann, 1988.

[Prz91] Teodor Przymusinski. Stable Semantics for Disjunctive Programs. *New Generation Computing Journal*, 9:401–424, 1991.

[Prz94] Teodor Przymusinski. Static Semantics For Normal and Disjunctive Logic Programs. *Annals of Mathematics and Artificial Intelligence*, Special Issue on Disjunctive Programs: to appear, 1995.

[SS94] Chiaki Sakama and Hirohisa Seki. Partial Deduction of Disjunctive Logic Programs: A Declarative Approach. In *Fourth International Workshop on Logic Program Synthesis and Transformation (LOPSTR '94)*. Lecture Notes in Computer Science, Springer-Verlag, July 1994.

Game Characterizations of Logic Program Properties

Howard A. Blair[1]

School of Computer and Information Science

Syracuse University

Syracuse, New York 13244-4100

blair@top.cis.syr.edu

Abstract

A family of simple two-player games will be presented which vary by
how play passes between players and by what constraint must be maintained by the players in order to avoid losing. The players are repre-
sentable as interacting almost independent logic programs. A correspondence between winning strategies, well-founded dependencies, constructive ordinals and hyperarithmetic sets is presented. Complexity results
can be obtained for logic program properties in a uniform way. This
paper demonstrates the technique as applied to two apparently divers
properties, each of a very high degree of undecidability.

1 Introduction

There are numerous results describing the degrees of unsolvability of various
properties of and relations among logic programs, as well as the degree of expressive power of various classes of logic programs. For example, Definite clause
programs define, via their maximum supported Herbrand models, the Σ_1^1 sets
[Bl82]. There are the more recent results [BMS92] that locally stratified logic
programs define precisely the hyperarithmetic sets via their unique stable models, and [CB93] that the class of locally stratified programs is complete Π_1^1. Since
there is about as much variety in the proof techniques that have been used to
obtain these results as there is in the results themselves, it would be clarifying
to have a reasonably uniform means of obtaining them. We introduce a game-
theoretic approach through which many results having to do with complexity,
degrees of unsolvability and expressive power of logic programs can be obtained
in a strikingly uniform way. We introduce a certain class of two-player games.
Individual games in the class are determined by two parameters, one being a

---

[1] Research partially supported by the U.S. Army Research Office through the Mathematical Sciences Institute of Cornell University.

number that amounts to an initial play of the game, the other being a binary relation that constitutes a constraint that must be maintained by the players in order to avoid losing. The games we define are examples of infinite-alphabet Gurevich-Harrington games, [GH82, YY93].

The game has the important feature that winning plays for one of the players (player 0) correspond to well-founded sequences of dependencies among atoms in the ground-instantiated version of the program. It will be seen that it is easy to control the degree of unsolvability of the class of winning plays available to one of the players by adjusting the parameters of the game. By using the correspondence and varying the logical connection between the players, and hence varying the type of dependency relation embodied by the overall program, the complexity of various properties of the program can be read off.

We will apply the games to give two results, one previously known, but with a new and much simpler proof, and the other result not previously published, about the degree of undecidability of certain program properties.

The main contribution of this paper is to show that the game-theoretic approach taken here is a useful, unifying device for complexity investigations. The argument for this point is that two theorems, which at first sight appear to be quite different, the one having to do with models of definite clause programs (where negations do not occur in program clause bodies), the other having to do with the property of local stratification (which appears to be intrinsically about dependencies on negations within the program), are actually two instances of the same underlying theorem about the degree of undecidability of the class of winning strategies for the games.

In the next section we define the games. Then, since what is logically expressed by a program's clauses is to be closely related to the program's dependency relation, the third section discusses converting a definite clause program into a binary definite clause program, and it is shown how to represent the game's players as definite clause programs. These programs describe the action of the game's players on each turn. These player programs can then be connected through calls each other in a way that reflects how the play passes from one player to another. By varying certain computable parameters within the player programs and by varying the manner in which the player programs are connected, we will be able to read off diverse results having to do with complexity and degrees of unsolvability associated with logic programs. It will become completely clear that the various manners in which programs may be connected are simple and do not have to hide encodings of complex properties. In particular, we will show that two quite distinct complete Π_1^1 properties of logic programs, namely unique fixed points of Horn clause programs, and local stratification, owe their high degree of unsolvability to the same underlying property.

2 The Game

We present an exact definition of the games in terms of two parameters, $n \in N$, where N is a fixed set of *nodes* and $R \subseteq N \times N$. We then illustrate the play of the games with an example in which R is represented as a finite directed graph, and discuss the nature of the games' winning strategies. Subsequently, we will be interested in games for which the underlying fixed set of nodes is (countably) infinite. Hence we will then identify N with the set of natural numbers $\mathbf{N}$.

Definition 2.1: $\Gamma(R, x_0)$ is played as follows:
Initially, set $z := x_0$. Player 0 *moves* first. The players alternate moves until one of them wins or loses. A *play* of the game is either a finite sequence of moves beginning with player 0's first move and ending with a move of player 1 resulting in a win for one of the players, or an infinite sequence of moves beginning with player 0's first move. The command `choose z` chooses a natural number.

player 0 executes: ($\mathtt{x} := \mathtt{z}$; `choose z`)
player 1 executes:
  if $R(\mathtt{x}, \mathtt{z})$ **then** (`choose z`; **if not** $R(\mathtt{x}, \mathtt{z})$ **then** player 0 wins)
        else (`choose z`; (**if** $R(\mathtt{x}, \mathtt{z})$ **then** player 1 wins **else** player 0 wins))

□

Example 2.1: Let R be the binary relation on the set of nodes {a,b,c,d,v,w,x, y,z} where $R(\mathrm{a,b})$, $R(\mathrm{a,x})$, $R(\mathrm{b,c})$, $R(\mathrm{c,v})$, $R(\mathrm{c,d})$, $R(\mathrm{d,y})$, $R(\mathrm{d,b})$, $R(\mathrm{v,w})$, $R(\mathrm{w,x})$, $R(\mathrm{w,y})$, $R(\mathrm{y,z})$ We will refer to this example in subsequent examples and remarks.

By definition, and consistent with the definition of *winning* in Gurevich-Harrington (GH) games which inspired these considerations, we define infinite plays that are not one by player 0 as *winning* plays for player 1.

We introduce some terminology for binary relations that will allow us to be more concise in describing winning strategies for the games $\Gamma(R, x_0)$. With the terminology made precise in the next definition we can say that the strategy for player 1 to avoid losing is for player 1 to avoid crossing the boundary of a *well* in R.

Definition 2.2: A binary relation R on a set A is *well-founded* iff there is no sequence $\{a_n\}_{n=0}^{\infty}$ of elements of A such that

$$R(a_1, a_0), \; R(a_2, a_1), \; \ldots, \; R(a_i, a_{i-1}), \; R(a_{i+1}, a_i), \; \ldots$$

R is *Noetherian* (terminology borrowed from the literature of term-rewriting systems, *cf.* [Hu80]) iff the converse of R is well-founded.

A *path* in R from a_0 to a_n is a finite sequence

$$a_0, a_1, \; \ldots, \; a_{n-1}, a_n \quad (n > 0)$$

of elements of A such that

$$R(a_{i-1}, a_i), \text{ for all } i = 0, \ldots, (n-1)$$

Thus, R is Noetherian iff from any element a of A, every path in R that starts from a is finite, where a path in R is a path in the usual sense when R is regarded as a directed graph.

We borrow terminology from graph theory via the following notation and terminology: The *trace* $R_{A'}$ of R on a subset A' of A is defined by

$$R_{A'}(x, y) \text{ iff } x \in A', \; y \in A' \text{ and } R(x, y).$$

A subrelation R' of R is *full* iff R' is the trace of R on a subset A' of A.

We say that R' is a *well* in R iff R' is maximal in the set of full Noetherian subrelations of R. The idea is that there are no paths leading out of wells, and one cannot move along a path in a well indefinitely. We take maximal relations of this kind because players of the games are interested in boundaries of wells.

The *field* of R, denoted by $\mathrm{fld}(R)$, is defined by

$$\mathrm{fld}(R) = \{x \mid \exists y \, R(x, y)\} \cup \{x \mid \exists y \, R(y, x)\}$$

Let R' be a subrelation of R. Then the *boundary* of R' is the set of elements a of A such that $\{a \notin \mathrm{fld}(R') \mid \exists y \in \mathrm{fld}(R') \, R(a, y)\}$.

A path in R from a to b *crosses* the boundary of a well W in R iff a is not in the well but b is in the well. (The last element of A in the path from a to b that is not in $\mathrm{fld}(W)$ is on the boundary of the well.) Note that a path does not terminate within the field of a well iff the path can be properly extended to a path with the same property. This completes definition 2.2. □

Example 2.2: In the directed graph corresponding to the relation R of example 2.1 the relation $R_{\mathbf{v}}$ is the trace of the relation R on the nodes $\mathbf{v}$, $\mathbf{w}$, $\mathbf{x}$, $\mathbf{y}$ and $\mathbf{z}$. $R_{\mathbf{v}}$ is a well whose boundary consists of the nodes $\mathbf{a}$, $\mathbf{c}$ and $\mathbf{d}$. □

Proposition 2.1: Player 0 wins $\Gamma(R, x)$ iff x is within the field of a well in R.
□

What makes $\Gamma(R, x_0)$ interesting for investigations of degrees of unsolvability is that computable R can be easily chosen to make the set of all x_0 such that player 0 wins $\Gamma(R, x_0)$ complete Π_1^1. We make this precise with lemma 2.1.

Definition 2.3: Let Σ be a possibly infinite set. We call Σ an *alphabet*. A 2-player Σ-*game* G is specified by a *game tree* and a set of winning plays for one of the players. A game tree $\mathcal{T}_G$ is a tree with nodes that are finite sequences of elements of Σ, also called *words* over Σ. Formally, $\mathcal{T}_G$ is a set of words over Σ which is closed under the operation of taking prefixes. The root of the tree is the empty word. In particular, an edge in $\mathcal{T}_G$ exists between any pair of nodes w and wa, that are both in $\mathcal{T}_G$. A node in $\mathcal{T}_G$ is also called a *position*. The infinite paths in $\mathcal{T}$ are the *plays* of G. The *available moves* of G in a position γ in $\mathcal{T}_G$ are those moves σ such that $\gamma \cdot \sigma$ is in $\mathcal{T}_G$. To specify G a triple $\langle \mathcal{T}, p, W \rangle$ is given in which $\mathcal{T}$ is a game tree with respect to Σ, p is a player and W is the set of plays won by player p. It is assumed that all plays in $\mathcal{T}$ which are not in W are won by the other player $1 - p$.
□

Generally, unless otherwise specified, positions in which player 0 moves, the collection of which is denoted by $\text{Pos}_G(0)$, are of even length, and positions in which player 1 moves, the collection of which is denoted by $Pos_G(1)$, are of odd length.

Conceptually, a strategy is a means by which players can select moves. The next definition makes rigorous what is meant by a *deterministic strategy*.

Definition 2.4: Let $p \in \{0, 1\}$ be a player in game G. A *deterministic p-strategy* is a function $f : \text{Pos}_G(p) \longrightarrow \Sigma$ such that if $\alpha \in \text{Pos}_G(p)$ then $\alpha \cdot \text{pos}_G(p) \in \mathcal{T}_G$. The set of positions in $\mathcal{T}_G$ *consistent with* a deterministic p-strategy f is inductively defined by: i) the empty sequence Λ is consistent with f. ii) if α is consistent with f and $\alpha \in \text{Pos}_G(1 - p)$ then every child of p is consistent with f. iii) if α is consistent with f and $\alpha \in \text{Pos}_G(p)$ then $\alpha \cdot f(\alpha)$ is consistent with f. A play is *consistent* with f if every position in the play is consistent with f. A deterministic p-strategy *wins* $G = \langle \mathcal{T}_G, p, W \rangle$ if every play consistent with f is in W. Player p *wins* G if there is a winning deterministic p-strategy.
□

The choice of R, by which we obtain complete Π_1^1 sets of games winnable by player 0, is given in the next lemma.

First, some notation: $\langle x, y \rangle$ is the code number (using a bijective pairing function) of the pair (x, y). If $c = \langle x, y \rangle$, then $(c)_0 = x$ and $(c)_1 = y$. The function φ_z is the zth partial recursive function with respect to a fixed acceptable indexing. Equivalently, z is the index of W_z, the zth recursively enumerable subset of $\mathbf{N}$. The notation is as in [Ro67].

Lemma 2.1: Let $R(x, z) \leftrightarrow \varphi_{(x)_0}((z)_0)$ converges within $(z)_1$ steps. Then the set of all n_0 such that player 0 has a winning strategy in $\Gamma(R, n_0)$ is complete Π_1^1.

Proof: (Sketch. *cf.* [Ro67].) Let C be the productive center of the identity function. C is a complete Π_1^1 set. Player 0 has a winning strategy in $\Gamma(R, n)$ iff $(n)_0 \in C$. The strategy has player 0 always choose z such that $R(x, z)$ holds, unless $W_{(x)_0} = \emptyset$, in which case player 1 will lose on his next turn. Such a choice can always be made if play starts from $(n)_0 \in C$ because $(x)_0 \in C$ implies either $W_{(x)_0} = \emptyset$ or $\exists z \, [(z)_0 \in W_{(x)_0} \subseteq C]$. If player 0 chooses z by this strategy, then for player 1 to avoid losing, he must either confirm player 0's choice of z or choose $z' \neq z$ with the property that $(z')_0 \in W_{(x)_0} \subseteq C$. C is structured so that it has a well-ordered partition $C = \bigcup_{\gamma=0}^{\omega_1^{\text{ck}}} C_\gamma$ such that $a \in C$ implies $a \in C_{\alpha+1}$ for some $\alpha < \omega_1^{\text{ck}}$, which, in turn, implies $W_a \in C_\alpha$. (ω_1^{ck} is the least nonconstructive ordinal.) This property entails that eventually player 0 must be able to choose z such that $W_{(z)_0} = \emptyset$. $\qquad\square$

In order to exclude certain unwanted entailments within the logic program representations of the games $\Gamma(R, n_0)$, we will use the following variation of the preceding lemma.

Corollary 2.1: Let $\Phi(Y) = \{x \mid x \in W_y \text{ for some } y \in Y\}$. Then $\bigcup_{i=0}^\infty \Phi^i(\{n\})$ is recursively enumerable. Define f by $W_{f(n)} = \bigcup_{i=0}^\infty \Phi^i(\{n\})$ and let $R_{n_0}(x, z) \leftrightarrow \varphi_{(x)_0}((z)_0)$ and $\varphi_{f(n_0)}((x)_0)$ both converge within $(z)_1$ steps. Then the set of all n_0 such that player 0 has a winning strategy in $\Gamma(R_{n_0}, n_0)$ is complete Π_1^1. $\qquad\square$

It should be observed that a winning strategy for player 1 in $\Gamma(R, n_0)$, is in general complete Π_1^1. Player 0's winning strategy, if it exists, is at worst recursive in the halting problem, and the cost of complicating R a little, can be made recursive.

3 Binary Logic Programs

The players of $\Gamma(R, x_0)$ can be represented by binary logic programs. This is desirable because of the close relationship between entailment and dependency in binary programs. A difficulty with controlling dependencies in programs is due to the fact that if conjunctions in clause bodies are replaced by disjunctions, then the dependency relation of the resulting program is the same as that of the original, but the models of the resulting program are, in general, vastly different. Another way to look at the difficulty is by considering a clause such as p(x) ← q(x,y), r(y) contained in some program P which has a least model in which, for example, q(a,b) is false. p(a) still depends on r(b), but this was perhaps not intended. We would like to control dependencies through the semantics of the program. This is achievable by converting a program P to a binary program which has the same least model as P with respect to the predicates defined in P.

Definition 3.1: Let P be a normal logic program *cf.* [Ll87], and let grd(P) be the set of ground clauses which are instances of clauses in P. The relations *refers positively to* and *refers negatively to* are defined by: A *refers positively* [*negatively*] *to* B iff there is a clause $A \leftarrow L_1, \ldots, L_n \in$ grd(P) such that B [$\neg B$] is L_i for some $i \in \{1, \ldots, n\}$. Define the *depends positively on* relation to be the reflexive transitive closure of the *refers positively to* relation, and let the *depends negatively on* relation be

(*depends positively on*)* $\circ$ (*refers negatively to*) $\circ$ (*depends depends positively on*)*

where $\circ$ denotes composition of relations. □

Definition 3.2: Let ground atom A depend positively on ground atom B with respect to program P. Then the pair (A, B) is said to be a *logical dependency* iff $P \cup \{B\} \models A$. A program is *dependency sound* if every pair of ground atoms in the positive dependency relation of P is a logical dependency. □

Definition 3.3: A *binary* logic program is a program where each program clause either has the form $A \leftarrow B$ or is a unit clause A, where A and B are atoms. □

The following proposition shows how binary programs "equate" entailment and dependency.

Proposition 3.1: Every binary program is dependency sound. □

Definition 3.4: Let L be a first order language and let P_1, P_2 be definite clause logic programs over L. Let L' be a sublanguage of L and suppose the restrictions of the least models of P_1 and P_2 to the Herbrand base of L' are the same. Then P_1 and P_2 are said to be *extensionally equivalent* with respect to L'. □

Definition 3.5 Let P be a definite clause program. Extend L to a language L' by adjoining a new function symbol f_p for each predicate symbol p in L other than the equality symbol, =. f_p has the same arity as p. Corresponding to each

atom $p(t_1, \ldots, t_n)$ of L, the *translation*, $f_p(t_1, \ldots, t_n)$ is a term of L'. In general, for each atom A of L, let t_A denote the translation of A. Corresponding to P the *binary extensional equivalent Q* of P is defined as follows. Extend L' by adjoining a new binary predicate symbol stack, a new binary function symbol cons and a new constant symbol nil. Corresponding to each program clause
$A \leftarrow B_1, \ldots, B_n$ of P, form the clause
$\mathtt{stack}(\mathtt{cons}(t_A, \mathtt{Y}), \mathtt{Z}) \leftarrow \mathtt{stack}(\mathtt{cons}(t_{B_1}, \mathtt{cons}(t_{B_2}, \ldots, cons(t_{B_n}, \mathtt{Y}) \ldots)), \mathtt{Z})$.
Q also contains a *bridging clause* for each predicate symbol p:
$p(\mathtt{X1}, \ldots, \mathtt{Xn}) \leftarrow \mathtt{stack}(\mathtt{cons}(f_p(\mathtt{X1}, \ldots, \mathtt{Xn}), \mathtt{nil}), f_p(\mathtt{X1}, \ldots, \mathtt{Xn}))$.
Finally, Q contains the *terminating* clause: $\mathtt{stack}(\mathtt{nil}, \mathtt{Z})$. $\qquad\square$

Occasionally, it will be convenient to be able to ensure that the *depends on* relation within binary definite clause programs is Noetherian. In the case of binary extensional equivalent programs, it will suffice to add a step-counter argument to the stack predicate.

Definition 3.6: The *step-counter augmentation* of a binary extensional equivalent program Q has clauses of the form

$\mathtt{stack}(\mathtt{s}(\mathtt{S}), \mathtt{cons}(t_A, \mathtt{Y}), \mathtt{Z}) \leftarrow$
$\qquad \mathtt{stack}(\mathtt{S}, \mathtt{cons}(t_{B_1}, \mathtt{cons}(t_{B_2}, \ldots, cons(t_{B_n}, \mathtt{Y}) \ldots)), \mathtt{Z})$.
$p(\mathtt{X1}, \ldots, \mathtt{Xn}) \leftarrow \mathtt{stack}(\mathtt{S}, \mathtt{cons}(f_p(\mathtt{X1}, \ldots, \mathtt{Xn}), \mathtt{nil}), f_p(\mathtt{X1}, \ldots, \mathtt{Xn}))$.
$\mathtt{stack}(\mathtt{0}, \mathtt{nil}, \mathtt{Z})$.

and is obtained by adding a step-counter argument to each of the clauses in Q.
$\qquad\square$

Proposition 3.2: The binary extensional equivalent of P is extensionally equivalent to P with respect to the language of P.

Proof: Use the bridging clauses. $\qquad\square$

Proposition 3.3: Let Q be the binary extensional equivalent of P and let A and B be ground atoms in the language of P. Then
A depends on $\mathtt{stack}(\mathtt{nil}, t_B)$ iff $Q \models A$ and B is A. $\qquad\square$

The following proposition will be convenient when we come to considering programs with unique fixed points.

Proposition 3.4 Suppose P is a binary program without unit clauses, and therefore with an empty least model. Then P has no nonempty supported models iff the *depends on* relation of P is Noetherian.

Proof: If the *depends on* relation of P is not Noetherian then there is an infinite sequence of ground atoms

$$A_0, \ldots, A_n, \ldots$$

such that A_i depends on A_{i+1} for all $i \in \mathbf{N}$. Since P is binary, $\mathbf{T}_P(\{A_{i+1}\})$ contains A_i, for each i. Let I be the set of atoms in the above sequence. Then $I \subseteq \mathbf{T}_P(I)$. Hence, since $\mathbf{T}_P$ is monotonic, there is a fixed point of $\mathbf{T}_P$ above I, which is, a fortiori, nonempty. Conversely, if $\mathbf{T}_P$ has a nonempty fixed point then we have immediately that the *depends on* relation is not Noetherian since P has no unit clauses. $\qquad\square$

4 Representing Players

Note that, informally, the players in $\Gamma(R, x_0)$ nondeterministically map $\mathbf{N}$ to $\mathbf{N}$. In the following definition P_0 and P_1 are intended to be executed when combined with others clauses, as will become clear in the sequel.

Definition 4.1: Let P_0 be the program consisting of only the unit clause p0(X, Z0, WinLoss). Informally, Z0 is the new value chosen by player 0. WinLoss records whether player 0 wins or loses in a finite number of moves.
Let P_1 be the program

$$\mathrm{p1}(\mathrm{X}, \mathrm{Z}, \mathrm{Z0}, 0, 0) \leftarrow \mathrm{p_R}(\mathrm{X}, \mathrm{Z}, \mathrm{s}(0)) \wedge \mathrm{p_R}(\mathrm{X}, \mathrm{Z0}, 0).$$
$$\mathrm{p1}(\mathrm{X}, \mathrm{Z}, \mathrm{Z0}, \mathrm{WinLoss}, \mathrm{s}(0)) \leftarrow \mathrm{p_R}(\mathrm{X}, \mathrm{Z}, \mathrm{s}(0)) \wedge \mathrm{p_R}(\mathrm{X}, \mathrm{Z0}, \mathrm{s}(0)).$$
$$\mathrm{p1}(\mathrm{X}, \mathrm{Z}, \mathrm{Z0}, \mathrm{WinLoss}, 0) \leftarrow \mathrm{p_R}(\mathrm{X}, \mathrm{Z}, 0) \wedge \mathrm{p_R}(\mathrm{X}, \mathrm{Z0}, \mathrm{WinLoss}).$$

$\mathrm{p_R}$ computes the characteristic function of relation R. The fifth argument of p1 is intended to record that play should continue when the second clause succeeds.

Suppose that R is a recursive relation. Let P_R be a definite clause program that computes the characteristic function of R using the predicate symbol $\mathrm{p_R}$. The notion of *compute* here is the obvious one in terms of least models, and was formalized in *cf.* [Bl87, Ap90].

Definition 4.2: Let the signature of L' be a subset of the signature of L. Let R be an n-ary relation over the Herbrand universe $U_{L'}$ of L' and let S be the relation computed by (P, p). Then (P, p) *computes R with respect to L'* iff $S \cap (U_{L'})^n = R$. $\qquad\square$

Computing with respect to L' is useful when we do not want to have to keep track of the extra tuples computed by a program due to the introduction of cons, nil, and the various f_p function symbols introduced in the construction of binary extensional equivalent programs.

Assume that there are no predicate symbols that occur in both programs P_0 and P_1. Assume also that the only predicate symbol that occurs in both programs P_1 and P_R is p_R. Further assume that p_0, the predicate symbol in the head of P_0 does not occur in either of the programs P_0 or P_R. The nonintersection of the sets of predicate symbols occurring in these programs can easily be arranged without loss of generality by renaming predicate symbols as necessary. (That a program to compute the characteristic function of R using p_R can be constructed from an explicit definition of R can be established by a variety of techniques; in particular, see [NS93].) The game has to get started. For this purpose we introduce the following definition.

Definition 4.3: An *initializing clause* a clause of the form

$$\texttt{start}(s^y(0), \texttt{WinLoss}) \leftarrow \texttt{p0}(s^y(0), \texttt{Z0}, \texttt{WinLoss}).$$

for some $y \in \mathbf{N}$. □

We will set up the program corresponding to $\Gamma(R, x_0)$ in two stages. In the first stage we define the player programs assuming that the relation R is recursive and that the corresponding program P_R is at hand. In the second stage we show how to connect the player programs together so that play may pass between them. The means of connection will be regarded as an adjustable parameter involving the presence or absence of negation signs. We also want to have that the *depends on* relations with respect to each of the player programs, respectively, are Noetherian. We do this by adding a step-counter to the programs representing the players.

Definition 4.4: Let Q_0 be the binary extensional equivalent of P_0 and let $Q_{R,1}$ be the binary extensional equivalent of $P_1 \cup P_R$. Let **player$_0$** and **player$_{R,1}$** be the step-counter augmentations of Q_0 and $Q_{R,1}$, respectively. The predicate symbol **stack** in each of the two programs is assumed to be renamed so that the programs have no predicate symbols in common. (One might imagine the **stack** symbols of **player$_0$** to be colored red and the **stack** symbol of **player$_{R,1}$** to be colored blue.) We also further assume, without loss of generality, that the only function symbols occurring in program P_R are the unary symbol s and the constant symbol 0. □

Hereafter, we will refer to the programs P_0 and P_1 as the *prototype player* programs, and the programs **player$_0$** and **player$_{R,1}$** as the **player** programs.

The next proposition informally says that **player$_0$** and **player$_{R,1}$** are correct implementations of player 0 and player 1, respectively, in the game $\Gamma(R, x_0)$.

Proposition 4.1: Let L' be the sublanguage of L whose function symbols are the unary function symbol s and constant 0.
1) Using predicate symbol p0, **player$_0$** computes with respect to L' the relation P consisting of all tuples $(\mathtt{s}^x(\mathtt{0}), \mathtt{s}^z(\mathtt{0}), \mathtt{s}^w(\mathtt{0}))$.
2) Using p1, **player$_{R,1}$** computes with respect to L' the relation Q where $Q_1(\mathtt{s}^x(\mathtt{0}), \mathtt{s}^z(\mathtt{0}), \mathtt{s}^{z_0}(\mathtt{0}), \mathtt{s}^w(\mathtt{0}))$ holds iff āny of the following conditions hold: (i) $R(x, z)$ and $\neg R(x, z_0)$ and $w = 0$, (ii) $R(x, z)$ and $R(x, z_0)$, (iii) $\neg R(x, z)$ and $R(x, z_0)$ and $w = 1$, or (iv) $\neg R(x, z)$ and $\neg R(x, z_0)$ and $w = 1$.

Proof: By proposition 3.2, it suffices to show that the prototype player programs P_0 and P_1 compute the relations P and Q, given in the proposition, using p0 and p1, respectively. This is nearly immediate. □

We now show how to connect the **player** programs. This will be done by replacing the empty bodies of the *terminating clauses* in the **player** programs by calls to instances of p0 and p1 literals.

Definition 4.5: The clauses (1) - (4), below, are called *connecting* clauses.

(1) $\mathtt{stack(0, nil}, f_{\mathtt{p0}}(\mathtt{X, Z, WinLoss})) \leftarrow \mathtt{p1(X, Z, Z1, WinLoss, s(0))}$.
(2) $\mathtt{stack(0, nil}, f_{\mathtt{p0}}(\mathtt{X, Z, WinLoss})) \leftarrow \neg\,\mathtt{p1(X, Z, Z1, WinLoss, s(0))}$.
(3) $\mathtt{stack(0, nil}, f_{\mathtt{p1}}(\mathtt{X, Z, Z0, WinLoss, s(0)})) \leftarrow \mathtt{p0(Z0, Z1, WinLoss)}$.
(4) $(\mathtt{X, Z, Z0, WinLoss, s(0)})) \leftarrow \neg\,\mathtt{p0(Z0, Z1, WinLoss)}$.

Connecting clauses (1) and (3) are said to be *positive*; connecting clauses (2) and (4) are *negative*. A *connection* is any one of the four programs consisting of two connecting clauses obtained by selecting *one* of the two clauses (1) and (2) and by selecting *one* of the two clauses (3) and (4). A *game program* consists of the initializing clause, and the clauses for the **player** programs but where the terminating clauses of the **player** programs are replaced by a connection. □

5 Unifying Two Theorems

In this section we show that two theorems that give the degree of unsolvability of two distinctly different classes of normal logic programs are actually two manifestations of the same underlying complexity of the dependency relations determined by the programs in these classes. This complexity is determined by lemma 2.1, above.

By a *sufficiently large* language we mean a language with at least one constant and one nonconstant function symbol and at least one binary predicate symbol or one binary function symbol. By independent means the following two theorems can be established.

Theorem 5.1: If L is a sufficiently large language, the set of normal logic programs over L that are locally stratified is complete Π_1^1. □

Theorem 5.2: If L is a sufficiently large language, the set of definite clause programs over L with a unique supported Herbrand model is complete Π_1^1. □

The first of these theorems is proved in [BMS92]. The second is contained in an unpublished technical report, [Bl86]. In this section we observe that both theorems are obtainable by essentially the same proof using lemma 2.1. The point is that the lemma is very generic, and the two theorems follow nearly immediately by the same short routine line of reasoning about game programs. We now prove both of these theorems together.

Proof of Theorems 5.1 and 5.2: Form two programs, Q^+ and Q^- as follows. First, choose $y \in \mathbf{N}$ and form the **player** programs using relation R_y where R_y is as in corollary 2.1. Next, connect the **player** programs by replacing their terminating clauses by a connection consisting of the positive connecting clauses in forming Q^+ and the negative connecting clauses in forming Q^-. Include the initializing clause

$$\texttt{start}(\texttt{s}^y(0), \texttt{WinLoss}) \leftarrow \texttt{p0}(\texttt{s}^y(0), \texttt{Z0}, \texttt{WinLoss}).$$

in Q^+ and Q^-. This completes the construction of Q^+ and Q^-. We now have the following claims.

claim 1: Q^- is locally stratified iff player 0 has a winning strategy in $\Gamma(R_y, y)$.
claim 2: Q^+ has a unique supported Herbrand model (which is empty) iff player 0 has a winning strategy in $\Gamma(R_y, y)$.
Proof of claims 1 and 2: We prove claim 1 first. A proof of claim 2 will then be at hand almost immediately. A program is locally stratified iff the *depends negatively on* relation is Noetherian. We have the following chain of equivalences: The *depends negatively on* relation (with respect to Q^-) is not Noetherian IFF there is an infinite sequence of ground atoms $A_0, \ldots, A_n, \ldots$ such that A_i depends negatively on A_{i+1} for all $i \in \mathbf{N}$ IFF (see the remark immediately following the proof) there is a sequence $\texttt{p0}(\texttt{s}^{k_0}(0), \texttt{s}^{k_1}(0), \texttt{s}(0))$, $\texttt{p1}(\texttt{s}^{k_0}(0), \texttt{s}^{k_1}(0), \texttt{s}^{k_1'}(0), \texttt{s}(0), \texttt{s}(0))$, $\texttt{p0}(\texttt{s}^{k_1'}(0), \texttt{s}^{k_2}(0), \texttt{s}(0))$, $\texttt{p1}(\texttt{s}^{k_1'}(0), \texttt{s}^{k_2}(0), \texttt{s}^{k_2'}(0), \texttt{s}(0), \texttt{s}(0))$, $\ldots$ $\texttt{p0}(\texttt{s}^{k_n'}(0), \texttt{s}^{k_{n+1}}(0), \texttt{s}(0))$,

$\text{p1}(s^{k'_n}(0), s^{k_n+1}(0), s^{k'_n+1}(0), s(0), s(0)), \ldots$ of atoms such that each atom in the sequence *depends negatively on* the succeeding atom IFF there is an infinite sequence $k_0, k_1, k'_1, k_2, k'_2, k_3, \ldots, k_n, k'_n, k_{n+1}, \ldots$ such that $R_y(k_0, k_1), R_y(k_0, k'_1)$ and for each $i \in \mathbf{N}$: $R_y(k'_i, k_{i+1})$ and $R_y(k'_i, k'_{i+1})$. IFF Player 0 does not have a winning strategy for the game $\Gamma(R_y, y)$.

This completes the proof of claim 1. To prove claim 2, replace *depends negatively on* by *depends positively on* in the above argument. The new argument goes through because the *depends positively on* relation, with respect to each of the player programs separately is always Noetherian. □

The reader may wonder whether, in the preceding proof, dependencies between, for example, atoms of the form $\text{p0}(t_1, t_2, t_3)$ and $\text{p1}(u_1, u_2, u_3, u_4, u_5)$ are relevant when the terms $t_1, t_2, t_3, u_1, u_2, u_3, u_4, u_5$ may contain occurrences of function and constant symbols other than s or 0. Such dependencies are relevant. However, by replacing every term v, where v has the form $s^k(v')$ and the principal function symbol of v' is neither s nor 0, with v^*, where v^* is $s^k(0)$, we obtain a dependency between $\text{p0}(t_1^*, t_2^*, t_3^*)$ and $\text{p1}(u_1^*, u_2^*, u_3^*, u_4^*, u_5^*)$.

6 Conclusions

The aim of this paper has been to show the utility of viewing interacting logic program modules as players in a game. The utility of this approach is that a variety of results about the degrees of unsolvability of logic program properties and expressive power can be established by representing various phenomena as plays in a game tree which in turn are represented as dependencies, and subsequently as entailments, with respect to various logic programs. The representations of the players of the games are both direct and uniform.

We speculate that the game-tree technique presented here will also be useful in investigations of subrecursive complexity properties of function-symbol-free programs, provided the stack-machine aspect of a binary extensional equivalent of a function-symbol-free program is kept clearly separate from the relations defined by the program itself. The reduction of programs to binary programs given in this paper also suggests that properties of programs closely related to dependencies between atoms should be further investigated by studying dependencies between finite *sets* of atoms. The techniques presented here also suggest that reversing the direction of application of the games may aid future investigation of GH-games. We stress that even where considerations about games are not essential, such considerations are illuminating.

Acknowledgements: The author has benefited from discussions with Anil Nerode, Jeffrey Remmel, Victor Marek, Alexander Yakhnis and Vladimir Yakhnis.

References

[Ap90]   Apt, K. R. "Logic Programming" in *Handbook of Theoretical Computer Science*, J. van Leeuwen, ed., Elsevier, 1990, pp. 494-574.

[Bl82]   Blair, H. A. "The Recursion-Theoretic Complexity of the Semantics of Predicate Logic as a Programming Language." *Information and Control*, July-August, 1982, pp. 25–47.

[Bl86]   Blair, H. A. *Decidability in the Herbrand Base.* (Manuscript) Workshop on Deductive Databases and Logic Programming, Washington D.C. Aug 18-22, 1986. Syracuse University Logic Programming Research Group Technical Report LPRG-TR88-13.

[Bl87]   Blair, H. A. "Canonical Conservative Extensions of Logic Program Completions". *IEEE Symposium on Logic Programming*, San Francisco, August, 1987. pp. 154-161.

[BMS92]   Blair, H.A., Marek, V.W. and Schlipf, J.S. *The Expressiveness of Locally Stratified Programs.* Technical Report, Mathematical Sciences Institute, Cornell University. To appear in *Fundamenta Informaticae.*

[CB93]   Cholak, P. and Blair, H.A. "The Complexity of Local Stratification", *Fundamenta Informaticae.* (To appear.)

[GH82]   Gurevich, Yuri & Harrington, Leo, "Trees, Automata and Games", *Proceedings of the 14th Annual ACM Symposium on Theory of Computing*, 1982, pp. 60-65.

[Hu80]   Huet, Gérard. "Confluent Reductions: Abstract Properties and Applications to Term Rewriting Systems", *JACM* Vol. 27, no. 4 (October, 1980), pp. 797-821.

[Ll87]   Lloyd, J.W. *Foundations of Logic Programming*, (2nd. ed.) Springer-Verlag, 1987.

[NS93]   Nerode, Anil & Shore, Richard, *Logic for Applications*, Springer-Verlag, 1993.

[Ro67]   Rogers, H. *Theory of Recursive Functions and Effective Computability.* McGraw-Hill, New York, 1967.

[YY93]   Yakhnis, A. & Yakhnis, V. "Extension of Gurevich-Harrington's Restricted Memory Determinacy Theorem: A Criterion for the Winning Player and an Explicit Class of Winning Strategies", *Annals of Pure and Applied Logic*, Vol. 48, 1990, pp.277-297.

Computing the Well-Founded Semantics Faster

Kenneth A. Berman, John S. Schlipf, and John V. Franco

University of Cincinnati, Department of ECE&CS, USA

Abstract. We address methods of speeding up the calculation of the well-founded semantics for normal propositional logic programs. We first consider two algorithms already reported in the literature and show that these, plus a variation upon them, have much improved worst-case behavior for special cases of input. Then we propose a general algorithm to speed up the calculation for logic programs with at most two positive subgoals per clause, intended to improve the *worst case* performance of the computation. For a logic program $\mathcal{P}$ in atoms $\mathcal{A}$, the speed up over the straight Van Gelder alternating fixed point algorithm (assuming worst-case behavior for both algorithms) is approximately $(|\mathcal{P}|/|\mathcal{A}|)^{(1/3)}$. For $|\mathcal{P}| \geq |\mathcal{A}|^4$, the algorithm runs in time linear in $|\mathcal{P}|$.

1 Introduction

Logic programming researchers have, over the last several years, proposed many logic-based declarative semantics for various sorts of logic programming. The hope is that, if they can be efficiently implemented, these semantics can restore the separation between logic and implementation that motivated the developers of Prolog but that Prolog did not achieve. In the last few years, several projects have begun to produce working implementations of some of these semantics.

A major difficulty with all such approaches is the complexity of the calculations, and various approaches have been tried to speed the calculations up. We feel that serious investigation of new algorithms to compute these semantics may have significant practical importance at this time.

Two quite popular semantics for the class of normal logic program are the stable [4] and the well-founded [14]. For propositional logic programs, computing inferences under the stable semantics is known to be co-NP-complete [7]; computing inferences under the well-founded is known to be quadratic time (folklore), and no faster algorithm is known. Work in [8] (on questions of updating accessibility relations in graphs) suggests that it may be quite difficult, using standard techniques, to break the quadratic-time bound on the well-founded semantics. We are concerned here with speeding up the computation of the well-founded semantics; in particular, we find a large class of propositional logic programs for which we can break the quadratic time bound. As an additional application, we also note that, if the calculation of the well-founded semantics can be made sufficiently fast, that calculation may also prove a highly useful subroutine in computing under the stable semantics.

The standard calculation methods for the well-founded semantics are based upon the alternating fixed-point algorithm of Van Gelder [13]. Several researchers have observed that, for propositional logic programs, the (worst case) time taken by that algorithm to compute the well-founded partial model is quadratic in the size of the program. More specifically, for a propositional logic program $\mathcal{P}$ with atom set $\mathcal{A}$, the complexity is $O(|\mathcal{A}||\mathcal{P}|)$: Van Gelder's algorithm can make at most $|\mathcal{A}|$ passes before reaching a fixed point (plus another to verify that a fixed point has been reached), and each pass consists of finding the least models of two Horn clause programs derived (essentially via the Gelfond-Lifschitz Transform) from $\mathcal{P}$. Since finding the least model of a Horn clause program is linear in the size of the program [2, 6, 10], this gives the stated complexity.

Some techniques are know for speeding up the calculation on "nice" programs. We investigate two of them and prove that they do achieve optimal speed-up on certain classes of programs. We then turn to improving the worst-case behavior. Our main result is Theorem 10: We show that our Algorithm 6 computes the well-founded partial model of any program $\mathcal{P}$ in proposition letters $\mathcal{A}$ in time $O(|\mathcal{P}| + |\mathcal{A}|^2 + |\mathcal{P}|^{\frac{2}{3}}|\mathcal{A}|^{\frac{4}{3}})$, so long as $\mathcal{P}$ has at most two positive subgoals per rule. Thus when $\mathcal{P}$ is sufficiently larger than $\mathcal{A}$, this new algorithm will thus have a noticably better worst-case time than Van Gelder's original construction. Thus also, $|\mathcal{A}|^4 \in O(|\mathcal{P}|)$, our algorithm takes time linear in $|\mathcal{P}|$, which is clearly optimal.

Restrictions

We shall be discussing only (finite) propositional logic programs. Now for many practical purposes, propositional logic programs are not especially interesting by themselves, but semantics for first order logic programs (a.k.a. intensional databases, or IDB's) over (finite) extensional databases (EDB's) are generally defined by translating, via Herbrand expansions, these logic programs and databases into (finite) propositional logic programs.

In our principal algorithm, Algorithm 6, we shall consider only logic programs with at most two positive subgoals per rule. In some circumstances, the restriction to two positive subgoals per rule is no particular hindrance; it is well known that, given any logic program $\mathcal{P}$ in proposition letters $\mathcal{A}$, one can easily construct a logic program $\mathcal{P}^+$ in proposition letters $\mathcal{A}^+ \supseteq \mathcal{A}$ which is a conservative extension of $\mathcal{P}$ under many logic programming semantics, including the well-founded, the stable, and Fitting's Kripke-Kleene (3-valued program completion) semantics [3]. However, in general $\mathcal{A}^+$ will be substantially larger than $\mathcal{A}$, making $|\mathcal{A}^+||\mathcal{P}^+|$ substantially larger than $|\mathcal{A}||\mathcal{P}|$ — and thus possibly negating all benefits of our algorithm.

Nevertheless, we feel Algorithm 6 may prove fairly widely useful. For example, there are very natural conditions on IDB's that guarantee that $c\sqrt[3]{(|\mathcal{A}|/|\mathcal{P}|)} < 1$ for all large enough EDB's, and for which the standard reduction to a program with two positive subgoals per rule, done in the IDB and thus automatically extended to the propositional program, leaves $|\mathcal{A}^+| \in O(|\mathcal{A}|)$.

Of course, in actual practice, computing the well-founded semantics for a particular IDB over various EDB's by first translating each IDB/EDB pair into a propositional logic program is generally not particularly efficient, since uniformities over the variables are lost. Accordingly, the utility of the methods we use here will be determined in part by whether they can be incorporated in faster methods for handling first order logic programs. This we intend to pursue in future research.

2 Terminology and Notation

A normal propositional logic program $\mathcal{P}$ over a set of atoms $\mathcal{A}$ consists of a finite set of *rules* of the form

$$a \leftarrow \beta_1 \wedge \beta_2 \wedge \cdots \wedge \beta_i,$$

where each β_i is a literal, i.e., a proposition letter in $\mathcal{A}$ or its negation. We shall use these meanings of $\mathcal{P}$ and $\mathcal{A}$ henceforth in the paper.

Definition 1. 1. Traditionally associated with a logic program $\mathcal{P}$ is a set of binary dependency relations on the atoms of the programs. For proposition letters a, b:
  - $b <_{pos} a$ (*a depends directly positively* upon b) if there is a rule in $\mathcal{P}$ with head a and subgoal b.
  - $b <_{neg} a$ (*a depends directly negatively* upon b) if there is a rule in $\mathcal{P}$ with head a and subgoal $\neg b$.

2. Two further dependency relations are defined from the above relations, using the obvious regular-set type notation, with sequence meaning concatenation:
  - $<_{dep} = <_{pos} \cup <_{neg}$. Relation $<_{dep}$ is called the *direct dependency relation*.
  - $<_{1neg} = <_{neg} <_{pos}^{*}$. Relation $<_{1neg}$ is the relation of *dependency through exactly one negation* (and any number of positives).

3. A logic program $\mathcal{P}$ is *stratified* if its relation $<_{1neg}$ is acyclic (i.e, if its transitive closure is irreflexive).

4. A logic program $\mathcal{P}$ is *positive-acyclic* if its relation $<_{pos}$ is acyclic.

For both stratified and positive-acyclic logic programs, it turns out that the well-founded semantics can be found in time linear in the size of the logic program.

In this paper we make some fairly standard assumptions about how complexity is measured. We assume that every propositional logic program has proposition letters $x_1, \ldots, x_n$, for some n. We assume that computations are performed on a machine with random-access memory, and, in particular, that accessing array positions, following pointers, and copying pointers, can done in constant time. (Similar assumptions were made, for example, in [2]). Finally, we assume that certain arithmetic operations (initialization to 0, adding, dividing, copying, incrementation, decrementation, and comparison) on natural numbers used for indices and reference counts can be done in unit time.

3   A Hypergraph Representation of Programs

To picture algorithms to compute the well-founded semantics, we frequently think of logic programs as representing directed hypergraphs in a fairly straightforward way:

Definition 2. Let $\mathcal{P}$ by a logic program over atoms set $\mathcal{A}$. With only trivial rewriting, we may assume there is exactly one rule $s \leftarrow$ with no positive subgoals, and for that s, $\neg s$ does not appear in $\mathcal{P}$. (If necessary, add such a proposition letter to a. If any other rule has no positive subgoals, add in s as a positive subgoal.) We shall call s the *source*. We shall ignore rule $s \leftarrow$ in building the hypergraph, instead treating the atom s specially.

Set $\mathcal{A}$ is the set of vertices of the hypergraph. First group together rules that have the same heads and positive subgoals: group rules

$$a \leftarrow b_1 \wedge \cdots \wedge b_i \wedge \neg c_1^1 \wedge \cdots \wedge \neg c_{j_1}^1$$
$$a \leftarrow b_1 \wedge \cdots \wedge b_i \wedge \neg c_1^2 \wedge \cdots \wedge \neg c_{j_2}^2$$
$$\vdots$$
$$a \leftarrow b_1 \wedge \cdots \wedge b_i \wedge \neg c_1^h \wedge \cdots \wedge \neg c_{j_h}^h$$

into "rules" the form

$$a \leftarrow b_1 \wedge \cdots \wedge b_i \wedge ((\neg c_1^1 \wedge \cdots \neg c_{j_1}^1) \vee (\neg c_1^2 \wedge \cdots \neg c_{j_2}^2) \vee \cdots \vee (\neg c_1^h \wedge \cdots \neg c_{j_h}^h))$$

The head and positive subgoals of the "rule" will be considered to be a *directed hypergraph edge* $\langle a, \{b_1, \ldots, b_i\}\rangle$, directed from the subgoals to the head. The remaining conjunct,

$$((\neg c_1^1 \wedge \cdots \wedge \neg c_{j_1}^1) \vee (\neg c_1^2 \wedge \cdots \wedge \neg c_{j_2}^2) \vee \cdots \vee (\neg c_1^h \wedge \cdots \wedge \neg c_{j_h}^h)),$$

will be called the *presupposition* of the edge. (Intuitively, the rule is applied only if the presupposition is either known to be true or not known to be false, depending upon circumstances.) Variable a will also be called the *head* of the edge, and $b_1, \ldots, b_i$ will be called the *tails* of the edge.

The *degree* $d(a)$ of any $a \in \mathcal{A}$ is the sum of the numbers of tails on all hyperedges with head a.

We shall use both hypergraph-theoretic language and logic-program language in this paper, whichever seems more obvious at any point.

Definition 3. Given a partial function $\tau : \mathcal{A} \to \{T, F\}$ we can easily assign an interpretation $\tau(p)$ for p the presupposition of any rule:

$$\tau(((\neg c_1^1 \wedge \cdots \wedge \neg c_{j_1}^1) \vee (\neg c_1^2 \wedge \cdots \wedge \neg c_{j_2}^2) \vee \cdots \vee (\neg c_1^h \wedge \cdots \wedge \neg c_{j_h}^h))$$

is true if every conjunct of some disjunct is true in τ, false if some conjunct of every disjunct is false in τ, and undefined otherwise. Moreover, this can be computed by substituting T literals true in τ and F for literals false in τ and doing standard simplifications.

The algorithm embodied in Van Gelder's alternating fixed point definition of the well-founded semantics [13] translates almost immediately into the following algorithm on their corresponding hypergraphs:

Algorithm 1. (Van Gelder) We are given a program $\mathcal{P}$ and its hypergraph interpretation $\mathcal{H}$. We maintain a partial truth assignment τ, consisting of the truth values already inferred, and two approximations to $\mathcal{H}$, an over-approximation $\overline{\mathcal{H}}$ and an under-approximation $\underline{\mathcal{H}}$. Initially, $\tau = \emptyset$ and $\overline{\mathcal{H}} = \mathcal{H}$, and $\underline{\mathcal{H}}$ is the set of hyperedges of $\mathcal{H}$ with empty presuppositions. We maintain two loop invariants: $\underline{\mathcal{H}}$ is the set of hyperedges of $\mathcal{H}$ with presuppositions p where $\tau(p) = T$, and $\overline{\mathcal{H}}$ is the set of hyperedges of $\mathcal{H}$ with presuppositions p where $\tau(p) \neq F$.

Repeat until no changes are made during an entire pass:
1. **Van Gelder edge deletion step:**
     Find all vertices $a \in \mathcal{A}$ accessible from s in hypergraph $\underline{\mathcal{H}}$.
     For every vertex a accessible in $\underline{\mathcal{H}}$
        set $\tau(a) = T$.
     For every hyperedge e of $\overline{\mathcal{H}} - \underline{\mathcal{H}}$ with presupposition p containing $\neg a$,
        if $\tau(p) = F$
           remove h from $\overline{\mathcal{H}}$.
2. **Van Gelder edge addition step:**
     Find all vertices $a \in \mathcal{A}$ accessible from s in hypergraph $\overline{\mathcal{H}}$.
     For every vertex a not accessible in $\overline{\mathcal{H}}$
        set $\tau(a) = F$.
     For every hyperedge e of $\overline{\mathcal{H}} - \underline{\mathcal{H}}$ with presupposition p containing $\neg a$,
        if $\tau(p) = T$
           add h to $\underline{\mathcal{H}}$.

The well-founded partial model of $\mathcal{P}$ is the final partial interpretation τ.

4   Elaborating Upon Known Speed-Ups

"Best-case" and informal "Average-case" speedup techniques for the well-founded semantics have been fairly widely observed.[1] The algorithmic methods of this section are already known; what is new here is (1) a new algorithm, which explicitly combines two older ideas, and (2) some propositions about these algorithms.

The first speedup we first heard from Subrahmanian [11]. Fitting's Kripke-Kleene semantics [3] (a.k.a. 3-valued program completion semantics) for logic programs can be calculated much as is the well-founded semantics, but with Van Gelder's search for inaccessible nodes of $\overline{\mathcal{H}}$ replaced by searches for nodes (other than the source s) with in-degree 0. (Essentially this algorithm is in [3].)

---

[1] We say "informal Average-case" because there is no accepted distribution on which to base the average and because, so far as we know, the conclusions are based only upon experimentation.

Proposition 4. (folklore) *The above-sketched computation of the Kripke-Kleene semantics for a propositional logic program P can be done in $O|P|$ time.*

Idea of Proof. The proof depends upon the fact that finding whether a node is inaccessible from s takes a search, while a node with in-degree 0 can be identified in constant time with the right data structure. ∎

Essentially the same proof gives that

Proposition 5. *Suppose an algorithm is of the following form, and a logic program P is given:*

Set $\tau = \emptyset$
Repeat until a fixed point is reached
  Use Fitting's algorithm to extend τ
  Set $\tau(a) = F$ for some proposition letters a previously not in the domain(τ).

Then the total time taken by all iterations of the Fitting algorithm is $O(|P|)$.[2]

Algorithm 2. (Subrahmanian et. al.)

Initialize $\tau = \emptyset$, $\overline{\mathcal{H}} = \mathcal{H}$, and $\underline{\mathcal{H}} = $ the hyperedges of $\mathcal{H}$ with empty
        presuppositions.
Repeat until no edges are deleted in step 2:
  1. **Fitting Algorithm:**
      Repeat until no changes are made during an entire pass:
        For every vertex a where $\tau(a)$ is undefined and for some hyperedge
            h of $\underline{\mathcal{H}}$, for all tails b of h, $\tau(b) = T$,
          set $\tau(a) = T$.
        For every vertex a where $\tau(a)$ is undefined and $d(a) = 0$ in $\overline{\mathcal{H}}$
          set $\tau(a) = F$.
        For every hyperedge e of $\overline{\mathcal{H}}$ with presupposition p,
          if $\tau(p) = F$
            remove h from $\overline{\mathcal{H}}$.
        For every hyperedge e of $\overline{\mathcal{H}} - \underline{\mathcal{H}}$ with presupposition p,
          if $\tau(p) = T$
            add h to $\underline{\mathcal{H}}$.
  2. Make one pass of Van Gelder edge addition step

The well-founded partial model of P is the final partial interpretation τ.

Subrahmanian noted that Algorithm 2 is faster than Algorithm 1 on most of the logic programs his group randomly generated. We find a partial explanation for this in the following proposition:

---

[2] We have sloughed over an important detail here since it does not affect this paper: we may have to use a four-valued logic, with the four truth values T, F, undefined, and contradictory.

Proposition 6. *Let P be a logic program.*

1. *If P is positive-acyclic, then the Kripke-Kleene semantics for P agrees with the well-founded semantics, and hence Algorithm 2 finds the well-founded semantics in time $O(|P|)$.*

2. *If P consists of a positive-acyclic program plus some additional rules giving only k hyperedges (possibly with very long presuppositions), then Algorithm 2 finds the well-founded semantics in time $O(k|P|)$.*

Proof. 1. In [9] we showed that, for a logic program with no positive subgoals, the Kripke-Kleene and well-founded semantics agree. It is routine to extend that proof to all positive-acyclic logic programs.

2. The difference between the Kripke-Kleene semantics and the well-founded semantics is that the well-founded semantics identifies positive dependency cycles. Since the first part of each pass through the outside loop of the algorithm is Fitting's algorithm, and since A is finite, each atom found by step 2 must be in a positive cycle, all of whose atoms are inaccessible. Thus one of the k special edges (the "back-edges") must be in this cycle. Accordingly, at least one back-edge will be added to $\overline{\mathcal{H}}$ in each pass. Thus, after k passes, no additional cycles in $\overline{\mathcal{H}}$ remain to be discovered, and all additional inferences may be made by the Fitting semantics. ∎

A second speed-up is motivated by the fact that calculating the perfect model of a stratified logic program (which is also the well-founded partial model and the unique stable model) can be done in linear time. Consider $\langle A, <_{dep} \rangle$ as a directed graph. In linear time, form the strongly connected components of the graph as in [12], and topologically sort the components. Let P_1 be the set of rules of P whose heads occur in the first component of the graph. P_1 turns out to be pure Horn, so its perfect model is its minimal model M_1, which can be computed in time linear in $|P_1|$ [2, 6, 10]. Let P_2 be the set of rules of P whose heads occur in the second component. Substitute into P_2 the truth values computed for variables in the first component and simplify. The result is a pure Horn program, so compute its minimal model, in time linear in $|P_2|$, and the perfect model of $P_1 \cup P_2$ is $M_1 \cup M_2$. Continue this way through all the strongly connected components. The result is the perfect model, which has been constructed in time linear in $|P|$.

Now for the well-founded and Kripke-Kleene semantics, the same partitioning into strongly-connected components, decomposition of the program by the components and construction of the semantics for each piece separately correctly constructs the semantics. We shall call the levels $P_1, P_2, \ldots$ *strata*, or *modules*. (These were investigated in the current context in [9] and [1].) Unlike with stratified programs, there is no guarantee that the strata are Horn. However, for programs with more than one strongly connected component, this gives a divide-and-conquer approach which can sometimes, as with stratified programs, reduce the complexity of finding the well-founded model. (The same technique has been used [5] to speed up the search for individual stable models, though the technique does not work directly for computing the intersection of stable models [9].)

We can combine this technique with Algorithm 2, giving what appears to be a fairly useful algorithm. We have chosen to start first with the Fitting step since, for example, calculations from non-stratified IDB's over nice EDB's can reduce to stratified programs after only one or two passes of the Fitting method, e.g., for modularly stratified programs.

In the following algorithm, we refer to the direct dependency relation $<_{dep}$ on $\overline{\mathcal{H}}$. This is the analogue of the logic programming definition: $a <_{dep} b$ if there is a hyperedge in $\overline{\mathcal{H}}$ with head a and with either b in its tail or $\neg b$ appearing in the presupposition for the hyperedge. Since the algorithm is merely a combination of previous pieces, we summarize the algorithm. The algorithm is initially called with $\tau = \emptyset$ and $\overline{\mathcal{H}}$ and $\underline{\mathcal{H}}$ as in Algorithms 1 and 2. The details of partitioning into strata are can be filled in easily by the reader.

Algorithm 3.

Initialize $\tau = \emptyset$.
Perform the Fitting step, updating τ and *all* current subgraphs of $\underline{\mathcal{H}}$ and $\overline{\mathcal{H}}$.
Decompose $\mathcal{A}$ into its strongly connected components $\mathcal{A}_1, \ldots, \mathcal{A}_j$ under $<_{dep}$
                (defined from $\overline{\mathcal{H}}$).
Topologically sort the components, $\mathcal{A}_1, \mathcal{A}_2, \ldots, \mathcal{A}_n$.
Construct the subprograms $\mathcal{P}_1, \mathcal{P}_2, \ldots, \mathcal{P}_n$.
For $i = 1$ to n:
     Construct the subhypergraphs $\underline{\mathcal{H}}_i$ and $\overline{\mathcal{H}}_i$.
     Perform one Van Gelder edge addition step on $\mathcal{P}_i$,
            updating τ and *all* current subgraphs of $\underline{\mathcal{H}}$.
     If any changes were made by the Van Gelder step
         Call Algorithm 3 recursively on $\mathcal{P}_i$.

The well-founded partial model of $\mathcal{P}$ is the final partial interpretation τ.

Proposition 7. *Consider the class of programs $\mathcal{P}$ consisting of strata $\mathcal{P}_1, \ldots, \mathcal{P}_i$ where each stratum is either Horn or positive-acyclic. Algorithm 3 finds the well-founded partial model of any such $\mathcal{P}$ in time* $O(|\mathcal{P}|)$.

5 Improving the Quadratic Bound on the Worst Case

The techniques of the previous section often speed up logic program evaluation, but they do not improve on the worst-case behavior of Van Gelder's algorithm. Moreover, writing an IDB for testing some property P of interest which, for all interesting EDB's, the properties of Proposition 7 hold, may be very difficult for some properties P. We would like a far more general speedup technique.

Van Gelder's algorithm repeatedly deletes edges from $\overline{\mathcal{H}}$ and rechecks for accessibility from s. Reif [8] has studied algorithms for updating vertex accessibility information in directed graphs during dynamic edge deletion. His work suggests it may in general be very difficult to speed up in the worst case past the size of the graph times the number of cycles of deletions. Thus his result seems to suggest that generally fast well-founded semantics algorithms might be difficult or impossible to find, even for logic programs with *at most one* positive subgoal per rule. This same lower bound applies to all algorithms for the well-founded semantics which we have previously seen, arousing our interest in the question of whether that quadratic time bound can be beaten. In what follows, we do improve in this bound, albeit, as previously noted, only for logic programs with at most two positive subgoals per rule and where $c \sqrt[3]{(|\mathcal{A}|/|\mathcal{P}|)} < 1$ for some constant c. The heart of the difference between Reif's work and ours is that we are looking for a fixed point of an operator and may stop as soon as one is reached.

From now on we limit attention to the class of propositional logic programs $\mathcal{P}$ over atoms $\mathcal{A}$ with at most 2 positive subgoals per clause.

Observe that, with Van Gelder's algorithm, it takes a relatively large amount of work to find an unfounded set: an entire depth-first search of the hypergraph $\overline{\mathcal{H}}$. If the unfounded set is large, there is a fairly good payoff for the work of searching, but if the unfounded set is small, there has been a high cost to identify a few truth values. It is fairly easy to construct examples where all the algorithms of the previous section take $|\mathcal{A}|$ passes to discover unfounded sets, each of small size. Our goal here is to find somewhat faster ways to find small unfounded sets.

The basic idea is this: Proceed as in Algorithms 2 and 3. Usually, before doing the Van Gelder edge addition step or attempting to decompose the program into strata (modules) (the two time-consuming steps), first do a depth-first search of a "small" approximation $\overline{\mathcal{H}'}$ to $\overline{\mathcal{H}}$, an approximation chosen so that any atom inaccessible in $\overline{\mathcal{H}'}$ is also inaccessible in $\overline{\mathcal{H}}$; since $\overline{\mathcal{H}'}$ is smaller than $\overline{\mathcal{H}}$, the depth-first search will proceed faster. If any inaccessible nodes are found, adjust τ and hypergraph $\underline{\mathcal{H}}$ as before and go back to repeat the Fitting computation, having found inaccessible nodes more quickly than the with the other algorithms. In general, only if no inaccessible nodes are found do we do a full depth-first search. If no inaccessible nodes are found by the full depth-first search either, then the algorithm is finished, with just an additional $O(|\mathcal{P}|)$ steps used in checking that we're done. If, on the other hand, some inaccessible nodes are found, i.e., a non-empty unfounded set is found, we shall show that a "large" number of inaccessible nodes must be found. Thus the cost of the depth-first search, averaged over all the inaccessible nodes found, is not too high.

If e_1, e_2 are hypergraph edges with the same head, and if every tail of e_1 is also a tail of e_2, then say that e_1 is a *subedge* of e_2, and e_2 *extends* e_1.

We do the approximation to hypergraph $\overline{\mathcal{H}}$ described above in two steps. We start with a subalgorithm that applies only to hypergraphs H where each subedge $\langle a, \{b\} \rangle$ occurs in $\leq \mu$ edges of H for some μ. It will use a parameter T to be optimized later.

Algorithm 4. Let $X = \{a \in \mathcal{A} : a$ is the head of $\geq T$ edges of $\mathcal{H}\}$.
Let $\mathcal{H}'' = \{\langle a, \{s\}\rangle : a \in X\} \cup \{e \in H : \text{head}(e) \notin X\}$.

Do a depth-first search of $\mathcal{H}''$ to find all inaccessible nodes.
If no inaccessible nodes are found,
    perform a depth-first search of $\mathcal{H}$ to find inaccessible nodes.

Lemma 8. *1. If a vertex is inaccessible in $\mathcal{H}''$, it is inaccessible in $\mathcal{H}$.*
2. If an inaccessible node is found in the first search, then the total cost of searching is $O(T|\mathcal{A}|)$.
3. The total cost of searching, if the second search is called, is $O(|\mathcal{H}|)$.
4. If no inaccessible node is found in the first search but an inaccessible node is found in the second one, then $\geq \frac{T}{\mu}$ inaccessible nodes are found. Thus the total cost of searching per inaccessible node found is $O(\frac{\mu|\mathcal{H}|}{T})$.

Proof. 1. Obvious.
2. $|\mathcal{H}''| \in O(T|\mathcal{A}|)$ since there are at most T edges in $\mathcal{H}''$ into each node.
3. Suppose a is inaccessible in $\mathcal{H}''$ but accessible in $\mathcal{H}$. Then $a \in X$, so there are $\geq T$ edges with head a. Thus at least one tail on each of these edges must be inaccessible in $\mathcal{H}$, since otherwise a would be accessible. Suppose $< \frac{T}{\mu}$ inaccessible nodes are found. Then, by the pigeon-hole principle, one of these inaccessible nodes must occur on $> T/(\frac{T}{\mu}) = \mu$ edges, contradicting the assumption on μ.
4. Immediate from the previous part since the cost of one depth-first search is $O(|\mathcal{H}|)$. ∎

To get a near-optimal value for T above, equate $T|\mathcal{A}|$ and $(\mu|\mathcal{H}|)/T$. Solving for T we get $T^2 = \mu|\mathcal{H}|/|\mathcal{A}|$. This gives an average cost of searching (per inaccessible vertex found) of $\sqrt{\mu|\mathcal{H}||\mathcal{A}|}$.

If in the original hypergraph $\mathcal{H}$ the value of μ (defined just before Algorithm 4) is small, the total cost of the algorithm is good. But there is, in general, no reason to expect that value of μ to be small. So we perform another approximation.

Algorithm 5. Let $\mathcal{H}$ be a hypergraph, and let B be a number. Let

$$\mathcal{H}_1 = \{\langle a, \{b_1\}\rangle : \text{for} \geq B \text{ vertices } b_2, \langle a, \{b_1, b_2\}\rangle \in \mathcal{H}\}$$
$$\mathcal{H}_2 = \{e \in \mathcal{H} : \text{no subedge of } e \text{ is in } \mathcal{H}_1\}$$
$$\mathcal{H}' = \mathcal{H}_1 \cup \mathcal{H}_2$$

Call Sublgorithm 4 on $\mathcal{H}'$ with $T = \sqrt{B|\mathcal{H}|/|\mathcal{A}|}$.
If no inaccessible nodes are found above,
    perform a depth-first search of $\mathcal{H}$.

Lemma 9. *Let* $\mathcal{H}$, B, *and* $\mathcal{H}'$ *be as in Sublgorithm 5.*

1. *If some vertex* $a \in \mathcal{A}$ *is inaccessible from* s *in* $\mathcal{H}'$, *it is inaccessible from* s *in* $\mathcal{H}$.
2. *No edge* $\langle a, \{b\} \rangle$ *is a subedge of* $\geq B$ *edges of* $\mathcal{H}'$ *(i.e, the property of Algorithm 4 holds with* $\mu = B$).
3. *If every vertex* $a \in \mathcal{A}$ *is accessible from* s *in* $\mathcal{H}'$, *but some vertex* $a \in \mathcal{A}$ *is inaccessible from* s *in* $\mathcal{H}$, *then* $\geq B$ *vertices of* $\mathcal{A}$ *are inaccessible in* $\mathcal{H}$.
4. $|\mathcal{H}'| \leq |\mathcal{H}|$. *Thus: (i) The cost to perform the search in this algorithm is in* $O(T|\mathcal{A}|)$ *if inaccessible nodes are found by the first step of Algorithm 4. (ii) If no inaccessible nodes are found by the first step of Algorithm 4 but some inaccessible nodes are found by the second step, the cost to perform the search is in* $O(|\mathcal{H}'|) \subseteq O(|\mathcal{H}|)$. *(iii) If no inaccessible nodes are found by Algorithm 4, the total cost to perform the search is* $O(|\mathcal{H}|)$.
5. *If this subalgorithm performs the search of* $\mathcal{H}$ *and inaccessible nodes are found, then the average cost of searching for each inaccessible node found is* $O(|\mathcal{H}|/B)$.

Proof.

1. Obvious.
2. The subedges in $\mathcal{H}_1$ have no tails in common, and no edge in $\mathcal{H}_1$ is a subedge of any edge in $\mathcal{H}_2$. If any B edges in $\mathcal{H}_2$ had a subedge in common, that subedge would have been put into $\mathcal{H}_1$.
3. Suppose a is accessible in $\mathcal{H}'$ but not in $\mathcal{H}$. Then there must be an edge $\langle a, \{b\} \rangle \in \mathcal{H}' - \mathcal{H}$ where b is accessible in $\mathcal{H}$ but a is not. By definition of $\mathcal{H}'$, there must be $\geq B$ edges $\langle a, \{b, b_2\} \rangle \in \mathcal{H}$. Since a is inaccessible in $\mathcal{H}$, none of these $\geq B$ b_2's can be accessible in $\mathcal{H}$ either.
4. That $|\mathcal{H}'| \leq |\mathcal{H}|$ is obvious. Statements (i) and (ii) then follow from Lemma 8. Statement (iii) follows from the fact that the searching cost is dominated by the cost of the search of $\mathcal{H}'$ followed by the search of $\mathcal{H}$.
5. Follows from the two previous steps. ∎

A near-optimal value for B can be computed by equating the two previous average costs, $\sqrt{B|\mathcal{H}||\mathcal{A}|}$ and $\frac{|\mathcal{H}|}{B}$. This gives $B = \sqrt[3]{|\mathcal{H}|/|\mathcal{A}|}$, yielding a final average cost of $O(|\mathcal{A}|^{(1/3)}|\mathcal{P}|^{(2/3)})$. The total search cost must be at most $|\mathcal{A}|$ times this — thus $O(|\mathcal{A}|^{(4/3)}|\mathcal{P}|^{(2/3)})$. The speedup is on the order of $\sqrt[3]{|\mathcal{P}|/|\mathcal{A}|}$, as stated earlier.

We put our modifications together into the following algorithm. In order to simplify the bookkeeping, we divide the program into strata only once, after the initial application of Fitting's algorithm.

Algorithm 6.

Initialize $\tau = \emptyset$. Initialize $\underline{\mathcal{H}}$ and $\overline{\mathcal{H}}$ as usual.
Perform the Fitting step, updating $\tau, \underline{\mathcal{H}}$ and $\overline{\mathcal{H}}$.
Decompose A into its strongly connected components $A_1, \ldots, A_j$ under $<_{dep}$
        (defined from $\overline{\mathcal{H}}$).
Topologically sort the components, $A_1, A_2, \ldots, A_n$.
Construct the subprograms $\mathcal{P}_1, \mathcal{P}_2, \ldots, \mathcal{P}_n$.
For $i = 1$ to n:
    Construct the subhypergraphs $\underline{\mathcal{H}}_i, \overline{\mathcal{H}}_i$ and construct $\overline{\mathcal{H}}_i'$ (as in Algorithm 5).
    Repeat until no inaccessible edges are found in the next step:
        Call Algorithm 5 on $\overline{\mathcal{H}}_i$, $\overline{\mathcal{H}}_i'$;
        If any inaccessible vertices are found,
            Update τ and *all* $\underline{\mathcal{H}}_j$'s, $\overline{\mathcal{H}}_j$'s, and $\overline{\mathcal{H}}_j'$'s for $j \geq i$.
            Perform the Fitting step, updating τ and all $\underline{\mathcal{H}}_j$'s, $\overline{\mathcal{H}}_j$'s, $\overline{\mathcal{H}}_j'$'s for $j \geq i$.

The well-founded partial model of $\mathcal{P}$ is the final partial interpretation τ.

It turns out that all the bookkeeping and graph construction and update for the above algorithm can be done in time $|\mathcal{P}| + |A|^2$, as we shall comment upon in the next section. Combining this with the preceding remarks, we have:

Theorem 10. *Algorithm 6 computes the well-founded semantics for any propositional logic program $|\mathcal{P}|$ in proposition letters A with ≤ 2 positive subgoals per clause in time $O(|\mathcal{P}| + |A|^2 + |A|^{\frac{4}{3}}|\mathcal{P}|^{\frac{2}{3}})$.*

Since we expect that in common examples $|A|^{(4/3)}|\mathcal{P}|^{(2/3)} > |\mathcal{P}| + |A|^2$, we have simplified the result above to saying that generally our speed-up over the straight Van Gelder algorithm is $O(|\mathcal{P}|^{(1/3)}|A|^{-(1/3)})$. Obviously, if $|A|^4 \in O(|\mathcal{P}|)$, the algorithm performs in time $O(\mathcal{P})$, which is best possible.

In practice, it would seem wise to optimize the algorithm by, approximately once every $\sqrt[3]{|\mathcal{P}|/|A|}$ iterations of the outer repeat loop, replacing the call to Algorithm 5 with a full depth-first search of $\overline{\mathcal{H}}_i$. Except for $\mathcal{P}$ much larger than A, this would not hurt the worst-case performance of the algorithm, and it might help if the algorithm is finding many small unfounded sets when finding some large unfounded set would substantially reduce the size of $\overline{\mathcal{H}}$.

Implementation

What we have left to do is to show that all the necessary bookkeeping for the algorithm can be done in time $O(|\mathcal{P}|)$. The basic work is in choosing the appropriate data structures and the right counts to store and in doing enough work in preprocessing, leaving mostly updating of the data structures and the counts for run time (plus, of course, many depth-first searches). All this can be done; we just sketch the construction below.

For each $a \in \mathcal{A}$ we store 3 doubly-linked lists of pointers: pointers to hyperedges with head a, hyperedges with a in the tail, and hyperedges with $\neg a$ in the presupposition. (In fact, we have to store the presuppositions themselves as doubly linked lists, simplifying according to the usual logic rules each time an atom is first set to T of F.) We store each $\mathcal{H}_i$, $\overline{\mathcal{H}}_i$, and $\overline{\mathcal{H}}_i'$ and update them dynamically; thus we actually need one set of the 3-doubly linked lists of pointers above for each of the three graphs. For this we need a pointer from each edge in $\overline{\mathcal{H}}_i$ to its subedge (if any) in $\overline{\mathcal{H}}_i'$, a counter on the subedge telling how many edges point to it (so we can tell when to remove the subedge from $\overline{\mathcal{H}}'$ and reinstall the original edges), and doubly-linked lists going from the subedge the edges from which it is derived to allow us to reinstall the original edges quickly. (Importantly, each reinstallation is done only once.) We store and dynamically update an $|\mathcal{A}|$-length array giving the number of rules of $\overline{\mathcal{H}}_i'$ with head each variable a; this allows us to create the approximations $\overline{\mathcal{H}}_i''$ of Algorithm 4 quickly. In the construction we use an $|\mathcal{A}| \times |\mathcal{A}|$ array giving the number of times each atom appears as a positive subgoal in a rule with head each $a' \in \mathcal{A}$.

6 Future Work

There are two clear open directions left by this work, both of which we intend to pursue. One is to relax the restriction to two positive subgoals per rule. We expect this can be done, but it is not clear how high the cost is going to be; we do not expect to be able to achieve the same speedup we have here. (By analogy, a better speedup can be achieved if we restrict attention to programs with only one positive literal per clause.) The other direction, as noted before, is to integrate an algorithm such as ours with actual deductive-database inference techniques.

The authors thank Fred Annexstein and R.P. Saminathan for helpful suggestions on this work.

The research of the second and third authors on this paper was partially supported by grant grant N00014-94-1-0382 from the Office of Naval Research.

Authors' address: Department of Electrical and Computer Engineering and Computer Science, The University of Cincinnati, Cincinnati, OH 45221-0008, USA.

References

1. J. Dix. A classification theory of semantics of normal logic programs: II. Weak properties. To appear in *JCSS*.
2. W. F. Dowling and J. H. Gallier. Linear time algorithms for testing the satisfiability of propositional Horn formulae. *Journal of Logic Programming* 1 (1984), 267–284.
3. M. Fitting. A Kripke-Kleene semantics for logic programs. *Journal of Logic Programming*, 2(4):295–312, 1985.
4. M. Gelfond and V. Lifschitz. The stable model semantics for logic programming. In *Proc. 5th Int'l Conf. Symp. on Logic Programming*, 1988.
5. V. Lifschitz and H. Turner. Splitting a logic program. Preprint.
6. A. Itai and J. Makowsky, On the complexity of Herbrand's theorem. Technical Report No. 243, Department of Computer Science, Israel Institute of Technology, Haifa (1982).
7. W. Marek and M. Truszczyński. Autoepistemic logic. *Journal of the ACM* 38(3), pages 588-619, 1991.
8. J. Reif. A topological approach to dynamic graph connectivity. *Information Processing Letters* 25(1), pages 65-70.
9. J. S. Schlipf. The expressive powers of the logic programming semantics. To appear in *JCSS*. A preliminary version appeared in *Ninth ACM Symposium on Principles of Database Systems*, pages 196-204, 1990. Expanded version available as University of Cincinnati Computer Science Technical Report CIS-TR-90-3.
10. M. G. Scutellà. A note on Dowling and Gallier's top-down algorithm for propositional Horn satisfiability. *Journal of Logic Programming* 8, pages 265-273, 1990.
11. V. S. Subrahmanian, personal communication.
12. R. Tarjan, "Depth first search and linear graph algorithms," *SIAM Journal on Computing* 1 (1972), 146-160.
13. A. Van Gelder. The alternating fixpoint of logic programs with negation. In *Eighth ACM Symposium on Principles of Database Systems*, pages 1-10, 1989. Available from UC Santa Cruz as UCSC-CRL-88-17.
14. A. Van Gelder, K. A. Ross, and J. S. Schlipf. The well-founded semantics for general logic programs. *Journal of the ACM* 38(3), pages 620-650, 1991.

Loop Checking
and the Well-Founded Semantics

Vladimir Lifschitz[1] and Norman McCain[1]
Department of Computer Sciences
University of Texas at Austin
Austin, TX 78712, USA
{vl,mccain}@cs.utexas.edu

Teodor C. Przymusinski[2]
Department of Computer Science
University of California at Riverside
Riverside, CA 92521, USA
teodor@cs.ucr.edu

Robert F. Stärk
Department of Mathematics
Stanford University
Stanford, CA 94305, USA
staerk@gauss.stanford.edu

Abstract. Using a calculus of goals, we define the success and failure of a goal for propositional programs in the presence of loop checking. The calculus is sound with respect to the well-founded semantics; for finite programs, it is also complete. A Prolog-style proof search strategy for a modification of this calculus provides a query evaluation algorithm for finite propositional programs under the well-founded semantics. This algorithm is implemented as a meta-interpreter.

1   Introduction

A "loop check" in a logic programming system is a mechanism that allows the system to avoid some infinite loops. For instance, the execution of the program

$$p \leftarrow not\ q, not\ r,$$
$$q \leftarrow r, p, \qquad\qquad\qquad (1)$$
$$r \leftarrow q, p$$

under Prolog would lead to an infinite loop, because Prolog would attempt to "derive" q from r and then r from q. A system with a loop check may be able to recognize this fact and to conclude that the goals q and r fail, and, accordingly, the goal p succeeds.

In this paper, we make the idea of loop checking precise and define mathematically what is meant by the success and failure of a goal in the presence of

---

[1] Partially supported by the National Science Foundation grant #IRI-9306751.
[2] Partially supported by the National Science Foundation grant #IRI-9313061.

loop checking. On the basis of this definition, we prove the soundness and completeness of query evaluation with loop checking for finite propositional programs under the well-founded semantics [Van Gelder *et al.*, 1990], [Przymusinski, 1991]. It follows that loop checking is also sound for the answer set ("stable model") semantics [Gelfond and Lifschitz, 1991].

The mathematical theory of loop checking in logic programming is the subject of a recent dissertation [Bol, 1991]. In this note, we limit attention to a particularly simple case, propositional programs; the problems investigated by Bol have to do, for the most part, with nonground queries and thus do not even arise in the propositional case. On the other hand, Bol's results are restricted to (locally) stratified programs, and in this respect our theory is more general. For example, it applies to program (1), which is not stratified.

Our definition of loop checking is in terms of a "calculus of goals," and in this sense we extend here the earlier work that relates Prolog to the "SLDNF calculus" [Lifschitz, 1995]. The view of Prolog developed in that paper provides a clean separation between the inference rules and the search strategy, which makes the mathematics of Prolog considerably simpler.[3] For instance, the soundness of the calculus from [Lifschitz, 1995] with respect to the completion semantics of [Clark, 1978] can be verified by straightforward induction, and then the soundness of Prolog with respect to that semantics will immediately follow.

The situation with the soundness of loop checking is similar, with an important difference. The existing approaches to the semantics of logic programming can be loosely divided into two categories, depending on how they treat cycles in the program. Consider the simplest cycle: $p \leftarrow p$. Adding a rule like this can affect the meaning of a program under the completion semantics [Clark, 1978] and some of its variants, whereas the well-founded semantics, the answer set semantics and other more recent semantical theories treat rules of this kind as irrelevant. Loop checking is a better match for the second group of theories than for the first. Accordingly, we establish the soundness of loop checking relative to the well-founded semantics and the answer set semantics — representatives of the second group.

In Section 2 we review the SLDNF calculus as defined in [Lifschitz, 1995], and then introduce its modification that is more convenient for our present purposes. The calculus for loop checking and its properties are discussed in Section 3. The calculus turns out to be sound and complete for finite programs under the well-founded semantics. In Section 4, we show how this calculus leads to a query evaluation method for the well-founded semantics, and how this method can be implemented as a meta-interpreter. Section 5 contains proof outlines.

---

[3] The SLDNF calculus is a reformulation of the inductive characterization of the success and failure of goals under SLDNF given in [Kunen, 1989], and is similar to the calculus proposed earlier in [Mints, 1986]. The idea of separating the search space from the search strategy underlies also the analysis of tabulation techniques in [Bol and Degerstedt, 1993].

2 The SLDNF Calculus

A *rule element* is a (propositional) atom possibly preceded by the negation as failure symbol *not*. [4] A *rule* is a pair *Head* ← *Body*, where *Head* is an atom and *Body* is a finite set of rule elements. A *program* is a set of rules (not necessarily finite).

2.1 First Formulation

A *goal* is a finite set of rule elements. In the *SLDNF calculus of the first kind* corresponding to a program Π [Lifschitz, 1995], the derivable objects are expressions of the forms $\models G$ ("*G* succeeds") and $\dashv G$ ("*G* fails"), where *G* is a goal. The only axiom is

$$\models \emptyset.$$

In the inference rules, the following abbreviation is used: For any atom A, *Bodies*(A) is the set of the bodies of all rules in Π with the head A. There are four inference rules:

$$(SP^1) \quad \frac{\models G \cup B}{\models G \cup \{A\}} \quad \text{where } B \in \textit{Bodies}(A)$$

$$(FP^1) \quad \frac{\dashv G \cup B \quad \text{for all } B \in \textit{Bodies}(A)}{\dashv G \cup \{A\}}$$

$$(SN^1) \quad \frac{\models G \quad \dashv \{A\}}{\models G \cup \{not\ A\}}$$

$$(FN^1) \quad \frac{\models \{A\}}{\dashv G \cup \{not\ A\}}$$

In the names of the inference rules, S stands for *success*, F for *failure*, P for *positive*, and N for *negative*. Note that the number of premises of (FP^1) equals the cardinality of *Bodies*(A). In particular, it can be zero or infinite.

As an example, consider the program

$$\begin{array}{l} p, \\ q \leftarrow p, not\ r, \\ q \leftarrow r, not\ p, \\ r \leftarrow p, not\ s. \end{array} \qquad (2)$$

Here is a derivation of $\dashv \{q\}$ in the corresponding SLDNF calculus of the first kind. To simplify notation, we drop the braces in expressions of the forms $\models \{\ldots\}$ and $\dashv \{\ldots\}$.

---

[4] We do not use here the term "literal," in order to avoid conflict with the terminology of [Gelfond and Lifschitz, 1991] and [Kowalski and Sadri, 1991], where literals are formed using classical (explicit) negation.

$$
(SN^1) \ \cfrac{\models \emptyset \qquad (FP^1)\ \cfrac{\ }{\ \dashv s}}{\models not\ s}
$$

$$
(SP^1)\ \cfrac{\ }{\models p,\ not\ s}
$$

$$
(SP^1)\ \cfrac{\ }{\models r} \qquad (SP^1)\ \cfrac{\models \emptyset}{\models p}
$$

$$
(FN^1)\ \cfrac{\ }{\dashv p,\ not\ r} \qquad (FN^1)\ \cfrac{\ }{\dashv r,\ not\ p}
$$

$$
(FP^1)\ \cfrac{\ }{\dashv q}
$$

Since the SLDNF calculus is sound relative to each of the three semantics of negation as failure mentioned in the introduction, this derivation shows that, under any of them, program (2) makes q false.

As observed in [Lifschitz, 1995], Prolog can be viewed as proof search in the SLDNF calculus of the first kind.

2.2 Second Formulation

In the *SLDNF calculus of the second kind* corresponding to a program Π, the derivable objects are expressions of the forms $\models E$ and $\dashv E$, where E is a rule element. There are no axioms.

In order to describe the inference rules, we need the following definition. For any atom A, a *cover* of A is a set of rule elements that has a nonempty intersection with each set in $Bodies(A)$. It is clear that a superset of a cover is a cover; we will be interested in the covers of A that are minimal relative to set inclusion. The set of such covers will be denoted by $Covers(A)$. For instance, in the case of program (2),

$$
\begin{aligned}
&Covers(q) = \{\{p,r\}, \{p,\ not\ p\}, \{not\ r,r\}, \{not\ r,\ not\ p\}\}, \\
&Covers(p) = \emptyset.
\end{aligned}
$$

The calculus has four inference rules:

$$
(SP^2)\ \cfrac{\models E \qquad \text{for all } E \in B}{\models A} \qquad \text{where } B \in Bodies(A)
$$

$$
(FP^2)\ \cfrac{\dashv E \qquad \text{for all } E \in C}{\dashv A} \qquad \text{where } C \in Covers(A)
$$

$$
(SN^2)\ \cfrac{\dashv A}{\models not\ A}
$$

$$
(FN^2)\ \cfrac{\models A}{\dashv not\ A}
$$

Note that any application of (SP^2) has finitely many premises, possibly zero. The number of premises of (FP^2) can be infinite. It is clear that, for any atom A, $\models not\ A$ is derivable in this calculus iff $\dashv A$ is derivable.

Here is a derivation of $\dashv q$ in the SLDNF calculus of the second kind for program (2):

$$
\cfrac{
  \cfrac{
    \cfrac{
      (SP^2)\,\rule{1cm}{0.4pt}
    }{\models p}
    \qquad
    \cfrac{
      (SN^2)\,\cfrac{(FP^2)\,\rule{1cm}{0.4pt}}{\dashv s}
    }{\models not\ s}
  }{
    (FN^2)\,\cfrac{(SP^2)\,\rule{3cm}{0.4pt}}{\models r}
  }
  \qquad
  \cfrac{(SP^2)\,\rule{1cm}{0.4pt}}{\models p}
}{}
$$

We will denote the two forms of the SLDNF calculus for a program Π by $SLDNF^1_\Pi$ and $SLDNF^2_\Pi$. The two formulations are equivalent:

Theorem 1. *For any program Π and any goal G,*

- *expression $\models G$ is derivable in $SLDNF^1_\Pi$ iff for every $E \in G$, $\models E$ is derivable in $SLDNF^2_\Pi$,*
- *expression $\dashv G$ is derivable in $SLDNF^1_\Pi$ iff for some $E \in G$, $\dashv E$ is derivable in $SLDNF^2_\Pi$.*

The second formulation is more economical, in the sense that its derivable expressions have a simpler structure. On the other hand, it is not clear how to extend it to programs with variables.

The calculus of goals for loop checking described in Section 3.1 below is analogous to the second form of the SLDNF calculus.

2.3 Prolog

As we mentioned in Section 2.1, Prolog can be viewed as a proof search strategy for the SLDNF calculus of the first kind. Alternatively, we can think of Prolog as a proof search strategy for the second formulation, as follows.

Consider a finite program Π and a rule element E. We would like to "evaluate" E, that is, to find a derivation of one of the expressions $\models E$, $\dashv E$ in the calculus $SLDNF^2_\Pi$. The process is recursive, and it terminates only if $\models E$ or $\dashv E$ is derivable.

Case 1: E is an atom A. Consider, one by one, the sets $B \in Bodies(A)$, and evaluate the elements of each, until a set B is found such that $\models F$ is derived for every element F of B; whenever a set B is determined not to have this property, note for which element F_B of B the expression $\dashv F_B$ is derived. If such a B is found, then $\models A$ can be derived by (SP^2). If all sets $B \in Bodies(A)$ have been checked and none turned out to have this property, then an element F_B is selected in each $B \in Bodies(A)$. Then the set $\{F_B \mid B \in Bodies(A)\}$ is a cover of A, and $\dashv A$ can be derived by (FP^2).

Case 2: E has the form *not* A. Then evaluate A. If $\models A$ is derived, then $\dashv E$ can be derived by (FN^2). If $\dashv A$ is derived, then $\models E$ can be derived by (SN^2).

Note that the objects that we operate with in the process of search are rule elements E, rather than derivable expressions $\models E$, $\dashv E$. Which of these two expressions will be eventually derived is not determined until the search is completed. This is what we call "Prolog-style" search. This organization of proof search is possible because of a close similarity between the inference rules for success (SP^2), (SN^2) on the one hand, and for failure (FP^2), (FN^2) on the other. The same similarity between success and failure rules will be found in the calculi for loop checking described in Sections 3.1 and 4.1, and for this reason we will be able to extend to those systems the idea of Prolog-style search.

3 Loop Checking

3.1 Calculus for Loop Checking

We will now define the calculus for loop checking corresponding to a given program Π. The derivable expressions of this calculus have the forms $\models E$ and $\dashv E\,[X]$, where E is a rule element and X a finite set of atoms. We will drop the annotation $[X]$ in $\dashv E\,[X]$ if X is empty. Intuitively, the expression $E\,[X]$ represents the task of "establishing" E by the rules of the program without having established any of the elements of X in the process — without "passing through" X. Consider, for instance, the program

$$p,$$
$$q \leftarrow p.$$

One can perform the task represented by q, but not the task represented by $q\,[p]$.

The axioms of the calculus are the expressions

$$\dashv A\,[X] \qquad (A \in X).$$

(If $A \in X$, then it is impossible to establish A without having established at least one element of X.) The inference rules are:

$$(SP^L)\ \frac{\models E \qquad \text{for all } E \in B}{\models A} \qquad \text{where } B \in Bodies(A)$$

$$(FP^L)\ \frac{\dashv E\,[X \cup \{A\}] \qquad \text{for all } E \in C}{\dashv A\,[X]} \qquad \text{where } C \in Covers(A)$$

$$(SN^L)\ \frac{\dashv A}{\models not\ A}$$

$$(FN^L)\ \frac{\models A}{\dashv not\ A\,[X]}$$

We will denote this calculus by L_Π.

Note that the success rules (SP^L) and (SN^L) are the same as (SP^2) and (SN^2). Rule FP^L says that it is impossible to establish A without passing through X if it is impossible to establish all elements of a cover of A without passing through $X \cup \{A\}$. Here, by adding A to X, we eliminate loops in the process of establishing A.

In the following example, Π is program (1). Here is a derivation of $\models p$:

$$
(SP^L) \cfrac{(SN^L) \cfrac{(FP^L) \cfrac{(FP^L) \cfrac{\dashv q\,[q,r]}{\dashv r\,[q]}}{\dashv q}}{\models not\ q} \qquad (SN^L) \cfrac{(FP^L) \cfrac{(FP^L) \cfrac{\dashv r\,[q,r]}{\dashv q\,[r]}}{\dashv r}}{\models not\ r}}{\models p}
$$

To simplify notation, we drop the braces in expressions of the form $[\{\ldots\}]$.

3.2 Soundness and Completeness Theorems

The calculus L_Π is sound for the well-founded semantics:

Theorem 2. *For any program Π and any atom A,*

- *if $\models A$ is derivable in L_Π then A belongs to the well-founded model of Π,*
- *if $\dashv A$ is derivable in L_Π then $\neg A$ belongs to the well-founded model of Π.*

It follows that L_Π is "consistent":

Corollary 1. *For any program Π, there is no rule element E such that both $\models E$ and $\dashv E$ are derivable in L_Π.*

It is possible, of course, that, for some atom A, neither $\models A$ nor $\dashv A$ is derivable in L_Π. Theorem 2 shows that this happens whenever the truth value of A in the well-founded semantics of Π is *undefined*. The simplest example is

$$p \leftarrow not\ p. \tag{3}$$

In the calculus corresponding to this program, one can derive neither $\models p$ nor $\dashv p$.

Since the answer set semantics is stronger than the well-founded semantics [Przymusinski, 1990], Theorem 2 implies that L_Π is sound for the answer set semantics also:

Corollary 2. *For any program Π and any atom A,*

- *if $\models A$ is derivable in L_Π then A belongs to all answer sets for Π,*
- *if $\dashv A$ is derivable in L_Π then A does not belong to any of the answer sets for Π.*

If Π is finite then L_Π is complete for the well-founded semantics:

Theorem 3. For any finite program Π and any atom A,

- expression $\models A$ is derivable in L_Π iff A belongs to the well-founded model of Π,
- expression $=\!\mid A$ is derivable in L_Π iff $\neg A$ belongs to the well-founded model of Π.

For infinite programs, the assertion of Theorem 3 is generally invalid. Consider, for instance, the program

$$p_n \leftarrow p_{n+1} \qquad (n \geq 0).$$

The expression $=\!\mid p_0$ is not derivable in the corresponding calculus for loop checking, although $\neg p_0$ belongs to the well-founded model of the program.

3.3 Query Evaluation in the Calculus for Loop Checking

The Prolog proof search strategy (Section 2.3) can be extended to the calculus L_Π as follows.

Consider a finite program Π, a rule element E, and a finite set X of atoms. To "evaluate" the expression $E\,[X]$ means now to find a derivation of one of the expressions $\models E$, $=\!\mid E\,[X]$ in the calculus L_Π.

Case 1: E is an atom A. If $A \in X$ then $=\!\mid A\,[X]$ is an axiom. Otherwise, consider, one by one, the sets $B \in Bodies(A)$, and, for the successive elements F of each of these sets, evaluate $F\,[X \cup \{A\}]$, until a set B is found such that $\models F$ is derived for every element F of B; whenever a set B is determined not to have this property, note for which element F_B of B the expression $=\!\mid F_B\,[X \cup \{A\}]$ has been derived. If such a B is found, then $\models A$ can be derived by (SP^L). If all sets $B \in Bodies(A)$ have been checked and none turned out to have this property, then an element F_B is selected in each $B \in Bodies(A)$. Then the set $\{F_B \mid B \in Bodies(A)\}$ is a cover of A, and $=\!\mid A\,[X]$ can be derived by (FP^L).

Case 2: E has the form *not* A. Then evaluate A. If $\models A$ is derived, then $=\!\mid E\,[X]$ can be derived by (FN^L). If $=\!\mid A$ is derived, then $\models E$ can be derived by (SN^L).

For example, when this process is applied to program (1) as Π, the atom p as E, and the empty set as X, the result is the derivation of $\models p$ shown at the end of Section 3.1.

As another example, consider the program

$$p \leftarrow p,$$
$$p.$$

We want to evaluate p. Using the first rule, we are led to evaluate the expression $p\,[p]$, and observe that $=\!\mid p\,[p]$ is an axiom. Hence, we go on to try the second rule of the program. Since the body of the rule is empty, $\models p$ is derivable by (SP^L) from the empty set of premises. The following derivation is found:

$$(SP^L)\; \frac{}{\models p}$$

It is interesting to observe that this figure is also a derivation in the SLDNF calculus of the second kind, but Prolog will not find this derivation. The reason is that Prolog will loop on the first rule of the program, and the second rule will never be tried.

Like Prolog, this query evaluation procedure does not necessarily terminate. For instance, it does not terminate whenever applied to an atom A (and the empty X) such that neither $\models A$ nor $=\!\mid A$ is derivable in L_Π; program (3) is an example. Moreover, it is possible that $\models A$ is derivable but the procedure does not terminate anyway. This is illustrated by the program

$$
\begin{aligned}
p &\leftarrow not\ q, \\
q &\leftarrow not\ p, \\
q &\leftarrow not\ r.
\end{aligned}
\tag{4}
$$

The expression $\models q$ is derivable in the corresponding calculus for loop checking:

$$(FP^L)\ \underline{\hspace{3cm}}$$
$$=\!\mid r$$

$$(SN^L)\ \underline{\hspace{3cm}}$$
$$\models not\ r$$

$$(SP^L)\ \underline{\hspace{3cm}}$$
$$\models q$$

However, given the atom q as input, the evaluation strategy described above will not find this derivation. The reason is that the algorithm loops infinitely through the following sequence of expressions:

$$q,\ not\ p,\ p,\ not\ q,\ q,\ not\ p, \ldots .$$

4 A Prolog-Style Evaluation Algorithm for the Well-Founded Semantics

We have seen that, for any finite program Π, the calculus L_Π is sound and complete with respect to the well-founded semantics of Π, but Prolog-style query evaluation in this calculus does not terminate in some cases. For this reason, Prolog-style proof search in L_Π does not provide a query evaluation algorithm for finite propositional programs under the well-founded semantics.

In this section we describe a simple modification of L_Π that has essentially no effect on its deductive capabilities but causes the evaluation method to always terminate. The algorithm based on the modified calculus is implemented below as a meta-interpreter.

4.1 Modified Calculus

The new calculus corresponding to a given program Π will be denoted by $\widetilde{L}_\Pi$. Its derivable expressions have the forms $\models E\ [Y]$ and $=\!\mid E\ [X;Y]$, where E is a rule element and X, Y are finite sets of atoms. We will drop the braces around the elements of X and Y, and will drop the annotations $[\emptyset]$ and $[\emptyset; \emptyset]$ altogether.

The axioms of $\widetilde{L}_\Pi$ are the expressions

$$\dashv A \, [X; Y] \qquad (A \in X).$$

The inference rules are:

$$(SP^{\widetilde{L}}) \quad \frac{\models E \, [Y] \quad \text{for all } E \in B}{\models A \, [Y]} \qquad \text{where } B \in Bodies(A)$$

$$(FP^{\widetilde{L}}) \quad \frac{\dashv E \, [X \cup \{A\}; Y] \quad \text{for all } E \in C}{\dashv A \, [X; Y]} \qquad \text{where } C \in Covers(A)$$

$$(SN^{\widetilde{L}}) \quad \frac{\dashv A \, [\emptyset; Y \cup \{A\}]}{\models not \, A \, [Y]} \qquad \text{where } A \notin Y$$

$$(FN^{\widetilde{L}}) \quad \frac{\models A \, [Y \cup \{A\}]}{\dashv not \, A \, [X; Y]} \qquad \text{where } A \notin Y$$

The restriction $A \notin Y$ in the last two rules guarantees that, in any branch of a derivation, these rules will not be applied to the same A more than once.

Here is a derivation in the modified calculus for program (1):

$$
\begin{array}{cc}
& \dashv q \, [q, r; q] & & \dashv r \, [q, r; r] \\
(FP^{\widetilde{L}}) \, \rule{3cm}{0.4pt} & & (FP^{\widetilde{L}}) \, \rule{3cm}{0.4pt} \\
& \dashv r \, [q; q] & & \dashv q \, [r; r] \\
(FP^{\widetilde{L}}) \, \rule{3cm}{0.4pt} & & (FP^{\widetilde{L}}) \, \rule{3cm}{0.4pt} \\
& \dashv q \, [; q] & & \dashv r \, [; r] \\
(SN^{\widetilde{L}}) \, \rule{3cm}{0.4pt} & & (SN^{\widetilde{L}}) \, \rule{3cm}{0.4pt} \\
& \models not \, q & & \models not \, r \\
(SP^{\widetilde{L}}) \, \rule{8cm}{0.4pt} & & \\
& \models p &
\end{array}
$$

To simplify notation, we drop the braces in expressions of the form $[\{\ldots\}; \{\ldots\}]$.

Theorem 4. *For any program Π and any rule element E, each of the expressions $\models E$, $\dashv E$ is derivable in $\widetilde{L}_\Pi$ iff it is derivable in L_Π.*

In this sense, the calculus $\widetilde{L}_\Pi$ is equivalent to L_Π. Consequently, L_Π can be replaced by $\widetilde{L}_\Pi$ in the statements of all theorems and corollaries in Section 3.2.

4.2 Query Evaluation in the Modified Calculus

Consider a finite program Π, a rule element E, and finite sets X, Y of atoms. To "evaluate" the expression $E \, [X; Y]$ means now to find a derivation of one of the expressions $\models E \, [Y]$, $\dashv E \, [X; Y]$ in the calculus $\widetilde{L}_\Pi$ or to determine that neither expression is derivable. Thus the evaluation of an expression can have one of three possible outcomes (rather than two, as in the algorithms presented in Sections 2.3 and 3.3).

Case 1: E is an atom A. *Case 1.1:* $A \in X$. Then $\dashv A \ [X;Y]$ is an axiom. *Case 1.2:* $A \notin X$. Then consider, one by one, the sets $B \in Bodies(A)$, and, for the successive elements F of each of these sets, evaluate $F \ [X \cup \{A\};Y]$, until a set B is found such that $\models F \ [Y]$ is derived for every element F of B; whenever a set B is determined not to have this property, continue the process of evaluation of the expressions $F \ [X \cup \{A\};Y]$ for the elements F of B until an element F_B of B is found for which $\dashv F_B \ [X \cup \{A\};Y]$ is derived. *Case 1.2.1:* Such a B is found. Then $\models A \ [Y]$ can be derived by $(SP^{\widetilde{L}})$. *Case 1.2.2:* There is no such B, and F_B is selected in every $B \in Bodies(A)$. Then the set $\{F_B \mid B \in Bodies(A)\}$ is a cover of A, and $\dashv A \ [X;Y]$ can be derived by (FP^L). *Case 1.2.3:* There is no such B, and for at least one B, F_B could not be selected. Then neither $\models E \ [Y]$ nor $\dashv E \ [X;Y]$ is derivable.

Case 2: E has the form *not A*. *Case 2.1:* $A \in Y$. Then neither $\models E \ [Y]$ nor $\dashv E \ [X;Y]$ is derivable. *Case 2.2:* $A \notin Y$. Then evaluate $A \ [\emptyset;Y \cup \{A\}]$. *Case 2.2.1:* $\models A \ [Y \cup \{A\}]$ is derived. Then $\dashv E \ [X;Y]$ can be derived by $(FN^{\widetilde{L}})$. *Case 2.2.2:* $\dashv A \ [\emptyset;Y \cup \{A\}]$ is derived. Then $\models E \ [Y]$ can be derived by $(SN^{\widetilde{L}})$. *Case 2.2.3:* Neither $\models A \ [Y \cup \{A\}]$ nor $\dashv A \ [\emptyset;Y \cup \{A\}]$ is derivable. Then neither $\models E \ [Y]$ nor $\dashv E \ [X;Y]$ is derivable.

In Case 1.2, we could have included an obvious optimization: Once it is determined that an element F_B cannot be selected in some $B \in Bodies(A)$, there is no need to continue looking for such elements in the other sets from $Bodies(A)$. This optimization is not implemented in the meta-interpreter presented in Section 4.3.

This algorithm always terminates. Indeed, define the *rank* of an expression $E \ [X;Y]$ to be the ordinal $\omega(n - |Y|) + (n - |X|)$, where n is the number of atoms in Π. In every recursive call of the algorithm, the rank of the expression to be evaluated goes down: In Case 1.2, $|Y|$ stays the same and $|X|$ increases; in Case 2.2, $|Y|$ increases. (In the definition of rank, ω can be replaced by $n + 1$.)

From Theorems 3 and 4 we see that the three possible answers produced by this algorithm applied to an atom A with empty X and Y exactly correspond to the three possible truth values of A in the well-founded semantics.

Consider, for instance, the application of this algorithm to program (3) with the expression p as E and with empty X and Y. (The rank of $E \ [X;Y]$ is $\omega + 1$.) First, we are in Case 1.2 and are led to evaluate *not p* $[p;]$ (rank ω). This brings us to Case 2.2 and to the task of evaluating $p \ [;p]$ (rank 1). Now we are in Case 1.2 and turn to the evaluation of *not p* $[p;p]$ (rank 0). This is Case 2.1; neither $\models$ *not p* $[p]$ nor $\dashv$ *not p* $[p;p]$ is derivable. Consequently, neither $\models p \ [p]$ nor $\dashv p \ [;p]$ is derivable (Case 1.2.3). Consequently, neither $\models$ *not p* nor $\dashv$ *not p* $[p;]$ is derivable (Case 2.2.3). Consequently, neither $\models p$ nor $\dashv p$ is derivable (Case 1.2.3). This result tells us that the truth value of p in the well-founded semantics is *undefined*.

4.3 Implementation

An implementation of the evaluation strategy described in the previous section is given below. It is implemented in Quintus Prolog as the meta-interpreter solve/4. The procedure wf/2 returns the value of an atom in the well-founded semantics.

The programs executed by the meta-interpreter are composed of ground Prolog clauses with purely conjunctive bodies. The symbol not is used for negation as failure.

```
:- op(600,fx,'not').
:- use_module(library(basics)).

wf(A,Val) :- solve(A,[],[],Val).

solve(not A,_X,Y,Ans) :-
        member(A,Y), !, Ans = und.                    % 2.1
solve(not A,_X,Y,Ans) :-
        !, solve(A,[],[A|Y],An1),                     % 2.2
        neg(An1,Ans).                                 % 2.2.1-2.2.3
solve(A,X,_Y,Ans) :-
        member(A,X), !, Ans = false.                  % 1.1
solve(A,X,Y,Ans) :-
        findall(B,clause(A,B),Bs),
        solve_bodies(Bs,[A|X],Y,Ans).                 % 1.2

solve_bodies([B|Bs],X,Y,Ans) :-
        solve_elements(B,X,Y,An1),
        ( An1 = true -> Ans = true
        ; solve_bodies(Bs,X,Y,An2),
          or(An1,An2,Ans) ).                          % 1.2.1-1.2.3
solve_bodies([],_X,_Y,false).

solve_elements((F,Fs),X,Y,Ans) :-
        !, solve(F,X,Y,An1),
        ( An1 = false -> Ans = false
        ; solve_elements(Fs,X,Y,An2),
          and(An1,An2,Ans) ).
solve_elements(F,X,Y,Ans) :-
        F = true -> Ans = true ; solve(F,X,Y,Ans).

and(false,_U,false).
and(und,U,V) :-
        U = false -> V = false ; V = und.
and(true,U,U).

or(false,U,U).
```

```
or(und,U,V) :-
        U = true -> V = true ; V = und.
or(true,_U,true).

neg(false,true).
neg(und,und).
neg(true,false).
```

5 Proofs

5.1 Proof of Theorem 1

Lemma 1. *For any program Π and any goals G_1, G_2, if $\models G_1$ and $\models G_2$ are derivable in $SLDNF^1_\Pi$ then so is $\models G_1 \cup G_2$.*

The proof is by induction on the derivation of $\models G_1$.

Lemma 2. *For any program Π and any goals G_1, G_2 such that $G_1 \subset G_2$, if $\dashv G_1$ is derivable in $SLDNF^1_\Pi$ then so is $\dashv G_2$.*

The proof is by induction on the derivation of $\dashv G_1$.

Proof of Theorem 1. Left-to-right, both parts of the theorem can be proved simultaneously by induction on the given derivation in $SLDNF^1_\Pi$. Right-to-left, both parts can be proved simultaneously by induction on the given derivation in $SLDNF^2_\Pi$, using Lemmas 1 and 2.

5.2 A Characterization of the Well-Founded Model

For a set X of atoms, define

$$\neg X = \{\neg A \ : \ A \in X\},$$
$$not(X) = \{not\ A \ : \ A \in X\}.$$

Every set of rule elements can be represented as $Pos \cup not(Neg)$, where Pos and Neg are sets of atoms.

The proofs of Theorems 2 and 3 are based on the following characterization of the well-founded semantics.

Lemma 3. *The well-founded model of a program Π is the set $I \cup \neg J$, where I and J are the least sets of atoms with the following two properties:*

(WF1) *If $A \leftarrow Pos \cup not(Neg)$ is a rule of Π such that $Pos \subset I$ and $Neg \subset J$ then $A \in I$.*

(WF2) *There exists a mapping ν from atoms to natural numbers such that for each atom $A \notin J$ there exists a rule $A \leftarrow Pos \cup not(Neg)$ in Π for which*
   (a) *for all $B \in Pos$, $\nu(B) < \nu(A)$ and $B \notin J$,*
   (b) *$Neg \cap I = \emptyset$.*

Proof. The well-founded model of a program can be described as the least fixpoint of an operator Ψ associated with this program ([Przymusinski, 1994], Theorem 13). If I and J satisfy (WF1) and (WF2) then $I \cup \neg J$ is a pre-fixpoint of Ψ. If $I \cup \neg J$ is a fixpoint of Ψ then I and J satisfy (WF1) and (WF2).

5.3 Proofs of Theorems 2 and 3

Proof of Theorem 2. Let I, J and ν be as in (WF1) and (WF2). Then the following can be shown by induction on the derivation:

1. If $\models A$ is derivable in L^{Π} then $A \in I$.
2. If $\models not\ A$ is derivable in L^{Π} then $A \in J$.
3. If $\dashv A\ [X]$ is derivable in L^{Π} and $\nu(A) < \nu(B)$ for all $B \in X$ then $A \in J$.
4. If $\dashv not\ A\ [X]$ is derivable in L^{Π} then $A \in I$.

Now the assertion of the theorem follows by Lemma 3.

Proof of Theorem 3. In this proof, by an atom we mean an atom occurring in Π. Let n be the number of atoms. Let I be the set of atoms A such that $\models A$ is derivable in L_{Π}. Let J be the set of atoms A such that $\dashv A$ is derivable in L_{Π}. By Lemma 3, it is sufficient to check that I and J satisfy (WF1) and (WF2).

(WF1) Assume that $A \leftarrow Pos \cup not(Neg)$ is a rule of Π such that $Pos \subset I$ and $Neg \subset J$. From the definitions of I and J we see that $\models A$ can be derived by rule (SP^L). Thus $A \in I$.

(WF2) For every atom $A \notin J$, let X_A be a set of maximal cardinality such that $\dashv A\ [X_A]$ is not derivable. For every atom $A \notin J$, define $\nu(A) = n - |X_A|$; for $A \in J$, set $\nu(A) = 0$. Take any $A \notin J$. Then $\dashv A\ [X_A]$ is not derivable. In particular, this expression is not an axiom, so that $A \notin X_A$. Moreover, it cannot be derived by (FP^L), that is, there exists a rule $A \leftarrow Pos \cup not(Neg)$ in Π such that $\dashv E\ [X_A \cup \{A\}]$ is not derivable for each $E \in Pos \cup not(Neg)$. (a) Take any $B \in Pos$. Then $\dashv B\ [\emptyset]$ is not derivable and thus $B \notin J$. We need to show that $\nu(B) < \nu(A)$. Since $\dashv B\ [X_A \cup \{A\}]$ is not derivable, it follows by the definition of X_B that $|X_A \cup \{A\}| \leq |X_B|$. Since $A \notin X_A$, we conclude that $|X_A| < |X_B|$ and thus $\nu(B) = n - |X_B| < n - |X_A| = \nu(A)$. (b) Take any $B \in Neg$. Then $\dashv not\ B$ is not derivable, and therefore $\models B$ is not derivable either, so that $B \notin I$. Thus $Neg \cap I = \emptyset$.

5.4 Proof of Theorem 4

Consider a derivation of $\models E$ or $\dashv E$ in the calculus $\tilde{L}_{\Pi}$. Replace every derived expression of the form $\models F\ [Y]$ in this derivation by $\models F$, and every derived expression of the form $\dashv F\ [X; Y]$ by $\dashv F\ [X]$. The result is a derivation in the calculus L_{Π}.

To prove the second half of the theorem, we need the following definition. A derivation in the calculus L_{Π} is *minimal* if any two different nodes in the same branch of this derivation contain different derived expressions. It is clear that every expression derivable in L_{Π} has a minimal derivation.

Consider a minimal derivation of $\models E$ or $\dashv E$ in the calculus L_Π. For any node α in this derivation, let Y_α be the set of atoms A such that one of the expressions $\models not\ A$, $\dashv not\ A\ [X]$ for some X occurs in the derivation under the node α. Replace every derived expression of the form $\models F$ in every node α by $\models F\ [Y_\alpha]$, and every derived expression $\dashv F\ [X]$ in a node α by $\dashv F\ [X; Y_\alpha]$. We claim that the result is a derivation in the calculus $\tilde{L}_\Pi$.

To prove this, we need to check that, for any node α of the given derivation that contains the conclusion $\models not\ A$ of (SN^L) or the conclusion $\dashv not\ A\ [X]$ of (FN^L), $A \notin Y_\alpha$. Assume that $A \in Y_\alpha$. Then some node α' under α contains either $\models not\ A$ or $\dashv not\ A[X']$ for some X'. Consider the premise of the rule leading to α and the premise of the rule leading to α'. Each of them is either $\models A$ or $\dashv A$. By the minimality of the derivation, they cannot be equal. Consequently, one of the expressions is $\models A$ and the other is $\dashv A$. But this is impossible, by Corollary 1 of Theorem 2.

6 Further Work

A derivation in the calculus L_Π can justify the assertion that an atom A is true in the well-founded semantics (if the derived expression is $\models A$) or that it is false (if the derived expression is $\dashv A$). In a forthcoming paper, we introduce a calculus with additional types of derived expressions that correspond to the properties of being not false (true or undefined) and not true (false or undefined). This calculus, like L_Π, is sound and complete for finite propositional programs.

We plan to extend this work in several directions. First, it needs to be generalized to programs with variables. Second, we would like to describe tabulation [Chen and Warren, 1993] in terms of a calculus of goals. Third, it is interesting to know how such calculi can be made complete relative to the answer set semantics.

References

[Bol and Degerstedt, 1993] Roland Bol and Lars Degerstedt. The underlying search for magic templates and tabulation. In *Logic Programming: Proceedings of the Tenth Int'l Conf. on Logic Programming*, pages 793–811, 1993.

[Bol, 1991] Roland Bol. *Loop Checking in Logic Programming*. PhD thesis, University of Amsterdam, 1991.

[Chen and Warren, 1993] Weidong Chen and David Warren. Query evaluation under the well founded semantics. In *The Twelfth ACM Symposium on Principles of Database Systems*, 1993.

[Clark, 1978] Keith Clark. Negation as failure. In Herve Gallaire and Jack Minker, editors, *Logic and Data Bases*, pages 293–322. Plenum Press, New York, 1978.

[Gelfond and Lifschitz, 1991] Michael Gelfond and Vladimir Lifschitz. Classical negation in logic programs and disjunctive databases. *New Generation Computing*, 9:365–385, 1991.

[Kowalski and Sadri, 1991] Robert Kowalski and Fariba Sadri. Logic programs with exceptions. *New Generation Computing*, 9:387–400, 1991.

[Kunen, 1989] Kenneth Kunen. Signed data dependencies in logic programs. *Journal of Logic Programming*, 7(3):231–245, 1989.

[Lifschitz, 1995] Vladimir Lifschitz. SLDNF, constructive negation and grounding. In *Proc. of ICLP-95*, 1995. To appear.

[Mints, 1986] Grigori Mints. A complete calculus for pure Prolog. *Proc. Academy of Sciences of Estonian SSR*, 35(4):367–380, 1986. In Russian.

[Przymusinski, 1990] Teodor Przymusinski. The well-founded semantics coincides with the three-valued stable semantics. *Fundamenta Informaticae*, pages 445–464, 1990.

[Przymusinski, 1991] Teodor Przymusinski. Stable semantics for disjunctive programs. *New Generation Computing*, 9:401–424, 1991.

[Przymusinski, 1994] Teodor Przymusinski. Well-founded and stationary models of logic programs. *Annals of Mathematics and Artificial Intelligence*, 12:141–187, 1994.

[Van Gelder et al., 1990] Allen Van Gelder, Kenneth Ross, and John Schlipf. The well-founded semantics for general logic programs. *Journal of ACM*, pages 221–230, 1990.

Annotated Revision Specification Programs

Melvin Fitting
mlflc@cunyvm.cuny.edu

Dept. Mathematics and Computer Science
Lehman College (CUNY), Bronx, NY 10468

Abstract. Marek and Truszczyński have introduced an interesting mechanism for specifying revisions of knowledge bases by means of logic programs. Here we extend their idea to allow for confidence factors, multiple experts, and so on. The appropriate programming mechanism turns out to be *annotated logic programs* and the appropriate semantic tool, *bilattices*. This may be the first example of a setting in which both notions arise naturally, and complement each other. We also show that several of the results of Marek and Truszczyński turn out to be essentially algebraic, once the proper setting has been formulated.

1   Introduction

In [8] Marek and Truszczyński provided an interesting formalism for specifying revisions in databases, knowledge bases, and belief sets. The specifications could be quite complex, containing conditionals (e.g., if this and that is present, so-and-so must be absent). And they presented an ingenious test for determining whether a candidate for an update is one in fact. This test was suggested by the definition of stable model and, in a sense, properly extends the notion.

We extend their work in the following ways. First, we generalize their formalism to systems that allow confidence factors, evidence, multiple experts, and other similar notions. The (fairly obvious) tool for this is an *annotated* version ([7]) of a revision program. To give semantic meaning to such annotated programs, *bilattices* are a natural tool. (This seems to be the first joint exploitation of annotations and bilattices together.) We show that several of the results of Marek and Truszczyński extend naturally to this broader setting. Finally, we show that simple algebraic bilattice calculations yield interesting results about the update mechanism. The ultimate goal (the present paper representing an early stage of the work) is to extract away all the details of the update formalism to reveal the abstract structure underneath. Work on this continues.

2   The Original Version

The following is a brief sketch of revision specification programs as presented in [8]. First, we have an underlying set U whose elements are not analysed further — they are called *atoms*. These can be thought of as things that can appear in a knowledge base. Next, there are two special unary operators, in and out, which apply to atoms. The idea is, in(a) says the atom a is (or ought to be) in a knowledge base, while out(a) says a is (or ought to be) out. We will call in(a) and out(a) *revision atoms*. A *revision rule* is an expression of one of the following forms:

$$\text{in}(p) \leftarrow \text{in}(q_1), \dots, \text{in}(q_m), \text{out}(s_1), \text{out}(s_n)$$
$$\text{out}(p) \leftarrow \text{in}(q_1), \dots, \text{in}(q_m), \text{out}(s_1), \text{out}(s_n)$$

The first kind are called *in-rules* and the second are *out-rules*. Bodies are allowed to be empty. Finally, a *revision program* is a collection of revision rules. (Note that, as defined here, a revision program is propositional. A formulation allowing variables is certainly possible, but let's keep things as simple as possible for the time being.) Now, the idea is to determine when a revised knowledge base B_R (a set of atoms) is a justified revision of an initial knowledge base B_I (another set of atoms) according to conditions specifiec by a revision program $\mathcal{P}$.

In order to do this, Marek and Truszczyński introduce a fairly elaborate process, inspired by the notion of stable model. First, some useful terminology. If B is a knowledge base, in(a) is *satisfied in* B if $a \in B$. Similarly out(a) is satisfied in B if $a \notin B$. Now let $\mathcal{P}$ be a revision program, and B_I and B_R be two knowledge bases. The process begins with the notion of the *reduct* of $\mathcal{P}$ with respect to (B_I, B_R), which is defined as follows.

1. Remove from $\mathcal{P}$ every rule whose body contains a revision atom that is not satisfied in B_R.
2. From the body of each remaining rule, delete any revision atom that is satisfied in B_I.

The notation Marek and Truszczyński use for this revised program is $\mathcal{P}_{B_R}|B_I$.

The revised program $\mathcal{P}_{B_R}|B_I$ is a logic program in the usual sense (without negations or function symbols) and so has a well-behaved minimal model. This minimal model is called the *necessary change basis*, and is denoted by $NCB(\mathcal{P}_{B_R}|B_I)$. It, in turn, determines the *necessary change* for $\mathcal{P}_{B_R}|B_I$ — this is the pair (I, O), where $I = \{a \mid \text{in}(a) \in NCB(\mathcal{P}_{B_R}|B_I)\}$ and $O = \{a \mid \text{out}(a) \in NCB(\mathcal{P}_{B_R}|B_I)\}$. If $I \cap O = \emptyset$ the revised program is called *coherent*.

Finally, suppose $\mathcal{P}_{B_R}|B_I$ is coherent, and (I, O) is the necessary change for $\mathcal{P}_{B_R}|B_I$. Then $(B_I \cup I)\backslash O$ is well-defined (add the members of I to B_I and remove the members of O). If it turns out that $B_R = (B_I \cup I)\backslash O$ then B_R is a $\mathcal{P}$-*justified revision of* B_I.

The intuitions about knowledge base revision that the notions above are intended to capture are these. First, a revised knowledge base should satisfy all the constraints imposed by the revision program. And second, each insertion or deletion performed in order to convert B_I into B_R must be justified by some rule in the revision program. See [8] for a discussion of these ideas, with several motivating examples. In this paper we simply take for granted that the intentions are correctly captured by the Marek, Truszczyński process, and note that their intuitions carry over to the setting of this paper. We concentrate on technical details.

3 Annotated Logic Programs

Generalized Annotated Programs are fully treated in [7], to which we refer you for details. In the interests of simplicity, here we use only a much stripped-down version, with no annotation variables, no function symbols, and more than minimal assumptions about the space of annotations. Given these simplifications, what we need of annotated logic programming can be easily sketched.

Let T be a complete distributive lattice, of truth values. These will be used as annotations in what follows. In addition we will assume T has a *de Morgan complement*, a one-one mapping that is order inverting, of period two. We denote the de Morgan complement of a by $\bar{a}$. It follows easily that de Morgan complementation satisfies the usual de Morgan laws. (In [7] annotations are only required to be an upper semilattice.)

Examples The following are typical and should be kept in mind for what follows.

1. $T = \{false, true\}$, with $false < true$. The de Morgan complement is the usual negation operation. This is the grand-daddy of all examples, of course.
2. T is the collection of all subsets of some set S, with $\subseteq$ as the ordering relation. The de Morgan complement is set-theoretic complement. Think of the members of S as experts, and a member A of T as a set of experts who assert the truth of some proposed fact.
3. $T = [0, 1]$, with the usual ordering. The de Morgan complement of a is $1 - a$. Think of a member of T as a degree of confidence in some proposition.
4. T is the set of all functions from some set S to the unit interval. The ordering is the pointwise one: $f \leq g$ provided $f(a) \leq g(a)$ for all $a \in S$. The de Morgan complement of f is the function given by $\bar{f}(a) = 1 - f(a)$. This combines the two previous examples. Think of $f \in T$ as assigning to each member of a set of experts a confidence factor.

An *annotated revision atom* is either $(\text{in}(a) : \alpha)$ or $(\text{out}(a) : \alpha)$ where a is an atom and α is an annotation — that is, a member of T. Think of $(\text{in}(a) : \alpha)$ as asserting that there is at least α reason to assume that a is (or should be) in the knowledge base. Similarly for $(\text{out}(a) : \alpha)$. For instance, using the first example above, $(\text{in}(a) : true)$ simply says a is in, while $(\text{in}(a) : false)$ gives no information. Using the second example above, $(\text{in}(a) : A)$, where A is a set of experts, can be thought of as saying the members of A assert that a should be in. Continuing this example, $(\text{in}(a) : \emptyset)$ gives no information. There are similar readings that can be supplied for the other examples.

Now, an *annotated revision rule* is an expression of the form $p \leftarrow q_1, \ldots, q_n$ where $p, q_1, \ldots, q_n$ are annotated revision atoms, and an *annotated revision program* is a set of annotated revision rules.

Example A Suppose there are two independent experts, or sources of information, p and q. Let T be the collection of all possible sets of experts: $\{\emptyset, \{p\}, \{q\}, \{p, q\}\}$, ordered by inclusion. This is our space of annotations, with complementation as the de Morgan complement. Also, assume there are three atoms a, b and c that are candidates for inclusion in a knowledge base. Now, here is an annotated revision program, which we will continue to discuss throughout the paper.

$$(\mathbf{in}(a) : \{p\}) \leftarrow (\mathbf{out}(b) : \{p\})$$
$$(\mathbf{in}(b) : \{p\}) \leftarrow (\mathbf{out}(a) : \{p\})$$
$$(\mathbf{out}(a) : \{q\}) \leftarrow (\mathbf{in}(a) : \{p\})$$
$$(\mathbf{in}(c) : \{p, q\}) \leftarrow (\mathbf{in}(a) : \{p\})$$
$$(\mathbf{out}(c) : \{q\}) \leftarrow (\mathbf{out}(a) : \{q\})$$

Since there are no variables or function symbols, either in atoms or in annotations, a fixpoint semantics is easy to describe. First, a $\mathcal{T}$-*valuation* is a mapping v from revision atoms to $\mathcal{T}$. $\mathcal{T}$-valuations are given the pointwise ordering, which makes the space of $\mathcal{T}$-valuations into a complete lattice itself. The $\mathcal{T}$-valuation v *satisfies* the annotated revision atom $(\mathbf{in}(a) : \alpha)$ provided $v(\mathbf{in}(a)) \geq \alpha$ — v satisfies $(\mathbf{out}(a) : \alpha)$ if $v(\mathbf{out}(a)) \geq \alpha$. We say v satisfies a list of annotated revision atoms if it satisfies each member of the list. As "intermediate" notation, let $t_{\mathcal{P}}(v)$ be the set of all annotated revision atoms that occur as the head of a clause in $\mathcal{P}$ whose body is satisfied by v. Finally, let $T_{\mathcal{P}}$ be the mapping from $\mathcal{T}$-valuations to $\mathcal{T}$-valuations given by:

$$T_{\mathcal{P}}(v)(A) = \bigvee \{\alpha \mid (A : \alpha) \in t_{\mathcal{P}}(v)\}.$$

It is straightforward to check that $T_{\mathcal{P}}$ is monotonic on the space of $\mathcal{T}$-valuations, hence has a smallest fixed point. This is the intended meaning of the program. Since negations are not present, the intended meaning is quite unproblematic, and appropriate computational mechanisms can be specified (assuming the lattice $\mathcal{T}$ itself has computble operations).

If we use the first example above, identify $(\mathbf{in}(a) : true)$ with $\mathbf{in}(a)$, ignore $(\mathbf{in}(a) : false)$, and treat $\mathbf{out}(a)$ similarly, revision programming is easily seen to be a special case of annotated revision programming.

4 Bilattices

Atoms can be in a knowledge base, or out of it, and each with a degree of confidence, an annotation. A natural semantic tool for representing this situation is a *bilattice*, as we will see shortly. We begin with a brief abstract presentation, then provide a concrete representation which should make their utility in this context clear. Bilattices were introduced in [6], and have turned out to be a useful tool for investigating logic programming semantics [1, 2], and stable model semantics in particular [3, 5]. We refer you to these papers for a fuller treatment.

Definition 1. A *pre-bilattice* is a structure $\langle \mathcal{B}, \leq_t, \leq_k \rangle$ where $\mathcal{B}$ is a non-empty set and $\leq_t$ and $\leq_k$ are each partial orderings giving $\mathcal{B}$ the structure of a lattice with a top and a bottom. We call $\mathcal{B}$ *complete* if each of the two lattices is complete in the usual sense that all meets and joins exist.

A pre-bilattice has a *negation* if there is a mapping $\neg$ from $\mathcal{B}$ to itself that is an involution, reverses the $\leq_t$ ordering, and preserves the $\leq_k$ ordering. Likewise it has a *conflation* if there is a mapping $-$ that is an involution, reverses the $\leq_k$ ordering, and preserves the $\leq_t$ ordering. If both a negation and a conflation exist, we generally will require that they commute with each other.

Definition 2. In a pre-bilattice $\langle \mathcal{B}, \leq_t, \leq_k \rangle$, meet and join under $\leq_t$ are denoted $\wedge$ and $\vee$, and meet and join under $\leq_k$ are denoted $\otimes$ and $\oplus$. Top and bottom under $\leq_t$ are denoted *true* and *false*, and top and bottom under $\leq_k$ are denoted $\top$ and $\bot$. If the pre-bilattice is complete, infinitary meet and join under $\leq_t$ are denoted $\bigwedge$ and $\bigvee$, and infinitary meet and join under $\leq_k$ are denoted $\prod$ and $\sum$.

Definition 3. A *distributive bilattice* is a pre-bilattice $\langle \mathcal{B}, \leq_t, \leq_k \rangle$ in which all 12 distributive laws connecting $\wedge$, $\vee$, $\otimes$ and $\oplus$ hold. An *infinitely distributive bilattice* is a complete pre-bilattice in which all infinitary, as well as all finitary, distributive laws hold.

It is an easy consequence that in a distributive bilattice, each of the lattice operations, $\wedge$, $\vee$, $\otimes$, $\oplus$, is monotone with respect to both orderings. These are generally called the *interlacing conditions*. If the pre-bilattice is infinitely distributive, each of the infinitary meet and join operations is monotone with respect to both orderings — the *infinitary interlacing conditions*.

Our interests here are confined to complete, infinitely distributive bilattices with negation and conflation that commute. There is a standard way of constructing such structures, which we now sketch. Let $\mathcal{T}$ be a complete distributive lattice with a de Morgen complement, such as could serve as a space of annotations above. We use it to construct a bilattice, which we denote $\mathcal{T} \odot \mathcal{T}$, as follows. The domain is $\mathcal{T} \times \mathcal{T}$. (Think of a member of the domain $\langle \alpha, \beta \rangle$ as saying, there is α evidence that some atom is in a knowledge base, and β evidence that it is out.) The two orderings have the following characterization. $\langle \alpha_1, \beta_1 \rangle \leq_k \langle \alpha_2, \beta_2 \rangle$ if $\alpha_1 \leq \alpha_2$ and $\beta_1 \leq \beta_2$. (Intuitively, knowledge goes up if all evidence, both for and against, increases.) $\langle \alpha_1, \beta_1 \rangle \leq_t \langle \alpha_2, \beta_2 \rangle$ if $\alpha_1 \leq \alpha_2$ and $\beta_1 \geq \beta_2$. (Intuitively, degree of truth goes up if evidence for increases and evidence against decreases.) It is not hard to check that this gives $\mathcal{T} \times \mathcal{T}$ the structure of a complete, infinitely distributive bilattice. Next, negation simply switches around the roles of for and against: $\neg \langle \alpha, \beta \rangle = \langle \beta, \alpha \rangle$. The intuition here is quite clear. Finally, conflation is a little more complicated, and involves the de Morgan complementation operation of $\mathcal{T}$. $-\langle \alpha, \beta \rangle = \langle \overline{\beta}, \overline{\alpha} \rangle$. (Intuitively, the conflation of a member b of $\mathcal{T} \times \mathcal{T}$ counts as evidence for inclusion in a knowledge base whatever b did not count as evidence against, and similarly the other way around.) Again it is easy to check that this does give a conflation that commutes with the negation operation.

We have sketched a method of constructing bilattices $\mathcal{T} \odot \mathcal{T}$ from lattices $\mathcal{T}$. The method is entirely general, in the sense that every complete, infinitely distributive bilattice with a negation and a conflation that commute is isomorphic to $\mathcal{T} \odot \mathcal{T}$ for some complete lattice $\mathcal{T}$ with a de Morgan complement. A proof of this, generalizing a representation theorem of Ginsberg, can be found in [1, 4].

5 Annotated Revision Programs and Bilattices

In section 3, $\mathcal{T}$-valuations were defined. It will be more convenient to work with valuations in a bilattice, and connections are easy to make. First, by a $(\mathcal{T} \odot \mathcal{T})$-*valuation* we mean a mapping from *atoms* (not revision atoms) to members of the bilattice $\mathcal{T} \odot \mathcal{T}$. If the atom a maps to $\langle \alpha, \beta \rangle$ under some $(\mathcal{T} \odot \mathcal{T})$-valuation, think of this as saying there is α reason to have a in a knowledge base and β reason to have a out. $(T \odot T)$-valuations are given two orderings, denoted $\leq_k$ and $\leq_t$, in the obvious pointwise way. Likewise the operations of negation, $\neg$, and conflation, $-$, lift pointwise to the space of $(T \odot T)$-valuations. It is easy to check that the space of $(T \odot T)$-valuations is, again, a complete infinitely distributive bilattice with negation and conflation that commute. As it happens, in this paper the $\leq_k$ ordering will play the primary role, with almost no role given to $\leq_t$.

There is an obvious correspondence between T-valuations and $(T \odot T)$-valuations — we denote this correspondence by θ. Suppose v is a T-valuation, mapping revision atoms to members of T. Associate with this the $(T \odot T)$-valuation $\theta(v)$ defined by: $\theta(v)(a) = \langle \alpha, \beta \rangle$ where $v(\text{in}(a)) = \alpha$ and $v(\text{out}(a)) = \beta$. θ is easily seen to be $1 - 1$ and onto, and so has an inverse, characterized by: if w is a $(T \odot T)$-valuation, $\theta^{-1}(w)$ is the T-valuation given by: $\theta^{-1}(w)(\text{in}(a)) = \alpha$ and $\theta^{-1}(w)(\text{out}(a)) = \beta$ provided $w(a) = \langle \alpha, \beta \rangle$.

The mapping θ is order-preserving, in the following sense. Suppose v and w are T-valuations, and $v \leq w$ in the ordering of T. Then $\theta(v) \leq_k \theta(w)$. This is easy to see, as is the fact that θ^{-1} is also order-preserving in the other direction.

Let $\mathcal{P}$ be an annotated revision program. In section 3 we associated with it a mapping $T_{\mathcal{P}}$ on the space of T-valuations. This induces a bilattice mapping which we denote $T_{\mathcal{P}}^b$ on the space of $(T \odot T)$-valuations, in a direct way:

$$T_{\mathcal{P}}^b(v) = (\theta T_{\mathcal{P}} \theta^{-1})(v).$$

We noted earlier that $T_{\mathcal{P}}$ is monotonic in the space of T-valuations. It follows from this, and the order-preserving properties of θ and θ^{-1}, that $T_{\mathcal{P}}^b$ is monotonic in the space of $(T \odot T)$-valuations, with respect to the $\leq_k$ ordering. From now on we generally confine our work to the bilattice $T \odot T$, and use the $T_{\mathcal{P}}^b$ mapping, pushing T and $T_{\mathcal{P}}$ into the background.

6 Program Transformation

Now we carry over directly to the present setting the program transformation of Marek and Truszczyński. If v is a $(T \odot T)$-valuation, we say v *satisfies* an annotated revision atom if the T-valuation $\theta^{-1}(v)$ satisfies it, as defined in section 3.

Definition 4. Let $\mathcal{P}$ be an annotated revision program and let B_I and B_R be $(T \odot T)$-valuations. We define the reduct of $\mathcal{P}$ with respect to (B_I, B_R) as follows.

1. First, remove from $\mathcal{P}$ every rule whose body contains an annotated revision atom that is not satisfied in B_R.
2. Second, from the body of each remaining rule delete any annotated revision atom that is satisfied in B_I.

The resulting program is denoted $\mathcal{P}_{B_R} | B_I$.

Now we use the notion of program reduct to define an operator on bilattices, as follows.

Definition 5. Let B_I, B_R, and v be three $(\mathcal{T} \odot \mathcal{T})$-valuations, and let $\mathcal{P}$ be an annotated revision program. The mapping $\mathfrak{R}_\mathcal{P}$ is given by the following.

$$\mathfrak{R}_\mathcal{P}(B_I, B_R, v) = T^b_{\mathcal{P}_{B_R}|B_I}(v).$$

The T notation is somewhat hair-raising, which is partly why we have introduced the $\mathfrak{R}$ (for "revision") notation. The idea, nonetheless, is straightforward. To compute $\mathfrak{R}_\mathcal{P}(B_I, B_R, v)$, begin with the annotated revision program $\mathcal{P}$, carry out step 1 of the reduction process using B_R; next carry out step 2 using B_I. This yields another annotated revision program $\mathcal{P}_{B_R}|B_I$. Apply the "single-step" T operator for this program, using v as input, after translating v from the bilattice $\mathcal{T} \odot \mathcal{T}$ to the underlying truth-value space $\mathcal{T}$, then translating the result back to the bilattice setting. The outcome is the value of $\mathfrak{R}_\mathcal{P}(B_I, B_R, v)$.

As an operator on $\mathcal{T} \odot \mathcal{T}$, $\mathfrak{R}_\mathcal{P}(B_I, B_R, v)$ has several nice properties. First, it is obviously monotonic in v, in the $\leq_k$ ordering (since T^b operators are). Next, it is also monotonic in B_I in the $\leq_k$ ordering (because if B_I is increased, more annotated revision atoms will be satisfied, so more parts of clause bodies will be deleted in step 2 of the reduction process, yielding a program whose clause bodies are more easily satisfied). Finally, it is even monotonic in B_R in the $\leq_k$ ordering (because if B_R is increased, fewer annotated revision atoms will be unsatisfied, so fewer clauses will be deleted in step 1 of the reduction process, again yielding a program whose clause bodies are more easily satisfied, since there are more of them).

Example A continued (Example A began in section 3.) Let B_I be given by:

$$B_I(a) = \langle \{q\}, \{p, q\} \rangle$$
$$B_I(b) = \langle \emptyset, \{p, q\} \rangle$$
$$B_I(c) = \langle \emptyset, \{q\} \rangle$$

And let B_R be given by:

$$B_R(a) = \langle \{p\}, \{q\} \rangle$$
$$B_R(b) = \langle \emptyset, \{p, q\} \rangle$$
$$B_R(c) = \langle \{p, q\}, \{q\} \rangle$$

Now, $(\text{out}(a) : \{p\})$ is not satisfied by B_R, but all other annotated revision atoms in clause bodies of $\mathcal{P}$ are, so only the second clause of $\mathcal{P}$ is deleted in step 1 above. Next, both $(\text{out}(b) : \{p\})$ and $(\text{out}(a) : \{q\})$ are satisfied by B_I so these are deleted from clause bodies in step 2. The resulting program, $\mathcal{P}_{B_R}|B_I$, is:

$$(\text{in}(a) : \{p\}) \leftarrow$$
$$(\text{out}(a) : \{q\}) \leftarrow (\text{in}(a) : \{p\})$$
$$(\text{in}(c) : \{p, q\}) \leftarrow (\text{in}(a) : \{p\})$$
$$(\text{out}(c) : \{q\}) \leftarrow$$

Notice, in this example, that according to B_I, expert q is acting inconsistently, asserting that a should both be present and absent. In the original Marek and Truszczyński setting such inconsistencies were not allowed — coherency conditions were explicitly imposed. We find it more natural to allow inconsistencies, and simply record their presence. This is especially useful in a setting like that above, where q is being inconsistent with regard to a, but p is not.

7 How To Make Change

Now that the notion of reduced program has been introduced in our setting, we can use it to compute the necessary change basis, and consequently the necessary change. Then we have the problem of how to carry out this change. But first things first.

Thought of as an annotated logic program, a reduced program $\mathcal{P}_{B_R}|B_I$ has a minimal model — a $\mathcal{T}$-valuation. Equivalently, because of the correspondence between $\mathcal{T}$-valuations and $(\mathcal{T} \odot \mathcal{T})$-valuations, we can take the least fixed point, in the $\leq_k$ ordering, of the mapping:

$$(\lambda v)\Re_P(B_I, B_R, v)$$

We call the least fixed point of this mapping the *necessary change* for $\mathcal{P}_{B_R}|B_I$.

Example A continued In the previous section the revised program $\mathcal{P}_{B_R}|B_I$ was given. It is easy to see that the minimal model for this is the set consisting of: $(\text{in}(a) : \{p\})$, $(\text{out}(a) : \{q\})$, $(\text{in}(c) : \{p,q\})$, and $(\text{out}(c) : \{q\})$. Corresponding to this is the $(\mathcal{T} \odot \mathcal{T})$-valuation C, the *necessary change* for $\mathcal{P}_{B_R}|B_I$, as follows:

$$C(a) = \langle \{p\}, \{q\} \rangle$$
$$C(b) = \langle \emptyset, \emptyset \rangle$$
$$C(c) = \langle \{p,q\}, \{q\} \rangle$$

Now that we have said what necessary change amounts to in the present context, we must say how to effectuate it. For this, we propose the following.

Definition 6. The result of applying change C to an initial knowledge base B_I is:

$$(B_I \otimes -C) \oplus C.$$

In this, $\otimes$ and $\oplus$ are the bilattice meet and join with respect to the $\leq_k$ ordering, and $-$ is the conflation operation. To get a feeling for how this works, suppose $\mathcal{T}$ is the lattice of subsets of the three-expert set $\{p,q,r\}$, $B_I = \langle \{p,q\}, \{r\} \rangle$, and $C(a) = \langle \emptyset, \{p\} \rangle$. Essentially $C(a)$ amounts to: put p in the "against" column, as far as a is concerned, and don't change anybody else. For this choice of $\mathcal{T} \odot \mathcal{T}$, conflation is given by $-\langle X, Y \rangle = \langle \overline{Y}, \overline{X} \rangle$, where $\overline{X}$ denotes complement. Then $-C(a) = \langle \{q,r\}, \{p,q,r\} \rangle$. Also, for $\mathcal{T} \odot \mathcal{T}$, $\langle X_1, X_2 \rangle \otimes \langle Y_1, Y_2 \rangle$ is $\langle X_1 \cap Y_1, X_2 \cap Y_2 \rangle$ and $\langle X_1, X_2 \rangle \oplus \langle Y_1, Y_2 \rangle$ is $\langle X_1 \cup Y_1, X_2 \cup Y_2 \rangle$. Then $(B_I \otimes -C)(a) = \langle \{q\}, \{r\} \rangle$. Notice that this amounts to the removal of p from the "for" side. Next, $((B_I \otimes -C) \oplus C)(a) = \langle \{q\}, \{p,r\} \rangle$, and indeed p has been withdrawn as "for" and added as "against." We leave it to you to try out examples involving inconsistent or incomplete information. We also note that if we restrict the setting to the original one of Marek and Truszczyński, the present definition is equivalent to theirs. (Recall, their setting is the simplest non-trivial bilattice, the four-element one, with valuations never taking on either $\bot$ or $\top$ as values.)

Example A continued We have computed the necessary change for Example A. Now, $-C$ is as follows:

$$-C(a) = \langle \{p\}, \{q\} \rangle$$
$$-C(b) = \langle \{p,q\}, \{p,q\} \rangle$$
$$-C(c) = \langle \{p\}, \emptyset \rangle$$

Continuing,

$$((B_I \otimes -C) \oplus C)(a) = \langle \{p\}, \{q\} \rangle$$
$$((B_I \otimes -C) \oplus C)(b) = \langle \emptyset, \{p,q\} \rangle$$
$$((B_I \otimes -C) \oplus C)(c) = \langle \{p,q\}, \{q\} \rangle$$

Definition 7. Extending the terminology of Marek and Truszczyński, we say B_R is a *$\mathcal{P}$-justified revision of B_I* if $B_R = (B_I \otimes -C) \oplus C$.

Notice that in Example A, B_R is in fact a $\mathcal{P}$-justified revision of B_I. We leave it to you to verify that the following is also a $\mathcal{P}$-justified revision of B_I:

$$B_R(a) = \langle \{q\}, \{p,q\} \rangle$$
$$B_R(b) = \langle \{p\}, \{q\} \rangle$$
$$B_R(c) = \langle \emptyset, \{q\} \rangle$$

8 Elementary Results

The notion of applying a necessary change to a knowledge base, introduced above in a general context, has several nice features, and some of these can be established by essentially algebraic methods.

Why only two values? We begin by showing why the full structure of a bilattice never arose in the original treatment of Marek and Truszczyński. As we noted earlier, their setting can be thought of as the four-element bilattice, but restricted to the portion consisting of only *false* and *true*. Ruling out $\top$ is done explicitly: they say a change is not well-defined if it is not coherent. Ruling out $\bot$ is done implicitly: all their models are considered totally defined. Now, how can such restrictions be extended to more general bilattices?

We showed in [2] that the conflation operator is the key. Suppose we call a member A of $\mathcal{T} \odot \mathcal{T}$ *exact* if $A = -A$. In the four-element bilattice, the exact members are simply *false* and *true*. In the setting of Example A, the exact values are those $\langle A, B \rangle$ for which B is exactly the complement of A (with respect to $\{p,q\}$). In general the exact members of $\mathcal{T} \odot \mathcal{T}$ always possess many important properties of the classical truth values (such as closure under $\wedge$, $\vee$, $\neg$, $\bigwedge$, and $\bigvee$) and can be considered a reasonable generalization. Further, suppose we call A *consistent* if $A \leq_k -A$. In the four-element setting, the consistent members are *false*, *true*, and $\bot$, and if the $\leq_t$ operations are restricted to these values, Kleene's strong three-valued logic results. In the setting of Example A, the consistent values are those $\langle A, B \rangle$ for which A and B do not overlap. In general, the consistent members of $\mathcal{T} \odot \mathcal{T}$ always constitute a complete semi-lattice with respect to $\leq_k$ and are closed under the operations of $\leq_t$, and thus naturally generalize Kleene's logic. With all this in mind, we have the following simple result.

Theorem 8. *Working in $T \odot T$, suppose an initial knowledge base B_I is exact and the necessary change C is consistent. Then the result of applying the change, $(B_I \otimes -C) \oplus C$, is exact.*

Proof. Let $D = (B_I \otimes -C) \oplus C$. We show $-D = D$, assuming that $-B_I = B_I$ and $C \leq_k -C$. Note, from the latter it follows that $C \otimes -C = C$. Now, using distributive laws and the de Morgan properties of conflation, we have the following calculation.

$$
\begin{aligned}
-D &= -((B_I \otimes -C) \oplus C) \\
&= -((-B_I \otimes -C) \oplus C) \\
&= (B_I \oplus C) \otimes -C \\
&= (B_I \otimes -C) \oplus (C \otimes -C) \\
&= (B_I \otimes -C) \oplus C \\
&= D
\end{aligned}
$$

Undoing changes. We wish to investigate under what circumstances changes can be undone. More specifically we will show that, at least sometimes, if C changes A to B, $\neg C$ will change B back to A. Now we should not expect this under all circumstances. For instance, if C says to add p to a knowledge base, and p is already in A, applying C to A yields A again. But $\neg C$ will tell us to remove p, and this certainly does not leave A unchanged. The problem here is that C is telling us to do something unnecessary.

Consider an example using the annotation space T of Example A. Suppose $C(a) = \langle \{p\}, \{q\} \rangle$ and $A(a) = \langle \alpha, \beta \rangle$. In order for C to be telling us to do something that actually needs doing, we must have $p \notin \alpha$ and $q \notin \beta$, or equivalently, $p \in \overline{\alpha}$ and $q \in \overline{\beta}$. But $-\neg \langle \alpha, \beta \rangle = \langle \overline{\alpha}, \overline{\beta} \rangle$, so we can state our requirements quite simply: we want $\langle \{p\}, \{q\} \rangle \leq_k -\neg \langle \alpha, \beta \rangle$. This leads us to the following notion.

Definition 9. We say C is an *essential change* with respect to A if $C \leq_k -\neg A$.

Theorem 10. *Suppose that T is not just a complete lattice with a de Morgan complement, but that it is also a Boolean algebra. Working in the bilattice $T \odot T$, suppose C is an essential change with respect to A, and A is classical. If C changes A into B, then $\neg C$ changes B into A. More precisely, if $B = (A \otimes -C) \oplus C$, then $A = (B \otimes -\neg C) \oplus \neg C$.*

Proof. If we assume the complement operation of T satisfies the Boolean algebra law, $x \wedge \overline{x} = \bot$, it follows easily that in $T \odot T$ we have (1) $X \otimes -\neg X = \bot$. Applying conflation to both sides of this, we also have (2) $-X \oplus \neg X = \top$. Since C is an essential change with respect to A, $C \leq_k -\neg A$, or equivalently, (3) $A \leq_k -\neg C$, or

$\neg C \leq_k -A$. Finally A is classical, so (4) $-A = A$. Now, the argument is as follows.

$$(A \otimes -C) \oplus C = B \qquad \text{Assumption}$$
$$[(A \otimes -C) \oplus C] \otimes -\neg C = B \otimes -\neg C$$
$$(A \otimes -C \otimes -\neg C) \oplus (C \otimes -\neg C) = B \otimes -\neg C \qquad \text{Distributive Law}$$
$$A \otimes -C \otimes -\neg C = B \otimes -\neg C \qquad \text{By (1)}$$
$$A \otimes -C = B \otimes -\neg C \qquad \text{By (3)}$$
$$(A \otimes -C) \oplus \neg C = (B \otimes -\neg C) \oplus \neg C$$
$$(A \oplus \neg C) \otimes (-C \oplus \neg C) = (B \otimes -\neg C) \oplus \neg C \qquad \text{Distributive Law}$$
$$A \oplus \neg C = (B \otimes -\neg C) \oplus \neg C \qquad \text{By (2)}$$
$$-A \oplus \neg C = (B \otimes -\neg C) \oplus \neg C \qquad \text{By (4)}$$
$$-A = (B \otimes -\neg C) \oplus \neg C \qquad \text{By (3)}$$
$$A = (B \otimes -\neg C) \oplus \neg C \qquad \text{By (4)}$$

Notice, incidentally, that the multiple expert examples do satisfy the Boolean algebra requirement of the Theorem above, but the unit interval example does not.

Dual revision programs. In [8] the notion of a *dual* revision program was introduced. $\mathcal{P}^D$ is the dual of $\mathcal{P}$ if every occurrence of **out** is replaced with an occurrence of **in**, and conversely. In fact the behavior of a dual program is easily characterized in bilattice terms — the program operator is the dual operator in the most straightforward sense. Suppose, for notational simplicity, we let $\mathcal{Q}$ be the dual of $\mathcal{P}$. Then it is quite easy to see that:

$$T_{\mathcal{Q}}^b(v) = \neg T_{\mathcal{P}}^b(\neg v).$$

It follows immediately that if C is a fixed point of $T_{\mathcal{P}}^b$, then $\neg C$ is a fixed point of $T_{\mathcal{Q}}^b$, because $T_{\mathcal{Q}}^b(\neg C) = \neg T_{\mathcal{P}}^b(\neg\neg C) = \neg T_{\mathcal{P}}^b(C) = \neg C$. It now follows from the previous discussion that dual programs can be used, under appropriate circumstances, to compute inverse changes.

Preprocessing. Both the original Marek and Truszczyński version and the present generalization only serve to verify that a candidate B_R for a revision of B_I in accordance with $\mathcal{P}$ really is one. No method is provided for computing a revision — indeed it is not clear what such a method would compute, since revisions are not generally unique, as Example A shows. What we consider now is whether there is some technique for computing at least that portion of the necessary changes that are common to all changes leading to correct revisions. We propose a simple method that is, in a sense, correct but not complete. Exactly what this means will become clear after the method has been presented.

Recall that for a given annotated revision program $\mathcal{P}$d an operator $\mathfrak{R}_{\mathcal{P}}(B_I, B_R, v)$ was defined to be $T_{\mathcal{P}_{B_R|B_I}}^b(v)$, and we observed that this was monotonic in each of B_I, B_R, and v, with respect to the $\leq_k$ ordering. Then we defined the necessary change for $\mathcal{P}_{B_R}|B_I$ to be the least fixed point of $(\lambda v)\mathfrak{R}_{\mathcal{P}}(B_I, B_R, v)$. It is convenient now to turn this notion itself into an operator.

$$\mathfrak{R}_{\mathcal{P}}'(B_I, B_R) = \text{the least fixed point of } (\lambda v)\mathfrak{R}_{\mathcal{P}}(B_I, B_R, v).$$

Then $\mathfrak{R}_{\mathcal{P}}'(B_I, B_R)$ is the necessary change (candidate) for revising B_I into B_R. It is not hard to verify, from general lattice properties, that $\mathfrak{R}_{\mathcal{P}}'(B_I, B_R)$ itself is monotonic in both B_I and B_R, in the $\leq_k$ ordering.

Definition 11. Let C_{B_I} be the least fixed point of $(\lambda w)\mathfrak{R}'_p(B_I, w)$.

The utility of this notion is simple: it yields a $\mathcal{T} \odot \mathcal{T}$-valuation that is compatible with the necessary change for revising B_I, *no matter what the candidate for revision might be*. More precisely, we have the following.

Theorem 12. *Suppose B_R is a $\mathcal{P}$-justified revision of B_I, and C is the necessary change for $\mathcal{P}_{B_R}|B_I$. Then $C_{B_I} \leq_k C$.*

Proof. Since C is the necessary change for $\mathcal{P}_{B_R}|B_I$, $\mathfrak{R}'_p(B_I, B_R) = C$. Also since B_R is a $\mathcal{P}$-justified revision of B_I, $B_R = (B_I \otimes -C) \oplus C$, so $C \leq_k B_R$. Since $\mathfrak{R}'_p(x, y)$ is monotonic in both inputs, $\mathfrak{R}'_p(B_I, C) \leq \mathfrak{R}'_p(B_I, B_R) = C$. It follows that the least fixed point of $(\lambda w)\mathfrak{R}'_p(B_I, w)$ is $\leq_k C$, or $C_{B_I} \leq_k C$.

What this means is, if we are given $\mathcal{P}$ and an initial knowledge base B_I, we can compute at least some of the necessary change, C_{B_I}, before we have a candidate B_R for a revised knowledge base. In a sense, C_{B_I} constitutes the *uniform* part of the change, that is, the part whose justification will be the same for every B_R. An example should help clarify this.

Example B Let $\mathcal{P}$ be the following annotated revision program, where the space of annotations is the collection of subsets of $\{p, q\}$.

$$(\mathbf{out}(a) : \{p\}) \leftarrow (\mathbf{in}(b) : \{p\})$$
$$(\mathbf{out}(b) : \{p\}) \leftarrow (\mathbf{in}(a) : \{p\})$$
$$(\mathbf{in}(c) : \{p\}) \leftarrow (\mathbf{in}(a) : \{p\})$$
$$(\mathbf{in}(c) : \{p\}) \leftarrow (\mathbf{in}(b) : \{p\})$$
$$(\mathbf{in}(a) : \{q\}) \leftarrow$$
$$(\mathbf{out}(b) : \{q\}) \leftarrow (\mathbf{in}(a) : \{q\})$$

Also let B_I be the following.

$$B_I(a) = \langle \{p\}, \emptyset \rangle$$
$$B_I(b) = \langle \{p\}, \emptyset \rangle$$
$$B_I(c) = \langle \emptyset, \emptyset \rangle$$

We can then calculate C_{B_I} by the usual technique of starting with $\perp$ and iterating operator application.

$$C_{B_I}(a) = \langle \{q\}, \emptyset \rangle$$
$$C_{B_I}(b) = \langle \emptyset, \{q\} \rangle$$
$$C_{B_I}(c) = \langle \emptyset, \emptyset \rangle.$$

Next, consider the following, both of which are, in fact, $\mathcal{P}$-justified revisions of B_I.

$$B_R^1(a) = \langle \{p, q\}, \emptyset \rangle$$
$$B_R^1(b) = \langle \emptyset, \{p, q\} \rangle$$
$$B_R^1(c) = \langle \{p\}, \emptyset \rangle$$

$$B_R^2(a) = \langle \{q\}, \{p\} \rangle$$
$$B_R^2(b) = \langle \{p\}, \{q\} \rangle$$
$$B_R^2(c) = \langle \{p\}, \emptyset \rangle$$

For B_R^1 the necessary change turns out to be the following.

$$C^1(a) = \langle \{q\}, \emptyset \rangle$$
$$C^1(b) = \langle \emptyset, \{p, q\} \rangle$$
$$C^1(c) = \langle \{p\}, \emptyset \rangle$$

And for B_R^2 the necessary change is the following.

$$C^2(a) = \langle \{q\}, \{p\} \rangle$$
$$C^2(b) = \langle \emptyset, \{q\} \rangle$$
$$C^2(c) = \langle \{p\}, \emptyset \rangle$$

As predicted, $C_{B_I} \leq_k C^1$ and $C_{B_I} \leq_k C^2$. Note, however, that if C is any change that effects a $\mathcal{P}$-justified revision of B_I, it must be the case that $\langle \{p\}, \emptyset \rangle \leq_k C(c)$, though $\langle \{p\}, \emptyset \rangle \not\leq_k C_{B_I}(c)$. Essentially this is because the reasons p has for asserting c are not uniform in all models. In B_R^1, p asserts c because p asserts a, while in B_R^2 it is because p asserts b. Nonetheless, use of C_{B_I} should allow a certain amount of preprocessing, and thus improve efficiency somewhat.

9 Conclusion

We believe the understanding of revision specification programs is still in early stages. Algorithms for computing revisions are missing. Non-trivial examples to which the theory applies would be useful. Still we believe generalizing as we have done does not obfuscate, but clarify — serving to bring out the essential algebraic structure underneath. We intend to continue with further investigations. We urge others to join in.

References

1. FITTING, M. C. Bilattices in logic programming. In *The Twentieth International Symposium on Multiple-Valued Logic* (1990), G. Epstein, Ed., IEEE, pp. 238–246.
2. FITTING, M. C. Bilattices and the semantics of logic programming. *Journal of Logic Programming 11* (1991), 91–116.
3. FITTING, M. C. The family of stable models. *Journal of Logic Programming 17* (1993), 197–225.
4. FITTING, M. C. Kleene's three-valued logics and their children. *Fundamenta Informaticae 20* (1994), 113–131.
5. FITTING, M. C. On prudent bravery and other abstractions. Submitted, 1994.
6. GINSBERG, M. L. Multivalued logics: a uniform approach to reasoning in artificial intelligence. *Computational Intelligence 4* (1988), 265–316.
7. KIFER, M., AND SUBRAHMANIAN, V. S. Theory of generalized annotated logic programming and its applications. *Journal of Logic Programming 12* (1992), 335–367.
8. MAREK, V. W., AND TRUSZCZYŃSKI, M. Revision specifications by means of programs. Presented at LPNMR workshop, Lexington, KY, 1994., 1994.

Update by Means of Inference Rules

Teodor C. Przymusinski[1]
Department of Computer Science
University of California
Riverside, CA 92521, USA
(teodor@cs.ucr.edu)

Hudson Turner[2]
Department of Computer Sciences
University of Texas at Austin
Austin, TX 78712 USA
(hudson@cs.utexas.edu)

1 Introduction

Katsuno and Mendelzon [KM91] have distinguished two abstract frameworks for reasoning about change: theory revision and theory update. *Theory revision* involves a change in knowledge or belief with respect to a static world. For example, suppose you are booked on a flight, but told only that your destination is either Australia or Europe, i.e., that *australia* ∨ *europe* holds. If sometime later you learn, in addition to what you already know, that you weren't booked on a flight to Europe, i.e., that ¬*europe* holds, then you are likely to conclude that your destination is Australia, i.e., that *australia* holds.

By contrast, *theory update* involves a change of knowledge or belief in a changing world. Again suppose you are booked on a flight, and told only that your destination is either Australia or Europe, i.e., that *australia* ∨ *europe* holds. Suppose you later learn that the situation has changed and all flights to Europe have just been cancelled, i.e., that ¬*europe* holds. Under these circumstances, you are not likely to conclude that you are going to Australia, i.e., that *australia* holds. In fact, it may be that your flight has just been cancelled.

In this paper we are concerned with theory update. A key insight into the nature of update is due to Winslett [Win88] who showed that reasoning about the effects of actions should be done "interpretation by interpretation". This insight is apparent in the general definition of theory update due to Katsuno and Mendelzon, which can be stated as follows. Let Γ and T be sets of propositional formulae. A set T' of formulae is a "theory update" of T by Γ if

$$Models(T') = \{ \, I' : \exists I \in Models(T) \, . \, I' \text{ is "an update of } I \text{ by } \Gamma" \, \} \, .$$

According to this definition, in order to determine "theory update", it suffices to define when an interpretation I' is an update of an interpretation I by a theory Γ. Accordingly, in this paper we focus exclusively on "interpretation update". However, by contrast to the work cited above, we investigate a more general case of update by means of sets $\mathcal{R}$ of inference rules, instead of sets Γ of formulae.

[1] Partially supported by National Science Foundation grant #IRI-9313061.
[2] Partially supported by an IBM Graduate Fellowship and NSF grant #IRI-9306751.

The first part of the paper is devoted to the study of *revision programs*, introduced by Marek and Truszczyński [MT93, MT94, MT95a] to formalize interpretation update in a language similar to the language of logic programming. Revision programs are essentially sets of logic program rules, which can be interpreted as inference rules and used to update interpretations. Marek and Truszczyński proved that *logic programs with stable semantics* are embeddable into revision programs. We show that, conversely, there is a simple embedding of revision programs into logic programs with stable semantics.[3] Thus the two formalisms are equivalent. We demonstrate that various properties of revision programs are easily derived from this translation and from known properties of logic programs. Moreover, the embedding of revision programs into logic programs suggests a new definition for the more general class of *disjunctive revision programs*, obtained by translating such programs into disjunctive logic programs.

While revision programs provide a useful and natural definition of interpretation update, they are limited to a fairly restricted set of update rules and thus are not sufficiently expressive to capture more complex interpretation updates which may be described by arbitrarily complex formulae, or, more generally, by arbitrary inference rules. Accordingly, in the second part of the paper we introduce the notion of *rule update* — interpretation update by arbitrary sets of inference rules. The proposed formalism is not only more general and expressive than revision programming, but also has a very simple and natural definition.

We show that Winslett's [Win88] approach to update by means of arbitrary sets of formulae corresponds to a simple subclass of rule update. More generally, we investigate how the "directionality" of inference rules contributes to the expressiveness of rule update. Finally, we specify a remarkably simple embedding of rule update into *default logic*, obtained by augmenting the original update rules with inertia axioms analogous to those used in the translation of revision programs into logic programs. The translation into default logic provides a bridge between our newly introduced formalism and a well-known non-monotonic formalism. It also suggests a possible generalization of rule updates to "disjunctive rule updates" based on *disjunctive default logic* [GLPT91].

The introduction of rule update provides a new framework for *interpretation updates* and thus also for *theory updates*. In spite of its great simplicity, rule update constitutes a powerful and expressive mechanism which determines updates of theories by arbitrarily complex sets of inference rules and is applicable to various knowledge domains. For example, in [MT95b] McCain and Turner apply rule update to the problem of reasoning about the effects of actions.

Preliminary definitions appear in Section 2. In Sections 3 & 4 we specify a simple embedding of revision programming into logic programming and show that basic results obtained by Marek and Truszczyński for revision programs are easily deduced from this embedding, using known properties of logic programs. In Sections 5–7 we define rule update, compare it to update by means of propositional formulae, investigate how the "directionality" of inference rules influences rule update, and specify a simple embedding of rule update into default logic.

---

[3] Chitta Baral [Bar94] independently found a somewhat more complex embedding.

2 Preliminary Definitions

Given a propositional language $\mathcal{K}$, we represent an interpretation of $\mathcal{K}$ as a maximal consistent set of literals from $\mathcal{K}$. For any set Γ of formulae from $\mathcal{K}$, by $Cn(\Gamma)$ we denote the least set of formulae from $\mathcal{K}$ that contains Γ and is also closed under propositional logic.

Inference rules over $\mathcal{K}$ will be written as expressions of the form

$$\frac{\phi}{\psi}$$

where ϕ and ψ are formulae from $\mathcal{K}$.[4] Let $\mathcal{R}$ be a set of inference rules over $\mathcal{K}$, and let Γ be a set of formulae from $\mathcal{K}$. We write

$$\Gamma \vdash_{\mathcal{R}} \phi$$

if ϕ is a formula from $\mathcal{K}$ belonging to the least set of formulae containing Γ that is closed under propositional logic and also closed with respect to the inference rules in $\mathcal{R}$. We say that Γ is *closed under* $\mathcal{R}$ if $\Gamma = \{\phi : \Gamma \vdash_{\mathcal{R}} \phi\}$.

A *default rule* over $\mathcal{K}$ is an expression of the form

$$\frac{\alpha : \beta_1, \ldots, \beta_n}{\gamma} \tag{1}$$

where all of $\alpha, \beta_1, \ldots, \beta_n, \gamma$ are formulae from $\mathcal{K}$. For a default rule r as in (1), we define $prerequisite(r) = \alpha$, $justifications(r) = \{\beta_1, \ldots, \beta_n\}$, and $consequent(r) = \gamma$. If $justifications(r)$ is empty, we identify r with the corresponding inference rule $\frac{\alpha}{\gamma}$. If in addition $prerequisite(r) = True$, we sometimes simply write γ.

A *default theory* over $\mathcal{K}$ is a set of default rules over $\mathcal{K}$. Let D be a default theory over $\mathcal{K}$ and let E be a set of formulae from $\mathcal{K}$. We define the *reduct of D with respect to E*, denoted by D^E, as follows.

$$D^E = \left\{ \frac{prerequisite(r)}{consequent(r)} : \exists r \in D \, . \, \forall \beta \in justifications(r) \, . \, \neg\beta \notin E \right\}$$

We say that E is an *extension* of D if E is the least set closed under D^E.[5]

A *logic program* over $\mathcal{K}$ is a set of *logic program rules* which are expressions of the form

$$A \leftarrow B_1, \ldots, B_m, \text{not } C_1, \ldots, \text{not } C_n \tag{2}$$

where A, B_i and C_j are atoms from $\mathcal{K}$ and $0 \leq m, n$. If $n = 0$ for all program rules, then the program is called *positive*. By interpreting lists of atoms in rule

---

[4] At times we will find it convenient to identify a propositional formula ϕ with the inference rule $\frac{True}{\phi}$.

[5] The definition of a default theory is due to Reiter [Rei80]. The definition of an extension given above follows [GLPT91], and is equivalent to Reiter's definition.

bodies as conjunctions of atoms, we can identify a positive logic program with a propositional Horn theory.[6]

Definition 1. (Stable Models) [GL88] Let P be a logic program over a language $\mathcal{K}$ and let M be an interpretation of $\mathcal{K}$. By the *quotient of P modulo M* we mean the positive logic program $\frac{P}{M}$ obtained from P by:

- removing from P all rules which contain a negative premise "not C" such that C is true in M, and
- deleting all negative premises "not C" from all the remaining rules of P.

Since program $\frac{P}{M}$ is a Horn theory, it has a unique least model $Least(\frac{P}{M})$. The interpretation M is called a *stable model* of the program P if $M = Least(\frac{P}{M})$. □

Definition 2. (Extended Logic Programs) [GL90] (see also [Prz94]) Let $\mathcal{K}$ be a propositional language. Let $\mathcal{K}^*$ be an *extended propositional language* obtained from $\mathcal{K}$ by augmenting it with new propositional letters $\sim A$, for some (or all) propositional letters A in $\mathcal{K}$. The new propositional symbols $\sim A$ are called *strong* (or "classical") negation of A. A logic program P over the extended language $\mathcal{K}^*$, which contains at least one of the atoms $\sim A$, is called an *extended logic program*. We call a stable model of P over $\mathcal{K}^*$ an *extended stable model* of P if there is no atom $A \in \mathcal{K}$ such that both A and $\sim A$ are true in M. □

3 Embedding Revision Programs into Logic Programs

Revision programs were introduced by Marek and Truszczyński in a series of papers [MT93, MT94, MT95a] in order to formalize interpretation update in a language similar to the language of logic programming. In [MT93] they showed that logic programs with stable semantics are embeddable into revision programs, and wrote: "The question whether there is a simple representation of revision programming in terms of logic programming remains open."[7] In this section we specify a remarkably simple embedding of revision programs into logic programs with stable semantics, and into extended logic programs with stable semantics. Consequently, the two formalisms are precisely equivalent. In the next section we demonstrate how one can easily derive various properties of revision programs from this translation and from known properties of logic programs.

---

[6] Before continuing, we recall the fact that propositional programs and default theories can be viewed as instantiated versions of programs and theories with variables. Thus the results in this paper apply to the general case.

[7] Somewhat paradoxically, they also said "revision programming is significantly more expressive. While logic programs do not allow one to state that an atom must be absent from a model, revision programs explicitly talk about deletions" [MT93].

3.1 Revision Programs

We first recall the definition of revision programs. Following [MT94] we fix a countable set U.

Definition 3. (Revision Programs) [MT94] A revision in-rule or, simply, an in-rule, is any expression of the form

$$in(p) \leftarrow in(q_1), \ldots, in(q_m), out(s_1), \ldots, out(s_n), \tag{3}$$

where p, q_i, $1 \leq i \leq m$, and s_j, $1 \leq j \leq n$, are all in U and m, $n \geq 0$. A revision out-rule or, simply, an out-rule, is any expression of the form

$$out(p) \leftarrow in(q_1), \ldots, in(q_m), out(s_1), \ldots, out(s_n), \tag{4}$$

where p, q_i, $1 \leq i \leq m$, and s_j, $1 \leq j \leq n$, are all in U and m, $n \geq 0$.

A collection of in-rules and out-rules is called a revision program. Any subset B of U is called a knowledge base. □

Clearly, revision programs can be syntactically viewed as positive logic programs (or as propositional Horn theories). However, as we will see below, they are given a special revision semantics which differs significantly from the least model semantics of positive logic programs. We first need the definition of the necessary change determined by a revision program.

Definition 4. (Necessary Change) [MT94] Let P be a revision program with least model M. The necessary change determined by P is the pair (I, O), where $I = \{q : in(q) \in M\}$ and $O = \{q : out(q) \in M\}$. The revision program is called coherent if $I \cap O = \emptyset$. □

Now we are ready to define the so-called P-justified revisions.

Definition 5. (P-Justified Revision) [MT94] Suppose that P is a revision program, B_I is the initial knowledge base and B_R is the revised knowledge base. The reduct of P with respect to (B_I, B_R) is defined[8] as the revision program $P_{B_R}|B_I$ obtained from P by:

- removing from the body of each rule all atoms $in(a)$ such that $a \in B_I \cap B_R$, and all atoms $out(a)$ such that $a \notin B_I \cup B_R$;
- removing from P every rule of type (3) or (4) such that $q_i \notin B_R$, for some i, $1 \leq i \leq m$, or, $s_j \in B_R$, for some j, $1 \leq j \leq n$.

If (I, O) is the necessary change determined by $P_{B_R}|B_I$ and $B_R = (B_I \cup I) - O$ and $P_{B_R}|B_I$ is coherent, then B_R is called a P-justified revision of B_I. □

---

[8] Although the definition given below differs slightly from the one given in [MT94] it is easily seen to be equivalent.

3.2 Translating revision programs into logic programs

We first show how to embed revision programs into logic programs with stable semantics and then we produce an equivalent embedding into extended logic programs. In order to translate revision programs into logic programs we first consider a propositional language $\mathcal{K}$ whose set of propositional letters consists of $\{in(q) : q \in U\} \cup \{out(q) : q \in U\} \cup \{in_I(q) : q \in U\} \cup \{out_I(q) : q \in U\}$.

Definition 6. (Translating revision programs into logic programs) The translation of the revision program P and the initial knowledge base B_I into a logic program is defined as the logic program $\mathcal{P}(P, B_I) = P_I \cup P_N \cup P$ over $\mathcal{K}$ consisting of the following three subprograms:

Initial Knowledge Rules P_I: All $q \in B_I$ are initially in and all $s \notin B_I$ are initially out:

$$in_I(q) \leftarrow \tag{5}$$
$$out_I(s) \leftarrow \tag{6}$$

for all $q \in B_I$ and all $s \notin B_I$.

Inertia Rules P_N: If q was initially in (respectively, out) then after revision it remains in (respectively, out) unless it was forced out (respectively, in):

$$in(q) \leftarrow in_I(q), \text{not } out(q) \tag{7}$$
$$out(q) \leftarrow out_I(q), \text{not } in(q) \tag{8}$$

for all $q \in U$.

Revision Rules P: All the in-rules and out-rules that belong to the original revision program P.

A stable model M of $\mathcal{P}(P, B_I)$ is called *coherent* if it does not contain both $in(q)$ and $out(q)$, for any $q \in U$. $\qquad\square$

Observe that the above translation is quite simple. It preserves the original revision program P and adds to it the set of facts representing the initial state B_I and two simple inertia axiom schemas stating that things do not change from one state to another unless they are forced to.[9]

Example 1. Consider the revision program

$$P = \{ out(a) \leftarrow in(b) \}$$

and the initial knowledge base $B_I = \{a, b\}$. Its translation $\mathcal{P}(P, B_I)$ into a logic program consists of P together with the initial conditions and inertia axioms[10]:

$$in_I(a)$$
$$in_I(b)$$
$$in(a) \leftarrow in_I(a), \text{not } out(a)$$
$$in(b) \leftarrow in_I(b), \text{not } out(b). \qquad\square$$

---

[9] By comparison, the translation in [Bar94] is complicated by the introduction of an auxiliary abnormality predicate.

[10] Notice that inertia axioms for $out()$ can be skipped.

Example 2. Consider now the revision program P:

$$in(a) \leftarrow out(b)$$
$$in(b) \leftarrow out(a)$$

and the initial knowledge base $B_I = \{\}$. Its translation $\mathcal{P}(P, B_I)$ into a logic program consists of P together with the initial conditions and inertia axioms[11]:

$$out_I(a)$$
$$out_I(b)$$
$$out(a) \leftarrow out_I(a), \text{not } in(a)$$
$$out(b) \leftarrow out_I(b), \text{not } in(b). \qquad \square$$

We now prove that the translation specified in Definition 6 indeed yields an embedding of revision programming into logic programming under the stable semantics.

Theorem 7. (Embedding of revision programs into logic programs) *Let P be a revision program and B_I be the initial knowledge base. There is a one-to-one correspondence between P-justified revisions of B_I and coherent stable models of its translation $\mathcal{P}(P, B_I)$ into a logic program.*

More precisely, to every P-justified revision B_R of B_I there corresponds a unique stable model M of $\mathcal{P}(P, B_I)$ such that:

$$B_R = \{q : in(q) \in M\} \qquad (9)$$
$$U - B_R = \{q : out(q) \in M\} \qquad (10)$$

and, conversely, for every coherent stable model M of $\mathcal{P}(P, B_I)$ the set $B_R = \{q : in(q) \in M\}$ is a P-justified revision of B_I. $\qquad \square$

Proof. ($\Leftarrow$) Suppose that M is a coherent stable model of $\mathcal{P}(P, B_I)$ and let $B_R = \{q : in(q) \in M\}$. We have to show that B_R is a P-justified revision of B_I.

By Definition 1, M is the least model of the positive logic program $Q = \frac{\mathcal{P}(P,B_I)}{M}$, namely the quotient of $\mathcal{P}(P, B_I)$ modulo M. Since both the initial knowledge rules P_I and the original revision rules P in $\mathcal{P}(P, B_I) = P_I \cup P_N \cup P$ are positive, only the inertia rules P_N :

$$in(q) \leftarrow in_I(q), \text{not } out(q) \qquad (11)$$
$$out(q) \leftarrow out_I(q), \text{not } in(q) \qquad (12)$$

will be affected by the quotient transformation and therefore $Q = P_I \cup \frac{P_N}{M} \cup P$.

Define $B'_R = \{q : out(q) \in M\}$ and let $B'_I = U - B_I$. According to Definition 1, in order to construct the quotient $\frac{P_N}{M}$ we have to remove from P_N all the inertia clauses (11) such that $q \in B'_R$ and all inertia clauses (12) such that $q \in B_R$. Subsequently, we have to remove all the negative premises from the

---

[11] Notice that now the inertia axioms for $in()$ can be skipped.

remaining clauses of P_N. As a result of the quotient transformation we obtain therefore the program $\frac{P_N}{M}$ consisting of rules:

$$in(q) \leftarrow in_I(q), \quad \text{for all } q \notin B'_R \tag{13}$$

$$out(q) \leftarrow out_I(q), \quad \text{for all } q \notin B_R. \tag{14}$$

Let us now observe that $B'_R = U - B_R$. Indeed, since M is coherent, the sets B_R and B'_R are disjoint. Suppose that there is a q such that $q \notin B_R$ and $q \notin B'_R$. Then both clauses $in(q) \leftarrow in_I(q)$ and $out(q) \leftarrow out_I(q)$ belong to the quotient program Q. Since we must either have $q \in B_I$ or $q \in B'_I$ and since M is the least model of Q we conclude that either $in(q)$ or $out(q)$ must belong to M, which contradicts our assumption that $q \notin B_R$ and $q \notin B'_R$.

We can also ignore all clauses whose premises are false in M and remove premises which are already known to be true in M. Consequently, the quotient $\frac{P_N}{M}$ of the set of inertia rules effectively consists of all clauses (facts) of the form:

$$in(q) \leftarrow, \quad \text{for all } q \in B_R \cap B_I \tag{15}$$

$$out(q) \leftarrow, \quad \text{for all } q \in B'_R \cap B'_I. \tag{16}$$

In addition, the quotient program Q contains the initial knowledge rules P_I:

$$in_I(q) \leftarrow, \quad \text{for all } q \in B_I \tag{17}$$

$$out_I(s) \leftarrow, \quad \text{for all } s \in B'_I \tag{18}$$

and all the original revision program in-rules (3) and out-rules (4) in P.

We now show that B_R is a P-justified revision of B_I. According to Definition 5, in order to compute the reduct $P_{B_R}|B_I$ of P we first have to remove from the body of each revision rule in P all atoms $in(q)$, such that $q \in B_I \cap B_R$, and all atoms $out(s)$, such that $s \in B'_I \cap B'_R$. Notice, that these are precisely the atoms that must be true in M due to the rules (15) and (16).

Subsequently, we remove from the (already reduced) revision program every rule of type (3) or (4) such that $q_i \notin B_R$, for some i, $1 \leq i \leq m$, or, $s_j \in B_R$, for some j, $1 \leq j \leq n$ thus obtaining the reduct $P_{B_R}|B_I$. Notice that by doing so we are removing from P those rules whose premises are false in M. As a result, the stable model M remains the least model of the reduced quotient program $Q^* = P_I \cup \frac{P_N}{M} \cup P_{B_R}|B_I$.

Let M_0 be the least model of the reduct $P_{B_R}|B_I$ and let (I, O) be the necessary change of $P_{B_R}|B_I$, i.e. $I = \{q : in(q) \in M_0\}$ and $O = \{q : out(q) \in M_0\}$. The program $Q^* = P_I \cup \frac{P_N}{M} \cup P_{B_R}|B_I$ consists of three independent parts: the initial knowledge rules (17) and (18), the (reduced) inertia axioms (15) and (16) and the reduct $P_{B_R}|B_I$ which no longer contains any premises from the other two parts. Consequently, the set of atoms that belong to the stable model M, which is the least model of this reduced quotient program Q^*, consists of:

- $\{in_I(q) : q \in B_I\} \cup \{out_I(q) : q \in B'_I\}$,
- $\{in(q) : q \in B_R \cap B_I\} \cup \{out(q) : q \in B'_R \cap B'_I\}$,
- $\{in(q) : q \in I\} \cup \{out(q) : q \in O\}$.

This shows that:

$$B_R = \{q : in(q) \in M\} = (B_R \cap B_I) \cup I, \tag{19}$$
$$B'_R = U - B_R = \{q : out(q) \in M\} = (B'_R \cap B'_I) \cup O. \tag{20}$$

Since $P_{B_R}|B_I$ is coherent by assumption, in order to verify that B_R is a P-justified revision of B_I it suffices to establish that $B_R = (B_I \cup I) - O$. Since B_R is disjoint from O, we obtain $B_R = (B_R \cap B_I) \cup I = ((B_R \cap B_I) \cup I) - (B_R \cap O) \subseteq (B_I \cup I) - O$. To see that $B_R \supseteq (B_I \cup I) - O$ it suffices to note that if $q \in (B_I \cup I) - O$ then either $q \in I$ or $q \in B_I - O$ which, in view of (19) and (20), implies that $q \in B_R$.

($\Rightarrow$) The proof in the opposite direction is exactly analogous. The steps of the above proof can be reversed to establish the converse implication. $\square$

Example 3. Consider the revision program P discussed in Example 1 and its translation $P' = \mathcal{P}(P, B_I)$ into a logic program. One easily checks that P' has a unique stable model, which contains only the atoms:[12]

$$\{out(a), \ in(b)\}$$

and therefore corresponds to the unique P-justified revision $B_R = \{b\}$. $\square$

Example 4. Consider now the revision program P discussed in Example 2 and its translation $P' = \mathcal{P}(P, B_I)$ into a logic program. One easily checks that P' has two stable models M_1 and M_2, which contain only the atoms[13]:

$$M_1 = \{out(a), \ in(b)\}$$
$$M_2 = \{in(a), \ out(b)\}$$

and therefore correspond to the two P-justified revisions $\{b\}$ and $\{a\}$. $\square$

3.3 Translating revision programs into extended logic programs

We now observe that the translation described above can be easily expressed in the language of extended logic programs with strong (or "classical") negation. According to Definition 2, in order to translate revision programs into extended logic programs we first consider a propositional language $\mathcal{K}^*$, whose set of propositional letters contains $\{in(q) : q \in U\} \cup \{in_I(q) : q \in U\}$ as well as their strong negations $\{\sim in(q) : q \in U\} \cup \{\sim in_I(q) : q \in U\}$.

Definition 8. (Translating revision programs into extended logic programs) The translation of the revision program P and the initial knowledge base B_I into an extended logic program is defined as the logic program $\mathcal{P}^*(P, B_I)$ over $\mathcal{K}^*$ consisting of the following three sets of rules:

---

[12] In addition to the initial state atoms $in_I(a)$, $in_I(b)$.
[13] In addition to the initial state atoms $out_I(a)$, $out_I(b)$.

Initial Knowledge Rules: All $q \in B_I$ are initially in and all $s \notin B_I$ are initially not in:

$$in_I(q) \leftarrow \tag{21}$$

$$\sim in_I(s) \leftarrow \tag{22}$$

for all $q \in B_I$ and all $s \notin B_I$.

Inertia Rules: If q was initially in (respectively, not in) then after revision it remains in (respectively, not in) unless forced to be not in (respectively, in):

$$in(q) \leftarrow in_I(q), \text{not} \sim in(q) \tag{23}$$

$$\sim in(q) \leftarrow \sim in_I(q), \text{not}\, in(q) \tag{24}$$

for all $q \in U$.

Revision Rules: All the in-rules and out-rules of the original revision program P, with propositions $out(q)$ replaced everywhere by $\sim in(q)$. □

The only difference between the two embeddings is that we are using the strong negation $\sim in()$ of $in()$ instead of $out()$. Revision programs are embeddable into extended logic programs with strong (or "classical") negation.

Theorem 9. (Embedding of revision programs into extended logic programs) *Let P be a revision program and B_I be the initial knowledge base. There is a one-to-one correspondence between P-justified revisions of B_I and extended stable models of its translation $\mathcal{P}^*(P, B_I)$ into an extended logic program.*

More precisely, to every P-justified revision B_R of B_I there corresponds a unique extended stable model M of $\mathcal{P}^*(P, B_I)$ such that:

$$B_R = \{q : in(q) \in M\} \tag{25}$$

$$U - B_R = \{q : \sim in(q) \in M\} \tag{26}$$

and, conversely, for every extended stable model M of $\mathcal{P}^*(P, B_I)$ the set $B_R = \{q : in(q) \in M\}$ is a P-justified revision of B_I. □

Theorem 9 follows immediately from Theorem 7 and Definition 2.

Example 5. The translation $P'' = \mathcal{P}^*(P, B_I)$ of the revision program P discussed in Example 2 into an extended logic program is given by:

$$in(a) \leftarrow \sim in(b)$$

$$in(b) \leftarrow \sim in(a)$$

together with the initial conditions and inertia axioms[14]:

$$\sim in_I(a)$$

$$\sim in_I(b)$$

$$\sim in(a) \leftarrow \sim in_I(a), \text{not}\, in(a)$$

$$\sim in(b) \leftarrow \sim in_I(b), \text{not}\, in(b).$$

---

[14] Notice that now the inertia axioms for $in()$ can be skipped.

One easily checks that P'' has two stable models

$$M_1 = \{\sim in(a),\ in(b),\ \sim in_I(a),\ \sim in_I(b)\}$$
$$M_2 = \{in(a),\ \sim in(b),\ \sim in_I(a),\ \sim in_I(b)\}$$

which correspond to the two P-justified revisions $\{b\}$ and $\{a\}$. $\qquad \square$

4 Properties of Revision Programs

Many of the results involving revision programs obtained by Marek and Truszczyński in [MT93, MT94, MT95a] become simple consequences of the embeddability of revision programs into logic programs with stable semantics. For example, the complexity results obtained in [MT95a, Theorem 4.2], stating the NP-completeness of some problems involving the computation of P-justified revisions, can be easily seen to follow from similar results already known about the computation of stable models and from the equivalence of the two formalisms. Below we give some other examples illustrating this claim. We also show that embeddability of revision programs into logic programs leads to a natural extension of revision programming to the *disjunctive* case.

First, let us observe that the fact that the translation $\mathcal{P}(P, B_I)$ of a revision program is completely symmetric with respect to the atoms *in* and *out* immediately yields the following result:

Theorem 10. *[MT94] Let P be a revision program and let B_I be a knowledge base. Then B_R is a P-justified revision of B_I if and only if $U - B_R$ is a P^D justified revision of $U - B_I$ where P^D is a dual of the program P obtained by simultaneously replacing everywhere in by out and vice versa.* $\qquad \square$

The following result from [MT94] stating that logic programs with stable semantics are embeddable into revision programs becomes an easy consequence:

Theorem 11. *[MT94] Let P be a logic program consisting of rules:*

$$p \leftarrow q_1, \ldots, q_m, \text{not } s_1, \ldots, \text{not } s_n$$

and let $R(P)$ be the revision program obtained by replacing each rule of P with the in-rule:

$$in(p) \leftarrow in(q_1), \ldots, in(q_m), out(s_1), \ldots, out(s_n).$$

Then M is a stable model of P if and only if its set of atoms B_R is an $R(P)$-justified revision of $B_I = \emptyset$. $\qquad \square$

Proof. By Definition 6, the translation $\mathcal{P}(R(P), \emptyset)$ of the revision program $R(P)$ into a logic program consists of $R(P)$ itself and the initial knowledge rules:

$$out_I(q) \leftarrow, \text{ for all } q \in U$$

together with the inertia rules[15]:

$$out(q) \leftarrow \text{not } in(q), \text{ for all } q \in U.$$

After performing a single step of partial evaluation on the premises $out(s_j)$ of rules from $R(P)$ (by using the above inertia rules), the rules of $R(P)$ become equivalent to:

$$in(p) \leftarrow in(q_1), \ldots, in(q_m), \text{not } in(s_1), \ldots, \text{not } in(s_n)$$

and thus they are equivalent (up to renaming) to the rules of the original program P. Theorem 7 now easily implies the equivalence between stable models of P and justified revisions of $P(R)$. □

The embeddability of revision programs into default logic follows from Theorem 9, given the well-known equivalence between extended logic programming (under the stable semantics) and the corresponding subset of default logic [GL90]. Finally, since revision programming is equivalent to logic programming under stable semantics, computational methods developed for the stable semantics (or, perhaps, for its approximations, such as the well-founded semantics) can be used to provide a query answering mechanism for revision programming.

4.1 Extending revision programming to disjunctive programs

In [MT94] the authors propose an extension of revision programming to *disjunctive* revision programs consisting of rules of the following form.

$$in(p_1) \vee \cdots \vee in(p_k) \vee out(r_1) \vee \cdots \vee out(r_l) \leftarrow in(q_1), \ldots, in(q_m), out(s_1), \ldots, out(s_n)$$

However, the proposed definition exhibits what may be undesirable behavior. For example, given the disjunctive revision program $P = \{in(a) \vee in(b)\}$ and initial knowledge base $B_I = \{a\}$ we obtain $\{a, b\}$ as one of two P-justified revisions of B_I, which seems to violate the principle of minimization of updates.[16]

However, it is easy to define a natural semantics for disjunctive revision programs P by first translating them, using Definition 6, into logic programs and then applying a suitable semantics to the resulting disjunctive logic program $\mathcal{P}(P, B_I)$. In particular, one can use the *disjunctive stable semantics*, originally introduced in [GL90, Prz91].

By utilizing the translation of revision programs into logic programs, default logic or some other non-monotonic formalism, we eliminate the need for the introduction of an entirely new formalism for revision programs but rely instead on already well-established and thoroughly investigated non-monotonic formalisms.

---

[15] Notice that we can skip the rules for *in* and remove the premises $out_I(q)$.

[16] Mirek Truszczyński has agreed (personal communication) that this example shows that the original definition fails to capture the intended intuition.

5 Rule Update

The revision programs discussed in the first part of the paper are sets of revision in-rules and out-rules, which can be interpreted as inference rules and are used to update interpretations. In the second part of the paper, we introduce a more general approach to update by means of inference rules, called rule update. Rule update has a simple fixpoint definition which not only extends revision programming, but also includes as a special case the approach to update by means of formulae introduced by Winslett in [Win88]. Furthermore, rule update has a remarkably simple embedding into default logic, using essentially the same inertia rules used in Section 3 to embed revision programming into logic programming.

Definition 12. (Rule Update) Let $\mathcal{R}$ be a set of inference rules. Let I, I' be interpretations. We say that I' is an *update of I by $\mathcal{R}$* if

$$I' = \{L : I \cap I' \vdash_{\mathcal{R}} L\}$$

where L ranges over literals. □

It follows that I' is an update of I by $\mathcal{R}$ if and only if the following two conditions are met: (i) $\forall L \in I'$, $I \cap I' \vdash_{\mathcal{R}} L$ and (ii) $Cn(I')$ is closed under $\mathcal{R}$.

Example 6. Consider the following.

$$I_1 = \{a, b, c\} \quad \mathcal{R}_1 = \left\{ \frac{a}{\neg b \vee \neg c} \right\} \quad I_2 = \{a, \neg b, c\}$$

First we will show that I_2 is an update of I_1 by $\mathcal{R}_1$. Notice that $I_1 \cap I_2 = \{a, c\}$ and that $I_1 \cap I_2 \vdash_{\mathcal{R}_1} \neg b$. So for all literals $L \in I_2$, $I_1 \cap I_2 \vdash_{\mathcal{R}_1} L$. And since $Cn(I_2)$ is closed under $\mathcal{R}_1$, we have shown that I_2 is an update of I_1 by $\mathcal{R}$. A symmetric argument shows that the interpretation $\{a, b, \neg c\}$ is also an update of I_1 by $\mathcal{R}$. On the other hand, if we take $I_3 = \{\neg a, b, c\}$, then we have $I_1 \cap I_3 = \{b, c\}$; and we see that $I_1 \cap I_3 \not\vdash_{\mathcal{R}_1} \neg a$. So I_3 is not an update of I_1 by $\mathcal{R}_1$. One can similarly show that the interpretation $\{a, \neg b, \neg c\}$ is not an update of I_1 by $\mathcal{R}$. □

The following theorem shows that revision programming can be seen as the special case of rule update in which every rule in $\mathcal{R}$ has the form

$$\frac{L_1 \wedge \cdots \wedge L_n}{L_0}$$

where each L_i $(0 \leq i \leq n)$ is a literal.

Theorem 13. (Embedding revision programming in rule update) *Let P be a revision program, with initial and revised knowledge bases B_I and B_R. Let $\mathcal{K}$ be the propositional language with atoms U. Let $\mathcal{R}$ be the set of inference rules over $\mathcal{K}$ obtained by replacing each in-rule (3) in P with the corresponding inference rule $\frac{q_1 \wedge \cdots \wedge q_m \wedge \neg s_1 \wedge \cdots \wedge \neg s_n}{p}$, and similarly replacing each out-rule (4) in P with the corresponding inference rule $\frac{q_1 \wedge \cdots \wedge q_m \wedge \neg s_1 \wedge \cdots \wedge \neg s_n}{\neg p}$. Let I, I' be the interpretations of $\mathcal{K}$ such that $I \cap U = B_I$ and $I' \cap U = B_R$. Then B_R is a P-justified revision of B_I if and only if I' is an update of I by $\mathcal{R}$.* □

A proof sketch for Theorem 13 appears in Section 7.

6 Properties of Rule Update

In this section we show that rule update includes as a special case the approach to update by means of formulae introduced by Winslett [Win88].[17] More generally, we investigate the role of the "directionality" of inference rules in rule update.

Definition 14. (Formula-update) Given interpretations I, I', I'', we say that I' is *closer to I than I'' is* if $I'' \cap I$ is a proper subset of $I' \cap I$.

Let Γ be a set of formulae. Let I, I' be interpretations. We say that I' is a *formula-update of I by Γ* if I' is a model of Γ such that no model of Γ is closer to I than I' is. $\qquad\square$

In order to compare formula-update and rule update, we introduce the following definition.

Definition 15. Given a set $\mathcal{R}$ of inference rules, we define a corresponding set of formulae $Theory(\mathcal{R})$ as follows.

$$Theory(\mathcal{R}) = \left\{ \phi \supset \psi : \frac{\phi}{\psi} \in \mathcal{R} \right\}$$

$\qquad\square$

Thus, for example, $Theory(\mathcal{R}_1) = \{a \supset \neg b \vee \neg c\}$.

Let $\mathcal{R}$ be a set of inference rules and I an interpretation. Notice that $Cn(I)$ is closed under $\mathcal{R}$ if and only if I is a model of $Theory(\mathcal{R})$. Thus, every update of I by $\mathcal{R}$ is a model of $Theory(\mathcal{R})$. In fact, we have the following stronger result.

Proposition 16. *Let $\mathcal{R}$ be a set of inference rules. Let I be an interpretation. Every update of I by $\mathcal{R}$ is a formula-update of I by $Theory(\mathcal{R})$.* $\qquad\square$

Proof. Assume that I' is an update of I by $\mathcal{R}$. So I' is a model of $Theory(\mathcal{R})$. Let I'' be a model of $Theory(\mathcal{R})$ such that $I' \cap I \subseteq I'' \cap I$. We need to show that $I'' = I'$. Since I' and I'' are both interpretations, it's enough to show that $I' \subseteq I''$.

$$
\begin{aligned}
I' &= \{L : I \cap I' \vdash_{\mathcal{R}} L\} && \{ \text{ } I' \text{ is an update of } I \text{ by } \mathcal{R} \text{ } \}\\
&\subseteq \{L : I \cap I'' \vdash_{\mathcal{R}} L\} && \{ \text{ } I' \cap I \subseteq I'' \cap I \text{ } \}\\
&\subseteq \{L : I'' \vdash_{\mathcal{R}} L\} && \{ \text{ } I'' \cap I \subseteq I'' \text{ } \}\\
&= I'' && \{ \text{ } I'' \text{ is a model of } Theory(\mathcal{R}) \text{ } \}
\end{aligned}
$$

$\qquad\square$

The converse of Proposition 16 doesn't hold in general. For instance, in Example 6 we saw that I_3 is not an update of I_1 by $\mathcal{R}_1$, and yet it is easy to verify that I_3 is a formula-update of I_1 by $Theory(\mathcal{R}_1)$.

On the other hand, the following proposition shows that if every inference rule in $\mathcal{R}$ has the form $\frac{True}{\phi}$ then the updates of I by $\mathcal{R}$ will be exactly the formula-updates of I by $Theory(\mathcal{R})$. Notice that this result implies that rule update generalizes formula-update.

---

[17] McCain and Turner [MT95b] discuss this comparison at some length, in the framework of reasoning about action. Propositions 16–18 below are essentially identical to Propositions 2–4 from [MT95b].

Proposition 17. *Let $\mathcal{R}$ be a set of inference rules, each of the form $\frac{True}{\phi}$. Every formula-update of an interpretation I by Theory($\mathcal{R}$) is an update of I by $\mathcal{R}$.* □

Proof. Assume that I' is a formula-update of I by *Theory*($\mathcal{R}$). Let I'' be a model of $(I \cap I') \cup$ *Theory*($\mathcal{R}$). Of course I'' is a model of *Theory*($\mathcal{R}$). Also $I' \cap I \subseteq I''$, so $I' \cap I \subseteq I'' \cap I$. Since no model of *Theory*($\mathcal{R}$) is closer to I than I' is, we can conclude that $I'' = I'$. So we have shown that I' is the only model of $(I \cap I') \cup$ *Theory*($\mathcal{R}$). Thus, $I' = \{L : (I \cap I') \cup$ *Theory*($\mathcal{R}$) $\vdash L\}$. Due to the special form of the rules in $\mathcal{R}$, it is clear that for every formula ϕ, we have $(I \cap I') \cup$ *Theory*($\mathcal{R}$) $\vdash \phi$ if and only if $I \cap I' \vdash_\mathcal{R} \phi$. So $I' = \{L : I \cap I' \vdash_\mathcal{R} L\}$. □

The following straightforward proposition shows that rules of the form $\frac{\phi}{False}$ only eliminate updates. This property reflects the influence of the "directionality" of inference rules in rule update.

Proposition 18. *Let $\mathcal{R}$ be a set of inference rules. Let I, I' be interpretations. For any formula ϕ, I' is an update of I by $\mathcal{R} \cup \left\{ \frac{\phi}{False} \right\}$ if and only if I' is an update of I by $\mathcal{R}$ such that $I' \not\models \phi$.* □

In fact, we see that if every rule in $\mathcal{R}$ has the form $\frac{\phi}{False}$, then I' is an update of I by $\mathcal{R}$ if and only if I is a model of *Theory*($\mathcal{R}$) and $I' = I$. Intuitively speaking, this is the most extreme example of the effect of the directionality of inference rules. At the other extreme, we have seen that if every rule in $\mathcal{R}$ has the form $\frac{True}{\phi}$, then I' is an update of I by $\mathcal{R}$ if and only if I' is a formula-update of I by *Theory*($\mathcal{R}$). Thus, in such cases, the directionality of rules has no effect at all. We might say, informally speaking: the greater the difference between update with respect to $\mathcal{R}$ and formula-update with respect to *Theory*($\mathcal{R}$), the greater the effect of directionality. We explore this remark below.

Definition 19. *Let $\mathcal{R}, \mathcal{R}'$ be sets of inferences rules. $\mathcal{R}'$ is as strong as $\mathcal{R}$ if for all sets Γ of formulae, if Γ is closed under $\mathcal{R}'$ then Γ is closed under $\mathcal{R}$.* □

It is clear that if R' is as strong as $\mathcal{R}$, then for any formula ϕ, $\Gamma \vdash_{\mathcal{R}'} \phi$ whenever $\Gamma \vdash_\mathcal{R} \phi$. We use this fact in the proof of the following proposition.

Proposition 20. *Let $\mathcal{R}, \mathcal{R}'$ be sets of inferences rules, with $Cn(Theory(\mathcal{R})) = Cn(Theory(\mathcal{R}'))$. Let I be an interpretation. If $\mathcal{R}'$ is as strong as $\mathcal{R}$, then an interpretation I' is an update of I by $\mathcal{R}'$ whenever it is an update of I by $\mathcal{R}$.* □

Proof. Assume that $\mathcal{R}'$ is as strong as $\mathcal{R}$ and that I' is an update of I by $\mathcal{R}$. Since $Cn(Theory(\mathcal{R})) = Cn(Theory(\mathcal{R}'))$ and $Cn(I')$ is closed under $\mathcal{R}$, we can conclude that $Cn(I')$ is closed under $\mathcal{R}'$. Let L be a literal in I'. We already know that $I \cap I' \vdash_\mathcal{R} L$. Since $\mathcal{R}'$ is as strong as $\mathcal{R}$, it follows by previous observation that $I \cap I' \vdash_{\mathcal{R}'} L$. □

Now we define an ordering on inference rules that, intuitively, allows us to compare the degree of directionality in (otherwise similar) rules.

Definition 21. Let ϕ, ϕ', ψ, ψ' be propositional formulae.

$$\frac{\phi}{\psi} \preceq \frac{\phi'}{\psi'} \quad \textit{iff} \quad \phi \vdash \phi' \quad \text{and} \quad \vdash (\phi \supset \psi) \equiv (\phi' \supset \psi')$$

□

Example 7. By the preceding definition, we have the following.

$$\frac{a \wedge b \wedge c}{False} \preceq \frac{a \wedge b}{\neg c} \preceq \frac{a}{\neg b \vee \neg c} \preceq \frac{True}{\neg a \vee \neg b \vee \neg c}$$

Roughly speaking, the idea behind this ordering of rules is that, as we move from left to right, the degree of directionality in the rule is lessened, which makes the rule "stronger". Below, we make this claim precise. □

Let $\mathcal{R}$ be a set of inferences rules, and let r and r' be inference rules such that $r \preceq r'$. It is clear that $Cn(Theory(\mathcal{R} \cup \{r\})) = Cn(Theory(\mathcal{R} \cup \{r'\}))$. Moreover, it follows easily from the definitions that $\mathcal{R} \cup \{r'\}$ is as strong as $\mathcal{R} \cup \{r\}$. Thus, we have the following corollary to Proposition 20.

Corollary 22. *Let $\mathcal{R}$ be a set of inferences rules and let r, r' be inference rules such that $r \preceq r'$. Let I be an interpretation. Every update of I by $\mathcal{R} \cup \{r\}$ is also an update of I by $\mathcal{R} \cup \{r'\}$.* □

7 Embedding Rule Update in Default Logic

In this section we specify a simple embedding of rule update into default logic, using essentially the same inertia rules used in Section 3 to embed revision programs into logic programs. The resulting default theories represent the principle of inertia by means of normal defaults, much as Reiter [Rei80] originally hoped.

We begin by extending a propositional language $\mathcal{K}$ in the following manner. For every atom A of $\mathcal{K}$, let A_0 be a new atom. By $\mathcal{K}'$ we denote the language obtained by extending $\mathcal{K}$ with all such new atoms. For any literal L of $\mathcal{K}$, by L_0 we denote the literal obtained from L by replacing the atom A that occurs in L by the new atom A_0. For any interpretation I of $\mathcal{K}$, let $I_0 = \{L_0 : L \in I\}$.

Definition 23. (**Translating rule update into default logic**) Let $\mathcal{R}$ be a set of inference rules over $\mathcal{K}$. Let I be an interpretation of $\mathcal{K}$. By $\mathcal{D}(\mathcal{R}, I)$ we denote the default theory over $\mathcal{K}'$ that consists of the following three sets of rules.

Initial Knowledge Rules: I_0
Inertia Rules: All rules of the form $\frac{L_0 : L}{L}$ where L is a literal of $\mathcal{K}$
Update Rules: $\mathcal{R}$ □

Theorem 24. (**Embedding rule update in default logic**) *Let $\mathcal{R}$ be a set of inference rules over $\mathcal{K}$, with I an interpretation of $\mathcal{K}$. The following hold.*

- *An interpretation I' of $\mathcal{K}$ is an update of I by $\mathcal{R}$ if and only if $Cn(I_0 \cup I')$ is an extension of the default theory $\mathcal{D}(\mathcal{R}, I)$.*

- If E is a consistent extension of $\mathcal{D}(\mathcal{R}, I)$, then there is an update I' of I by $\mathcal{R}$ such that $E = Cn(I_0 \cup I')$. $\quad\square$

Corollary 25. Let $\mathcal{R}$ be a set of inference rules over $\mathcal{K}$. Let I, I' be interpretations of $\mathcal{K}$. Let Lit denote the set of all literals in $\mathcal{K}$. The following hold.

- I' is an update of I by $\mathcal{R}$ if and only if there is an extension E of $\mathcal{D}(\mathcal{R}, I)$ such that $I' = E \cap Lit$.
- For all consistent extensions E of $\mathcal{D}(\mathcal{R}, I)$, $E \cap Lit$ is an update of I by $\mathcal{R}$. $\quad\square$

Proof of Theorem 24. For the first part, let I' be an interpretation of $\mathcal{K}$. Take $E = Cn(I_0 \cup I')$. We have

$$\mathcal{D}(\mathcal{R}, I)^E = \mathcal{R} \cup \left\{ \frac{L_0}{L} : L \in I' \right\} \cup I_0 .$$

Take

$$D' = \mathcal{R} \cup (I \cap I') \cup I_0 .$$

It's not hard to see that E is the extension of $\mathcal{D}(\mathcal{R}, I)^E$ if and only if E is the extension of D'. Furthermore, since I_0 and $\mathcal{R} \cup (I \cap I')$ have no atoms in common, it's clear that E is the extension of D' if and only if $Cn(I')$ is the extension of $\mathcal{R} \cup (I \cap I')$. So we have shown that E is an extension of $\mathcal{D}(\mathcal{R}, I)$ if and only if $Cn(I')$ is the extension of $\mathcal{R} \cup (I \cap I')$. We use this fact in the last step below.

I' is an update of I by $\mathcal{R}$ iff $I' = \{L : I \cap I' \vdash_\mathcal{R} L\}$
     iff $Cn(I') = \{\phi : I \cap I' \vdash_\mathcal{R} \phi\}$
     iff $Cn(I')$ is the extension of $\mathcal{R} \cup (I \cap I')$
     iff $Cn(I_0 \cup I')$ is an extension of $\mathcal{D}(\mathcal{R}, I)$

For the second part, assume that E is a consistent extension of $\mathcal{D}(\mathcal{R}, I)$. We need to show that there is an interpretation I' of $\mathcal{K}$ such that $E = Cn(I_0 \cup I')$. Suppose otherwise; so there is an atom A of $\mathcal{K}$ such that $A \notin E$ and $\neg A \notin E$. But D includes the following two inertia rules.

$$\frac{A_0 : A}{A} \qquad \frac{\neg A_0 : \neg A}{\neg A}$$

It follows that $\mathcal{D}(\mathcal{R}, I)^E$ includes the following two rules.

$$\frac{A_0}{A} \qquad \frac{\neg A_0}{\neg A}$$

Since neither A nor $\neg A$ belong to E, we can conclude that neither A_0 nor $\neg A_0$ belong to I_0, which contradicts the fact that I is an interpretation of $\mathcal{K}$. $\quad\square$

We are now ready to sketch an easy proof of Theorem 13 from Section 5, showing that rule update extends revision programming.

Proof of Proposition 13 (sketch). By Theorem 24, I' is an update of I by $\mathcal{R}$ if and only if $Cn(I_0 \cup I')$ is an extension of $\mathcal{D}(\mathcal{R}, I)$. Take $M = \{in_I(A) : A \in I \cap U\} \cup \{\sim in_I(A) : A \in U \setminus I\} \cup \{in(A) : A \in I' \cap U\} \cup \{\sim in(A) : A \in U \setminus I'\}$. According to Theorem 9, M is an extended stable model of $\mathcal{P}^*(P, I \cap U)$ if and only if $I' \cap U$ is a P-justified revision of $I \cap U$. It remains only to show that $Cn(I_0 \cup I')$ is an extension of $\mathcal{D}(\mathcal{R}, I)$ if and only if M is an extended stable model of $\mathcal{P}^*(P, I \cap U)$. This is reasonably straightforward, using the well-known equivalence between extended logic programming (under the stable semantics) and the corresponding subset of default logic [GL90]. □

7.1 Extending rule update via disjunctive default logic

It is possible to extend the definition of rule update so that it encompasses the disjunctive revision programs considered in Section 4.1. We briefly sketch this possibility below.

In Section 4.1 we considered the fact that disjunctive revision programs can be given a natural semantics based on disjunctive logic programming under the stable semantics. But the treatment of disjunction under this approach differs markedly from the treatment of disjunction in rule update. For instance, we can compare the disjunctive revision program $P = \{\ in(a) \vee out(a)\ \}$ with update by $\mathcal{R} = \{\ a \vee \neg a\ \}$. According to the definition considered in Section 4.1, both $\emptyset$ and $\{a\}$ would be P-justified revisions of $\{a\}$, whereas $\{a\}$ is the only update of $\{a\}$ by $\mathcal{R}$.

Using *disjunctive default logic* [GLPT91], it is possible to incorporate these two different treatments of disjunction into a single definition of update. Disjunctive default logic is an extension of default logic in which the symbol "|" is introduced to denote a second kind of disjunction, corresponding essentially to the treatment of disjunction in disjunctive logic programming. Thus, disjunctive logic programming under the stable semantics is equivalent to a subclass of disjunctive default logic [GLPT91]. We rely on this fact below.

Let $\mathcal{R}$ be a justification-free disjunctive default theory. So $\mathcal{R}$ is a set of rules of the form

$$\frac{\phi}{\psi_1 \mid \ldots \mid \psi_n}$$

where all of $\phi, \psi_1, \ldots, \psi_n$ are propositional formulae. Let I, I' be interpretations. We can say that I' is a *disjunctive update of I by $\mathcal{R}$* if $Cn(I')$ is an extension of the disjunctive default theory $(I \cap I') \cup \mathcal{R}$.

It is straightforward to verify that this definition would extend rule update. Furthermore, if we were to translate "disjunctive rule update" into disjunctive default logic, as in Definition 23, we would obtain a correct embedding of disjunctive update into disjunctive default logic.[18] Finally, it would also be easy to extend Corollary 13 to show that "disjunctive rule update" indeed extends the notion of disjunctive revision programming considered in Section 4.1.

---

[18] The proof of Theorem 24 is easily adapted to show correctness in this more general setting.

Acknowledgements

The authors wish to thank Vladimir Lifschitz, Norman McCain and Halina Przymusinska for many helpful discussions and suggestions. The authors are also grateful for comments from Chitta Baral, Enrico Giunchiglia and G. N. Kartha.

References

[Bar94] Chitta Baral. Rule-based updates on simple knowledge bases. In *Proc. AAAI-94*, pages 136–141, 1994.

[GL88] M. Gelfond and V. Lifschitz. The stable model semantics for logic programming. In R. Kowalski and K. Bowen, editors, *Proceedings of the Fifth Logic Programming Symposium*, pages 1070–1080, Cambridge, Mass., 1988. Association for Logic Programming, MIT Press.

[GL90] M. Gelfond and V. Lifschitz. Logic programs with classical negation. In *Proceedings of the Seventh International Logic Programming Conference, Jerusalem, Israel*, pages 579–597, Cambridge, Mass., 1990. Association for Logic Programming, MIT Press.

[GLPT91] Michael Gelfond, Vladimir Lifschitz, Halina Przymusińska, and Mirosław Truszczyński. Disjunctive defaults. In James Allen, Richard Fikes, and Erik Sandewall, editors, *Principles of Knowledge Representation and Reasoning: Proc. of the Second Int'l Conf.*, pages 230–237, 1991.

[KM91] Hirofumi Katsuno and Alberto O. Mendelzon. On the difference between updating a knowledge base and revising it. In James Allen, Richard Fikes, and Erik Sandewall, editors, *Principles of Knowledge Representation and Reasoning: Proc. of the Second Int'l Conf.*, pages 387–394, 1991.

[MT93] W. Marek and M. Truszczyński. Revision programming. Research report, University of Kentucky, 1993.

[MT94] W. Marek and M. Truszczyński. Revision specifications by means of revision programs. In *Logics in AI. Proceedings of JELIA '94*. Lecture Notes in Artificial Intelligence. Springer-Verlag, 1994.

[MT95a] W. Marek and M. Truszczyński. Revision programming, database updates and integrity constraints. In *Proceedings of the 5th International Conference on Database Theory — ICDT 95*, pages 368–382. Springer-Verlag, 1995.

[MT95b] Norman McCain and Hudson Turner. A causal theory of ramifications and qualifications (extended abstract). In *Working Notes: AAAI Spring Symposium on Extending Theories of Action*, 1995. To appear.

[Prz91] T. C. Przymusinski. Stable semantics for disjunctive programs. *New Generation Computing Journal*, 9:401–424, 1991. (Extended abstract appeared in: Extended stable semantics for normal and disjunctive logic programs. *Proceedings of the 7-th International Logic Programming Conference, Jerusalem*, pages 459–477, 1990. MIT Press.).

[Prz94] T. C. Przymusinski. Static semantics for normal and disjunctive logic programs. *Annals of Mathematics and Artificial Intelligence*, 1994. (in print).

[Rei80] Raymond Reiter. A logic for default reasoning. *Artificial Intelligence*, 13(1,2):81–132, 1980.

[Win88] Marianne Winslett. Reasoning about action using a possible models approach. In *Proc. AAAI-88*, pages 89–93, 1988.

A Sphere World Semantics
for Default Reasoning

João C. P. da Silva[1]   Sheila R. M. Veloso[2]

Programa de Engenharia de Sistemas e Computação
COPPE, Universidade Federal do Rio de Janeiro
Caixa Postal 68511
21945-970 Rio de Janeiro Brazil
E-mail: joaoc@cos.ufrj.br
sheila@cos.ufrj.br

Abstract. The purpose of this paper is to show that we can consider an extension of a Reiter's default theory (W,Δ) as the expansion of the (belief) set W by some maximal set D of consequences of defaults in Δ. We will use the model of revision functions proposed by Grove [13] to characterize the models of the extensions in Reiter's default logic [24], showing that the class of models we obtain in the special case when a revision is an expansion (i.e., a new sentence A is added to a belief set K and no sentence in K is deleted), is the class of models of some extension in Reiter's default logic. Furthermore, we will show that the class of models in Poole's system for default reasoning can be characterized in the same way.

1 - Introduction

The study of belief revision and non-monotonic logic has shown that there is a close relation between such approaches. As Makinson and Gärdenfors [20] pointed out, this appears more clearly when we compare the general conditions on nonmonotonic inference operations to the theory revision operations, although on the level of specific constructions such correspondence is not always exact. In [20], they showed that the relationship between Poole's default logic [23] and the full meet revision of Alchourron and Makinson [2] is very close to identity.

Other results about such relationship can be found in Boutilier's work ([6] and [7]). He presented a family of modal logics for reasoning about belief revision in which the process of revising a knowledge base by some sentence is represented with a conditional connective. The modal framework used in there allows to demonstrate the connections between revision, default reasoning and autoepistemic logic.

Reiter's default logic [24] is an example of nonmonotonic logic without a counterpart in the logic of theory of change. A Reiter's default theory is defined as the pair (W,Δ) where W is a set of facts and Δ is a set of defaults. Applying a default (A :

---

[1]The author is sponsered by CAPES fellowship.
[2]The author is partially sponsored by the Brazilian National Research Agency (CNPq).

B / C) ∈ Δ to the set W means that we add the sentence C to the set W under some conditions (i.e., if A is true and B is consistent with some extension of the set W). Remember that no sentence in the set W should be given up after adding C to it.

In belief revision, when we add a sentence to a belief set without giving up any other sentences in there, we are just *expanding* this belief set by such a sentence. So, we can consider that applying a default is the same as expanding a belief set by a sentence when some conditions are satisfied.

Thus, the purpose of this paper is to show that we can see an extension of a Reiter's default theory (W,Δ) as the expansion of the (belief) set W by some maximal set D of consequences of defaults in Δ. To do this, we will use the model of revision functions proposed by Grove [13] to characterize the models of the extensions in Reiter's default logic [24]. We show that the class of models we obtain in the special case when a revision is an expansion (i.e., a new sentence A is added to a belief set K and no sentence in K is deleted), is the class of models of some extension in Reiter's default logic. Moreover we show that Poole's system for default reasoning [23] can be characterized in the same way.

This work is organized as follows : in the next section, we briefly review the AGM model [1] of belief revision. In section 3, we present Grove's systems of sphere [13], and in sections 4 and 5 we determine, respectively, the relationship between this model and Reiter's default logic [24] and Poole's default system [23].

2 - The AGM Model

Belief sets ([1],[11]) are sets of sentences closed under (classical) logical consequences, used to model the statics of epistemic states. For a consistent belief set K and any sentence A, we can have one of the following epistemic attitudes :

(a) A ∈ K (i.e., A is accepted);

(b) ¬A ∈ K (i.e., A is rejected);

(c) A ∉ K and ¬A ∉ K (i.e., A is undetermined).

These epistemic states can be changed one into another, which characterizes an update on the epistemic states. We can have the following updates :

- *Expansion* : represented by K^+_A. A new sentence A (and its logical consequences) is added to the belief set K. The epistemic attitude (c) is changed into (a) or (b).
- *Contraction* : represented by K^-_A. A sentence in K is retracted. No sentence is added to the belief set K. The epistemic attitude (a) or (b) is changed into (c).
- *Revision* : represented by K^*_A. A new sentence, inconsistent with the belief set K, is added to it. In order to maintain consistency, it is necessary to remove some sentences from K. The epistemic attitude (a) (resp. (b)) is changed into (b) (resp. (a)).

To remove some sentences from the belief set K when it is reviewed by a sentence A means that we want to give up sentences that are inconsistent with the new sentence A retaining as much as possible of old beliefs in K. So, it is important to know which beliefs will continue in K after revision and which beliefs will not. For this, the AGM theory uses the following set of postulates that any revision function * should satisfy :

(K*1) For any sentence A and any belief set K, K^*_A is a belief set ;

(K*2) $A \in K^*_A$, the new sentence A will belong to the reviewed belief set ;

(K*3) $K^*_A \subseteq K^+_A$,
(K*4) If $\neg A \notin K$ then $K^+_A \subseteq K^*_A$
       Postulates (K*3) and (K*4) entail that expansion is a special case of revision
       except when $K^*_A = K_\perp$, where $K_\perp$ is the set of all sentences.

(K*5) $K^*_A = K_\perp$ iff $\vdash \neg A$
       K^*_A is a consistent belief set unless $\neg A$ is logically necessary.

(K*6) If $\vdash A \leftrightarrow B$, then $K^*_A = K^*_B$.
       Equivalent sentences yield identical changes in a belief set

(K*7) $K^*_{A \wedge B} \subseteq (K^*_A)^+_B$.
(K*8) If $\neg B \notin K^*_A$, then $(K^*_A)^+_B \subseteq K^*_{A \wedge B}$
       The postulates (K*7) and (K*8) are generalizations of (K*3) and (K*4) that
are applied to iterated changes of belief.

As we saw in the postulates (K*3) and (K*4), expansion is a special case of
revision. Revision and contraction functions are related through the following
identities :

$$\text{Levi}: \quad K^*_A = (K^-_{\neg A})^+_A$$
$$\text{Harper}: K^-_A = K \cap K^*_{\neg A}$$

3 - Grove's System of Spheres

This is a model of revision functions proposed by Grove [13] that uses a
system of spheres similar to the "sphere" semantics for counterfactuals presented by
Lewis [15]. Let M_L be the set of all possible worlds that can be describe in L. So, a
belief set K can be represented by the subset [K] of M_L consisting of all maximal sets
where all the sentences in K are satisfied : $[K] = \{m \in M_L : m \models K$, m maximal$\}$.

If the belief set K is inconsistent, then $[K] = \varnothing$. For any formula A, the set
[A] is formed by elements in M_L where A is satisfied. Conversely, for any non empty
subset $S \neq \varnothing$ of M_L, the set $K_S = \cap$ (m $\in$ S), of formulas included in all elements of S
is a belief set. If $S = \varnothing$, define $K_S = K_\perp$.

Definition 1 (Grove [13])

A *system of spheres centered on* [K] is a collection S of subsets of M_L that
satisfies the following conditions :
       (S1) S is totally ordered by $\subseteq$;
       (S2) [K] is the $\subseteq$-minimum of S ;
       (S3) M_L is in S ;
       (S4) If A is a sentence and there is a sphere in S intersecting [A], then there
            is a smallest sphere in S intersecting [A].■

M_L

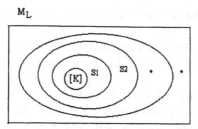

Fig. 1 - System of spheres centered on [K]

For any sentence A, (S4) ensure that if [A] intersects any sphere in S, then there is a smallest sphere S_A in S that is intercepted by [A]. If [A] does not intercept any sphere (by S3, since M_L is in S then [A] = $\varnothing$), then $S_A = M_L$. The set C(A) = [A] $\cap S_A$ is the set of the "closest" elements in M_L to [K] of which A is an element.

M_L

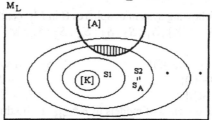

Fig. 2 - A possible world representation of the set C(A)
(hatched area)

Grove [13] shows the following theorems :

Theorem 1 (Grove[13])

Let S be any system of spheres in M_L centered on [K] for some belief set K. If, for any A, K^*_A is defined to be $K_{C(A)}$, then the resulting revision function satisfies (K*1)-(K*8).■

Theorem 2 (Grove[13])

Let * any revision function satisfying (K*1)-(K*8). Then for any (fixed) belief set K there is a system S of spheres that is centered on [K] and that satisfies $K^*_A = K_{C(A)}$ for all sentences A.■

So, in Grove's models, the revision K^*_A of the belief set K by A can be represented by the set C(A) of "worlds". A revision function can be represented by a family of systems of spheres, one for each belief set K.

Example 1

Consider L = {A,B} and K = {A, A→B}. Then :

$$M_L = \{(A,B), (\neg A,B), (A,\neg B), (\neg A,\neg B)\} \quad \text{and} \quad [K] = \{(A,B)\}$$

Consider the following system of spheres S in M_L centered on [K] :

$$S = \{ S_0, S_1, S_2, S_3 \}$$

where $S_0 = [K]$; $S_1 = \{(A,B), (\neg A,B)\}$; $S_2 = \{(A,B), (\neg A,B), (A,\neg B)\}$; $S_3 = M_L$. Note that $S_0 \subseteq S_1 \subseteq S_2 \subseteq S_3$.

Suppose we want to determine the model of the belief set $K^*_{\neg B}$. To do this, we have :

$[\neg B] = \{(A,\neg B), (\neg A,\neg B)\}$, i.e., a maximal set where $\neg B$ is satisfied.

$S_{\neg B} = S_2 = \{(A,B), (\neg A,B), (A,\neg B)\}$, i.e., the smallest sphere in S intercepted by $[\neg B]$.

So, $C(\neg B) = [\neg B] \cap S_{\neg B} = \{(A,\neg B)\}$, and then, the set of models of $K^*_{\neg B}$ is $\{(A,\neg B)\}$. Note that we can not have $K^*_{\neg B} = \{A \rightarrow B, \neg B\}$ because in this case, the world $\{(A,\neg B)\}$ does not satisfy $\{A \rightarrow B, \neg B\}$. ■

Example 2

Consider $L = \{A,B,C\}$ and $K = \{A, A \rightarrow B, B \rightarrow C\}$. Then :

$M_L = \{(A,B,C), (A,B,\neg C), (\neg A,B,C), (\neg A,B,\neg C), (A,\neg B,C), (A,\neg B,\neg C), (\neg A, \neg B,C), (\neg A,\neg B,\neg C)\}$ and

$[K] = \{(A,B,C)\}$

Consider the following system of spheres S in M_L centered on [K] :
$$S = \{ [K] ; S_1 ; S_2 ; S_3 ; S_4 ; S_5 ; S_6 ; M\}$$

where :

$S_1 = \{(A,B,C), (A,B,\neg C)\}$ $S_2 = S_1 \cup \{(\neg A,B,C)\}$

$S_3 = S_2 \cup \{(\neg A,B,\neg C)\}$ $S_4 = S_3 \cup \{(A,\neg B,C)\}$

$S_5 = S_4 \cup \{(A,\neg B,\neg C)\}$ $S_6 = S_5 \cup \{(\neg A,\neg B,C)\}$

Let's determine the class of models of $K^*_{\neg B}$. To do this, we have :

$[\neg B] = \{(A,\neg B,C), (A,\neg B,\neg C), (\neg A,\neg B,C), (\neg A,\neg B,\neg C)\}$

$S_{\neg B} = S_4 = \{(A,B,C), (A,B,\neg C), (\neg A,B,C), (\neg A,B,\neg C), (A,\neg B,C)\}$

So, $C(\neg B) = [\neg B] \cap S_{\neg B} = \{(A,\neg B,C)\}$, and $K^*_{\neg B} = K_{C(\neg B)} = Cn(A,\neg B, C)$, that is, $K^*_{\neg B}$ is formed by $\{A, B \rightarrow C\}$. Note that $K^*_{\neg B}$ can not be formed by neither $\{A, A \rightarrow B\}$ nor $\{A \rightarrow B, B \rightarrow C\}$ because the world $\{(A,\neg B,C)\}$ does not satisfy these sets. ■

4 - System of Spheres and Default Reasoning

In the revision process of a belief set K by a sentence A (K^*_A), sometimes we should give up some beliefs in K that contradicts the new sentence added in order to preserve consistency. As we saw in section 2, a particular case occurs if all sentences still belong to K when the new belief A is added to it. In this case, we say that the belief set K is *expanded* by A, represented by K^+_A, and we call this process an *expansion* of a belief set.

Using Grove's system of spheres, the revision K^*_A is an *expansion* ($K^*_A = K^+_A$) when [K] is the smallest sphere that [A] intercepts (condition S4) and so we have that $C(A) \subseteq [K]$. If $C(A) = [K]$ then we have $\neg A \notin K$ for A belongs to all worlds in [K] ($K^*_A = K^+_A$ by postulates (K*3) and (K*4)). In the case that $C(A) \subset [K]$, if we had $\neg A \in K$, we would have that $[K] \subseteq [\neg A]$. Then for all $m \in [K]$, $m \models \neg A$. But this contradicts the fact that $K \cup \{A\}$ is consistent (by $C(A) \subset [K]$), hence, $\neg A \notin K$ and $K^*_A = K^+_A$.

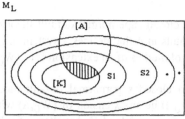

Fig. 3 - C(A) ⊂ [K]

Consider that we have a *closed supernormal default theory* (W,Δ), where Δ = {:A/A}. The default {:A/A} can be applied to W, that is, we can *add* the consequent of this default to W, when its justification is consistent with the sentences in the set W. In this case, as the justification and the consequent of the default are the same, we can add A to W whenever A is consistent with this set.

From the point of view of the Grove's model of revision functions, adding a sentence A to a belief set W without giving up any belief in this set means that the belief set W is *expanded* by the sentence A. Note that when {A} is inconsistent with W, it generates a revision on this set, and the smallest sphere S_i that [A] intercepts should be different from [W] (see figure 4). So, the default {:A/A} can be applied to W if, and only if, adding A to W causes only an expansion and not a revision on this set, that is, $[A] \cap [W] \neq \emptyset$.

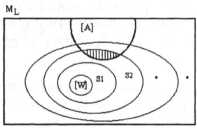

Fig. 4

Thus, we can determine the class of models of a *closed supernormal default theory* (W,Δ) using the Grove's sphere model as follows :

Definition 2

Let S be any system of sphere in M_L centered on [K], for some belief set K. A sentence A *expands* K when the set [A] intercepts the smallest sphere in the system S, i.e., $C(A) = [A] \cap [K]$.∎

By the remarks made above we have in this case that $K^+_A = K_{C(A)} = K^*_A$ and consequently $[K^+_A] = [K^*_A] = [A] \cap [K] \neq \emptyset$.

Definition 3

Let $(W_0,Δ)$ be a closed supernormal default theory. Let $<d_i> = <d_1, ..., d_n>$ be a sequence of defaults in Δ such that $d_j = : B_j / B_j$. Define a system of spheres S centered on $[W_n]$ as the set $\{M_L, [W_0], ..., [W_n]\}$ where for each j = 1 to n , $[W_j] = [W_{j-1}] \cap [B_j]$ if $[W_{j-1}] \cap [B_j] \neq \emptyset$. ∎

Note that $[W_1] = [W_0] \cap [B_1] = [W_0^+{}_{B_1}] = [W_0^*{}_{B_1}]$, ..., $[W_n] = [W_{n-1}] \cap [B_n] = [W_{n-1}^+{}_{B_n}] = [W_{n-1}^*{}_{B_n}] = [(...((W_0^+{}_{B_1})^+{}_{B_2})...)^+{}_{B_n})]$

Theorem 3 [3]

Let (W_0,Δ) be a closed supernormal default theory. Let $<d_1, \ldots, d_n>$ be a maximal sequence of defaults in Δ that defines a system of spheres S as in definition 3. Then $[W_n]$ is the class of models of an extension E of (W_0,Δ). ∎

Theorem 4

Let (W_0,Δ) be a closed supernormal default theory. Let $E = Th(W_0 \cup CONSEQUENT(GD(E,W_0)))$ be an extension of (W_0,Δ) with $GD(E,W_0) = \{d_1, d_2, \ldots, d_n\}$, and M the class of all models of E. A system of spheres S as in definition 3 can be defined, with $<d_i> = <d_{i1}, d_{i2}, \ldots, d_{in}>$ where $d_{ij} \in GD(E,W_0)$, such that for all $i \in [1,n] : [W_i] \neq \varnothing$ and $[W_n] = M$. ∎

Example 3 (Reiter [24])

Consider the closed default theory (W_0,Δ) where $W_0 = \{B \rightarrow \neg A \wedge \neg C\}$ and $\Delta = \{:A/A \, , :B/B \, , :C/C\}$. This theory has two extensions : $E_1 = Th(W_0 \cup \{A,C\})$ and $E_2 = Th(W_0 \cup \{B\})$ which have, respectively, the following class of models : $M_1 = \{(A,\neg B,C)\}$ and $M_2 = \{(\neg A,B,\neg C)\}$.

Using the definition 3 and the theorem 4, we can construct the following systems of spheres centered on M_1 and M_2, respectively :

- Considering the sequence of defaults $<:A/A \, , :C/C>$, we have S formed by :

$M_L = \{(A,B,C) \, , (A,B,\neg C) \, , (\neg A,B,C) \, , (\neg A,B,\neg C) \, , (A,\neg B,C) \, , (A,\neg B,\neg C), (\neg A, \neg B,C) \, , (\neg A,\neg B,\neg C)\}$,

$[W_0] = \{(\neg A,B,\neg C) \, , (A,\neg B,C) \, , (A,\neg B,\neg C) \, , (\neg A,\neg B,C) \, , (\neg A,\neg B,\neg C)\}$,

$[W_1] = [W_0] \cap [A] = \{(A,\neg B,C) \, , (A,\neg B,\neg C)\} = [W_0^{+}A] \neq \varnothing$,

$[W_2] = [W_1] \cap [C] = \{(A,\neg B,C)\} = [W_1^{+}C] = [(W_0^{+}A)^{+}C] \neq \varnothing$.

So, we have $M_1 = [W_2]$.

- Considering the sequence of defaults $<:B/B>$, we have S' formed by :

$M_L = \{(A,B,C) \, , (A,B,\neg C) \, , (\neg A,B,C) \, , (\neg A,B,\neg C) \, , (A,\neg B,C) \, , (A,\neg B,\neg C), (\neg A, \neg B,C) \, , (\neg A,\neg B,\neg C)\}$,

$[W_0] = \{(\neg A,B,\neg C) \, , (A,\neg B,C) \, , (A,\neg B,\neg C) \, , (\neg A,\neg B,C) \, , (\neg A,\neg B,\neg C)\}$,

$[W_1] = [W_0] \cap [B] = \{(\neg A,B,\neg C)\} = [W_0^{+}B] \neq \varnothing$,

So, we have $M_2 = [W_1]$.

Note that the sequence of defaults $<:A/A \, , :B/B \, , :C/C>$ can not generate a system of spheres as it is defined in definiton 3, since :

$M_L = \{(A,B,C) \, , (A,B,\neg C) \, , (\neg A,B,C) \, , (\neg A,B,\neg C) \, , (A,\neg B,C) \, , (A,\neg B,\neg C), (\neg A, \neg B,C) \, , (\neg A,\neg B,\neg C)\}$,

$[W_0] = \{(\neg A,B,\neg C) \, , (A,\neg B,C) \, , (A,\neg B,\neg C) \, , (\neg A,\neg B,C) \, , (\neg A,\neg B,\neg C)\}$,

$[W_1] = [W_0] \cap [A] = \{(A,\neg B,C) \, , (A,\neg B,\neg C)\} = [W_0^{+}A] \neq \varnothing$,

$[W_2] = [W_1] \cap [B] = \varnothing$, which does not satisfy the condition that $[W_1] \cap [B] \neq \varnothing$

Indeed, any sequence of defaults that includes all of them, can not generate a system of spheres that satisfies all the conditions we want. So, we have only two maximal sequences of defaults that satisfy the conditions on theorem 4. ∎

---

[3]For lack of space, we will omit the proof of all theorems. We will present some examples to illustrated them.

We can generalize the discussion above to closed normal default theories with prerequisite as follows :

Definition 4

Let (W_0,Δ) be a closed normal default theory. Let $<d_i> = <d_1, ..., d_n>$ be a sequence of defaults in Δ such that $d_j = A_j : B_j / B_j$. Define a system of spheres S centered on $[W_n]$ as the set $\{M_L,[W_0], ..., [W_n]\}$ where for each $j = 1$ to n , $[W_j] = [W_{j-1}] \cap [B_j]$ if $[W_{j-1}] \subseteq [A_j]$ and $[W_{j-1}] \cap [B_j] \neq \varnothing$. ∎

Theorem 5

Let (W_0,Δ) be a closed normal default theory. Let $<d_1, ..., d_n>$ be a maximal sequence of defaults in Δ that defines a system of spheres S as in definition 4. Then $[W_n]$ is the class of models of an extension E of (W_0,Δ). ∎

Theorem 6

Let (W_0,Δ) be a closed normal default theory. Let $E = Th(W_0 \cup CONSEQUENT(GD(E,W_0)))$ be an extension of (W_0,Δ), with $GD(E,W_0) = \{d_1, d_2, ..., d_n\}$ where $d_i = A_i : B_i /B_i$, and M the class of all models of E. A system of spheres S as in definition 4 can be defined, with $<d_i> = <d_{i1}, d_{i2}, ..., d_{in}>$ where $d_{ij} \in GD(E,W_0)$, such that for all $i \in [1,n]$: $[W_i] \neq \varnothing$ and $[W_n] = M$. ∎

Example 4 (Reiter [24])

Consider the default theory (W_0,Δ) where $W_0 = \{A\}$ and $\Delta = \{A:B/B$, $B:C/C\}$. Such theory has only one extension $E_1 = Th(W_0 \cup \{B,C\})$ which has the model $M_1 = \{A, B, C\}$.

Considering the sequence of defaults $<A:B/B , B:C/C>$, we have a system of spheres S formed by :

$M_L = \{(A,B,C) , (A,B,\neg C) , (\neg A,B,C) , (\neg A,B,\neg C) , (A,\neg B,C) , (A,\neg B,\neg C),$
$\qquad (\neg A,\neg B, C) , (\neg A,\neg B,\neg C)\}$,

$[W_0] = \{(A,B,C) , (A,B,\neg C) , (A,\neg B,C) , (A,\neg B,\neg C)\}$,

$[W_1] = [W_0] \cap [B] = \{(A,B,C) , (A,B,\neg C)\} = [W_0{}^+{}_B] \neq \varnothing$,

$[W_2] = [W_1] \cap [C] = \{(A,B,C)\} = [W_1{}^+{}_C] = [(W_0{}^+{}_B){}^+{}_C] \neq \varnothing$.

Note that in such system of spheres, we have that $[W_0] \subseteq [A]$ and $[W_1] \subseteq [B]$.

Thus, we have $M_1 = [W_2] = [(W_0{}^+{}_B){}^+{}_C]$.

Note that, with the sequence of defaults $<B:C/C , A:B/B>$, we can not construct a system of spheres since we do not have $[W_0] \subseteq [B]$. ∎

Finally, the case of closed non-normal default is considered :

Definition 5

Let (W_0,Δ) be a closed default theory where the defaults are of form $\{A_k : B_k / C_k\}$. Let $<d_i> = <d_1, ..., d_n>$ be a sequence of defaults in Δ such that $d_j = A_j : B_j / C_j$. Define a system of spheres S centered on $[W_n]$ as the set $\{M_L,[W_0], ..., [W_n]\}$ where for each $j = 1$ to n , $[W_j] = [W_{j-1}] \cap [C_j]$ if $[W_{j-1}] \subseteq [A_j]$ and $[W_{j-1}] \cap [B_j] \neq \varnothing$. ∎

Theorem 7

Let (W_0,Δ) be a closed default theory where the defaults are of form $\{A_k : B_k / C_k\}$. Let $<d_1, ..., d_n>$ be a maximal sequence of defaults in Δ that defines a system of spheres S as in definition 5 and such that for each $i \in [1,n]$, $[W_i] \neq \varnothing$ and $[W_n] \cap [B_i] \neq \varnothing$. Then $[W_n]$ is the class of models of an extension E of (W_0,Δ). ■

Theorem 8

Let (W_0,Δ) be a closed default theory where the defaults are of the form $\{A_k : B_k / C_k\}$. Let $E = Th(W_0 \cup CONSEQUENT(GD(E,W_0)))$ be an extension of (W_0,Δ), with $GD(E,W_0) = \{d_1, d_2, ..., d_n\}$ where $d_i = A_i : B_i / C_i$, and M the class of all models of E. A system of spheres S as in definition 5 can be defined, with $<d_i>$ $= <d_{i1}, d_{i2}, ..., d_{in}>$ where $d_{ij} \in GD(E,W_0)$ such that for all $i \in [1,n] : [W_i] \neq \varnothing$, $[W_n] \cap [B_i] \neq \varnothing$ and $[W_n] = M$, where $[W_n]$ is the innermost sphere of S. ■

Example 5

Consider the default theory (W_0,Δ) where $W_0 = \varnothing$ and $\Delta = \{:A/\neg A\}$. This theory has no extension. By theorem 7, there is no maximal sequence of defaults that can define a system of spheres as in definition 5. Indeed, note that $M_L = [W_0] = \{(A) , (\neg A)\}$ and, using the sequence of defaults $<:A/\neg A>$, we would have $[W_1] = [W_0] \cap [\neg A] = \{\neg A\} \neq \varnothing$, but $[W_1] \cap [A] = \varnothing$. So, we have that the justification A is inconsistent with the consequent $\neg A$ of the applied default and then the expansion of W_0 by $\neg A$ is not valid and then such theory does not have any extension. ■

Example 6 (Etherington [10])

Consider the default theory where $W_0 = \varnothing$ and $\Delta = \{: (A \wedge \neg B)/A ; : (B \wedge \neg C) /B; :(C \wedge \neg A)/C\}$. It is well known that this theory has no extension. Let us see what happens in the model of spheres.

We have that :

$$[W_0] = M_L$$

Suppose we want to construct a system of spheres using the following sequence of defaults : $<:(A \wedge \neg B)/A ; :(B \wedge \neg C)/B ; :(C \wedge \neg A)/C>$. Note that :

- since $[W_0] \cap [A \wedge \neg B] = \{(A,\neg B,C) , (A,\neg B,\neg C)\} \neq \varnothing$, then we have $[W_1] = [W_0] \cap [A] = \{(A,B,C) , (A,B,\neg C) , (A,\neg B,C) , (A,\neg B,\neg C)\}$;
- since $[W_1] \cap [B \wedge \neg C] = \{(A,B,\neg C)\} \neq \varnothing$, then we have $[W_2] = [W_1] \cap [B] = \{(A,B,C) , (A,B,\neg C)\}$; but
- $[W_2] \cap [C \wedge \neg A] = \varnothing$, and then, by definition 5, we can not consider the sphere $[W_3] = [W_2] \cap [C]$ in our system of spheres.

So, we have to consider only the sequence of defaults $< : (A \wedge \neg B)/A ; :(B \wedge \neg C) /B>$. But, in this case :

- $[W_2] \cap [A \wedge \neg B] = \varnothing$, that is, the consequence of the second default contradicts the justification of the first default applied, which contradicts one of the conditions of theorem 8.

Note that a sequence of defaults formed by only one default (in the example, $<:(A \wedge \neg B)/A>$) will be not maximal. Moreover, the same occurs if we begin with any other default in Δ. So, such default theory has no extension by theorem 7. ■

Example 7 (Brewka [8])

Consider the default theory (W_0,Δ) where $W_0 = \{B \vee C\}$ and $\Delta = \{: A \wedge \neg B / A ; : D \wedge \neg C /D\}$. This theory has one extension : $E_1 = Th(W_0 \cup \{A,D\})$ which has the following class of models $M = \{(A,B,C,D), (A,B,\neg C,D), (A,\neg B, C,D)\}$.

Consider the sequence of defaults $<: A \wedge \neg B/A ; : D \wedge \neg C /D>$. We have the system of spheres S formed by :

M_L

$[W_0] = \{(A,B,C,D), (A,B,C,\neg D), (A,B,\neg C,D), (A,B,\neg C,\neg D), (A,\neg B,C,D),$
$(A,\neg B,C,\neg D), (\neg A,B,C,D), (\neg A,B,C,\neg D), (\neg A,B,\neg C,D), (\neg A,B,\neg C,\neg D),$
$(\neg A,\neg B,C,D), (\neg A,\neg B,C,\neg D)\}.$

- Since $[W_0] \cap [A \wedge \neg B] \neq \emptyset$, we have $[W_1] = [W_0] \cap [A] = \{(A,B,C,D),$
  $(A,B,C,\neg D), (A,B,\neg C,D), (A,B,\neg C,\neg D), (A,\neg B,C,D), (A,\neg B,C,\neg D)\} \neq \emptyset.$
  So, $[W_1] = [W_0{}^+{}_A]$.

- Since $[W_1] \cap [D \wedge \neg C] \neq \emptyset$, we have $[W_2] = [W_1] \cap [D] = \{(A,B,C,D),$
  $(A,B,\neg C,D), (A,\neg B,C,D)\} \neq \emptyset.$ So, $[W_2] = [W_1{}^+{}_D] = [(W_0{}^+{}_A){}^+{}_D]$

Moreover, we have $[W_2] \cap [A \wedge \neg B] \neq \emptyset$ and $[W_2] \cap [D \wedge \neg C] \neq \emptyset.$ So, the consequents of the applied defaults do not contradict any justification of any other applied defaults. As we want, $M = [W_2]$. ∎

5 - Poole's System for Default Reasoning

In the simplest case of Poole's system for default reasoning ([23]), there are two sets of formulas to consider : a consistent set F formed by *facts* about the world, and a set Δ formed by the *possible hypotheses* one can assume. A *scenario* of (F,Δ) is a consistent set $F \cup D$ where D is a set of ground instances of elements of Δ. An *extension* of (F,Δ) is the set of logical consequences of a maximal scenario (under set inclusion) of (F,Δ).

Note that a scenario (or extension) $F \cup \{A_1, A_2, ..., A_n\}$ can be seen, from the point of view of the Grove's model of revision functions, as the expansion of the "belief set" F by all elements in $\{A_1, A_2, ..., A_n\}$ (i.e., $(...((W^+{}_{A1})^+{}_{A2})...)^+{}_{An} = (...((W^*{}_{A1})^*{}_{A2})...)^*{}_{An}$). So, the class of all models of a scenario (or an extension) $F \cup \{A_1, A_2, ..., A_n\}$ and the class of all models of the expansion of "belief set" F by $\{A_1, A_2, ..., A_n\}$ are the same. Indeed, this case of Poole's system corresponds to the case of supernormal default logic we have seen before (theorem 3).

A more interesting case occurs when we consider the introduction of a set of *constraints*. This set is introduced to avoid the undesirable use of the contrapositive of a default (blocking the application of a default under certain conditions), without any other effect. Thus, a *scenario* of (F,Δ,C) is a consistent set $F \cup D$ where D is a set of ground instances of elements of Δ such that $D \cup F \cup C$ is consistent. An *extension* of (F,Δ,C) is the set of logical consequences of a maximal scenario (under set inclusion) of (F,Δ, C).

Suppose that $\Delta = \{A\}$, $F \cup \{A\}$ and $F \cup \{A\} \cup C$ are consistent sets, that is, $F \cup \{A\}$ is an extension of (F,Δ,C). In the Grove's model of spheres, this means that $[F \cup C]$ is the smallest sphere that $[A]$ intercepts since if $[F \cup C] \cap [A] = \emptyset$ and $[F] \cap [A] \neq \emptyset$, we have that A is inconsistent with sentences in the set of constraints C. Thus, we must have a revision of the set $F \cup C$ by the sentence $\{A\}$ (see figures 5 and 6).

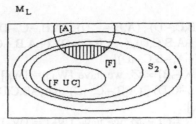

Fig. 5 : The set [F ∪ C] is reviewed by the sentence A

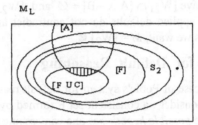

Fig. 6 : The set [F ∪ C] is expanded by the sentence A

Thus we have that :

Definition 6

Let (F_0,Δ,C) be a triple such that F_0 is a set of facts about the world, Δ is a set of possible hypotheses, and C is a set of constraints. Let $<h_i> = \{h_1, ... , h_n\}$ be a set of possible hypotheses in Δ such that $h_j = A_j$. Define a system of spheres **S** centered on $[F_n]$ as the set $\{M_L,[F_0], ... , [F_n]\}$ such that :

for each j = 1 to n , $[F_j] = [F_{j-1}] \cap [h_j]$ if $[F_{j-1} \cup C] \cap [h_j] \neq \varnothing$. ∎

Theorem 9

Let (F_0,Δ,C) be a triple such that F_0 is a set of facts about the world, Δ is a set of possible hypotheses, and C is a set of constraints. Let $\{h_1, ... , h_n\}$ be a (maximal) set of possible hypotheses in Δ that defines a system of spheres **S** as in definition 6 with $F_j \neq \varnothing$ for all j = 1,...,n. Then $[F_n]$ is the class of all models of some scenario (extension) of (F_0,Δ,C).∎

Theorem 10

Let (F_0,Δ,C) be a triple such that F_0 is a set of facts about the world, Δ is a set of possible hypotheses, and C is a set of constraints. Let $E = F \cup \{h_1, ... , h_n\}$ be a scenario (or an extension) of (F_0,Δ,C), and M the class of all models of E. Let **S** = $(\{[F_0]\} \cup \{[F_i] : \text{for all } i \in [1,n], [F_i] = [F_{i-1}] \cap [h_i]\} \cup M_L)$ be a system of spheres centered on $[F_n]$ as in definition 6. Then $[F_i] \neq \varnothing$ and $[F_n] = M$. ∎

Example 8 (Poole [23])

Consider the following sets of facts, possible hypotheses and constraint, respectively :

$F_0 = \{ \}$; $\Delta = \{A,B,D\}$; and $C = \{B \rightarrow \neg A , D \rightarrow \neg B , A \rightarrow \neg D\}$

In this case, Poole's system obtains three extensions (one for each case of assuming one of the defaults) whose classes of models are : $M_1 = \{(A,B,D) , (A,B,\neg D) , (A,\neg B,D), (A,\neg B,\neg D)\}$, $M_2 = \{(A,B,D), (A,B,\neg D), (\neg A,B,D), (\neg A,B,\neg D)\}$,

$M_3 = \{(A,B,D), (\neg A,B,D), (A,\neg B,D), (\neg A,\neg B,D)\}$. Let us see what happens in the model of spheres.

Consider the set of possible hypotheses $\{A\}$. We have that :
$$[F \cup C] = [C] = \{(A,\neg B,\neg D), (\neg A,B,\neg D), (\neg A,\neg B,D), (\neg A,\neg B,\neg D)\}$$
and the system of spheres S formed by :

- $[F_0] = M_L$;
- since $[F_0 \cup C] \cap [A] = \{(A,\neg B,\neg D)\} \neq \varnothing$, then $[F_1] = [F_0] \cap [A] = \{(A,B,D), (A,B,\neg D), (A,\neg B,D), (A,\neg B,\neg D)\} = M_1$.

Note that neither B nor D can belong to the set we are considering since :
$$[F_1 \cup C] \cap [B] = \varnothing \text{ and } [F_1 \cup C] \cap [D] = \varnothing.$$
So, considering the system of spheres $S = \{[F_0], [F_1]\}$, all conditions on theorem 10 are satisfied. We can use the same reasoning to prove that $\{B\}$ and $\{D\}$ are sets of possible hypotheses which can generate, respectively, systems of spheres S' and S'' such that $[F'_1] = M_2$ and $[F''_1] = M_3$. ∎

6 - Conclusion

In this work, we used Grove's system of spheres [13] for belief revision to characterize the models of extensions in Reiter's default logic [24] and Poole's system [23] for default reasoning. In both, an extension can be viewed as a belief set obtained through expanding, under certain conditions, the (belief) set W by the set of consequences of the defaults that generates such extension. Such conditions, that we have to satisfy in the construction of an extension in default logic, are expressed in terms of revisions and expansions in Grove's sphere models.

We can also use this relationship to semantically characterize extensions in other default logics as Lukaszewicz's [17], Brewka's [8] and Schaub's [25] default logics. Furthermore, the relationship between belief revision and default logic stated here and the connection between default logic and logic program stated in Bidoit and Froidevaux [4,5] and Marek and Truszczynski [22] suggest that Grove's models for belief revision could be used to define the meaning of logical databases and logic programs, stating a connection between belief revision and logic programming.

Acknowledgments

We would like to thank Gerson Zaverucha for his comments and suggestions. The first author would like to thank Stefan Brass, Miguel Dionísio, Udo Lipeck and Mark Ryan for the opportunity of showing part of this work and for their comments, and to Universidade Estácio de Sá for its partial financial support (Programa de Qualificação e Participação).

References

1. C.E. Alchourrón, P. Gärdenfors, D. Makinson: On the logic of theory change : partial meet functions for contraction and revision. Journal of Symbolic Logic 50, 510-530 (1985).
2. C.E. Alchourrón, D. Makinson: On the logic of theory change : contraction functions and their associated revision functions. Theoria 48, 14-37 (1982).

187

3. P. Besnard: An introduction to default logic. Berlin: Springer-Verlag 1989.
4. N. Bidoit and C. Froidevaux. General logical databases and programs: default logic semantics and stratification. Information and Computation 91, 15-54 (1991).
5. N. Bidoit and C. Froidevaux. Negation by default and unstratifiable logic programs. Theoretical Computer Science 78, 85-112 (1991).
6. C. Boutilier: Unifying default reasoning and belief revision in a modal framework. Artificial Intelligence 68, 33-85 (1994).
7. C. Boutilier:.: Conditional logics of normality : a modal approach. Artificial Intelligence 68, 87-154 (1994).
8. G. Brewka: Nonmonotonic reasoning : logical foundations of commonsense. Cambridge University Press, Cambridge 1991.
9. D. W. Etherington.: Reasoning with imcomplete information. Research Notes in Artificial Intelligence, Pitman, London 1987.
10. D. W. Etherington: Formalizing non-monotonic reasoning systems. Artificial Intelligence 31, 41- 85 (1987).
11. P. Gärdenfors: Nonmonotonic inference based on expectations. Artificial Intelligence 65, 197-245 (1994).
12. P. Gärdenfors: Knowledge in flux : modeling the dynamics of epistemic states. The MIT Press, Bradford Books, Cambridge, MA, 1988.
13. A. Grove: Two modellings for theory change. Journal of Philosophical Logic 17, 157-170 (1988).
14. R. Guerreiro and M.Casanova. An alternative semantics for default logic. The Third International Workshop on Nonmonotonic Reasoning, South Lake Tahoe, 1990.
15. D. Lewis: Counterfactuals. Harvard University Press, Cambridge, Mass. 1973.
16. V. Lifschitz. On open default. In J. Lloyd, editor, Cmputational logic. Symposium proceedings, pp. 80-95. ESPRIT Basic Research Series. Berlin: Springer-Verlag, 1990.
17. W. Lukaszewicz: Non-monotonic reasoning: Formalization of commonsense reasoning. Ellis Horwood Series in Artificial Intelligence (Ellis Horwood, Chichester,England) 1990.
18. W. Lukaszewicz: Two results on default logic. In: Proceedings IJCAI-85, Los Angeles, CA, 459-461 (1985).
19. D. Makinson: General theory of cumulative inference. In: M.Reinfrank,J. de Kleer, M.L. Ginsberg and Sandewall, eds., Non-monotonic Reasoning, Lecture Notes in Artificial Intelligence 346. Berlin: Springer 1989.
20. D. Makinson, P. Gärdenfors: Relations between the logic of theory change and nonmonotonic logic. In A. Fuhrmannand M. Morreau, eds., The Logic of Theory Change, Lecture Notes in Artificial Intelligence 465. Berlin: Springer 1991, pp. 185-205.
21. W. Marek and M. Truszczynski: Nonmonotonic logic - context-dependent reasoning. Berlin: Springer-Verlag 1993.
22. W. Marek and M. Truszczynski: Stable semantics for logic programs and default theories. In E. Lusk and R. Overbeek, editors, Proceedings of the North

American Conference on Logic Programming, pp. 243-256, Cambridge, MA, MIT Press, 1989.

23. D. Poole: A logical framework for default reasoning. Artificial Intelligence 36, 27- 47 (1988).

24. R. Reiter: A logic for default reasoning. Artificial Intelligence 13, 81-132 (1980).

25. M. Ryan: Ordered presentations of theories - default reasoning and belief revision. Ph. D. Thesis, 1992.

26. Schaub, T.H. , Considerations on default logic, Ph.D. Thesis, 1992.

Revision by Communication

program revision by consulting weaker semantics

Cees Witteveen[1] and Wiebe van der Hoek[2]

[1] Delft University of Technology, Dept of Mathematics and Computer Science,
P.O.Box 356, 2600 AJ Delft, The Netherlands, witt@cs.tudelft.nl
[2] Utrecht University, Dept of Computer Science, Utrecht, The
Netherlands,wiebe@cs.ruu.nl

Abstract. We deal with the problem of revising logic programs that, according to some non-monotonic semantics, do not have an acceptable model. We propose to study such revisions in a framework where a number of semantical agents is distinguished, each agent associated with a different semantics but all agents interpreting the same program. If an agent cannot find an acceptable model for the program, he has to perform program revision. For logic programs, the different agents can be partially ordered by inferential strength. We propose a revision framework where an agent may consult his weaker colleagues, adds the information they can infer to the program and tries to find an acceptable model for the expanded program. In this paper we will concentrate on the kind of information needed to find successful revisions of programs. We point out some parameters along which our framework can be analyzed and suggest some further research.

1 Introduction

In the logic programming community one usually distinguishes the set of *intended* or *acceptable* models of a program from its set of *classical* models, the former usually being a strict subset of the latter.

By using a set of intended models, one can obtain stronger (but defeasible) conclusions from a program than by using classical models. This increase in inferential power, however, does not hold for all programs: sometimes agents cannot establish an intended model for a program, even if the program is classically consistent. In this case, there is a need for program revision: the current program does not satisfy any interpretation acceptable to the agent.

For classical theories, there is a framework ([4]) for doing revision that essentially is *inconsistency-driven*. In order to remove inconsistencies, the theory is narrowed by giving up some existing piece of information.

In this paper we will not deal with classically inconsistent theories. Instead, we will investigate classically consistent theories which, under a given non-monotonic semantics *Sem*, do not have an acceptable model. Let us call such a theory *non-monotonically inconsistent* or *nm-inconsistent*. Solving nm-inconsistency is quite different from solving the classical inconsistency problem, for nm-inconsistent theories do have a classical model and therefore, retraction of statements is not a prerequisite to restore consistency.

Example 1. Suppose that you know that (i) under normal conditions it takes a person at most one hour to drive from A to your place B, (ii) you know that some person left A to drive to you more than an hour ago and (iii) she didn't arrive yet. The following program P describes this scenario:

$$one\_hour\_drive \leftarrow drive\_A\_to\_B, \neg abnormal.$$
$$drive\_A\_to\_B \leftarrow .$$
$$\bot \leftarrow one\_hour\_drive.$$

According to the dominant non-monotonic *stable model* semantics, P is nm-inconsistent: it does not have a stable model. There is, however, a model giving the intuitive meaning of this program by interpreting *abnormal* as true. Hence, there is no need to *retract* information from the program. On the contrary, if we *expand* the program, e.g., by adding the contrapositive of the first rule:

$$abnormal \leftarrow drive\_A\_to\_B, \neg one\_hour\_drive.$$

The expansion has a stable model in which *abnormal* is true.

As we have shown in [14], the reason why the stable model semantics fails for this program is clear: it lacks the possibility of modeling reasoning by *reductio ad absurdum*. In the example above, first we assume that everything is normal, then detect a contradiction and finally find out that the assumption does not hold, so we have to conclude that there is something abnormal. To enable a stable reasoner to draw such conclusions, adding information in the form of contrapositives may help him in finding a suitable model.

This idea of *revision by expansion* to restore nm-consistency has been applied in [5, 10, 11, 7, 14]. In all these approaches, however, the existence of different, but related, non-monotonic semantics for programs has not been recognized: only revision under the stable model semantics has been dealt with. Moreover, the only source of additional information was consulting *classical* models to add information to the program.

In this paper we will generalize this existing framework by paying attention to the interrelations between different non-monotonic semantics of programs. Since, at least for logic programs, these semantics can be *partially* ordered according to inferential strength, we will deal with the following questions:

- How can an nm-reasoner make use of information obtained from arbitrarily *weaker*, not necessarily classical, semantics?
- Given an nm-reasoner R and some weaker semantics R', what kind of information is necessary to expand the program successfully?

After we give some preliminaries in 1.1, we will present our general framework for revision by communication in Section 2. In Section 3, we will concentrate on one aspect of this framework: the role that different kinds of information play in revising programs. Then, in Section 4 we briefly address some refinements of the framework, discussing credulous variants of the skeptical approach taken before.

1.1 Preliminaries

We assume the reader to be familiar with some basic terminology used in logic programming, as for example in Lloyd [8]. A *normal program rule* r is a directed propositional clause of the form $A \leftarrow B_1, \ldots, B_m, \neg D_1, \ldots, \neg D_n$, where $A, B_1, \ldots, D_1, \ldots, D_n$ are all atomic propositions. The head of such a rule r is denoted by $hd(r)$ and its body by $body(r) = \alpha(r) \wedge \beta(r)$, where $\alpha(r)$ denotes the conjunction of the B_i atoms and $\beta(r)$ denotes the conjunction of the negative atoms $\neg D_j$. A program is called *positive* if for every rule r, $\beta(r) = \top$ (the empty conjunction). *Constraints* are special rules of the form $\perp \leftarrow body$, expressing that the conjunction of the literals occurring in $body$ cannot be true. We do not allow $\perp$ (*falsum*) to occur in the body of any rule r. If L is a set of literals over B_P, $Atoms(L)$ denotes the set of atoms $a \in B_P$ occurring positively or negatively in L and L^+ (L^-) denotes the subset of atoms occurring positively (negatively) in L. If l is a literal, $\bar{l}$ equals $\neg l$ if l is a positive atom and equals s if $l = \neg s$ for some $s \in B_P$. Likewise, $\neg L = \{\bar{l} \mid l \in L\}$.

As usual, interpretations and models of a program are denoted by maximal consistent subsets I of literals over the Herbrand base B_P of P, i.e., for every atom $a \in B_P$, either a or $\neg a$ occurs in an interpretation I. A *model* of a program P is an interpretation I satisfying all the rules and constraints of P. Here, a rule r is said to be satisfied by an interpretation I iff $I \models body(r)$ implies $I \models hd(r)$ and a constraint $\perp \leftarrow body$ is satisfied by I iff $I \not\models body$ or, equivalently, I is a model of the constraint as a rule and I^+ does not contain the special atom $\perp$. We denote the set of classical models of P by $Mod(P)$. The definitions of *minimal* model, *supported* model and *stable* model are assumed to be well-known. If M is both supported and minimal, M is called *positivistic* (see [1]).

To discuss properties of stable models, we use the Gelfond-Lifschitz reduction $G(P, M)$ of a program P with respect to a model M, where $G(P, M) = \{c \leftarrow \alpha \mid c \leftarrow \alpha \wedge \beta \in P, \; M \models \beta\}$.

We use $Mod(P), MinMod(P), Supp(P), Pos(P), Stable(P)$ to denote the class of classical models, minimal models, supported models, positivistic models and stable models, respectively. It is well-known that the following inclusion relations hold (cf [1]): For every logic program P, $Stable(P) \subseteq Pos(P) \subseteq Supp(P) \subseteq Mod(P)$ and $Pos(P) \subseteq MinMod(P) \subseteq Mod(P)$.

We use all these semantics in the *definite* sense (cf. [16]), i.e., If $Sem(P) = \emptyset$ then $Sem(P) \models \emptyset$ and $Sem(P) \models x$ iff $Sem(P) \neq \emptyset$ and every $M \in Sem(P)$ satisfies x for $Sem \in \{Mod, MinMod, Supp, Pos, Stable\}$.

In the subsequent sections, we will also need clause-representations of rules. If r is the rule $c \leftarrow a_1, \ldots, a_m, \neg b_1, \ldots, \neg b_n$, then the clause $C(r)$ *associated* with r is $C(r) = \{c, \neg a_1, \ldots, \neg a_m, b_1, \ldots, b_n\}$. If P is a logic program, $C(P) = \{C(r) \mid r \in P\}$ is the set of clauses associated with P. Conversely, if $\mathcal{C}$ is a set of clauses, $Normal(\mathcal{C})$ refers to the set of *normal* logic programs P such that $\mathcal{C} = C(P)$. Note that, since we include constraints, for every set of clauses $\mathcal{C}$, there exists such a normal program P.

A clause C is called *positive* if it contains only positive atoms. A clause C is said to be a *prime implicate* of a set of clauses Σ iff $\Sigma \models C$ and there is no proper subset $C' \subset C$ such that $\Sigma \models C'$.

Let us finally mention some results that will be used in subsequent sections.

Lemma 1 (Marek & Truszczyński [9]).
Let M be a model of a program P. Then $T_{G(P,M)} \uparrow \omega \subseteq M$.

The easy proof of the following result appears in [15]:

Lemma 2. *Let P be a program and $M_{neg} = \{\neg a \mid a \in B_P\}$ an interpretation of P. Then the following statements are equivalent:*
(i) M_{neg} is a model of P.
(ii) M_{neg} is a stable model of P.
(iii) the body of every rule r of P contains at least one positive literal.

Lemma 3 (Expansion Lemma).
Let P be a program, $P' \subset P$ and M a model of P. Then, if M is a (minimal, supported, positivistic, stable) model of P', M also is a (minimal, supported, positivistic, stable) model of P.

Proof. Let M be a minimal model of P' and a model of P. Then it is immediate that M is also a minimal model of P.

For supported models the lemma is also trivially true, since $M \in Sup(P')$ and $M \in Mod(P)$ implies that for all $a \in M$, there is at least one rule $r \in P' \subseteq P$ with $hd(r) = a$, such that $M \models hd(r) \wedge body(r)$. Therefore, M is also a supported model for P. Combining the results for M as a supported and minimal model is sufficient to prove the property for positivistic models.

For stable models, assume that M is a stable model of P' and a model of P. Then $M^+ = lfp\, T_{G(P',M)} \subseteq lfp(T_{G(P,M)})$, where the last inclusion holds, since $G(P', M)$ and $G(P, M)$ are positive programs and $G(P', M) \subseteq G(P, M)$. Since M is a model of P, by Lemma 1, it follows that $lfp(T_{G(P,M)}) \subseteq M^+$. So $M^+ = lfp(T_{G(P,M)})$, and therefore $M \in Stable(P)$. □

2 Revision by Communication

In this paper we will generalize the existing framework for revision by expansion in a significant way. Instead of using one particular non-monotonic semantics and the use of classical information to perform program revision, we propose to do program revision in a larger, *distributed*, framework.

In this set-up, several agents are associated with the different (non-monotonic) semantics and each agent interprets the same program. These agents are partially ordered according to their inferential strength. As soon as an agent cannot find an intended model for the program, he is allowed to consult one or more weaker agents in parallel. The information he obtains from them will be compared and the results can be added to the program. Then the agent may try to find an intended model for the *expanded* program. To keep notation simple, we will identify agents with the semantics $Sem \in \{Stable, Pos, Supp, MinMod, Mod\}$ they use to interpret the (same) program P. Then we say that an agent Sem'

is *weaker* than an agent Sem, denoted by $Sem' < Sem$ iff for every program P, $Sem(P) \subseteq Sem'(P)$ and for at least one program P a strict inclusion holds.

The idea of revision by communication essentially is that, if an agent Sem cannot find an acceptable model for a program P, he may consult a weaker colleague Sem'. This agent can tell him what he considers as true in his set of intended models. The agent Sem can add this information to the program and tries to find an acceptable model for the expanded program. So the intended models of P are equal to

$$Sem(P + \{x \in R \mid Sem'(P) \models x\})$$

where R is a set of statements that may be communicated to Sem, e.g. some subsets of atoms or rules. Note that this approach to revision is a *skeptical* one.

The framework sketched in [14] in fact was a starting point for this research: it was confined to stable agents using classical consequences only and the intended models of the revised theory were obtained as the stable models of classical expansions:

$$Stable(P + \{x \in R \mid Mod(P) \models x\})$$

In this paper we will only touch upon a few aspects of this new framework.

First, we will concentrate on the nature of the information to be provided if an agent Sem consults a weaker colleague Sem' and we will show that, in this respect, there exists a sharp distinction between *Stable* and the remaining semantics.

Note that we will discuss revision for *normal logic programs* and always assume that the programs to be revised are *classically consistent*.

3 Adding information to a program

Given some semantics $Sem > Sem'$ we would like to know which kind of information Sem needs from Sem' in order to revise a program P successfully, i.e., obtaining an expansion P' such that $Sem(P') \neq \emptyset$.

Definition 4. Let O be an expansion operation on programs P, such that $O(P) \supseteq P$. We say that O is *successful* with respect to semantics Sem, if $Sem(O(P)) \neq \emptyset$.

Since the information delivered by Sem' has to be added to the program, we can distinguish between *atomic* or factual information and *rule* information provided by Sem'. Then we can investigate which information can be used to construct successful expansion operations for various semantics Sem.

3.1 Adding atomic information

It is not difficult to show that in general, atomic information is not sufficient to construct successful expansion operations for an arbitrary non-monotonic reasoner[1].

---

[1] Atomic information is sufficient if we do not require the skeptical form of revision, see Section 4.

Lemma 5. *There are programs P, such that for every semantics Sem, Sem $>$ MinMod^2, the following holds:*

For every $Sem' < Sem$ and every $F \subseteq \{a \in B_P \mid Sem'(P) \models a\}$,
$Sem(P + F) = \emptyset$.

Proof. Consider the following program:

$$P: \quad a \leftarrow \neg a, b$$
$$b \leftarrow \neg c$$
$$c \leftarrow \neg b$$
$$d \leftarrow \neg d, c$$

Since P has no supported, positivistic or stable model, it is sufficient to prove that for $Sem' = MinMod$, the set $\{a \in B_P \mid MinMod(P) \models a\} = \emptyset$. This is easy to see, as P has exactly two complementary minimal models: $M_1^+ = \{a, b\}$ and $M_2^+ = \{c, d\}$. $\qquad\square$

3.2 Adding contrapositives

Since atomic information does not always help, we will try to establish results for agents also accepting rule information from their weaker colleagues.

Definition 6. If P is a program

- $Rules(B_P)$ denotes the set of all *normal rules* that can be constructed from atoms in B_P.
- $Contra(P)$ denotes the set of all logical contrapositives $Contra(r)$ of rules r, including r, such that $Contra(r)$ again is a normal rule.

So for example, if $r : a \leftarrow \neg b, c$ is a rule, $Contra(r) = \{a \leftarrow \neg b, c, \ , \ b \leftarrow \neg a, c \ , \ \bot \leftarrow \neg a, \neg b, c\}$. Observe that $Contra(P) \subseteq Rules(B_P)$.

Remark. There is a close correspondence between shift operations as defined in [2] and taking contrapositives. In fact, contrapositives can be considered as bidirectional shift operations where literals can be moved from the head of a rule to the body and vice-versa. Contrapositives also have been used to give a logical reconstruction of the idea of *dependency-directed backtracking* (see [3]), used in both logic programming and truth maintenance as a revision technique. In [13] we have shown that in a three-valued logic, adding contrapositives is successful in obtaining a stable model of the expanded program. For a two-valued semantics, we will show that this picture is different.

Before looking at the general case, we will try to establish the information content of contrapositives of rules.

Our first result is that even for positivistic reasoners it is sufficient to use contrapositive information:

---

^{2} Since P is a consistent normal program, $(Min)Mod(P) \neq \emptyset$, therefore, revision does not apply to $Sem \in \{Mod, MinMod\}$.

Lemma 7. *Let P be a program. Then $Pos(P + Contra(P)) \neq \emptyset$.*

Proof. Since P is assumed to be consistent, P has at least one minimal model M. Recall that $C(P)$ is the set of clauses associated with P. We distinguish the following cases:

1. M does not contain a positive literal.
   By Lemma 2, M is a stable model of P, and since $Stable(P) \subseteq Pos(P)$, $M \in Pos(P)$. By the Expansion Lemma, it follows that $M \in Pos(P + Contra(P))$.
2. M contains at least one positive literal.
   Let $M^+ = \{c_1, c_2, \ldots, c_m\}$. Since M is a minimal model, for every $c_i \in M^+$ there is at least one (not necessarily positive) clause $C_i = \{c_i, l_1, \ldots, l_m\}$. in $C(P)$ containing the atom c_i, such that M minimally satisfies C_i by making c_i true and making the remaining literals l_j, $j = 1, 2, \ldots, m$ false. For else, c_i could be safely removed from M^+, contradicting the fact that M is minimal. Then $M \models c_i$ and $M \models \neg l_j$ for $j = 1, \ldots m$, hence, c_i is supported by the following rule $r(c_i)$ associated with such a clause C_i:

$$r(c_i) = c_i \leftarrow \bar{l}_1, \bar{l}_2, \ldots, \bar{l}_m.$$

   Let $P' = \{r(c_i) \mid c_i \in M^+\}$. By definition, M is a supported model of P'. Hence, by the Expansion Lemma, M is also a supported model of $P + P'$. Since $M \models Contra(P)$ and $P + P' \subseteq P + Contra(P)$, again by the Expansion Lemma, it follows that $M \in Supp(P + Contra(P))$.
   Finally, since M is a minimal model of P and M is also a model of $P + Contra(P)$, M is also a minimal model of $P + Contra(P)$. Hence, M is a positivistic model of $P + Contra(P)$. $\qquad\square$

We state the following theorem as an easy consequence of Lemma 7 :

Theorem 8. *For every program P and semantics Sem such that Sem < Stable, $Sem(P + Contra(P)) \neq \emptyset$.*

In fact, it is not difficult to show that for $Sem \in \{Pos, Supp\}$, such semantics reduces to the minimal model semantics of P: $Sem(P + Contra(P)) = MinMod(P)$.

3.3 Adding contrapositives and condensations

In the case of the stable model semantics, adding contrapositives in general is *not* sufficient. We present a simple example:

Example 2. Take the program: $P = \{b \leftarrow \neg a,\ a \leftarrow b,\ b \leftarrow a\}$. *The unique (classical) model of P is $M = \{a, b\}$. Since $Mod(P) = Mod(P + F)$ for every $F \subseteq Contra(P)$, if there is a stable model of some $P + F$, it should be equal to M. Note, however, that for each such a program $P' = P + F$, we have $G(P', M) \subseteq \{a \leftarrow b, b \leftarrow a\}$. Hence, $lfp(T_{G(P',M)}) = \emptyset \neq M^+$ and M cannot be stable.*

Since adding contrapositives is not always sufficient to help stable agents, we will investigate the use of other rule information. We will now allow for the addition of contrapositives of program rules in which some literals may have been removed. Such *condensations*, however, of rules are only allowed if they appear as classical consequences of P.

We will then show that combining contrapositives and condensations is sufficient to allow a stable reasoner to find an acceptable model for every program. This immediately implies that such information is sufficient to provide every weaker reasoner with an intended model.

Definition 9. A set of clauses C' is said to be a *condensation*[3] of a set of clauses C iff C' is an inclusion minimal set of clauses such that

1. for every clause $C \in C$ there exists a clause $C' \in C'$ such that $C' \subseteq C$ and
2. $Mod(C) = Mod(C')$.

Analogously, we say that P' is a *condensation* of P, if $C(P')$ is a condensation of $C(P)$.

Definition 10. A program P' is said to be obtained from a program P by *contracondens* operations if P' is a condensation of P. $ContraCondens(P)$ denotes the set of rules that can be obtained from P by applying contracondens operations. More formally: $ContraCondens(P) = \bigcup_{P' \text{ is a condensation of } P} P'$.

We note that, in particular, $Contra(P) \subseteq ContraCondens(P)$.

It is often useful to consider *extreme* condensations of clauses and programs:

Definition 11. A set of clauses C' is a *maximal condensation* of a set of clauses C iff

1. C' is a condensation of C and
2. C' is *maximally condensed*, i.e. for every condensation C'' of C', $C'' = C'$.

Analogously, a program P is a *maximally* condensed program iff $C(P)$ is a maximally condensed set of clauses and $C'(P)$ is a maximal condensation of P if $C'(P)$ is a maximal condensation of $C(P)$.

Maximal condensations of P can be obtained by using prime implicates of $C(P)$:

Proposition 12.
Let $\Pi(P)$ denote an inclusion-minimal set of prime-implicates of $C(P)$, such that for every $C \in C(P)$ there is a prime implicate C' of $C(P)$ with $C' \subseteq C$ and $C' \in \Pi(P)$. Then $\Pi(P)$ is a maximal condensation of $C(P)$.

A nice property of maximal condensations we will need in the next section is that in every positive clause C of $\Pi(P)$ every atom $c \in C$ occurs in at least one minimal model M of P that also minimally satisfies C:

---

[3] This notion is related to the notion of a condensation of a clause, as discussed in [6], but not identical. In this latter article, a condensation relation is a relation between two clauses instead of sets of clauses.

Observation 13.

For every positive clause C in $\Pi(P)$ and for every atom c in C there exists at least one minimal model M of P such that M minimally satisfies C and M makes c true.

This property will be used in the next section to show that given an arbitrary consistent program P, we can construct a normal program P' from $\Pi(P)$ such that (i) $P' \subseteq ContraCondens(P)$ and (ii) $Stable(P') \neq \emptyset$.

3.4 Constructing stable models for maximally condensed programs

In this section we will show that expansion by contracondensing is successful for every consistent normal program with respect to the stable semantics.

The construction idea we will use can be summarized as follows:

1. We assume to be given a maximal condensation $\Pi = \Pi(P)$ of an arbitrary consistent program P.
2. From $\Pi(P)$ we construct a normal program P' and a stable model M'.
3. We show that $C(P') = \Pi$, i.e. the model-theoretical interpretations of P' and P are identical.

Note that for such an M' we also have $M' \models ContraCondens(P)$ and $P \subseteq ContraCondens(P)$. Since for every logic program P, there exists a maximal condensation $\Pi(P)$ of P, by the Expansion Lemma, for every program P there exists a program $P'' = P + ContraCondens(P)$ such that P'' has a stable model.

As an important corollary, we note that, with respect to maximally condensed logic programs, adding contrapositives is sufficient to obtain a stable model.

The method we will present constructs P' and M' in a finite number of stages $i = 0, 1, \ldots, n$. At every stage i, $i \geq 0$, the currently partial realization of P' is denoted as P^i and the (partial) model associated with P^i as M^i. To show that M' is a stable model of P' and can be constructed in finite time, it suffices to prove the following claims:

Claim 1 At every stage i, M^i will be a stable model of P^i and a partial model of Π, i.e. there exists at least one complete extension M of M^i which is a model of Π.   -

Claim 2 After a finite number of stages, $C(P^i) = \Pi$.

The stages themselves are defined as follows:

At Stage 0, let $P^0 = \emptyset$ and $M^0 = \emptyset$. Since M^0 is a stable model of P^0 and M^0 is extendible to a model M of Π, Claim 1 holds for Stage 0.

At Stage $i+1$, proceeding inductively, we have at our disposal a partial model M^i of Π, which is also a stable model of P^i, where $C(P^i) \subseteq \Pi$.

Then we consider a set of clauses $\Pi^{i+1} = R(\Pi, M^i)$, derived from Π and M^i as follows:

1. remove every *clause* C' from Π containing a literal c such that $M^i \models c$;

2. remove in the remaining clauses C', every *literal* c such that $M^i \models \neg c$;
3. let the resulting set of clauses be Σ^{i+1} and construct a set Π^{i+1} of maximally condensed clauses from Σ^{i+1}.

Note that, since M^i has an extension M satisfying Π, Σ^{i+1} and Π^{i+1} must be satisfiable.

We distinguish two cases:

Case 1. $\Pi^{i+1} = R(\Pi, M^i)$ does not contain any positive clause.
By Lemma 2, $M_{neg} = \neg(Atoms(\Pi) - Atoms(P^i))$, is a stable model of Π^{i+1}.
Note that $C(P^i) \subseteq \Pi$. Let $P'' \in Normal(\Pi - C(P^i))$ be an arbitrary normal program, $P^{i+1} = P^i \cup P''$ and $M^{i+1} = M^i \cup M_{neg}$.
Then it is not difficult to show that
(a) M^{i+1} is a stable model P^{i+1}
(b) $C(P^{i+1}) = \Pi$
This implies that Stage $i + 1$ is a final stage, $M' = M^{i+1}$ and $P' = P^{i+1}$, proving Claim 1+2.

Case 2. $\Pi^{i+1} = R(\Pi, M^i)$ contains at least one positive clause $C'' = \{c_1, \ldots, c_k\}$.
Since Π^{i+1} is a maximally condensed set, by Observation 13, there is a minimal model M_m of Π^{i+1}, making c_1 true and every c_j, $j \geq 2$ false.
Let $C' \supseteq C''$ be a clause in Σ^{i+1} from which C'' has been derived and

$$C = \{c_1, \ldots, c_k, c_{k+1}, \ldots, c_m\}$$

be the clause in Π from which C' has been derived.
Consider the following rule $r(C)$ to be added to P^i to form the program P^{i+1}:

$$r(C) = c_1 \leftarrow \bar{c}_2, \ldots, \bar{c}_k, \bar{c}_{k+1}, \ldots, \bar{c}_m$$

Let $M^{i+1} = M^i \cup \{c_1, \bar{c}_2, \ldots, \bar{c}_m\}$. It is easy to show that:
(a) M^{i+1} is a stable model of P^{i+1};
(b) there is at least one extension M of M^{i+1} such that M is a model of Π; finishing the proof of Claim 1.
Finally, note that at the end of Stage $i + 1$, either (i) $C(P^{i+1}) = \Pi$ or (ii) $C(P^i) \subset C(P^{i+1}) \subset \Pi$ and $C(P^{i+1}) - C(P^i) \neq \emptyset$. This means that after a finite number of stages, the construction of P' is completed, proving Claim 2.

From the properties of this construction, the following result can be easily derived:

Lemma 14.
For every consistent maximally condensed set of clauses Π, there is a normal program P' such that $C(P') = \Pi$ and $Stable(P') \neq \emptyset$.

The procedure given in Figure 1 is a succinct description of our method to find a stable model of an expansion of a normal program P.

We present an example to illustrate the application of the method described above.

input: the set $\Pi = \Pi(P)$.
output: a program $P' \subseteq ContraCondens(P)$ and a stable model M' of P'.
begin

    Let $j := 0$;
    Let $P^j := \emptyset$; let $M^j := \emptyset$; Let $\Pi^j := \Pi$;
    while $R(\Pi, M^j)$ contains a positive clause C containing c_1 derived from
        some $C^0 = \{c_1, c_2, \ldots, c_m\} \in \Pi$
        $r := c_1 \leftarrow \bar{c}_2, \ldots, \bar{c}_m$;
        $P^{j+1} := P^j + r$;
        $M^{j+1} := M^j \cup \{c_1, \bar{c}_2, \ldots, \bar{c}_m\}$;
        $j := j + 1$
    wend
    $M' := M^j \cup \neg(Atoms(\Pi) - Atoms(P^j))$;
    Let $P'' \in Normal(\Pi - C(P^j))$;
    $P' := P^j \cup P''$;
    return (P', M');
end;

Fig. 1. Finding a suitable transformation of a normal program P

Example 3.
Consider the following program:

$$P: \quad b \leftarrow \neg a$$
$$c \leftarrow \neg b$$
$$a \leftarrow \neg c$$

Since $Stable(P) = \emptyset$, we transform P into a set of clauses $C(P) = \{\{a, b\}, \{b, c\}, \{c, a\}\}$. Note that $\Pi = \Pi(P) = C(P)$. Π^0 contains a positive clause $\{c, a\}$; therefore, let P^1 contain the rule $r = c \leftarrow \neg a$ and $M^1 = \{c, \neg a\}$. Now the clauses $\{a, c\}$ and $\{b, c\}$ can be removed from Π^0 and a can be removed from $\{a, b\}$. Therefore, $\Pi^1 = \Sigma^1 = \{b\}$. Since $\{b\}$ has been derived from $\{a, b\}$, P^2 will include the rule $b \leftarrow \neg a$ and $M^2 = \{b, c, \neg a\}$.

Note that now $\Sigma^3 = \emptyset$ and therefore, does not contain a positive clause. Now $\Pi - C(P^2) = \{b, c\}$. So, let $P'' = \{c \leftarrow \neg b\}$. Then

$$P' = P^2 \cup \{c \leftarrow \neg b\}$$
$$= \{c \leftarrow \neg a, b \leftarrow \neg a, c \leftarrow \neg b\}$$

and $M' = \{b, c, \neg a\}$ is a stable model of P'.

Since $ContraCondens(P) \supseteq Contra(\Pi(P))$, from Lemma 14, by application of the Expansion Lemma, it follows immediately that

Theorem 15. *For every logic program P, $Stable(P + ContraCondens(P)) \neq \emptyset$.*

Since, for every $Sem' \geq Mod$, $Sem'(P) \models ContraCondens(P)$, applying the Expansion Lemma, we obtain a much stronger result for non-atomic information, stating that every nm-reasoner can use rule information from weaker reasoners:

Theorem 16. *For every program P and every pair of semantics $Sem > Sem'$, $Sem(P + \{r \in Rules(B_P) \mid Sem'(P) \models r\}) \neq \emptyset$.*

4 Exploring the Framework

We have only explored a part of the new framework. We will briefly address some more observations:

Disjunctive Programs We have concentrated on normal programs + constraints. It is not difficult to show, however, that (see [15]) *ContraCondens*-additions to general, disjunctive programs are sufficient to provide every (consistent) program with an acceptable model. In that sense, *ContraCondens* operations are more powerful than the shift-operations discussed in [12] to provide disjunctive programs with a (weakly)-stable model.

Skeptical versus credulous revision The framework offers possibilities to study both skeptical and credulous forms of revision. In this paper we have discussed the skeptical variant by requiring that information should be inferable in every intended model belonging to some class. It should also be possible to use the *credulous* form of revision by using one or a few intended models belonging to a weaker semantics.

In fact, is is not difficult to prove that for every credulous reasoner, atomic information from weaker reasoners suffices. If we decide to add rule information such as contrapositives, in a credulous approach we could choose to add those rules $r \in Contra(P)$ such that there exists a weaker model M of P and $M \models hd(r)$. Let us briefly elaborate on the following aspects of this type of credulous revision:

- *Reasoners do not always completely reduce to weaker reasoners.*
  Just like the positivistic semantics, it is not difficult to prove that in the skeptical approach, the stable semantics reduces to the minimal semantics: $Stable(P + ContraCondens(P)) = MinMod(P)$. It is an open problem how to characterize rule information sets R such that R is derived skeptically and e.g. $Stable(P + R) = Pos(P)$ or $Stable(P + R) = Supp(P)$.
  In a credulous approach these reduction effects do not need to occur: For example, take the following program, adapted from [1]:
  $P = \{a \leftarrow c, \neg b, \ c \leftarrow \neg a, \ c \leftarrow b, \ c \leftarrow c\}$
  This program does not have a stable or positivistic model. Its minimal models are $M_1^+ = \{a\}$, $M_2^+ = \{b, c\}$ and its single supported model is $M_3^+ = \{a, c\}$.

In order to find a positivistic model, we add those contrapositives r to P for which there is a supported model M such that $M \models hd(r)$. For P we have to add the rule $a \leftarrow \neg c$. Then the expanded program P' has a single positivistic model $\{a\} = M_1^+$. Hence, $Pos(P') \neq Supp(P)$ and $Pos(P') \neq MinMod(P)$.

– *Constrained communication versus unconstrained communication.*
In the skeptical approach, we used the set of contrapositives to communicate. In fact, this comes down to allowing agents to communicate with weakest (classical) agents in order to obtain additional information.

In a credulous approach, it really makes a difference whether or not we allow for unconstrained information: Take the last program P. If we only allow the positivistic reasoner to communicate with his strongest weaker neighbor, the supported agent, we obtain an expansion P' such that $Pos(P') = \{M_1\}$. If, however, we allow the positivistic reasoner to communicate with the minimal agents directly, the expanded program P'' would be equal to $P + \{a \leftarrow \neg c , b \leftarrow c, \neg a\}$. The result is that both minimal models are supported, hence $Pos(P'') = \{M_1, M_2\}$.

Note that unconstrained communication in this example will result in a loss of inferential power: $Pos(P') \models a$, while $Pos(P'') \not\models a$.

5 Conclusion

We have developed a new framework for doing revision of logic programs, stressing the relations between different non-monotonic semantics and the kind of information needed to find a successful revision of the program.

Our main results were:

1. pointing out an important difference between stable and positivistic (supported) reasoners in terms of the information they need to obtain from weaker reasoners in order to revise their programs.
2. a characterization of the kind of information needed to do successful revision for every nm-reasoner.
3. observing that in credulous revision, an agent A that consults a weaker agent B does not always derive exactly the conclusions of B; moreover, if $C < B < A$, there may be subtle differences between A consulting C directly, and A consulting B (who then may consult C).

Acknowledgments We would like to thank the anonymous referees for the careful reading of the submitted paper and for pointing out some significant omissions and mistakes. Their remarks were greatly appreciated.

References

1. N. Bidoit, Negation in rule-based database languages: a survey, *Theoretical Computer Science*, **78**, (1991), 3–83.
2. J. Dix, G. Gottlob, V. Marek, Causal Models of Disjunctive Logic Programs, in: *Proceedings of the Tenth International Conference on Logic Programming ICLP'94*, 1994.
3. Doyle, J., A Truth Maintenance System, *Artificial Intelligence* 12, 1979.
4. P. Gärdenfors, *Knowledge in Flux*, MIT Press, Cambridge, MA, 1988.
5. L. Giordano and A. Martelli, Generalized Stable Models, Truth Maintenance and Conflict Resolution, in: D. Warren and P. Szeredi, editors, *Proceedings of the 7th International Conference on Logic Programming*, pp. 427-441, 1990.
6. G. Gottlob, C. G. Fermüller, Removing Redundancy from a clause, *Artificial Intelligence*, **61** (1993), 263–289.
7. C. M. Jonker and C. Witteveen. Revision by expansion. In G. Lakemeyer and B. Nebel, editors, *Foundations of Knowledge Representation and Reasoning*, pp. 333-354. Springer Verlag, LNAI 810, 1994.
8. J. W. Lloyd, *Foundations of Logic Programming*, Springer Verlag, Heidelberg, 1987.
9. W. Marek, V.S. Subrahmanian, The relationship between stable, supported, default and auto-epistemic semantics for general logic programs, *Theoretical Computer Science* 103 (1992) 365–386.
10. L. M. Pereira, J. J. Alferes and J. N. Aparicio, Contradiction Removal within well-founded semantics. In: A. Nerode, W. Marek and V. S. Subrahmanian, editors, *First International Workshop on Logic Programming and Non-monotonic Reasoning*, MIT Press, 1991
11. L. M. Pereira, J. J. Alferes and J. N. Aparicio, The Extended Stable Models of Contradiction Removal Semantics. In: P. Barahona, L.M. Pereira and A. Porto, editors, *Proceedings -EPIA 91*, Springer Verlag, Heidelberg, 1991.
12. M. Schaerf, Negation and Minimality in Disjunctive Databases. In: C. Beeri (ed.), *Proceedings of the Twelfth Conference on Principles of Database Systems (PODS-93)*, pp. 147-157, ACM-Press, 1993.
13. C. Witteveen and G. Brewka, Skeptical Reason Maintenance and Belief Revision, *Artificial Intelligence*, **61** (1993) 1–36.
14. C. Witteveen, W. van der Hoek and H. de Nivelle. Revision of non-monotonic theories: Some postulates and an application to logic programming. In D. Pearce C. MacNish and L.M. Pereira, editors, *Logics in Artificial Intelligence, LNAI 838*, pp. 137-151, Springer Verlag, 1994.
15. C. Witteveen. Shifting and Condensing Logic programs. TWI-report 1994, Delft University of Technology, to appear.
16. D. Zacca, Deterministic and Non-Deterministic Stable Model Semantics for Unbound DATALOG queries. in: G. Gottlob, M. Vardi, editors, *Database Theory - ICDT'95*, Springer Verlag, 1995.

Hypothetical Updates, Priority and Inconsistency in a Logic Programming Language[*]

D. Gabbay[1], L. Giordano[2], A. Martelli[2], N. Olivetti[2]

[1] Department of Computing, Imperial College,
180 Queen's Gate - London SW7 2BZ, UK
e-mail: dg@doc.ic.ac.uk
[2] Dipartimento di Informatica, Università di Torino,
Corso Svizzera 185, 10149 Torino, Italy
e-mail: laura,mrt,olivetti@di.unito.it

Abstract. In this paper we propose a logic programming language which supports hypothetical updates together with integrity constraints. The language allows sequences of updates by sets of atoms and it makes use of a revision mechanism to restore consistency when an update violates some integrity constraint. The revision policy we adopt is based on the simple idea that more recent information is preferred to earlier one. This language can be used to perform several types of defeasible reasoning. We define a goal-directed proof procedure for the language and develope a logical characterization in a modal logic by introducing an abductive semantics.

1 Introduction

In [8] we have proposed a logic programming language which supports hypothetical updates together with integrity constraints. The language, called *CondLP* (*Conditional Logic Programming*), is an extension of N_Prolog [10, 7]. As in N_Prolog and in other similar logic programming languages [16, 15], Horn clause logic is extended to allow embedded implications both in goals and in clause bodies. In this paper we present a language, *CondLP+* which extends *CondLP* in many different ways.

Operationally, in N_Prolog an embedded implication $D \Rightarrow G$, where G is a goal and D is a set of clauses, succeeds from a program P if G succeeds from the enlarged program $P \cup D$. Hence, D can be regarded as an hypothetical update to the current program, that only affects the proof of G. As a difference with N_Prolog, in both *CondLP* and *CondLP+* a program may contain integrity constraints, and inconsistencies may arise, when updates are performed. In both languages we have a restriction on the allowed updates: in *CondLP* D is a single atom, while in *CondLP+* D is a set of atoms.

---

[*] This work has been partially supported by ESPRIT Basic Research Project 6471, Medlar II.

In *CondLP⁺* a program consists of a protected part and a removable part. The protected part contains clauses and integrity constraints of the form $G \rightarrow \perp$, whose meaning is that G must be false ($\perp$ is a distinguished symbol denoting falsity). Due to the presence of constraints, when an implication goal $D \Rightarrow G$ is proved, adding D to P may produce an inconsistency. When this happens, a *revision* of the program is needed in order to restore consistency. However, revision does not affect the protected part, which is permanent. The removable part of the program, consisting of a set of atomic formulas, is partitioned in different priority classes, and, when an inconsistency occurs, the atoms which are responsible for the inconsistency and have the least preference are overruled.

When new atoms are hypothetically added to the program, they are incorporated in its removable part with the highest priority. Thus, the removable part of the program consists of a sequence of sets of atoms which can change during the computation.

The problem of maintaining consistency in face of contradictory information has been widely studied in belief revision and in database theory (see [20] for a survey). Concerning logical databases, [17] provides a procedure for reasoning over inconsistent data, with the aim of supporting temporal, hypothetical and counter-factual queries. As in [17] we adopt the idea that earlier information is superseded by later information, but as a difference with respect to [17], we provide update operations in the language itself, by making use of embedded implications, on the line of [14, 6].

In *CondLP* [8] we focused on the case when atoms in the removable part of the program are totally ordered, and only atomic updates are allowed. In such a case, a unique revision of the program can always be found when an update is performed. On the contrary, in *CondLP⁺*, since the removable part of the program may contain more atoms with the same priority, there might be more than one revision of the program. In general, the problem is that of reasoning with multiple, possibly contradicting, sources of information ordered according to their reliability. Such a problem has been tackled by several authors (see, for instance, [1, 3]). In particular, in [3] it is considered the case when the knowledge sources are consistent sets of literals, and two possible attitudes towards the preference ordering are distinguished: the "suspicious attitude", which consists in rejecting all information provided by a knowledge source if a part of it is contradictory with some more reliable knowledge source; and the "trusting" attitude, which consists in rejecting just the information that contradicts the one contained in a more reliable source. In [4] logical systems are defined for reasoning according to each attitude.

In this paper we will adopt a "trusting attitude", though we can easily model the "suspicious" attitude behaviour by a proper formulation of knowledge (essentially, by regarding an update by a set of atoms as the update by the single formula which is the conjunction of all of them). In our work the knowledge sources are just sets of atoms, and inconsistencies among them may arise because of the presence of integrity constraints in the permanent part of the program. Differently from [4], our programs usually contain background knowledge common to all the sources (including integrity constraints), and we do not assume each single source to be consistent. As an example of use of the language for reasoning with multiple (inconsistent) sources of informations, consider the following one, which comes from [1] and has also been analyzed in [4].

Example 1. The example describes the knowledge of an inspector and of John and Bill, witnesses of a murder. The knowledge of the inspector consists of the following set Δ of clauses and constraints:

$orange\_coat \rightarrow light\_coat$
$black\_coat \rightarrow dark\_coat$
$light\_coat \wedge dark\_coat \rightarrow \perp$
$orange\_coat \wedge pink\_mercedes \rightarrow suspect\_Jeff$
$black\_coat \wedge pink\_mercedes \rightarrow suspect\_Ed$

The knowledge of John and Bill consists, respectively, of the following facts:

$John = \{orange\_coat, no\_hat, pink\_mercedes\}$
$Bill = \{black\_coat, no\_hat\}.$

Given the knowledge of the inspector, the information provided by John and Bill is contradictory. Let us assume that Bill is more reliable than John, since Bill was closer than John to the scene of the murder. The goal

$\{orange\_coat, no\_hat, pink\_mercedes\} \Rightarrow (\{ black\_coat, no\_hat\} \Rightarrow suspect\_Ed)$

succeeds from the program P, containing the clauses and constraints in Δ. On the other hand, the goal

$\{orange\_coat, no\_hat, pink\_mercedes\} \Rightarrow (\{ black\_coat, no\_hat\} \Rightarrow suspect\_Jeff)$

fails. In fact, the information *orange_coat* is not accepted, since it is in conflict with the more reliable information *black_coat*. Nevertheless, the inspector believes the information *pink_mercedes*, coming from the same source (John), since we adopt a trusting attitude.

Apart from the possibility of having sets of atoms with the same reliability, another difference between $CondLP^+$ and $CondLP$ in [8] is that, in $CondLP$, each time an update is performed in the proof of a given goal, the consistency of the new database is checked, and, when an inconsistency arises, the database is permanently modified by deleting some facts. By contrast, in $CondLP^+$, rather than removing an inconsistency as soon as it arises, the proof procedure maintains the consistency of logical consequences of the database by making use of the priority among facts. In this respect we follow the approach of [3, 17].

For instance, in the previous example, the information *orange_coat* is not accepted, since it is overridden by the more reliable information *black_coat*. However, *orange_coat* is not removed from the program. If, later on, a new witness Tom , which is more reliable than both John and Bill, came out and said that the coat was light, then Bill's information *black_coat* would be rejected and John's information *orange_coat* would be accepted in turn. The inspector could hence prove *suspect_Jeff*. This would be the effect of evaluating the hypothetical goal

$\{ orange\_coat, no\_hat, pink\_mercedes\} \Rightarrow (\{ black\_coat, no\_hat\} \Rightarrow$
$(light\_coat \Rightarrow suspect\_Jeff))$

from the above program P.

In section 2 we define a goal directed proof procedure for the language, and in section 3 we give some examples of its use. We also show that in this language we can easily represent normal logic programs with negation as failure. Since negation as failure is not explicitly provided by the language, we can say that the language follows the principle of "abstracting away NAF from the object level syntax of the representational language", as the languages in [12] and in [13]. In section 4 we provide a logic characterization of $CondLP^+$ by moving to a modal setting and representing updates by modal operators rather than hypothetical implications. Moreover, to account for the non-monotonic behaviour of the language, we introduce an abductive semantics for it, which has strong similarities with Eshghi and Kowalski's abductive semantics for negation as failure [5]. We give a soundness result for the proof procedure with respect to the abductive semantics. Furthermore, we sketch an abductive extension of our proof procedure, and finally, in section 5 we briefly discuss the predicative case.

2 The Operational Semantics

In this section we define the syntax of the language and its operational semantics. We will deal only with the propositional case. Let $true$ and $\perp$ be distinguished propositions (true and false) and let A denote atomic propositions different from $\perp$. The syntax of the language is the following:

G:= $true \mid A \mid G_1 \wedge G_2 \mid A_1, \ldots, A_n \Rightarrow G$
D:= $G \rightarrow A$
I:= $G \rightarrow \perp.$ [3]

In this definition G stands for a goal, D for a clause and I for an integrity constraint. Notice that a goal G may contain nested hypothetical implications as $a, b \Rightarrow (c \Rightarrow (a \wedge c))$, and that, hypothetical implications are allowed in the body of clauses and in constraints , as in the formulas $(a, b \Rightarrow (c \Rightarrow a)) \wedge (b \Rightarrow c) \rightarrow d$ and $(a \Rightarrow c) \wedge f \rightarrow \perp$.

A *program* P is defined as a set of clauses S, a set of integrity constraints IC, and a list of sets of atoms L (the removable part): $P = \Delta \mid L$, where $\Delta = S \cup IC$, and $L = S_1, \ldots, S_n$, each S_i being a set of atoms (facts). While the clauses and constraints in Δ cannot be removed, i.e., they are *protected*, the facts in L are *revisable*: each fact $A \in S_i$ in L can be used in a proof unless it is inconsistent with Δ together with some facts $B \in S_j$ (with $j > i$) with higher priority. Hence, we assume a total ordering among sets of atoms in the list L, and each S_i is less preferred than S_{i+1}. We assume that atoms in S_n have the highest priority, as Δ, and they cannot be removed.

We will now define a goal directed proof procedure for the language. The idea is that, during the computation of a given goal from a program P, the protected

---

[3] We will regard $\perp$ as a proposition without any special properties, since we will use $\perp$ to express integrity constraints in the program, and we do not want to derive everything from the violation of integrity constraints.

part of the program remains fixed, while the list of removable atoms changes, when updates are performed. When an atom from the list L is needed in a proof, it must be verified that it is not inconsistent with the atoms with equal or higher preference (i.e. it must be verified that, assuming that atom, an inconsistency can not be derived from the permanent database and the atoms with equal or higher preference).

The operational derivability of a given goal G from a context $\Delta \mid S_1, \ldots, S_n \mid \{H\}$ (written $\Delta \mid S_1, \ldots, S_n \mid \{H\} \vdash_o G$) is defined by the following proof rules. Each S_i is a set of atomic hypotheses. H represents a temporary hypothesis, which is checked for consistency, and it is virtually part of S_n. However, we keep it separate from S_n since when entering a new context, represented by a further update S_{n+1}, we do not want to assume it anymore.

1. $\Delta \mid S_1, \ldots, S_n \mid \{H\} \vdash_o true$;
2. $\Delta \mid S_1, \ldots, S_n \mid \{H\} \vdash_o G_1 \wedge G_2$ if $\Delta \mid S_1, \ldots, S_n \mid \{H\} \vdash_o G_1$ and
   $\Delta \mid S_1, \ldots, S_n \mid \{H\} \vdash_o G_2$;
3. $\Delta \mid S_1, \ldots, S_n \mid \{H\} \vdash_o S \Rightarrow G$ if $\Delta \mid S_1, \ldots, S_n, S \mid \{\} \vdash_o G$;
4. $\Delta \mid S_1, \ldots, S_n \mid \{H\} \vdash_o A$ (including the case $A = \bot$) if
   (i) there is a clause $G \rightarrow A \in \Delta$ such that $\Delta \mid S_1, \ldots, S_n \mid \{H\} \vdash_o G$, or
   (ii) $A \in S_n$ or $A = H$, or
   (iii) $A \in S_i$ $(i = 1, \ldots, n-1)$ and not $\Delta \mid S_i, \ldots, S_n \mid \{A\} \vdash_o \bot$.

Rule 3 says that, to prove a hypothetical implication $S \Rightarrow G$, the set of atoms S is added to the removable part of the program with the highest priority. Rule 4 defines under what conditions an atomic formula A succeeds: either by making use of a clause in the program (item (i)), or when A is the temporary hypothesis (H), or it belongs to the set of atoms on the top (item (ii)), or, finally, when it belongs to one of the other sets S_i in the list (item (iii)). In this last case, the atomic goal A in S_i is regarded as proved if it is not inconsistent with equally or more reliable atoms. To check this, A is assumed as temporary hypothesis, and a proof of $\bot$ is tried. For the atom A to be usable, $\bot$ is required not to be derivable. A natural choice in a logic programming context is to consider "non-provability" in the last item of rule (4) as finite failure to prove. Hence, we read not $\Delta \mid S_i, \ldots, S_n \mid \{A\} \vdash_o \bot$ as "$\bot$ *finitely fails* from $\Delta \mid S_i, \ldots, S_n \mid \{A\}$".

We say that a goal G is derivable from a program $P = \Delta \mid S_1, \ldots, S_n$ if $\Delta \mid S_1, \ldots, S_n, \{\} \vdash_o G$. The language equipped with the above operational semantics, is obviously non-monotoñic. Consider the following example.

Example 2. Let P be the program with $\Delta = \emptyset$ and $L = \epsilon$. The query $a \Rightarrow (b \Rightarrow (c \Rightarrow a))$ operationally succeeds from P with the following steps of derivation.

$\Delta \mid \epsilon \mid \{\} \vdash_o a \Rightarrow (b \Rightarrow (c \Rightarrow a))$
$\Delta \mid \{a\} \mid \{\} \vdash_o b \Rightarrow (c \Rightarrow a)$, by rule (3),
$\Delta \mid \{a\}, \{b\} \mid \{\} \vdash_o c \Rightarrow a$, by rule (3),
$\Delta \mid \{a\}, \{b\}, \{c\} \mid \{\} \vdash_o a$, by rule (3),
not $\Delta \mid \{a\}, \{b\}, \{c\} \mid \{a\} \vdash_o \bot$, by rule 4(iii)

The query $\perp$ fails from $\Delta \mid \{a\}, \{b\}, \{c\} \mid \{a\}$ since there is no constraint in the program to derive $\perp$. Hence, not $\Delta \mid \{a\}, \{b\}, \{c\} \mid \{a\} \vdash_o \perp$ succeeds, and so does the initial query. On the other hand, the query $a \Rightarrow (b \Rightarrow (c \Rightarrow a))$ fails from the program $P' = \Delta'$, obtained from P by adding the constraint $a \wedge b \rightarrow \perp$. The first part of the derivation is the same as before. However, in this case, the query $\Delta' \mid \{a\}, \{b\}, \{c\} \mid \{a\} \vdash_o \perp$ succeeds,

$$\Delta' \mid \{a\}, \{b\}, \{c\} \mid \{a\} \vdash_o \perp,$$
$$\Delta' \mid \{a\}, \{b\}, \{c\} \mid \{a\} \vdash_o a \wedge b,$$
$$\Delta' \mid \{a\}, \{b\}, \{c\} \mid \{a\} \vdash_o a \text{ and } \Delta' \mid \{a\}, \{b\}, \{c\} \mid \{a\} \vdash_o b,$$

since a succeeds by rule 4(ii), and b succeeds by rule 4(iii), since $\Delta' \mid \{b\}, \{c\} \mid \{b\} \vdash_o \perp$. Thus its negation fails, and so does the initial query.

If we now add the new constraint $b \wedge c \rightarrow \perp$ to P', we get a new program P'' from which the goal $a \Rightarrow (b \Rightarrow (c \Rightarrow a))$, succeeds. In fact, since b is overridden by c, which is preferred, a becomes again visible after the update by c.

This behaviour of the procedure above is different from that one of *CondLP* in [8]. In that case, the consistency check is done after each update, and the less preferred atoms responsible for inconsistencies are permanently removed. Thus, the goal $a \Rightarrow (b \Rightarrow (c \Rightarrow a))$, would fail in *CondLP* since, after the update with b, a is permanently removed from the current program.

3 Examples

Default reasoning

We start with a very simple instance of default reasoning.

Example 3. Let $P = \Delta \mid \{normal\_bird\}$ be a program with protected part
$\Delta = \{ bird \wedge normal\_bird \rightarrow fly \quad\quad true \rightarrow bird$
$\quad\quad\quad penguin \rightarrow notfly \quad\quad\quad fly \wedge notfly \rightarrow \perp\}$
Both the goals $G_1 = fly$ and $G_2 = penguin \Rightarrow notfly$ succeed from P. On the other hand, the goal $G_3 = penguin \Rightarrow fly$ finitely fails.

As it is clear from the above example, the most straightforward way to represent defaults in our language is to make use of the underlying belief revision machinery, and to regard atoms in the revisable list as normal defaults without prerequisite. For instance, in Example 3 we have encoded the default rule $: normal\_bird/normal\_bird$ by adding the atom $normal\_bird$ to the list.

This way of dealing with defaults is close to the approach proposed by Poole [18], where defaults are seen as *possible hypotheses*, which can be assumed if they are consistent. Actually, Poole shows the equivalence of his defaults with Reiter's normal defaults. Following Poole, we can give names to defaults and restrict hypotheses to default names without any loss of generality. In our example, $normal\_bird$ can be considered as the name of the default.

In general, there may be more than one default rule. Let us assume that we want to compute a goal G from a program containing a set of clauses Δ and a set of normal default rules : d_1/d_1, ..., : d_n/d_n . Then we can compute the goal G from the program $\Delta \mid \{d_1, d_2, \ldots, d_n\}$, where the d_i's are removable atoms. If two defaults, d_i and d_j are mutually inconsistent, our procedure may loop. In Section 4 we will discuss how to extend the procedure to overcome this drawback.

It has been claimed that normal defaults are not adequate to deal with exceptions. However, in our case, the expressiveness of defaults may be increased by means of priorities among assumptions.

Example 4. Let us consider the program P containing the following clauses in its protected part Δ:

$unemp\_if\_st \wedge unistudent \rightarrow unemployed$
$emp\_if\_ad \wedge adult \rightarrow employed$
$ad\_if\_st \wedge unistudent \rightarrow adult$
$unistudent$[4]
$unemployed \wedge employed \rightarrow \perp$

Typically, we want to give preference to the first default with respect to the second one, if both of them are applicable. Thus we consider the following program
$P = \Delta \mid \{emp\_if\_ad\}, \{unemp\_if\_st\}, \{ad\_if\_st\}$
The two assumptions emp_if_ad and $unemp\_if\_st$ are inconsistent, and the one with highest priority, i.e. $unemp\_if\_st$ is chosen. Thus the goal $unemployed$ succeeds from the program P, whereas $employed$ fails.

The problem of dealing with exceptions arises in many areas. In particular, in legal reasoning one has to reason about what can be derived from a set of possibly conflicting norms, i.e., general laws, exceptions to the laws and exceptions to the exceptions. We refer the reader to an extended version of the paper for an example from the legal reasoning domain.

Reasoning about actions

The update mechanism of this language provides a way to deal with the frame problem. As an example consider the following formulation of the *shooting problem*[5].

Example 5. Consider the program containing the set Δ of clauses and constraints:

$G \rightarrow holds(G, [])$
$(loaded \Rightarrow holds(G, L)) \rightarrow holds(G, [load|L])$
$loaded \wedge (dead, unloaded \Rightarrow holds(G, L)) \rightarrow holds(G, [shoot|L])$
$unloaded \wedge holds(G, L) \rightarrow holds(G, [shoot|L])$
$holds(G, L) \rightarrow holds(G, [wait|L])$
$dead \wedge alive \rightarrow \perp$
$loaded \wedge unloaded \rightarrow \perp,$

---

[4] In the following we will represent a clause $true \rightarrow A$ as A, as a shorthand.
[5] In this and in the following example we will make use of a first order language, though the discussion of the predicate case is deferred to section 5.

and the list of sets of atoms containing the only set *{alive, unloaded}*. The goal *holds(dead,[load,wait,shoot])* succeeds from the program, while the goal *holds(alive,[load,wait,shoot])* fails. In fact, when the action of shooting is performed, since the gun is loaded, the new fact *dead* is hypothetically added to the program. This produces an inconsistency with the fact *alive*, which is rejected.

Negation as failure

In our language we can easily represent normal logic programs with negation as failure. We model negation as failure by replacing each negative literal *not q*, with the goal $\bar{q} \Rightarrow (true \Rightarrow \bar{q})$, where $\bar{q}$ is a new symbol representing the negation of q. The idea is to add the atom $\bar{q}$ to the program, and then to check if it is consistent with it. The embedded implication with antecedent *true* is needed since atoms added to the program through the last update cannot be removed (i.e., $\bar{q} \Rightarrow \bar{q}$ always succeeds).

A normal program P containing clauses with negation can be represented in our language by a program $\bar{P}$ containing: *(a)* the clauses obtained from those of P by replacing each negative goal *not q* with the goal $\bar{q} \Rightarrow (true \Rightarrow \bar{q})$, and *(b)* for each new symbol $\bar{q}$, the constraint $q \wedge \bar{q} \rightarrow \bot$.

Hence, the program $\bar{P} = \Delta_P$ contains only permanent clauses and constraints, and no removable atoms. Let us see how the goal $\bar{q} \Rightarrow (true \Rightarrow \bar{q})$, is proved from $\bar{P}$:

$$\Delta_P \parallel \{\} \vdash_o \bar{q} \Rightarrow (true \Rightarrow \bar{q}), \qquad (*)$$
$$\Delta_P \mid \{\bar{q}\} \mid \{\} \vdash_o true \Rightarrow \bar{q}$$
$$\Delta_P \mid \{\bar{q}\}, \{true\} \mid \{\} \vdash_o \bar{q}$$
$$not \; \Delta_P \mid \{\bar{q}\}, \{true\} \mid \{\bar{q}\} \vdash_o \bot$$

which succeeds if the derivation of $\bot$ fails finitely from $\Delta_P \mid \{\bar{q}\}, \{true\} \mid \{\bar{q}\}$. Since Δ_P contains only constraints of the form $p \wedge \bar{p} \rightarrow \bot$, and no atom $\bar{p}$ is defined in Δ_P, the only rule applicable to derive $\bot$ is $q \wedge \bar{q} \rightarrow \bot$. Hence, (*) succeeds if $q \wedge \bar{q}$ fails from $\Delta_P \mid \{\bar{q}\}, \{true\} \mid \{\bar{q}\}$, that is, if q fails from that program.

Note that, if, in the proof of q, a new negative goal *not s* has to be proved, then the hypothesis $\bar{s}$ and *true* are added to the list of facts, with higher priority than $\bar{q}$, and $\bar{s}$ is checked from the new program. The hypothesis $\bar{q}$ is removed from the list when rule 4(iii) is applied to prove $\bar{s}$. Thus, no more than one hypothesis $\bar{q}$ can be used at the same time.

4   Logical Characterization of the Language

We present a logical characterization of our non-monotonic language in two steps. First, we introduce a monotonic modal logic which describes the monotonic part of the language. Then, we provide an *abductive semantics* to account for the non-monotonic behaviour of the language. An alternative account of the non-monotonic part of the language has been given in [9] by defining a completion construction, for the language with atomic updates.

We define a modal language L containing: a modal operator $[S]$, for each set of atomic propositions S and a modal operator $\Box$, which is used to denote those formulas that are permanent. In this modal language, we represent an hypothetical implication $D \Rightarrow G$, where D is a set of atoms, by the modal formula $[D]G$. Moreover, we regard the implication $\rightarrow$ in clauses and constraints as *strict implication*: a clause (or a constraint) $G \rightarrow A$ is regarded as a shorthand for $\Box(G \supset A)$, with the meaning that $G \supset A$ is permanently true.

The modalities $[S]$ are ruled by the axioms of the logic K, while the modality $\Box$ is ruled by the axioms of S4. Moreover, the axiom system for L contains the following *interaction axiom* schema: $\Box F \supset [S]\Box F$, where F stands for an arbitrary formula. In [2] such a modal language has been used to extend logic programming with module constructs. Note that the modality $\Box$, which is used to denote permanent information has some similarities with the well known *common knowledge* modality.

In our operational semantics, after any sequence of updates ending with the update S, all the atoms in S are true. We can represent this fact by introducing the additional axiom schema $\Box[S]p$, where S ranges on sets of atoms and p ranges on atomic propositions in S.

In order to account for the non-monotonic behaviour of the language, and for the interaction among subsequent updates, we introduce an abductive semantics in the style of Eshghi and Kowalski's abductive semantics for negation as failure [5]. In the abductive interpretation of "negation as failure" [5], *negative* literals are interpreted as abductive hypotheses that can be assumed to hold, provided they are consistent with the program and a canonical set of integrity constraints. Here, on the contrary, we make use of *positive* assumptions of the form $[S_1]\ldots[S_n]p$, with $p \in S_1$, whose meaning is that an atom p in S_1 holds after the sequence of updates $[S_1]\ldots[S_n]$. Assumptions of this kind are needed to model the non-monotonic part of the operational semantics, given by rule 4 (iii). Such an assumption is blocked whenever the negation of p, $\neg p$ [6], holds after the same sequence of updates (i.e. if p is overridden by atoms with a higher preference).

Actually, in the following we will make use of assumptions (abducibles) of the form $\Box[S_1]\ldots[S_n]p$ (with $p \in S_1$), whose meaning is that p is true after *any* sequence of updates ending with $S_1,\ldots,S_n$.

Let $\vdash_L$ be the derivability relation in the monotonic modal logic defined above. Let Δ be a set of clauses and constraints as defined above, and D be a set of *abducibles* of the form $\Box[S_1]\ldots[S_n]p$ (with $p \in S_1$). We say that D is an *abductive solution* for Δ if

(a) $D \cup \Delta \not\vdash_L \bot$; and
$D \cup \Delta \not\vdash_L [S_1]\ldots[S_n]\neg p \wedge [S_1]\ldots[S_n]p$,
(b) $D \cup \Delta \vdash_L [S_1]\ldots[S_n]\neg p$ or $\Box[S_1]\ldots[S_n]p \in D$,

for all sets of atoms $S_1,\ldots,S_n$ $(n \geq 1)$ and for all atoms $p \in S_1$.

Conditions (a) and (b) are quite similar to the conditions introduced by Eshghi and Kowalski [5] to define their abductive semantics for negation as failure,

---

[6] $\neg p$ is to be interpreted as $p \supset \bot$.

which they prove equivalent to the stable model semantics. The first one is a consistency condition, whereas the second one is a maximality condition which forces an assumption $[S_1]\ldots[S_n]p$ (with $p \in S_1$) to be made, if the formula $[S_1]\ldots[S_n]\neg p$ is not derivable from the current set of assumptions.

It has to be noted that if Δ is itself inconsistent (i.e., $\Delta \vdash_L \bot$), there is no abductive solution D for Δ. Moreover, a logic characterization of this kind is subject to all the problems of the stable model semantics. In particular, solutions do not always exist, and this not only because of the possibility of $\Delta \vdash_L \bot$, but also because of the second (totality) requirement in the definition of abductive solution, which forces either $[S_1]\ldots[S_n]p$ or $[S_1]\ldots[S_n]\neg p$ to hold in each abductive solution (for $p \in S_1$). This is a problem that also occurs in the case of normal logic programs, due to the presence of negative odd loops in the program, i.e., loops of the form $not\ p \supset p$. A program containing such a clause has no abductive solution in Eshghi and Kowalski's framework (and no stable model). To overcome this problem we would need to move to three-valued semantics, on the line of what has been done for normal logic programs.

Notice that it may happen that a program has more than one abductive solution. For instance, the program that is obtained by translating the normal clauses $not\ p \supset q$ and $not\ q \supset p$, according to the translation in section 3, has two abductive solutions, the one containing $not\ p$ and the other containing $not\ q$ (as usual in the stable model semantics).

Roughly speaking, we can say that the relation of this logical characterization with the proof procedure above is similar to the relation of stable models semantics with SLDNF. On one hand, we do not have completeness, since a goal G may loop in the proof procedure while being true in the semantics. On the other hand, we do not have correctness in general, since a goal may succeed in the procedure while there is no abductive solution at all. However, we can say that, the procedure is correct at least in the cases when an abductive solution exists.

Theorem 1. *Given a consistent set of clauses and constraints Δ, for all abductive solutions D for Δ, if $\Delta \mid S_1, \ldots, S_n \vdash_o G$ then $D \cup \Delta \vdash_L [S_1]\ldots[S_n]G$.*

Example 6. Consider again the program $P = \Delta \mid \{normal\_bird\}$ of Example 3. The goals $G_1 = fly$ and $G_2 = [penguin]not\,fly$ both succeed from P. Notice that the goal $[penguin]fly$ fails, and, correspondingly, there is no abductive solution D such that $D \cup \Delta \vdash_L [normal\_bird][penguin]fly$, since
$$D \cup \Delta \vdash_L [normal\_bird][penguin]\neg normal\_bird.$$
Thus, $\Box[normal\_bird][penguin]normal\_bird$ cannot be assumed.

The choice of an abductive semantics is motivated by the fact that, we are considering programs in which facts belonging to the removable part are not totally ordered. Equally reliable facts may be inconsistent with each other, so that, when the program is updated, more than one revision of it may be obtained, in general. In an abductive semantics, we can capture these alternative models through different abductive solutions.

Now that we have introduced an abductive semantics for our language, it would be quite natural to define an abductive procedure for it. The proof procedure in section 2 is not abductive, and is not able to compute alternative abductive solutions: goals proved by that procedure are true in all abductive solutions (if at least one solution exists). On the contrary, an abductive proof procedure, given a goal, searches an abductive solution in which the goal holds. Maybe there is more than one abductive solution, and the goal does not hold in all of them, but only in someone. Consider for instance the following diagnosis example, describing a simple circuit consisting of a battery connected with a bulb, with the constraint that the light cannot be on and off at the same time.

Example 7. Let Δ be the following set of clauses and constraints

(1) $normal\_battery \rightarrow voltage$
(2) $voltage \land normal\_bulb \rightarrow light\_on$
(3) $light\_on \land light\_off \rightarrow \bot$

The propositions $normal\_battery$ and $normal\_bulb$ represent normality of components. Let us consider the program $P = \Delta \mid \{normal\_battery, normal\_bulb\}$. From this program the query $light\_on$ succeeds, using the procedure in section 2. Assume that we have the observation that the light is off. Then we would expect that the query $light\_off \Rightarrow light\_on$ fails, for, when $light\_off$ is added to the database, one of the removable facts $normal\_battery$ or $normal\_bulb$ must be deleted. However, with the above operational semantics, the query loops (since the success of $normal\_battery$ depends on $normal\_bulb$ not being proved, and vice-versa). Since $normal\_battery$ and $normal\_bulb$ have the same preference, an abductive procedure is expected to find a solution in which $normal\_battery$ holds, and another one in which $normal\_bulb$ holds. But no solution in which both of them hold.

For lack of space we do not give the proof procedure which computes abductive explanations for a given goal. The procedure, extending the one in section 2, could be defined on the line of Eshghi and Kowalski's abductive procedure for logic programs with negation as failure [5]. It interleaves two different phases of computation: an abductive phase, in which we reason backward from the current goal and we collect the required abductive hypotheses; and a consistency phase, in which we check the consistency of the collected hypotheses. The procedure makes use of an auxiliary set D of abductive hypotheses, that is initially empty, and is augmented during the computation. At the end, if the query succeeds, this set represents the abductive solution in which the query holds. As a difference with Eshghi and Kowalski's procedure, in our case, the abductive hypotheses should be positive modalized atoms as introduced above.

5 The Predicate Case

In the previous sections we have defined and studied the language $CondLP^+$ only in the propositional case. In this section we briefly discuss the problems that arise when we move to the first order case. As for *CondLP* [8], the difficulty comes mainly from the fact that, during the proof of a goal, atomic formulas containing *free variables* (to be regarded as existentially quantified) can be introduced in the program through updates. Indeed, in order to prove a goal $D(X) \Rightarrow G(X)$, the clause $D(X)$ (containing the free variable X) has to be added to the program.

Example 8. Let Δ be the following set of clauses:

$r(X) \to q$

$r(Y) \wedge p(Y) \to \perp$

The goal $p(b) \Rightarrow q$ succeeds from the program $P = \Delta \mid \{s(b), r(a)\}$, while the goal $p(Z) \Rightarrow q$, where Z is regarded as existentially quantified, fails from P.

$\Delta \mid \{s(b), r(a)\} \mid \{\} \vdash_o p(Z) \Rightarrow q$

$\Delta \mid \{s(b), r(a)\}\{p(Z)\} \mid \{\} \vdash_o q$

$\Delta \mid \{s(b), r(a)\}\{p(Z)\} \mid \{\} \vdash_o r(X)$

not $\Delta \mid \{s(b), r(a)\}\{p(Z)\} \mid \{r(a)\} \vdash_o \perp$

which fails since the query $\perp$ succeeds from $\Delta \mid \{s(b), r(a)\}\{p(Z)\} \mid \{r(a)\}$ with computed answer $\{Z/a\}$ (by making use of the constraint in Δ).

The result is not the expected one. Here the problem is similar to that of floundering for negation as failure. While in NAF a floundering situation occurs when, during a computation, the goal to be solved only contains nonground negative literals, in this setting, a floundering situation occurs when in rule 4(iii) we check whether $\perp$ fails finitely from a non-closed program.

To avoid floundering, the introduction of existentially quantified variables in the program can be prevented by putting some syntactic restrictions on programs, which happens to be similar to the *allowedness condition* defined for normal logic programs. For a broader discussion on the subject we refer to [11, 8].

6 Conclusions and Related Work

In this paper we have defined a language with embedded implications which extends N_Prolog by allowing the definition of constraints and preference ordering among information. The language is able to perform defeasible and counterfactual reasoning: when new information is added to the program, some of the constraints may be violated, and a revision of the program may be required to restore consistency.

In section 4 we have provided a logical characterization of $CondLP^+$ by defining an abductive semantics based on a modal logic. Updates are represented by modal operators rather than hypothetical implications. Our use of the modal language somewhat resembles the language presented in [6], which contains a modal operator *assume[L]* (where L is a literal) to represent addition and deletion of atomic formulas. However, differently from [6] where a model-based approach to revision is adopted, here we have followed a *formula-based approach*, so that the semantics of an update depends on the syntax of the formulas present in the program (see [20]).

As mentioned in the introduction, our proposal can also be regarded as an extension of the proposal in [3], where a modal logic for reasoning with multiple knowledge sources is defined. In [3] a comparison with the belief revision problem is also provided. Similar results can be obtained for our language, in the case the protected part of the programs contains only clauses without updates, and the atoms in the revisable list are totally ordered.

The problem of combining theories and resolving inconsistencies among them has also been studied in [1]. In particular, [1] analyzes the problem of combining priorized (first order) theories, $T_1 \prec T_2 \prec \ldots \prec T_k \prec IC$ where the priority relation is a total order, and a set of integrity constraints IC has the highest priority. Two approaches are devised: *bottom-up* and *top-down*. With the bottom-up approach T_1 is combined with T_2, with preference to T_2; the result is combined with T_3, and so on, until the result is combined with T_k, and, finally, with IC. The operational semantics of *CondLP* presented in [8] is essentially based on this approach, though in *CondLP* the case is much simpler, since each theory contains a single atomic formula. The top-down approach amounts to start combining T_k with IC (with preference to IC), then combining T_{k-1} with the result (with preference for the last one), and so on, up to T_1. As observed in [1], top-down combining is more informative then bottom-up combining. The top-down approach is the one we have adopted in *CondLP+*.

In defining *CondLP+* we have chosen a specific policy to deal with updates. In particular, more recent updates are given higher priority with respect to previous ones. This has been motivated by some applications to counterfactual reasoning, and to planning (see [8]). In principle, however, other policies might be used. For instance, updates with lowest priority could be allowed, that could be useful to deal with default assumptions which are expected to have a lower priority than other kinds of assumptions in updates. Of course, these extensions of the language would require a modification of the logic on which the language is based, and, in particular, a modification of the conditions (a) and (b) in the definition of the abductive semantics, that have been defined ad hoc for the chosen policy. By contrast, allowing clauses in the revisable part of the program, would only require a minor extension of the language.

The language we have defined is somewhat related to other logic programming languages which deal with explicit negation and constraints and are faced with the problem of removing contradictions when they occur. In particular, Pereira et al. in [19] present a contradiction removal semantics and a proof-procedure (similar to an abductive procedure) for extended logic programs with explicit negation and integrity constraints. Furthermore, Leone and Rullo develop a language in which programs can be structured in partially ordered sets of clauses (with explicit negation). These languages [19, 13] do not support updates. On the other hand, dealing with partially ordered sets of clauses within our update mechanism would be a major extension of our language, and will be subject of further research.

References

1. C. Baral, S. Kraus, J. Minker, V. S. Subrahmanian. Combining knowledge bases consisting of first-order theories. *J.Automated Reasoning*, vol.8, n.1, pp. 45–71, 1992.

2. M. Baldoni, L.Giordano, and A.Martelli. A multimodal logic to define modules in logic programming. In *Proc. 1993 International Logic Programming Symposium*, pages 473–487, Vancouver, 1993.

3. L. Cholvy. Proving theorems in a multi-source environment. In *Proc. International Joint Conference on Artificial Intelligence*, pages 66–71, Chambery, 1993.

4. L. Cholvy. A logical approach to multi-sources reasoning. In *Knowledge Representation and Reasoning under Uncertainty–LNAI 808*, pages 183–196 , 1994.

5. K. Eshghi and R. Kowalski. Abduction compared with negation by failure. In *Proc. 6th Int. Conference on Logic Programming*, pages 234–254, Lisbon, 1989.

6. L. Fariñas del Cerro and A. Herzig. An automated modal logic for elementary changes. In P. Smets et al., editor, *Non-standard Logics for Automated Reasoning*. Academic Press, 1988.

7. D. M. Gabbay. NProlog: An extension of Prolog with hypothetical implications.II. *J.Logic Programming*, 2(4):251–283, 1985.

8. D. Gabbay, L. Giordano, A. Martelli, and N. Olivetti. Conditional logic programming. In *Proc. 11th Int. Conf. on Logic Programming*, Santa Margherita Ligure, pages 272–289, 1994.

9. D. Gabbay, L. Giordano, A. Martelli, and N. Olivetti. A language for handling hypothetical updates and inconsistency. *MEDLAR II Deliverable DII.5.2P*, 1994.

10. D. M. Gabbay and N. Reyle. NProlog: An extension of Prolog with hypothetical implications.I. *Journal of Logic Programming*, (4):319–355, 1984.

11. L.Giordano and N.Olivetti. Negation as failure in intuitionistic logic programming. In *Proc. Joint International Conference and Symposium on Logic Programming*, pages 431–445, Washington, 1992.

12. A.C. Kakas, P. Mancarella, P.M. Dung. The acceptability semantics for logic programs. In *Proc. 11th Int. Conf. on Logic Programming*, Santa Margherita Ligure, pages 504–519, 1994.

13. N. Leone and P. Rullo. Ordered logic programming with sets. *J. of Logic and Computation*, 3(6):621–642, 1993.

14. S. Manchanda and D.S. Warren. A logic-based language for database updates. In J. Minker, editor, *Foundation of Deductive Databases and Logic Programming*. Morgan-Kaufman, Los Alto, CA, 1987.

15. L. T. Mc Carty. Clausal intuitionistic logic. i. fixed-point semantics. *J. Logic Programming*, 5(1):1–31, 1988.

16. D. Miller. A theory of modules for logic programming. In *Proc. IEEE Symp. on Logic Programming*, pages 106–114, September 1986.

17. S. Naqvi and F. Rossi. Reasoning in inconsistent databases. In *Proc. of the 1990 North American Conf. on Logic Programming*, pages 255–272, 1990.

18. D. Poole A logical framework for default reasoning. *Artificial Intelligence*, 36 (1988) 27–47.

19. L.M. Pereira, J.J. Alferes, J.N. Aparicio Contradiction Removal within the Well Founded Semantics. *Logic Programming and Non-monotonic Reasoning Workshop*, (1991) 105–119.

20. M. Winslett. *Updating Logical Databases*. Cambridge University Press, 1990.

Situation Calculus Specifications for Event Calculus Logic Programs

Rob Miller

Department of Computing, Imperial College of Science, Technology and Medicine,
180 Queen's Gate, London SW7 2BZ, ENGLAND
email: rsm@doc. ic. ac. uk
WWW: http://laotzu.doc.ic.ac.uk/UserPages/staff/rsm/rsm.html

Abstract. A version of the Situation Calculus is presented which is able to deal with information about the actual occurrence of actions in time. Baker's solution to the frame problem using circumscription is adapted to enable default reasoning about action occurrences, as well as about the effects of actions. Two translations of Situation Calculus style theories into Event Calculus style logic programs are defined, and results are given on the soundness and completeness of the translations.

1 Introduction

This paper compares two formalisms and two associated default reasoning techniques for reasoning about action – the Situation Calculus [11], using a variant of Baker's circumscriptive solution to the frame problem [1], and the logic-programming based Event Calculus [8], in which default reasoning is realised through negation-as-failure. The version of the Situation Calculus used enables information about the occurrences of actions along a time line to be represented. A course of actions identified as actually occurring is referred to as a narrative, and this formalism is referred to as the Narrative Situation Calculus. Information about a narrative might be incomplete, so that default assumptions might be required. The circumscription policy incorporated in the Narrative Situation Calculus minimises action occurrences along the time-line. The original Event Calculus incorporates an analogous default assumption - that the only action occurrences are those provable from the theory.

The present paper shows that under certain circumstances the Narrative Situation Calculus may be regarded as a specification for Event Calculus style logic programs. The programs presented here are described as 'Event Calculus style' because of their use of *Initiates* and *Terminates* predicates to describe the effects of actions, because of the form of their persistence axioms, and because of the use of a time-line rather than the notion of a sequence or structure of situations. They differ from some other variants of the Event Calculus in that they do not assume complete knowledge of an initial state, and in that properties can hold (and persist) even if they have not been explicitly initiated by an action. Two classes of programs are discussed, both of which are "sound", for a wide class of domains, in that they only allow derivation of *Holds* information which is semantically entailed by their circumscriptive specifications. Programs of the second type, although more complex, have an advantage over those of the first in that they are also "complete" even where information is missing about the state of affairs before any action occurs.

Notation: Many-sorted first order predicate calculus together with parallel and prioritized circumscription is used to describe the Narrative Situation Calculus. Variable names begin with a lower case letter. All variables in formulas are universally quantified with maximum scope unless otherwise indicated. To simplify descriptions of the implementations, logic programs are written in a subset of the same language, supplemented with the symbol *not* (negation-as-failure). Metavariables are often written with Greek symbols, so that, for example, ζ might represent an arbitrary ground term of a particular sort. The parallel circumscription of predicates $\pi_1, \ldots, \pi_n$ in a theory T with $\xi_1, \ldots, \xi_k$ allowed to vary is written as

$$CIRC[T ; \pi_1, \ldots, \pi_n ; \xi_1, \ldots, \xi_k]$$

If $\rho_1, \ldots, \rho_m$ are also circumscribed, at a higher priority than $\pi_1, \ldots, \pi_n$, this is written as

$$CIRC[T ; \rho_1, \ldots, \rho_m ; \pi_1, \ldots, \pi_n, \xi_1, \ldots, \xi_k] \wedge$$
$$CIRC[T ; \pi_1, \ldots, \pi_n ; \xi_1, \ldots, \xi_k]$$

Justification for this notation can be found, for example, in [9].

One other piece of notation for specifying uniqueness-of-names axioms will be useful. $UNA[\phi_1, \ldots, \phi_m]$ represents the set of axioms necessary to ensure inequality between different terms built up from the (possibly 0-ary) function symbols $\phi_1, \ldots, \phi_m$. It stands for the axioms

$$\phi_i(x_1, \ldots, x_k) \neq \phi_j(y_1, \ldots, y_n)$$

for $i < j$ where ϕ_i has arity k and ϕ_j has arity n, together with the following axiom for each ϕ_i of arity $k > 0$

$$\phi_i(x_1, \ldots, x_k) = \phi_i(y_1, \ldots, y_k) \rightarrow [x_1 = y_1, \ldots, x_k = y_k]$$

2 A Narrative Situation Calculus

In this section an overview is given of the Narrative Situation Calculus employed here as a specification language. This work is presented more fully in [13]. A class of many sorted first order languages is defined, and the types of sentence which can appear in particular domain descriptions are then described. Finally, the circumscription policy is discussed.

Definition 1 (Narrative domain language). A *Narrative domain language* is a first order language with equality of four sorts; a sort $\mathcal{A}$ of actions with sort variables $\{a, a_1, a_2, \ldots\}$, a sort $\mathcal{F}_g$ of generalised fluents[1] with sort variables $\{g, g_1, g_2, \ldots\}$, a sort $\mathcal{S}$ of situations with sort variables $\{s, s_1, s_2, \ldots\}$, and a sort $\mathbb{R}$ of time-points with sort variables $\{t, t_1, t_2, \ldots\}$. The sort $\mathcal{F}_g$ has three sub-sorts; the sub-sort $\mathcal{F}^+$ of positive fluents with sort variables $\{f^+, f_1^+, f_2^+, \ldots\}$, the sub-sort $\mathcal{F}^-$ of negative fluents with sort variables $\{f^-, f_1^-, f_2^-, \ldots\}$, and the sub-sort $\mathcal{F}_f$ of fluents with sort variables $\{f, f_1, f_2, \ldots\}$, such that

$$\mathcal{F}^+ \cap \mathcal{F}^- = \emptyset, \qquad \mathcal{F}^+ \cup \mathcal{F}^- = \mathcal{F}_f, \qquad \mathcal{F}_f \subset \mathcal{F}_g$$

It has time-point constant symbols corresponding to the real numbers, a finite number of action and positive fluent constants, a single situation constant $S0$, and no negative or generalised fluent constants. It has five functions:

$$Neg : \mathcal{F}^+ \mapsto \mathcal{F}^- \qquad Sit : \mathcal{F}_g \mapsto \mathcal{S} \qquad State : \mathbb{R} \mapsto \mathcal{S}$$
$$Result : \mathcal{A} \times \mathcal{S} \mapsto \mathcal{S} \qquad And : \mathcal{F}_g \times \mathcal{F}_g \mapsto \mathcal{F}_g$$

and six predicates (other than equality): $Holds$ ranging over $\mathcal{F}_g \times \mathcal{S}$, Ab ranging over $\mathcal{A} \times \mathcal{F}_f \times \mathcal{S}$, $Absit$ ranging over $\mathcal{F}_g$, $<$ (infix) ranging over $\mathbb{R} \times \mathbb{R}$, $\leq$ (infix) ranging over $\mathbb{R} \times \mathbb{R}$, and $Happens$ ranging over $\mathcal{A} \times \mathbb{R}$.

Only models are considered in which the predicates $<$ and $\leq$ are interpreted in the usual way as the "less-than" and "less-than-or-equal-to" relationships between real numbers. $Happens(\alpha, \tau)$ represents that an action α occurs at time τ[2], and $State(\tau)$ represents the situation at time τ.

Several domain independent axioms will always appear in Narrative Situation Calculus theories. The following five axioms are taken from [1].

$$Holds(And(g_1, g_2), s) \leftrightarrow [Holds(g_1, s) \wedge Holds(g_2, s)] \tag{B1}$$

$$Holds(Neg(f^+), s) \leftrightarrow \neg Holds(f^+, s) \tag{B2}$$

$$Holds(g, Sit(g)) \leftarrow \neg Absit(g) \tag{B3}$$

$$Sit(g_1) = Sit(g_2) \rightarrow g_1 = g_2 \tag{B4}$$

$$[Holds(f, Result(a, s)) \leftrightarrow Holds(f, s)] \leftarrow \neg Ab(a, f, s) \tag{F1}$$

Axioms (B1)-(B4) are Baker's "existence-of-situations" axioms. Every combination of positive and negative fluents (negative fluent terms being those constructed with the Neg function) has at least one corresponding single "generalised fluent" term which can be constructed using the And function. Axioms

---

[1] Baker introduces generalised fluents in order to supply names to conjunctions of primitive fluents, so that for example the generalised fluent $And(Loaded, Neg(Alive))$ represents the joint property of being loaded and dead.

[2] Action occurrences are thus represented here as instantaneous. However the approach can easily be modified to represent actions with a duration (see [13]). The choice of the real numbers is also somewhat arbitrary – any ordered (or "non-converging" partially ordered) set of time-units would suffice (see [12] for further details).

(B1)-(B4), together with minimisation of *Absit*, ensure that, in each preferred model, for each consistent combination of fluents (characterised by some generalised fluent ϕ_g) there is at least one situation ($Sit(\phi_g)$) in which all of these fluents hold. Such situations are not characterised in the language by any actions which have led to them, but simply by the fluents that hold in them. Axiom (F1) is a frame axiom. *Ab* is minimised to represent the assumption that actions result only in changes demanded by the domain theory. Baker's minimisation policy avoids the "Yale Shooting Problem" by incorporating existence-of-situations axioms and by circumscribing *Ab* (at a lower priority than *Absit*) whilst allowing the *Result* function to vary. Varying *Result* ensures that for a given term $Result(\alpha, \sigma)$, the circumscription (and not the structure of the term) determines the set of possible situations to which it might refer. For any given model, inclusion in the language of terms of the form $Result(\alpha, \sigma)$ does not necessitate inclusion of extra situations in the universe of discourse, since there are no uniqueness-of-names axioms mentioning *Result*. Indeed, under the circumscription, the preferable models are generally those in which the denotation of each situation term of the form $Result(\alpha, \sigma)$ is the same as for some term of the form $Sit(\phi_g)$. (For further details consult [1].)

Two more domain independent axioms are included in the Narrative Situation Calculus, concerning properties of narratives and time-points:

$$State(t) = S0 \leftarrow \neg \exists a_1, t_1 [Happens(a_1, t_1) \wedge t_1 < t] \tag{N1}$$

$$\begin{aligned} State(t) = Result(a_1, State(t_1)) \leftarrow & \\ [Happens(a_1, t_1) \wedge t_1 < t \wedge & \\ \neg \exists a_2, t_2 [Happens(a_2, t_2) \wedge [a_1 \neq a_2 \vee t_1 \neq t_2] \wedge t_1 \leq t_2 \wedge t_2 < t]] \end{aligned} \tag{N2}$$

Axiom (N1) relates all time points before the first action occurrence to the initial situation $S0$, and Axiom (N2) says that if action α_1 happens at τ_1, τ_1 is before τ and no other action happens between τ_1 and τ, then the situation at time τ is equal to $Result(\alpha_1, State(\tau_1))$.

Several types of axioms are either required or allowed in Narrative Situation Calculus theories[3]. The following definitions specify the form of such sentences.

Definition 2 (Initial conditions description). A formula is an *initial conditions description* if it is of the form

$$Holds(\phi, S0) \quad \text{or} \quad Holds(Neg(\phi), S0)$$

where ϕ is a positive fluent constant.

Definition 3 (Action description). A formula is an *action description* if it is of one of the following two forms

---

[3] In this paper theories do not include domain constraints, since no translation of domain constraints into logic programs will be given. The Narrative Situation Calculus in [13] includes such constraints.

$$Holds(\phi, Result(\alpha, s))$$

$$Holds(\phi, Result(\alpha, s)) \leftarrow [Holds(\phi_1, s) \wedge \ldots \wedge Holds(\phi_n, s)]$$

where α is an action constant, $\phi, \phi_1, \ldots, \phi_n$ are (positive or negative) ground fluent terms, and for each i and j, $1 \le i, j \le n$, $\phi_i \ne Neg(\phi_j)$.

Definition 4 (Narrative domain description). Given a narrative domain language with positive fluent constants $\phi_1, \ldots, \phi_n$ and action constants $\alpha_1, \ldots, \alpha_m$, a formula N is a *narrative domain description* if it is a conjunction of action descriptions, initial conditions descriptions, occurrence descriptions, the frame axiom (F1), existence-of-situations axioms (B1)-(B4), axioms (N1) and (N2), uniqueness-of-names axioms

$$UNA[\phi_1, \ldots, \phi_n, And, Neg] \qquad UNA[\alpha_1, \ldots, \alpha_m]$$

and a domain closure axiom for fluents

$$f = \phi_1 \vee \ldots \vee f = \phi_n \vee f = Neg(\phi_1) \vee \ldots \vee f = Neg(\phi_n)$$

Although domain constraints have not been explicitly included in narrative domain descriptions, care must be taken, since domain constraints might be derived from pairs of action descriptions, together with Axiom (B2). For example, from the two action descriptions

$$Holds(F_1, Result(A, s)) \leftarrow Holds(F_2, s)$$

$$Holds(Neg(F_1), Result(A, s)) \leftarrow Holds(F_3, s)$$

the sentence $\neg[Holds(F_2, s) \wedge Holds(F_3, s)]$ can be derived. In fact, domain constraints are entailed only from pairs of action descriptions of this form (see [12]). Hence the following definition is included, of narrative domain descriptions with no implicit domain constraints.

Definition 5 (Fluent independence). A narrative domain description N is *fluent independent* if for every pair of action descriptions in N of the form

$$Holds(\phi, Result(\alpha, s)) \leftarrow [Holds(\phi_1, s) \wedge \ldots \wedge Holds(\phi_m, s)]$$

$$Holds(Neg(\phi), Result(\alpha, s)) \leftarrow [Holds(\phi_{m+1}, s) \wedge \ldots \wedge Holds(\phi_n, s)]$$

there is some i, $1 \le i \le m$, and some j, $m + 1 \le j \le n$, such that $\phi_i = Neg(\phi_j)$ or $\phi_j = Neg(\phi_i)$.

The following definition is also useful. It identifies narrative domain descriptions in which information about what holds in the initial situation is complete.

Definition 6 (Initially specified narrative domain description). A narrative domain description N is *initially specified* if for every positive fluent constant ϕ in the language, either $Holds(\phi, S0)$ or $Holds(Neg(\phi), S0)$ is an initial conditions description in N.

Two examples of fluent independent narrative domain descriptions are given below. The first is initially specified and the second is not. Only the domain-dependent axioms are given. Example 1 is a version of the Yale Shooting Problem, including a simple narrative in which a *Sneeze*[4] action occurs at time 1 followed by a *Shoot* action at time 3. The full narrative domain description is referred to as N_{YSP}.

Example 1 (The Yale Shooting Problem, N_{YSP}). Fluent constants: *Alive*, *Loaded*, Action constants: *Sneeze*, *Shoot*, Domain-specific axioms:

$$Holds(Neg(Alive), Result(Shoot, s)) \leftarrow Holds(Loaded, s) \tag{Y1}$$

$$Holds(Loaded, S0) \tag{Y2}$$

$$Holds(Alive, S0) \tag{Y3}$$

$$UNA[Alive, Loaded, And, Neg] \tag{Y4}$$

$$UNA[Sneeze, Shoot] \tag{Y5}$$

$$f = Alive \lor f = Loaded \lor f = Neg(Alive) \lor f = Neg(Loaded) \tag{Y6}$$

$$Happens(Sneeze, 1) \tag{Y7}$$

$$Happens(Shoot, 3) \tag{Y8}$$

Because it is not initially specified, Example 2 below is useful in illustrating that logic program translations cannot be used to derive *Holds* literals not warranted by their specifications (i.e. that they are "sound"). It concerns an electric gate, connected to a button which will open the gate when pressed, provided the system is connected to an electric supply. The gate is initially closed. There is no information as to whether the system is initially connected to an electrical supply, hence it is not possible to deduce that the gate will be either open or closed after the button has been pressed. The full narrative domain description is referred to as N_{GATE}.

Example 2 (N_{GATE}). Fluent constants: *Open*, *Connected*, Action constant: *Press*, Domain-specific axioms:

$$Holds(Open, Result(Press, s)) \leftarrow Holds(Connected, s) \tag{G1}$$

---

[4] The *Sneeze* action takes place of the *Wait* action in the original formulation. In the Narrative Situation Calculus, "waits" are more naturally represented by the absence of an action occurrence within some time interval.

$$Holds(Neg(Open), S0) \tag{G2}$$

$$UNA[Open, Connected, And, Neg] \tag{G3}$$

$$f = Open \lor f = Connected \lor f = Neg(Open) \lor f = Neg(Connected) \tag{G4}$$

$$Happens(Press, 1) \tag{G5}$$

For a given domain description D, Baker's original circumscription policy is

$$CIRC[D \; ; \; Absit \; ; \; Ab, Result, Holds, S0] \land CIRC[D \; ; \; Ab \; ; \; Result, S0]$$

The above policy will be referred to as $CIRC_b$. As regards the Yale Shooting Problem, Baker shows that

$$CIRC_b[N_{YSP}] \models Holds(Neg(Alive), Result(Shoot, Result(Sneeze, S0)))$$

The Narrative Situation Calculus introduces an extended circumscription policy representing the assumption that the only action occurrences are those explicitly described in the narrative description. The separation of sentences in theories into those which describe actions' effects and those which refer to the narrative allows this to be achieved in a natural way simply by circumscribing $Happens$ in parallel with Ab, while varying $State$ along with $Result$. As before, $Absit$ is circumscribed at a higher priority so as to ensure the existence of all consistent situations. Thus, given a narrative domain description N, the extended circumscription policy is

$$CIRC[N \; ; \; Absit \; ; \; Ab, Happens, Result, Holds, S0, State] \land$$
$$CIRC[N \; ; \; Ab, Happens \; ; \; Result, S0, State]$$

This circumscription policy is referred to as $CIRC_n$.

Three theorems are useful at this point. Full proofs of these can be found in [12]. The first two theorems show that minimisation of $Happens$ does not interfere with the minimisation used to solve the frame problem. The third theorem shows that, unsurprisingly, circumscribing $Happens$ has the same effect as forming its completion.

Theorem 7. *Let N be a narrative domain description. Then $CIRC_n[N] \models CIRC_b[N]$*

Theorem 8. *Let N be a narrative domain description, and let Φ be a sentence which does not contain the predicate symbol $Happens$ and does not contain the function symbol $State$. Then $CIRC_n[N] \models \Phi$ if and only if $CIRC_b[N] \models \Phi$*

Theorem 9. *Let N be a narrative domain description with k occurrence descriptions. If $k < 0$ and the set of occurrence descriptions is $\{Happens(\alpha_i, \tau_i) \mid 1 \leq i \leq k\}$ then*

$$CIRC_n[N] \models Happens(a, t) \leftrightarrow \bigvee_{i=1}^{k} [a = \alpha_i \land t = \tau_i]$$

and if $k = 0$ then $CIRC_n[N] \models \neg \exists a, t[Happens(a, t)]$

These theorems together with Axioms (N1) and (N2) allow the deduction of what fluents hold at different time points, i.e. they facilitate temporal projection. For example, in the Yale Shooting Problem it can be shown that

$$CIRC_n[N_{YSP}] \models Holds(Neg(Alive), State(5))$$

A derivation of this is given in [12].

3   A Translation into Logic Programs

For the purposes of deriving information about what holds along the time-line, the narrative domain descriptions of the previous section can be translated into Event Calculus style logic programs which do not contain situation terms or arguments. In this section, logic programs are defined which use the following predicate symbols: $HoldsAt$, $Initially$, $Initiates$, $Terminates$, $ClippedBetween$, $ClippedBefore$, $Happens$, $<$ and $\leq$[5]. Given a narrative domain description N, the aim is to define a logic program $EC[N]$ which facilitates temporal projection and which is "sound" in the following sense; for any ground fluent term ϕ and real number τ, the positive literal $HoldsAt(\phi, \tau)$ is derivable from $EC[N]$ only if $CIRC_n[N] \models Holds(\phi, State(\tau))$. The following definition of *fluent converses* will be useful.

Definition 10 (Converse of a fluent). Let ϕ be a ground fluent term. Then the *converse* of ϕ, written $\overline{\phi}$, is $Neg(\phi)$ if ϕ is a positive fluent constant, or ϕ' if ϕ is of the form $Neg(\phi')$ for some positive fluent constant ϕ'.

Occurrence descriptions are included directly in logic programs as conditionless clauses. The following definitions show how initial conditions descriptions and action descriptions are translated into domain-specific *Initially*, *Initiates* and *Terminates* clauses.

Definition 11 ($IN[IC]$). Let ϕ be a ground fluent term and let IC be an initial conditions description of the form $Holds(\phi, S0)$. The program clause $IN[IC]$ is defined as $Initially(\phi)$.

Definition 12 ($INIT[AD]$ and $TERM[AD]$). Let $\phi, \phi_1, \ldots, \phi_n$ be ground fluent terms, α an action constant, s a situation variable and t a time-point variable. Let AD be an action description of the form

$$Holds(\phi, Result(\alpha, s)) \leftarrow [Holds(\phi_1, s) \wedge \ldots \wedge Holds(\phi_n, s)]$$

The program clause $INIT[AD]$ is defined as

---

[5] It is assumed that $\leq$ and $<$ are defined in such a way that for all $\tau_1, \tau_2 \in \mathbb{R}$, goals of the form "$\tau_1 \leq \tau_2$" and "$\tau_1 < \tau_2$" succeed if and only if $\tau_1 \leq \tau_2$ and $\tau_1 < \tau_2$ respectively.

$$Initiates(\alpha, \phi, t) \leftarrow [HoldsAt(\phi_1, t) \land \ldots \land HoldsAt(\phi_n, t)]$$

and the program clause $TERM[AD]$ is defined as

$$Terminates(\alpha, \overline{\phi}, t) \leftarrow [not\ HoldsAt(\overline{\phi_1}, t) \land \ldots \land not\ HoldsAt(\overline{\phi_n}, t)]$$

In the definition above, although $Terminates$ clauses have a similar structure to $Initiates$ clauses, their bodies incorporate negated (not) literals with fluent converses as their first arguments. As will be seen from the examples below, this difference becomes important for domains where information about the initial situation is incomplete. Whereas $Initiates$ clauses describe the immediate effects of actions, $Terminates$ clauses play the role of domain-specific "abnormality" clauses which help determine those fluents which do not persist through an action occurrence. Where information about what holds at a time-point τ is incomplete, the sub-goal $not\ HoldsAt(\overline{\phi_i}, t)$ might succeed even when the sub-goal $HoldsAt(\phi_i, t)$ fails. Were it not for this use of negative literals, the definition of $Terminates$ would in general be incomplete, so that the default persistence of fluents would in general be incorrect. Similar techniques are used for example in [4] and [2]. (In fact, in the special case where domains are initially specified these domain-specific $Terminates$ clauses may be replaced by a general definition, expressing that an action terminates a fluent if and only if it initiates its converse.) The following definition gives a complete translation of a narrative domain description into a logic program.

Definition 13 ($EC[N]$). Given a narrative domain description N with action descriptions $AD_1, \ldots, AD_n$, initial condition descriptions $ID_1, \ldots, ID_m$ and occurrence descriptions $OD_1, \ldots, OD_k$ then the logic program $EC[N]$ is defined as

$$\{INIT[AD_1], \ldots, INIT[AD_n], TERM[AD_1], \ldots, TERM[AD_n],$$
$$IN[ID_1], \ldots, IN[ID_m], OD_1, \ldots, OD_k\}$$

together with the following domain-independent clauses

$$HoldsAt(f, t) \leftarrow [Initially(f) \land not\ ClippedBefore(f, t)] \qquad \text{(EC1)}$$

$$HoldsAt(f, t3) \leftarrow [Happens(a, t1) \land t1 < t3 \land Initiates(a, f, t1) \land \qquad \text{(EC2)}$$
$$not\ ClippedBetween(t1, f, t3)]$$

$$ClippedBefore(f, t) \leftarrow \qquad \text{(EC3)}$$
$$[Happens(a, t1) \land t1 < t \land Terminates(a, f, t1)]$$

$$ClippedBetween(t1, f, t3) \leftarrow \qquad \text{(EC4)}$$
$$[Happens(a, t2) \land t1 \leq t2 \land t2 < t3 \land Terminates(a, f, t2)]$$

Clauses (EC1) and (EC2) above are persistence axioms. Clause (EC1) states that a fluent ϕ holds at a time τ if it is initially true and it has not been cancelled ("clipped") before τ by some action. Clause (EC2) states that a fluent holds at a time τ_3 if it is initiated by an action occurrence at some previous time τ_1, and is not clipped in the meantime. Clauses (EC3) and (EC4) give the definitions for $ClippedBefore$ and $ClippedBetween$ in terms of $Happens$ and $Terminates$. Notice that there is no direct representation of Axiom (B2) in EC programs. No clause such as $HoldsAt(Neg(f),t) \leftarrow not\ HoldsAt(f,t)$ is included, since this would clearly cause unsoundness in cases where narrative domain descriptions are not initially specified (such as Example 2). In the method presented here, a distinction should be made between $EC[N] \nvdash_{SLDNF} HoldsAt(\phi, \tau)$ which should be interpreted as "it is not provable that ϕ holds at time τ", and $EC[N] \vdash_{SLDNF} HoldsAt(Neg(\phi), \tau)$ which should be interpreted as "ϕ does not hold at time τ".

As regards the Yale Shooting Problem (Example 1), the domain dependent clauses in $EC[N_{YSP}]$ are

$$Initiates(Shoot, Neg(Alive), t) \leftarrow HoldsAt(Loaded, t) \qquad (INIT[(Y1)])$$

$$\begin{aligned}Terminates(Shoot, Alive, t) \leftarrow \\ not\ HoldsAt(Neg(Loaded), t)\end{aligned} \qquad (TERM[(Y1)])$$

$$Initially(Loaded) \qquad (IN[(Y2)])$$

$$Initially(Alive) \qquad (IN[(Y3)])$$

$$Happens(Sneeze, 1) \qquad (Y7)$$

$$Happens(Shoot, 3) \qquad (Y8)$$

so that $EC[N_{YSP}] \vdash_{SLDNF} HoldsAt(Neg(Alive), 5)$.

In Example 2 (of the electric gate), the initial situation was not fully specified. $EC[N_{GATE}]$ consists of the clauses (EC1)-(EC4) together with the clauses

$$Initiates(Press, Open, t) \leftarrow HoldsAt(Connected, t) \qquad (INIT[(G1)])$$

$$\begin{aligned}Terminates(Press, Neg(Open), t) \leftarrow \\ not\ HoldsAt(Neg(Connected), t)\end{aligned} \qquad (TERM[(G1)])$$

$$Initially(Neg(Open)) \qquad (IN[(G2)])$$

$$Happens(Press, 1) \qquad (G5)$$

Example 2 illustrates the need for the use of negation-as-failure in the bodies of $Terminates$ clauses. Had $TERM[(G1)]$ simply been

$$Terminates(Press, Neg(Open), t) \leftarrow HoldsAt(Connected, t)$$

then the query $HoldsAt(Neg(Open), 2)$ would succeed, so that the program would be "unsound" in the sense described above. In fact, Theorems 14 and 15 below show that the translation method given is unsound only under two circumstances – (i) if the narrative domain description is not fluent independent, and (ii) if two or more different actions in the narrative occur simultaneously. If there are no simultaneous action occurrences and the narrative domain description is both fluent independent and initially specified, then the translations are both sound and complete.

Theorem 14. *Let N be a fluent independent narrative domain description, let τ be a real number, and let ϕ be a ground fluent term. Suppose that for all $\tau' \le \tau$ there is at most one occurrence description in N of the form $Happens(\alpha, \tau')$. Then*

$$CIRC_n[N] \models Holds(\phi, State(t)) \quad if \quad EC[N] \vdash_{SLDNF} HoldsAt(\phi, t)$$

Theorem 15. *Let N be an initially specified fluent independent narrative domain description, let τ be a real number, and let ϕ be a ground fluent term. Suppose that for all $\tau' \le \tau$ there is at most one occurrence description in N of the form $Happens(\alpha, \tau')$. Then*

$$CIRC_n[N] \models Holds(\phi, State(t)) \quad if \; and \; only \; if$$
$$EC[N] \vdash_{SLDNF} HoldsAt(\phi, t)$$

Space limitations do not permit full proofs to be given here; these can be found in [12]. However, the following remarks summarise the arguments. Theorems 14 and 15 rely on several intermediate propositions. In the proofs, an intermediate translation of circumscribed theories into Situation Calculus style logic programs is given, and soundness and/or completeness is shown for these programs. They are of independent interest because equivalents of Theorems 14 and 15 above hold for these programs even where simultaneous actions occur. To show this, several properties of the circumscriptions are first proved. In particular, it is shown that the circumscriptive formulation is, in Lin and Shoham's terms, epistemologically complete [10]. That is, given any situation in which all fluent values are known, then all fluent values will also be known in the situation resulting from a single action. Propositions are also proved which express limits to the possible extension of the predicate Ab, even in the general case where theories are not initially specified, in terms of syntactic properties of the theories. These results are used to show that, in this general case, it is possible to partially compute what holds in any situation named by sequence of actions by consideration of each action in turn, and that the nesting of the *not* operator in the Situation Calculus programs (similar to the nesting of *not* in the Event Calculus programs resulting from the *Terminates* literals in the bodies of clauses (EC3) and (EC4)) provides precisely the strength of default persistence required.

The intermediate logic programs which are of Situation Calculus style contain a frame axiom which is a clausal counterpart to Axiom (F1). They also contain domain-specific *Ab* clauses analogous to the *Terminates* clauses in the Event Calculus style programs. The essential difference between the two types of program is as follows. In the Situation Calculus implementations, persistence of properties along the time-line is derived indirectly by proving persistence one step at a time through a series of situations, which is in turn shown to correspond to the narrative in question. In the Event Calculus programs, default persistence of properties through a narrative is expressed more directly (clauses (EC1)–(EC4)). To show the correspondence between the Event Calculus style and Situation Calculus style programs (in the case where no simultaneous actions occur within the narrative), it is necessary to show that these two expressions of persistence are equivalent, and this can be done using induction on the number of occurrence descriptions within the narrative.

Where domains are not initially specified, program translations are "incomplete" because, in the specifications, Axiom (B2) allows the derivation of disjunctions of ground *Holds* literals. To take a trivial example, given a language including fluent constants F_1 and F_2 and action constant A, it is possible to derive $Holds(F_1, S0) \lor Holds(Neg(F_1), S0)$, so that the action descriptions and occurence description

$$Holds(F_2, Result(A, S0)) \leftarrow Holds(F_1, S0)$$

$$Holds(F_2, Result(A, S0)) \leftarrow Holds(Neg(F_1), S0)$$

$$Happens(A, 1)$$

entail $Holds(F_2, State(2))$. The corresponding *"EC"* logic program translation will not yield an SLDNF derivation of $HoldsAt(F_2, 2)$, since programs do not include a direct representation of Axiom (B2). However, since domains include only a finite number of fluents, it is not hard to design meta-level procedures or program enhancements which provide completeness in the more general case by testing *"EC"* programs with each possible initial situation consistent with the information in the specification. A more detailed discussion of this can be found in [12], but for the purposes of stating an appropriate theorem, a simple definition of such a program enhancement is given below:

Definition 16 ($EC^+[N]$). Let N be a narrative domain description written in a narrative domain language with positive fluent constants $\phi_1, \ldots, \phi_n$. Let l be a variable not used in $EC[N]$. Then the program $EC^+[N]$ is defined as $EC[N]$ with the variable l added as an extra first argument in all occurrences of *HoldsAt*, *Initiates*, *Terminates*, *ClippedBetween*, *ClippedBefore* and *Initially* literals, together with the following clauses:

$$Initially(l, f) \leftarrow On(f, l)$$

$$Fluent(\phi_1) \ \ldots \ Fluent(\phi_n)$$

$MholdsAt(f, t) \leftarrow Forall(PossibleFluentCombination(l), HoldsAt(l, f, t))$

$PossibleFluentCombination(l) \leftarrow$
$\quad [Setof(f, (Fluent(f) \wedge not\ Initially([], f) \wedge not\ Initially([], Neg(f))), p)$
$\quad \wedge Permutation(p, l)]$

$Permutation([], [])$

$Permutation([f|r1], [f|r2]) \leftarrow Permutation(r1, r2)$

$Permutation([f|r1], [Neg(f)|r2]) \leftarrow Permutation(r1, r2)$

The following theorem, the proof of which is given in [12], states the sense in which EC^+ programs[6] are sound and complete.

Theorem 17. *Let N be a fluent independent narrative domain description, let τ be a real number, and let ϕ be a ground fluent term. Suppose that for all $\tau' \leq \tau$ there is at most one occurrence description in N of the form $Happens(\alpha, \tau')$. Then*

$$CIRC_n[N] \models Holds(\phi, State(t)) \quad \text{if and only if}$$
$$EC^+[N] \vdash_{SLDNF} MholdsAt(\phi, t)$$

4 Related Work

This paper contributes to a growing body of research into the correspondence between different mechanisms for reasoning about actions, and on the use of logic programming in this respect. A recent result of Kowalski and Sadri [7] is closely related to the topic of this paper. This shows that, in the context of a class of logic programs for reasoning about actions and narratives, then all other aspects of programs being equal, the Situation Calculus type and Event Calculus type persistence clauses are interchangeable. Pinto and Reiter [14] also show how reasoning about a narrative may be accomplished with Situation Calculus style logic programs. The present paper addresses Pinto and Reiter's criticism of the original Event Calculus – that it "does not characterize a class of sound programs". Both Kowalski and Sadri and Pinto and Reiter use the Clark completion to give a "semantics" to programs, which can be regarded as their "specification" in the sense used here. Shanahan [15] shows how a circumscription policy related to Baker's may be used with Event Calculus style first order theories to model default reasoning.

---

[6] Strictly speaking, it is not possible to use SLDNF with such programs as they contain calls to the "second order" primitives *Forall* and *Setof* included here for readability. It is assumed that these primitives are appropriately re-interpreted. For definitions and practical details the reader may consult [16].

As regards the non-narrative aspects of reasoning about action, various results enable a network of correspondences between formalisms to be built up. For example, Kartha [6] shows a correspondence between Baker's formalism and the Language $\mathcal{A}$ introduced in [5]. Furthermore, Gelfond and Lifschitz [5], Dung [4], Baral and Gelfond [2], and Denecker and De Schreye [3] have each shown how the Language $\mathcal{A}$ can be used as a specification for various logic programming formulations. Like the work in this paper, most results are restricted to cases where theories are "fluent independent".

References

1. A. B. Baker, *Nonmonotonic Reasoning in the Framework of the Situation Calculus*, Artificial Intelligence, vol 49, page 5, 1991.
2. Chitta Baral and Michael Gelfond, *Representing Concurrent Actions in Extended Logic Programming*, Proceedings IJCAI 1993, Morgan Kaufmann, page 866, 1993.
3. Marc Denecker and Danny De Schreye, *Representing Incomplete Knowledge in Abductive Logic Programming*, Proceedings of the International Symposium on Logic Programming, 1993.
4. Phan Minh Dung, *Representing Actions in Logic Programming and its Applications in Database Updates*, Proceedings of the Tenth International Conference on Logic Programming, ed David S. Warren, MIT Press, pages 222-238, 1993.
5. Michael Gelfond and Vladimir Lifschitz, *Representing Actions in Extended Logic Programming*, Proceedings of the Joint International Conference and Symposium on Logic Programming, ed. Krzysztof Apt, MIT Press, page 560, 1992.
6. G. Neelakantan Kartha, *Soundness and Completeness Theorems for Three Formalizations of Action*, Proceedings IJCAI 1993, page 724, 1993.
7. R. A. Kowalski and F. Sadri, *The Situation Calculus and Event Calculus Compared*, Proceedings of the International Logic Programming Symposium, 1994
8. R. A. Kowalski and M. J. Sergot, *A Logic-Based Calculus of Events*, New Generation Computing, vol 4, page 267, 1986.
9. V. Lifschitz, *Circumscription*, in Handbook of Logic in A.I., ed.s D. Gabbay *et al.*, OUP, pages 297–352, 1995
10. Fangzhen Lin and Yoav Shoham, *Provably Correct Theories of Action*, Proceedings AAAI 1991, MIT Press, page 349, 1991.
11. J. McCarthy and P. J. Hayes, *Some Philosophical Problems from the Standpoint of Artificial Intelligence*, in Machine Intelligence 4, ed.s D. Michie and B. Meltzer, Edinburgh University Press, 1969.
12. R. S. Miller, *Narratives in the Context of Temporal Reasoning*, Imperial College Research Report DoC 94/3, available from Department of Computing, Imperial College, 180 Queen's Gate, London SW7 2BZ, 1994.
13. R. S. Miller and M. P. Shanahan, *Narratives in the Situation Calculus*, in Journal of Logic and Computation, Special Issue on Actions and Processes, vol 4 no 5, Oxford University Press, 1994.
14. J. Pinto and R. Reiter, *Temporal Reasoning in Logic Programming: A Case for the Situation Calculus*, Proceedings ICLP 93, page 203, 1993.
15. M. P. Shanahan, *A Circumscriptive Calculus of Events*, to appear in Artificial Intelligence, 1994.
16. Leon Sterling and Ehud Shapiro, *The Art of Prolog*, MIT Press, 1986.

On the Extension of Logic Programming with Negation through Uniform Proofs

Li-Yan Yuan* and Jia-Huai You

Department of Computing Science
University of Alberta
Edmonton, CANADA T6G 2H1
{yuan, you}@cs.ualberta.ca

Abstract. In the past, logic program semantics have been studied often separately from the underlying proof system, and this, consequently, leads to a somewhat confusing status of semantics. In this paper we show that elegant, yet natural semantics can be obtained by building a mechanism of justifying default assumptions on top of a proof system. In particular, we propose extended logic programming languages with negation through *uniform proofs*. The result is a very general framework, in which *any* abstract logic programming language can be extended to a nonmonotonic reasoning system, and many semantics, previously proposed and new, can be characterized and understood in terms of uniform proofs.

1   Introduction

There has been considerable attention to extending the logic programming paradigm either to include richer logics than Horn clauses or to incorporate default negation. Two notable examples are abstract logic programming languages for higher-order logics [8] and disjunctive logic programs extended with negation [5, 11].

Miller *et al.* [8] provide a theoretical foundation for logic programming based on the observation that logic programs are intended to specify search behavior and that the declarative meaning of a logic program, provided by provability in a logic system, should coincide with its operational meaning which is based on interpreting logical connectives as simple and fixed search instructions. Consequently, the notion of *uniform proof* was introduced to define abstract logic programming languages. Informally, an abstract logic programming language is a triple $\langle \mathcal{P}, \mathcal{G}, \vdash \rangle$ such that for any program $\Pi \subset \mathcal{P}$ and any query $G \in \mathcal{G}$, $\Pi \vdash G$ if and only if there exits a uniform proof from Π to G. A uniform proof is a proof in a sequent calculus without the cut inference rule, which guarantees proper answer substitutions. The class of such languages includes classical higher-order Horn clauses and the hereditary Harrop formulas.

Despite the fact that the essential aspect of logic programming is the coincidence of its declarative and operational semantics, declarative semantics of logic programs with negation have often been studied separately from its operational semantics. For example, many prominent semantics for disjunctive logic programs, including Przymusinski's *static semantics* and *perfect semantics* [10, 11], and Gelfond and

---

* Currently on leave at the ISIS, Fujitsu Labs, Numazu, Shizuoka, Japan.

Lifschitz's stable semantics [5], are based on a notion of minimal model in that the semantics of a disjunctive program with negation is determined by the minimal models of the program whose negation values have been determined. Because of lack of a clear understanding of what a proof system for minimal models is, this approach may cause confusion in searching for declarative semantics. Even for stratified disjunctive programs, for example, the static semantics and perfect semantics do not agree with each other (cf. Example 6). As a matter of fact, various alternatives to the minimal model approach have been reported [13].

In this paper, we extend abstract logic programming languages of Miller *et al.* to incorporate default negation, based on a very simple but quite different idea: The interpretation of default negation shall be determined by the underlying operational semantics.

Since uniform proofs serve not only as an operational semantics but also the theoretical foundation of abstract logic programming languages, the intended semantics of a program Π can be characterized naturally by Π, together with the appropriate values of default negations (also called assumptions) in Π, under the uniform proof system, while the appropriateness of default negations are justified on top of the same uniform proof system. This is illustrated below:

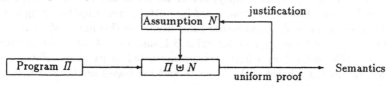

where $\Pi \uplus N$ denotes the GL-transformation.

By justifying default negations on top of uniform proofs, we not only characterize natural semantics of logic programming with negation, but also preserve the essential aspects of logic programming: the coincidence of its declarative and operational semantics, and the specification of search behavior. Furthermore, using one proof system to derive positive conclusions as well as to justify default negations will significantly simplify the implementation of logic programming languages with negation.

The plan of this paper is as follows. The next section introduces Miller's system of uniform proofs and the notion of abstract logic programming. In Section 3, we present the general idea of how to extend an abstract logic programming language into a nonmonotonic reasoning system, and discuss a number of particular semantics resulting from it. Section 4 formally defines extended logic programming languages with negation in terms of a *semantical transformation* and a number of common semantical properties that guarantee the well-behavedness property [3]. Section 5 is devoted to disjunctive programs and various semantics. In addition, we introduce a *program enhancement* and show how some of the previously proposed semantics are related to the uniform proof semantics under this enhancement.

2 Uniform Proof and Logic Programming

We consider here a first-order language with logic constants $\wedge, \vee, \subset, \forall$, and $\exists$, and shall use the following notations.

$\mathcal{P}$ A set of formulas that serves as possible program clauses.

$\mathcal{G}$ A set of formulas that serves as possible goals and queries.

A   An atomic formula, i.e., a formula without logic constants.

Π A subset of $\mathcal{P}$, referred to as a program.

G   A member of $\mathcal{G}$, referred to as a goal or query.

Given a program Π and a goal G, the most important relation that could be asked in a logic programming environment is the following

$$\text{Is it the case that } \Pi \vdash G?$$

Here, $\vdash$ denotes some notion of logical provability (such as classical provability). Furthermore, we are not merely interested in whether G can be derived or not, but also in which instances of G are derivable, commonly referred to as answer substitutions.[2]

It has been generally agreed that logic programming is based on a notion of *goal-directed* search and the goal-directed search has been formulated using the concept of *uniform proof*, as given below [8].

First, a brief introduction to *sequent-style* proof systems. A *sequent* is a pair $\langle \Gamma, G \rangle$, where Γ is a (possible empty) set of formulas and G a formula, which is written as $\Gamma \longrightarrow G$. Proofs are constructed by putting them together using inference rules. Figure 1 contains all the inference rules needed in this paper. We assume that c is not free in the lower sequent for rules $\exists$-L and $\forall$-R. A *proof* for the sequent $\Gamma \longrightarrow G$ is then defined as a finite tree, constructed using these inference rules, such that the root is labeled with $\Gamma \longrightarrow G$ and the leaves are labeled with *initial sequents*, i.e., sequents Identity and true-R.

$$\frac{}{\Gamma, G \longrightarrow G} \text{ Identity} \qquad\qquad \frac{}{\Gamma \longrightarrow \text{true}} \text{ true-R}$$

$$\frac{B, C, \Gamma \longrightarrow G}{B \wedge C, \Gamma \longrightarrow G} \wedge\text{-L} \qquad\qquad \frac{\Gamma \longrightarrow B \quad \Gamma \longrightarrow C}{\Gamma \longrightarrow B \wedge C} \wedge\text{-R}$$

$$\frac{B, \Gamma \longrightarrow G \quad C, \Gamma \longrightarrow G}{B \vee C, \Gamma \longrightarrow G} \vee\text{-L} \qquad \frac{\Gamma \longrightarrow B}{\Gamma \longrightarrow B \vee C} \vee\text{-R} \qquad \frac{\Gamma \longrightarrow C}{\Gamma \longrightarrow B \vee C} \vee\text{-R}$$

$$\frac{\Gamma \longrightarrow B \quad C, \Gamma \longrightarrow G}{B \supset C, \Gamma \longrightarrow G} \supset\text{-L} \qquad\qquad \frac{B, \Gamma \longrightarrow C}{\Gamma \longrightarrow B \supset C} \supset\text{-R}$$

$$\frac{[t/x]P, \Gamma \longrightarrow G}{\forall x P, \Gamma \longrightarrow G} \forall\text{-L} \qquad\qquad \frac{\Gamma \longrightarrow [c/x]P}{\Gamma \longrightarrow \forall x P} \forall\text{-R}$$

$$\frac{[c/x]P, \Gamma \longrightarrow G}{\exists x P, \Gamma \longrightarrow G} \exists\text{-L} \qquad\qquad \frac{\Gamma \longrightarrow [t/x]P}{\Gamma \longrightarrow \exists x P} \exists\text{-R}$$

Fig. 1. The inference rules

---

[2] However, such an answer substitution may not be always possible (cf. Subsection 2.1).

Goal-directed search is then characterized operationally by the bottom-up construction of proofs in which right-introduction rules are applied first and left introduction rules are applied only when the right-hand side is atomic. This means the logic connectives in a goal are decomposed uniformly and independently from the program and the program is only considered when the goal is atomic. More formally,

Definition 1. ([8]) A *uniform proof* is a proof using the given inference rules such that for each occurrence of a sequent $\Gamma \longrightarrow G$ in it, the following conditions are satisfied

- If G is $B \wedge C$ then that sequent is inferred by $\wedge$-R from $\Gamma \longrightarrow B$ and $\Gamma \longrightarrow C$.
- If G is $B \vee C$ then that sequent is inferred by $\vee$-R from either $\Gamma \longrightarrow B$ or $\Gamma \longrightarrow C$.
- If G is $\exists x P$ then that sequent is inferred by $\exists$-R from $\Gamma \longrightarrow [t/x]P$ for some term t.
- If G is $B \supset C$ then that sequent is inferred by $\supset$-R from $B, \Gamma \longrightarrow C$.
- if G is $\forall x P$ then that sequent is inferred by $\forall$-R from $\Gamma \longrightarrow [c/x]P$, where c is a parameter that does not occur in the given sequent. □

The notion of uniform proof reflects the search instructions associated with the logical connectives. We say goal G succeeds given program Π, denoted as $\Pi \vdash_U G$ if and only if there is a uniform proof of the sequent $\Pi \longrightarrow G$. For comparison, we denote classical provability as $\Pi \vdash_C G$, that is, $\Pi \vdash_C G$ if and only if $\Pi \vdash G$ under the classical first-order logic.

Now we present the definition of the abstract logic programming language.

Definition 2. ([8]) An abstract logic programming language is defined as a triple $\langle \mathcal{P}, \mathcal{G}, \vdash \rangle$, where $\vdash$ is a notion of provability, such that for all subsets Π of $\mathcal{P}$ and all formulas G of $\mathcal{G}$, $\Pi \vdash G$ if and only if $\Pi \vdash_U G$. □

Uniform proofs not just specify the goal-directed search but also guarantee proper answer substitutions for succeeded goals, a very important property of logic programming.

2.1 Restrictions of Abstract Logic Programming

Before we present some examples of abstract logic programming languages, we describe two logical systems, as given in [8], that are not abstract logic programming languages by our definition.

First, we consider disjunctive logic programs. Let A be a syntactic variable that ranges over atomic formulas. Further, let $\mathcal{G}_1$ be the collection of first-order formulas defined by the following inductive rule:

$$G := \top \mid A \mid G_1 \wedge G_2 \mid G_1 \vee G_2 \mid \forall x G \mid \exists x G,$$

where $\top$ stands for **truth**. Similarly, let $\mathcal{P}_1$ be the collection of first-order formulas defined by the following inductive rule:

$$D := A \mid G \supset A \mid D_1 \wedge D_2 \mid D_1 \vee D_2 \mid \forall x D,$$

where G is a formula in $\mathcal{G}_1$. The *disjunctive logic program system* is then defined as a logic system $\langle \mathcal{P}_1, \mathcal{G}_1, \vdash_C \rangle$. This system, contrary to one's expectation, is not an abstract logic programming language, according to the above definition, since

$$\{a \vee b; \ c \subset a; \ d \subset b\} \vdash_C c \vee d$$

but there is no uniform proof for $\{a \vee b; \ c \subset a; \ d \subset b\} \longrightarrow c \vee d$.

As another example, let us take for $\mathcal{P}_2$ the set of positive Horn clauses extended by permitting the antecedents of implication to contain negative literals, for $\mathcal{G}_2$ the existential closure of the conjunction of atoms, and for $\vdash$ the notion of classical provability. The resulting system, $\langle \mathcal{P}_2, \mathcal{G}_2, \vdash_C \rangle$, is also not an abstract logic programming language. Consider, for example, $\Pi = \{q(a) \subset p; \ q(b) \subset \neg p\}$. We know $\Pi \vdash_C \exists x q(x)$. However, there exists no term t such that $\Pi \vdash_C q(t)$. For this reason, there is no uniform proof for $\exists x q(x)$ from Π.

These two examples demonstrate that the "uniformed" proof is achieved by excluding *indefinite information*, which also explains why it guarantees proper answer substitutions. However, as we will show later, if we restrict to a special class of goals, the goals that do not contain disjunctive or existentially quantified queries, then we will have an abstract disjunctive logic programming language. For example, the reader may want to verify that there exist uniform proofs for the following two sequents: $\{a \vee b; \ c \subset a; \ c \subset b\} \longrightarrow c$ and $\{a \vee b; \ a \subset b; \ b \subset a\} \longrightarrow a$.

2.2 Sample Systems

Horn clauses are defined as the universal closures of disjunctions of literals that contains at most one positive literal. We present, following Miller's convention, an alternative definition for Horn clauses here. Let $\mathcal{G}_3$ be the collection of first order formulas defined by the following inductive rule:

$$G := \top \mid A \mid G_1 \wedge G_2 \mid G_1 \vee G_2 \mid \forall x G \mid \exists x G,$$

and $\mathcal{P}_3$ be the collection of formulas defined by the following inductive rule:

$$D := A \mid G \supset A \mid D_1 \wedge D_2 \mid \forall x G.$$

The *Horn programming system* is then defined as $\langle \mathcal{P}_3, \mathcal{G}_3, \vdash_C \rangle$.

Theorem 3. ([8]) *The Horn programming system $\langle \mathcal{P}_3, \mathcal{G}_3, \vdash_C \rangle$ is an abstract logic programming language.* □

Now we present an abstract logic programming language that makes stronger use of logical connectives. Specifically, it permits implication to occur in the body of a clause. Let $\mathcal{G}_4$ and $\mathcal{P}_4$ be defined by the following mutually recursive rules:

$$G := \top \mid A \mid G_1 \wedge G_2 \mid G_1 \vee G_2 \mid \forall x G \mid \exists x G \mid D \supset G$$
$$D := A \mid G \supset A \mid \forall x D \mid D_1 \wedge D_2.$$

A formula from $\mathcal{P}_4$ is called a *hereditary Harrop formula*. Then the *hereditary Harrop system* is defined as a triple $\langle \mathcal{P}_4, \mathcal{G}_4, \vdash_C \rangle$ [8].

Theorem 4. ([8]) *The hereditary Harrop system is an abstract logic programming language.* □

3 Logic Programs with Negation: Semantics

In this section, we discuss logic programs extended with negation from the semantical point of view.

Definition 5. Assume $\langle \mathcal{P}, \mathcal{G}, \vdash \rangle$ is an abstract logic programming language. Then $\mathcal{P}_N$ is defined as the collection of formulas defined by the following inductive rule:

$$C := D \mid \text{not}G \supset C,$$

where D and G are formulas in $\mathcal{P}$ and $\mathcal{G}$ respectively, and a *logic program* (extended with negation) of $\mathcal{P}_N$ is defined as a subset of $\mathcal{P}_N$. □

We call $\text{not}G$ an assumption (or an assumed negation). In the sequel, each assumed negation $\text{not}G$ is treated as a propositional symbol in any proof system.

Example 1. Consider the Horn programming system $\langle \mathcal{P}_3, \mathcal{G}_3, \vdash_C \rangle$ presented earlier. The following is a sample of *normal programs* which are defined as a subset of $\mathcal{P}_{3N}$.

$P(x) \subset Q(f(x)) \land \text{not}Q(x) \land \text{not}\forall y R(y)$
$Q(z) \subset \text{not}\exists y P(y).$

Note x and z are two free variables in their respective clauses. □

The *negation-base* of a logic program Π, denoted as $\mathcal{N}_\Pi$, is defined as the set of all assumed negations of formulas in $\mathcal{G}$ that are constructed from non-logical symbols in Π. That is, $\mathcal{N}_\Pi = \{\text{not}G \mid G \in \mathcal{G}$ and all non-logical symbols of G appear in $\Pi\}$. Then an *assumption set* (or *assumed negation set*) N is defined as a subset of $\mathcal{N}_\Pi$.

Definition 6. The GL-transformation of Π wrt N, denoted as $\Pi \uplus N$, is defined as

$$\Pi \uplus N = \{D \in \mathcal{P}_N \mid D \subset \text{not}G_1 \land \cdots \land \text{not}G_n \in INST(\Pi) \text{ and}$$
$$\{\text{not}G_1, \ldots, \text{not}G_n\} \subseteq N\},$$

where $INST(\Pi)$ is the set of all formulas instantiated from Π in every possible way over free variables in the Herbrand universe. □

Remark *It is not difficult to see that both $\mathcal{N}_\Pi$ and $\Pi \uplus N$ may be countably infinite even if both Π and N are finite. This shall not, however, affect our discussion since uniform proofs are defined regardless the size of the premise in each inference rule.*

Given an abstract logic programming language $\langle \mathcal{P}, \mathcal{G}, \vdash \rangle$, the intended meaning of logic program Π is determined by the provability $\vdash$ and the appropriate values of assumptions. In particular, the default assumptions should be justified using the same system of uniform proofs. So the question arises how to justify an assumption set. Though many approaches, based on various frameworks, have been proposed in searching for appropriate assumption sets, most, if not all, of them can be characterized by a simple justification principle, called the *justification rule* [18, 20].

Given a logic program Π, let $\mathcal{T}(\Pi)$ and $\mathcal{F}(\Pi)$ denote the sets of goals and assumed negations whose values are universally acceptable to any rational agent respectively. Clearly, we shall identify both $\mathcal{T}(\Pi)$ and $\mathcal{F}(\Pi)$ first.

Since the uniform proof is the sole proof system used in logic programming, we shall assume that

$$\mathcal{T}(\Pi) = \{G \mid \Pi \vdash G\}.$$

(Note for any abstract logic programming language, $F \vdash_U G$ if and only if $F \vdash G$.) The *unfounded set* of a program Π, defined by Van Gelder, Ross, and Sclipf [15] as the set of all goals that cannot be derived from Π **even if** all goals not in $T(\Pi)$ are assumed false, is the unique set of goals whose negations are assumed by all reasonable semantics. Naturally, we specify $\mathcal{F}(\Pi)$ as the unfounded set of Π, i.e.,

$$\mathcal{F}(\Pi) = \{\text{not}G | \Pi \uplus \{\text{not}B | \Pi \not\vdash B\} \not\vdash G\}.$$

With such a clear understanding of $\mathcal{F}(\Pi)$, the appropriateness of an assumption set can be easily justified. That is,

Definition 7. An assumption set N is said to be a *justified set* of Π if

$$N = \mathcal{F}(\Pi \uplus N). \qquad \square$$

Thus, N is **justified** just in case $\text{not}G \in N$ if and only if $\text{not}G \in \mathcal{F}(\Pi \uplus N)$.

Example 2. Consider the following normal program

$$\Pi = \{a \subset \text{not}b; \ b \subset \text{not}a \wedge \text{not}c; \ p \subset a \wedge \text{not}p; \ q \subset a; \ q \subset b\}.$$

Π has three justified sets, that is, $N_1 = \{\text{not}c\}$, $N_2 = \{\text{not}b, \text{not}c\}$, and $N_3 = \{\text{not}a, \text{not}c, \text{not}p\}$. $\qquad \square$

Among all justified sets, the following are of particular interest.

Definition 8. Let Π be a logic program and N a justified set. Then N is

1. a *stable set* of Π if $N = \{\text{not}G \mid \Pi \uplus N \not\vdash G\}$;
2. a *least justified set* of Π if N is a subset of any justified set of Π.
3. a *regularly-justified set* of Π if N is a maximal justified set, wrt set inclusion.
$\qquad \square$

As a matter of fact, define a transformation R as follows:

$$R(N) = \{\text{not}G \mid \Pi \uplus N \not\vdash G\}.$$

It is straightforward to show the following.

Theorem 9. *1. N is a stable set if and only if $N = R(N)$.*
2. N is a justified set if and only if $N \subseteq R(N)$ and $N = R(R(N))$. $\qquad \square$

Since R is not monotonic, a program may not have any stable set. It is easy to show that, however, R is anti-monotonic, that is, $R(N_1) \subseteq R(N_2)$ if $N_2 \subseteq N_1$ [1], and consequently, R^2 is monotonic. This leads to the following:

Theorem 10. *Any logic program has a least justified set.* $\qquad \square$

In the context of normal programs, it has been shown [16] that a stable set, a justified set, and a regularly-justified set correspond to a *stable model* [6], a *three-valued stable model* [12], and a *regular model* [17] (which is also equivalent to a *preferred extension* [4]), respectively.

Given a set of justified negation sets $N_1, \ldots, N_m$, there are different ways to use these sets. We may be interested in the intersection of N_i's, or we may want to treat every justified negation set N_i equally. In the following, for lack of appropriate terminologies, we use the term *skeptical* to characterize the former and *credulous* to characterize the latter.

Definition 11. Let Π be a logic program extended with negation. Then

1. the *skeptical well-founded semantics* of Π is characterized by $\Pi \uplus N_{s.wf}$, where $N_{s.wf}$ is the intersection of all justified sets of Π;
2. the *skeptical regular semantics* of Π is characterized by $\Pi \uplus N_{s.rg}$, where $N_{s.rg}$ is the intersection of all the regularly-justified sets of Π;
3. the *credulous regular semantics* of Π is characterized by $\bigvee(\Pi \uplus N_i)$, where N_i are all regularly-justified sets of Π; and
4. the *credulous stable semantics* of Π is characterized by by $\bigvee(\Pi \uplus N_i)$, where N_i are stable sets of Π.

Note that, according to Theorem 10, $N_{s.wf}$ is also the least justified set of Π. $\quad\square$

Example 3. Consider Π in Example 2 again. N_1 is the least justified set of Π, N_3 is the only stable set of Π, and both N_2 and N_3 are regularly-justified sets of Π. Thus, both the skeptical well-founded and skeptical regular semantics of Π are characterized by $\Pi \uplus N_1 = \{a \subset notb; \ b \subset nota; \ p \subset a \wedge notp; \ q \subset a; \ q \subset b\}$; the credulous regular semantics of Π is characterized by $(\Pi \uplus N_2) \vee (\Pi \uplus N_3) = \{a \vee b; \ a \subset notb; \ b \subset nota; \ p \subset a; \ q \subset a; \ q \subset b\}$; and the credulous stable semantics by $\Pi \uplus N_3 = \{b; \ a \subset notb; \ p \subset a; \ q \subset a; \ q \subset b\}$.

Therefore, no positive conclusions can be derived from both skeptical semantics, q can be derived from the credulous regular semantics, and the (credulous) stable semantics deduces both b and q. $\quad\square$

It is straightforward to find other ways of using different justified sets, e.g., the skeptical stable semantics. Of course, other mechanisms of making default assumptions, such as those used in the *O-semantics* [9] and the *acceptability semantics* [7], are also applicable, as long as the uniform proof is employed as the underline proof system.

The interest in these different ways of using justified sets lies in their extensibility from a monotonic system to a nonmonotonic system, which will be discussed in the next section.

4 Extended Logic Programming

In this section, we present a different point of view for declarative semantics of extended logic programming with negation: A declarative semantics is viewed as a program transformation within the same abstract logic programming language, which provides very much needed insights toward extending logic programming.

Definition 12. S is said to be a *semantical transformation* if there exist a set of negation sets $N_1, \ldots, N_m$ such that

$$S(\Pi) = \bigvee_{i=1}^{m} \Pi \uplus N_i.$$

S is a skeptical transformation if $m = 1$, and a credulous one if otherwise. $\quad\square$

It is essential to require that a declarative semantics of logic programs be characterized in terms of a semantical transformation. Otherwise, the notion of uniform proof and consequently answer substitutions disappear.

Example 4. Consider $\Pi = \{a \subset b;\ b \subset \text{not}a\}$ and semantics WFS^+, a semantics for normal programs defined by extending the well-founded semantics with an additional inference rule: Deriving D if both $D \subset a$ and $D \subset \text{not}a$ are derivable by uniform proofs [14, 3]. Under the WFS^+ semantics, a is **true**. There exists, however, no negation sets N_i such that $(\bigvee_{i=1}^{m} \Pi \uplus N_i) \vdash_U a$. Therefore, WFS^+ may be anything but a semantics for an extended logic programming language. □

Now we discuss semantical properties in terms of semantical transformation.

Definition 13. A semantical transformation is said to satisfy

1. the **N-cumulativity** if $S(S(\Pi)) = S(\Pi)$;
2. the **negative justification** if
   $S(\Pi) \vdash D$ if $D \subset \text{not}G \in INST(\Pi)$ and $\text{not}G \in \mathcal{F}(S(\Pi))$.
   where $INST(\Pi)$ is the instantiated program of Π.
3. the **simplicity** if $S(\Pi) \vdash G$ if $\Pi \vdash G$. □

The N-cumulativity (read cumulativity based on negation sets) expresses the idea that if the intended semantics of Π is to be characterized by $S(\Pi)$, then $S(\Pi)$ should have the same semantics as Π.[3] Since default assumptions are justified on top of the underlying proof system, $\text{not}G$ should be assumed if there exists no uniform proof from Π to G even if the negations of all non-derivable goals from $S(\Pi)$ are assumed. This is the negative justification. The simplicity requires that G should be **true** in the intended semantics of Π if there exists a uniform proof for $\Pi \longrightarrow G$, which is plain and simple. Note these three properties are a direct extension of those classified for normal programs [20], and, together with the relevance, they are essentially the same as the well-behaved semantical properties proposed by Dix [3].[4]

The following lemma explains how the N-cumulativity expresses the cumulativity based on negative assumptions.

Lemma 14. *Let Π be a program and $S(\Pi)$ a semantical transformation that satisfies the negative-justification. Then*

$$S(S(\Pi)) = S(\Pi) \text{ if and only if for any } N \subseteq \mathcal{F}(S(\Pi)),\ S(\Pi) = S(\Pi \uplus N).$$ □

Definition 15. Let $L = \langle \mathcal{P}, \mathcal{G}, \vdash \rangle$ be an abstract logic programming language. Then an *extended logic programming language* (with negation) of L is a quadruple

$$L_N = \langle \mathcal{P}_N, \mathcal{G}, \vdash, S \rangle$$

where $\mathcal{P}_N$ is defined by the inductive rule

$C := D \mid \text{not}G \supset C,$

where D and G are formulas in $\mathcal{P}$ and $\mathcal{G}$ respectively, and S is a function from $\mathcal{P}_N$ to $\mathcal{P}_N$ such that

1. $\langle \mathcal{P}_N, \mathcal{G}, \vdash \rangle$ is an abstract logic programming language, and

---

[3] Note the close relationship between this notion of cumulativity with that of Brewka [2].
[4] Since the property of relevance has no effect on the interpretation of default negation, it need not be included in extending logic programming languages.

2. S is a semantical transformation satisfying
the N-cumulativity, negative justification, and simplicity. □

Then for any program Π of $\mathcal{P}_N$ and query G, G is considered **true** under extended logic programming if and only if there exists a uniform proof for $S(\Pi) \longrightarrow G$.

The following theorem shows that the well-founded semantics is the lower semantical bound of any extended logic programming language.

Theorem 16. *Let $\langle \mathcal{P}_N, \mathcal{G}, \vdash, S \rangle$ be an extended logic programming language, $\Pi \subseteq \mathcal{P}_N$ a program, and $S_{s.wf}(\Pi) = \Pi \uplus N_{s.wf}$, where $N_{s.wf}$ is the least justified set of Π. Then*

$$S(\Pi) \vdash S_{s.wf}(\Pi).$$ □

Now we present some sample systems of extended logic programming languages.

Theorem 17. *Let $\langle \mathcal{P}, \mathcal{G}, \vdash \rangle$ be an abstract logic programming language, and S a skeptical transformation that satisfies the N-cumulativity, negative justification, and simplicity. Then $\langle \mathcal{P}_N, \mathcal{G}, \vdash, S \rangle$ is an extended logic programming language with negation.* □

This theorem, whose proof directly follows that of Lemma 18 below, shows that the skeptical semantics are well defined for any abstract logic programming languages.

Lemma 18. *All four semantical transformations outlined in Definition 11 satisfy the N-cumulativity, negative justification, and simplicity.*

Proof. It follows from the following two facts.

1. Let $\{N_1, \ldots, N_m\}$ be a set of assumed sets, and $\Pi_0 = \bigvee \Pi \uplus N_i$. For any $N \in \{N_1, \ldots, N_m\}$, N is a justified set of Π if and only if N is a justified set of Π_0.
2. N is a justification set of Π if and only if N is a justification set of $\Pi \uplus N'$, where $N' \subseteq N$. □

For any abstract logic programming language $\langle \mathcal{P}, \mathcal{G}, \vdash \rangle$, let us take $S_{s.wf}$ and $S_{s.rg}$ to be the two skeptical semantical transformations based on, respectively, the set of all justified sets and the set of all regularly-justified sets. Then both $\langle \mathcal{P}_N, \mathcal{G}, \vdash, S_{s.wf} \rangle$ and $\langle \mathcal{P}_N, \mathcal{G}, \vdash, S_{s.rg} \rangle$ are extended logic programming languages. The credulous semantics, however, does not enjoy such preservation, as demonstrated by the following example.

Example 5. Consider a normal program $\Pi = \{p(a) \subset \text{not} p(b); \ p(b) \subset \text{not} p(a)\}$. Since Π has two regularly-justified sets $N_1 = \{\text{not} p(a)\}$ and $N_2 = \{\text{not} p(b)\}$,

$$S_{c.rg}(\Pi) = \{p(a) \vee p(b); \ p(a) \subset \text{not} p(b); \ p(b) \subset \text{not} p(a)\}.$$

Though $S_{c.rg}(\Pi) \vdash_C \exists x p(x)$, there is no uniform proof for $S_{c.rg}(\Pi) \longrightarrow \exists x p(x)$. □

Obviously, the uniform provability was lost in the credulous regular semantics because of the multiple choices of regularly-justified negations, and therefore, further restrictions are inevitable if we wish to preserve the essential correspondence between the (classical) provability and the uniform provability.

An abstract logic programming language $\langle \mathcal{P}, \mathcal{G}, \vdash \rangle$ is said to be $\vee$-*extendable* if $\langle \mathcal{P}_\vee, \mathcal{G}, \vdash \rangle$ is also an abstract logic programming language, where $\mathcal{P}_\vee$ is the collection of formulas defined by

$C := D \mid C_1 \vee C_2,$ where D is a formula in $\mathcal{P}$.

Theorem 19. *Let $\langle \mathcal{P}, \mathcal{G}, \vdash \rangle$ be an $\vee$-extendedable abstract logic programming language and $S_{cr}(\Pi) = \bigvee_{i=1}^{m} \Pi \uplus N_i$, where N_i are justified negation sets of Π. Then*

$\langle \mathcal{P}_N, \mathcal{G}, \vdash, S_{cr} \rangle$ *is an extended logic programming language.* □

Again, the proof of this theorem follows that of Lemma 18.

From the discussion above, we can see that all the skeptical semantics based on the justification sets of any extended logic programming languages, and all the credulous semantics based on the justification sets of any extended logic programming languages with definite goals are also extended logic programming languages.

5 Disjunctive Logic Programs

Although the disjunctive programming system fails to satisfy our definition, as demonstrated in Section 2, we shall see that one of its subsystems is an abstract logic programming language.

Let $\mathcal{P}_5$ and $\mathcal{G}_5$ be the collection of first-order formulas defined by the following two inductive rules:

$G := \top \mid A \mid G_1 \wedge G_2 \mid \forall x G,$
$D := A \subset G \mid D_1 \wedge D_2 \mid D_1 \vee D_2 \mid \forall x D.$

The *sub-disjunctive logic programming system*, or just the *disjunctive logic program system* if no confusion arises, is then defined as a logic system $\langle \mathcal{P}_5, \mathcal{G}_5, \vdash_C \rangle$. A disjunctive program is then defined as a subset of $\mathcal{P}_5$. Note that the only restrictions are that a goal should be a definite formula.

The following theorem shows that the sub-disjunctive logic program system is an $\vee$-extendable abstract logic programming language.

Theorem 20. *The sub-disjunctive logic program system $\langle \mathcal{P}_5, \mathcal{G}_5, \vdash_C \rangle$ is an $\vee$-extendable abstract logic programming language.*

Proof. It follows the two basic facts that for any logic formula F, G_1 and G_2,

1. $F \vdash_C G_1 \wedge G_2$ if and only if $F \vdash_C G_1$ and $F \vdash_C G_2$;
2. $F \vdash_C \forall x G_1(x)$ if and only if $F \vdash_C G_1(c)$, assuming c is not free in F and G_1. □

The following theorem follows Theorems 17, 19 and 20.

Theorem 21. *Let $\langle \mathcal{P}_5, \mathcal{G}_5, \vdash_C \rangle$ be the sub-disjunctive logic program language, and*

1. $S_{s.wf}(\Pi) = \Pi \uplus N_{s.wf}$, *where $N_{s.wf}$ is the least justified set of Π, and*
2. $S_{c.rg}(\Pi) = \bigvee_{i=1}^{m} \Pi \uplus N_i$, *where N_i are all regularly-justified sets of Π.*

Then both $\langle \mathcal{P}_{5N}, \mathcal{G}_5, \vdash_C, S_{s.wf} \rangle$ and $\langle \mathcal{P}_{5N}, \mathcal{G}_5, \vdash_C, S_{c.rg} \rangle$ are extended logic programming languages. □

Example 6. Consider $\Pi_1 = \{a \vee b; \; d \subset notb\}$. Π_1 has a unique justified set $N = \{nota, notb\}$. Thus $S_{s.wf}(\Pi_1) = S_{c.rg}(\Pi) = \{a \vee b; \; d\}$, and d is derivable from $\{a \vee b; \; d\}$. The coincidence of the skeptical well-founded and credulous regular semantics of Π_1 is consistent with the fact that Π_1 is stratified.

On the other hand, both the *well-founded circumscriptive semantics* [19] and static semantics [11][5] of Π renders b **undefined**, and therefore, are characterized by $(a \vee b)$. Despite the fact that Π_1 is stratified, these two semantics of Π differ from the perfect semantics of Π_1 which yields $(a \vee b) \wedge (b \vee d)$. ☐

The question arises as whether there is some kind of relationship between some semantics in the literature and the ones defined in this paper. We answer this question positively by using a program enhancement technique.

Let us define Π_{shift} as the program obtained from Π by, for each program clause $A_1 \vee \cdots \vee A_m \subset body$ in Π, adding

$A_1 \subset body \wedge not A_2 \wedge \cdots \wedge not A_m$

$\cdots$

$A_m \subset body \wedge not A_1 \wedge \cdots \wedge not A_{m-1}$.

Example 7. Consider enhancing the program in Example 6. We thus have

$\Pi_2 = \Pi_{1shift} = \{a \vee b; \; a \subset notb; \; b \subset nota; \; d \subset notb\}$.

Then Π_2 has two regularly-justified sets, namely, $N_1 = \{nota, notd\}$ and $N_2 = \{notb\}$, and $S_{c.rg}(\Pi_2) = \{a \vee b; \; b \vee d; \; a \subset notb; \; b \subset nota; \; d \subset notb\}$. Therefore, under the credulous regular semantics, Π_2 implies $(a \vee b) \wedge (b \vee d)$, which coincides with the perfect and stable semantics of Π_1. ☐

This example shows that a number of semantics proposed for disjunctive programs, including the perfect, the stable, the well-founded circumscriptive, and the static semantics, are not characterized by justifying assumptions on top of the underlying proof system. We believe this is responsible for the confusing status facing disjunctive logic program semantics.

Since both perfect and stable semantics justify assumptions based on a "minimal model" approach and the minimal model approach corresponds to the program shifting, the program shifting does preserve such semantics. For example, it is not difficult to show that the stable semantics of Π coincides with the stable semantics of Π_{shift}. Furthermore, these semantics can also be characterized by our semantics in terms of program shifting, as shown below.

Theorem 22. *1. The perfect semantics [10] of stratified disjunctive program Π coincides with the credulous regular semantics of Π_{shift}.*

2. The stable semantics [5] of disjunctive program Π coincides with the credulous stable semantics of Π_{shift}.

Proof. (2) Let $C(N) = \{G \mid notG \notin N\}$ for any assumed set N. Then it is sufficient, and not difficult, to show that $C(N)$ is a minimal model of Π^N if and only if $C(N) = \{L | \Pi_{shift}{}^N \vdash_C L\}$, where Π^N denotes $\Pi \uplus N$ with all clauses with negations removed. (1) This follows (2) above and the fact that the perfect semantics coincides with the stable semantics for locally stratified disjunctive programs. ☐

---

[5] These two semantics are equivalent for the disjunctive programs.

6 Conclusions

Logic programming stands out, among others, as a powerful and convenient framework for non-monotonic reasoning, mainly because of its capability of specifying search behavior and of guaranteeing proper answer substitutions, both of which depend on uniform provability. Justifying default negation separately from uniform proofs, therefore, severely handicaps applications of logic programming in nonmonotonic reasoning.

By justifying default negation on top of uniform proofs, we not only extend many prominent semantics originally proposed for normal programs, such as the well-founded, stable, and regular semantics, into a much more general framework of extended logic programming languages, but also clarify some semantical confusion caused by separating default negation from the underlying proof system.

One obvious advantage of characterizing semantics in terms of uniform proofs is the preservation of specifying search behavior and of proper answer substitutions in the proposed extended logic programming languages.

Another advantage of the proposed approach is computational simplicity, mainly because both the positive conclusions and default negations are derived/justified using the same proof system. Such a simplicity, however, is not shared by many others. For example, the well-founded circumscriptive semantics for disjunctive logic programs and its equivalent static semantics have to use classical proofs for positive conclusions and circumscriptive derivations for justifying default negations.

Since the value of any assumed negation is determined by uniform proofs, default negation can also be viewed as a regular goal, and therefore, be quantified. This indicates that our approach may also be used to find intended interpretations for default assumptions with *quantifier-in*, a very difficult problem in nonmonotonic reasoning, as demonstrated in the following example.

Example 8. Consider the *hereditary Harrop system* $\langle \mathcal{P}_4, \mathcal{Q}_4, \vdash_C \rangle$, where $\mathcal{Q}_4$ and $\mathcal{P}_4$ are mutually defined by:

$$Q := \top \mid A \mid Q_1 \wedge Q_2 \mid Q_1 \vee Q_2 \mid \forall x Q \mid \exists x Q \mid P \supset Q,$$
$$P := A \mid Q \supset A \mid \forall x P \mid P_1 \wedge P_2.$$

Let's redefine $\mathcal{G}_4$ and $\mathcal{P}_{4N}$ as the set of all goals, possibly with negation, and the set of all program clauses, respectively, by the following two mutual rules.

$$G := \top \mid A \mid G_1 \wedge G_2 \mid G_1 \vee G_2 \mid \forall x G \mid \exists x G \mid D \supset G \mid \text{not} G$$
$$D := A \mid G \supset A \mid \forall x D \mid D_1 \wedge D_2.$$

Then it is not difficult to show that $\langle \mathcal{P}_{4N}, \mathcal{Q}_4, \vdash, S \rangle$, where S is a skeptical transformation similarly defined as $S_{s.wf}$, is an extended logic programming language. This system presents a natural semantics for extended logic programs with quantified negations. $\square$

Acknowledgments The authors would like to thank anonymous reviewers for constructive comments. This work is partially supported by grants from the National Science and Engineering Research Council of Canada. The work of the first author was performed while visiting the Institute of Social Information Science, Fujitsu Labs, Numazu, Shizuoka, Japan.

References

1. C. R. Baral and V. S. Subrahmanian. Stable and extension class theory for logic programs and default logics. *Journal of Automated Reasoning*, 8:345–366, 1992.
2. G. Brewka. Cumulative default logic: in defense of nonmonotonic inference rules. *Artificial Intelligence*, 50:183–205, 1994.
3. J. Dix. A classification theory of semantics of normal logic programs: II. weak properties. *to appear*, 1994.
4. P. M. Dung. Negations as hypotheses: an abductive foundation for logic programming. In *Proceedings of the 8th ICLP*, pages 3–17, 1991.
5. M. Gelfond and V. Lifschitz. Classical negation in logic programs and disjunctive databases. *New Generation Computing*, 9:365–386, 1991.
6. M. Gelfond and V. Lifschitz. The stable model semantics for logic programming. In *Proc. of the 5th ICLP*, pages 1070–1080, The MIT Press, 1988.
7. A. Kakas, P. Mancarella, and P.M. Dung. The acceptability semantics for logic programs. In *Proc. 11th ICLP*, pages 504–519, 1994.
8. D. Miller, G. Nadathur, F. Pfenning, and A. Scedrov. Uniform proofs as a foundation for logic programming. *Annals of Pure and Applied Logic*, 51:125–157, 1991.
9. L.M. Pereira, J.J. Alferes, C.V. Damasio, and J.N. Aparicio. Adding closed world assumptions to well-founded semantics. *Theoretical Computer Science*, 122:49–68, 1993.
10. T. C. Przymusinski. On the declarative semantics of deductive databases and logic programs. In J. Minker, editor, *Foundations of Deductive Databases and Logic Programming*, pages 193–216, Morgan Kaufmann Publishers, 1988.
11. T. C. Przymusinski. Static semantics for normal and disjunctive logic programs. *Annals of Mathematics and Artificial intelligence*, 1994. (To appear).
12. T. C. Przymusinski. Well-founded semantics coincides with three-valued stable semantics. *Foundamenta Informaticae*, 13:445–463, 1990.
13. C. Sakama and K. Inoue. An alternative approach to the semantics of disjunctive logic programs and deductive databases. *Journal of Automated Reasoning*, 13:145–172, 1994.
14. J.S. Schlipf. Formalizing a logic for logic programming. *Annals of Mathematics and Artificial Intelligence*, 5:279–302, 1992.
15. A. Van Gelder, K. Ross, and J. Schlipf. The well-founded semantics for general logic programs. *JACM*, 38:620–650, 1991.
16. J.-H. You and L. Y. Yuan. On the equivalence of semantics for normal logic programs. *Journal of Logic Programming*, 22(3):209-219, 1995.
17. J.-H. You and L. Y. Yuan. A three-valued semantics of deductive databases and logic programs. *Journal of Computer and System Science*, 49:334–361, 1994. The preliminary version appears in the Proc. of the 9th ACM PODS, page 171-182, 1990.
18. L. Y. Yuan. Autoepistemic logic of first order and its expressive power. *Journal of Automated Reasoning*, 13:69–82, 1994.
19. L. Y. Yuan and J.-H. You. Autoepistemic circumscription and logic programming. *Journal of Automated Reasoning*, 10:143–160, 1993.
20. L.Y. Yuan and J.-H. You. On coherence approach to logic program revision. In *Proc. of the 12th ICLP*, 1995.

Papers by You and Yuan are available at http://web.cs.ualberta.ca/~you and http://web.cs.ualberta.ca/~yuan respectively.

Default Consequence Relations as a
Logical Framework for Logic Programs

Alexander Bochman

E-mail: bochman@bimacs.cs.biu.ac.il

Abstract. We consider the use of default consequence relations suggested in [1,2] as a 'logical basis' for normal logic programs. We give a representation of major semantics for logic programs in this framework and study the question what kind of default reasoning is appropriate for them. It is shown, in particular, that default consequence relations based on three-valued inference are adequate for these semantics.

Introduction

We believe that modern logic programming is a sufficiently mature and 'self-conscious' theory to deserve a formal framework of its own. Moreover, we assume that this framework must be a kind of a *logical system* reflecting various types of reasoning underlying current declarative models of logic programs. Though logic programming has arisen as an implementation of logic as a programming language, it is pretty clear now that classical logic is inappropriate as a reasoning framework for logic programs involving 'negation as failure' operator.

It is widely acknowledged in this respect that the study of logic programs and their semantics is intimately connected with the field of nonmonotonic reasoning. This connection is established mainly in the form of translation of logic programs into different nonmonotonic formalisms, such as default logic, circumscription or modal nonmonotonic logics. However, we pursue here a different approach. What is suggested here is a kind of general formalism that encompasses (normal) logic programs and a number of nonmonotonic systems in a single framework. To be more exact, we propose to use a theory of default consequence relations, introduced in [1,2], as such a logical framework. Default consequence relations were used in these papers as a generalization of both default logic and modal nonmonotonic logics. Here we suggest to use this formalism as a logical basis of normal logic programs. One of the main advantages of such an approach is that many of the above mentioned translation results can be recast in the form of a simple equivalence between different nonmonotonic constructions.

Our basic objects will be default rules, or sequents, of the form

$$a : b \vdash A,$$

where a, b are finite sets of propositions. Propositions from a are called *positive premises*, while those from b—*negative premises*. On the intended interpretation, the role of negative premises consists in restricting applicability of the rules.

Accordingly, these rules could be assigned the following informal reading: "If a, then A, unless (one of) b".

The use of such rules for studying nonmonotonic reasoning is not new (see, e.g., [10,19]). What distinguishes our approach from other works in this area is an explicit and systematic use of 'metarules' allowing to infer new sequents from given ones. This feature of our system makes it more in accord with common monotonic inference systems. Moreover, the use of derived rules greatly simplifies matters and makes explicit many of the intuitions underlying different kinds of nonmonotonic reasoning.

As a first step, a program rule $A \leftarrow A_1, \ldots, A_n, \mathbf{not}\, B_1, \ldots, \mathbf{not}\, B_m$ will be identified with a default sequent

$$A_1, \ldots, A_n : B_1, \ldots, B_m \vdash A.$$

Default sequents representing program rules are considered then as 'axioms' of the default consequence relation corresponding to a program. Note that this representation presupposes that negated premises in the bodies of program clauses are treated simply as a new kind of atomic premises rather than, say, compound literals consisting of **not** and an atom. Nevertheless, we will see that this representation is sufficiently expressive to capture the essential aspects and properties of logic programs and their models.

It is interesting to note that a theory of default consequence relations that correspond in this way to logic programs is not simply a particular case of a theory of default consequence relations in general. Default consequence relations were defined in [1,2] for propositional languages that involve (at least) the language of classical logic and presuppose also a notion of deductive inference. Default sequents corresponding to program clauses, however, are just sequents that involve only atomic propositions. Consequently, a theory of default consequence relations restricted to such sequents turns out to be a *structural* theory of default consequence relations that does not depend on a particular language and propositional connectives involved, while more rich theories could be seen as specializations of this theory obtained by adding rules treating various language specific features and connectives. In this respect, existing semantic approaches in logic programming can contribute much to such a theory.

1 Default Consequence Relations and Logic Programs

A *default consequence relation* will be defined as a set of sequents with the form $a : b \vdash A$ (where a, b are finite) satisfying the following rule:

(*Monotonicity*) If $a : b \vdash A$ and $a \subseteq a'$, $b \subseteq b'$, then $a' : b' \vdash A$.

Monotonicity says, in effect, that default sequents are applicable in *all* contexts in which their premises hold. Note that this immediately distinguish such rules from, e.g., those used in preferential entailment theories [8], where the applicability of the rules is restricted to 'preferred' contexts in which the premises hold. However, the monotonicity condition seems to be in accord with the meaning of program clauses, since the latter are generally considered as rules allowing to infer their heads in all contexts that satisfy the bodies.

The above definition can be immediately extended to infinite sets of premises by stipulating that for any possibly infinite sets of propositions u and v,

(Compactness) $\qquad u : v \vdash A$ if and only if $a : b \vdash A$,

for some finite a, b such that $a \subseteq u$, $b \subseteq v$. This stipulation also ensures that default consequence relations will satisfy the *compactness property*.

In what follows $\mathbb{C}n(u : v)$ will denote the set of all consequences of a pair of sets (u, v), that is, $\{A \mid u : v \vdash A\}$.

The general notion of a default consequence relation is only a frame that can be 'filled' with additional rules that would provide a more tight description of our intuitions about nonmonotonic reasoning. As we will see, there is no single system that reflects adequately *all* our intuitions. In fact, different constructions and models admit different reasoning paradigms.

For a normal logic program P, let $\vdash_P$ denote the least default consequence relation containing all sequents corresponding to clauses from P and $\mathbb{C}n_P$ the associated provability operator. It is easy to see that $A \in \mathbb{C}n_P(u : v)$ if and only if there is a rule in P with the head A such that the positive premises in its body are included in u, while all the negated premises are from v. In other words, $\mathbb{C}n_P$ is just an 'immediate consequence' operator T_P corresponding to P (see, e.g., [20]). This observation is a key to the correspondence between the objects we are going to define and main constructs used for defining semantics of normal logic programs. $\overline{u}$ below denotes the complement of a set u.

Definition 1.1 (i) A pair of sets of propositions (u, v) will be called a *theory* in $\vdash$ if $\mathbb{C}n(u : v) \subseteq u$;

(ii) A pair (u, v) is *supported* if $u \subseteq \mathbb{C}n(u : v)$.

Pairs (u, v) can be seen as formal counterparts of *(Herbrand) interpretations* as they are used in logic programming, where u (v) corresponds to propositions that are true (resp., false) in an interpretation. A pair (u, v) is *consistent* if u and v are disjoint and *complete* if their union is a set of all propositions. Ordinary (two-valued) interpretations correspond to pairs that are both consistent and complete. Unfortunately, such a constraint on interpretations turns out to be too restrictive. However, an important condition on possible models is still that they should be closed with respect to applicability of the rules. This means that positive sets of interpretations must include the heads of all rules that are applicable with respect to them. This is reflected in our notion of a theory. The notion of a support, on the other hand, corresponds to the well-known requirement that any accepted proposition must be *demonstrably true* by being a head of some clause whose body is true with respect to an interpretation.

There is an additional constraint on interpretations implicit in current three-valued approaches to logic programming semantics. It amounts to the requirement that, even for inconsistent or incomplete interpretations, a pair $(\overline{v}, \overline{u})$ consisting of 'nonfalse' and 'nontrue' propositions should also be closed with respect to the program rules. This requirement is embodied in the notion of a t-theory given below. As we will see, such theories arise naturally if the program rules are seen as preserving the truth content, or t-order.

Definition 1.2 (i) (u, v) is a *t-theory* if both (u, v) and $(\overline{v}, \overline{u})$ are theories.
(ii) A t-theory (u, v) is *supported* if both (u, v) and $(\overline{v}, \overline{u})$ are supported.

It follows from the definition that supported t-theories are pairs satisfying the following two conditions:

$$u = \mathbb{C}\mathrm{n}(u : v) \qquad \overline{v} = \mathbb{C}\mathrm{n}(\overline{v} : \overline{u})$$

Note also that if a pair is complete and consistent, it can be identified with its positive component; the corresponding (t-)theories (we will call them *2-theories*) can be defined simply as sets u satisfying the condition $\mathbb{C}\mathrm{n}(u : \overline{u}) \subseteq u$, while supported 2-theories are characterized by the condition $u = \mathbb{C}\mathrm{n}(u : \overline{u})$.

Below is a list of correspondences between the above defined objects and well known logic programming constructs.

- 2-theories of $\vdash_P$ coincide with Herbrand models of P;
- Supported 2-theories correspond to models of $Comp(P)$;
- Consistent t-theories in general correspond to 3-valued models from [14,15];
- Consistent supported t-theories correspond to 3-valued models of $Comp(P)$ [4][1].

Now we will give a characterization of partial (3-valued) stable models from [15]. The notion of a strong support defined below stems from a requirement that demonstration should be 'positively minimal', taken together with the principle that only theories matter in determining the meaning of logic programs. To simplify the presentation, we will say that a set u is v-*closed* if (u, v) is a theory.

Definition 1.3 (i) A pair (u, v) is *strongly supported* if u is included in any v-closed set;
(ii) A t-theory (u, v) will be called *stable* if both (u, v) and $(\overline{v}, \overline{u})$ are strongly supported.

Note that an intersection of v-closed sets is also v-closed. Consequently, there always exists a least v-closed set. Moreover, we have

Lemma 1.1 (u, v) *is a stable theory iff u is the least v-closed set and $\overline{v}$ is the least $\overline{u}$-closed set.*

The following theorem shows that partial stable models of a program P coincide with consistent stable theories of $\vdash_P$.

---

[1] Fitting's operator Φ_P can be defined as follows: $\Phi_P(u, v) = (\mathbb{C}\mathrm{n}(u, v), \overline{\mathbb{C}\mathrm{n}(\overline{v} : \overline{u})})$. Fixed points of this operator are clearly supported t-theories in our sense.

Theorem 1.2 *A pair (u, v) is a partial stable model of P if and only if it is a consistent stable theory in $\vdash_P$.*

As is shown in [15], the well-founded semantics [20] coincides with the k-least partial stable model, while complete stable models coincide with stable models in the sense of Gelfond and Lifschitz [7]. Complete theories corresponding to the latter can be characterized as sets u such that u coincides with the least $\bar{u}$-closed set. We will call such theories *stable 2-theories*.

1.1 Iterative and Reflexive Consequence Relations

In order to give a more perspicuous description of the above models for logic programs, we need to develop our intuitions further. As was already hinted above, an additional principle that is implicit in all the above models is that only theories are essential for determining the meaning of logic programs. It turns out that this requirement can be expressed in the form of a condition that provable conclusions can be used as additional positive premises in the proof.

A default consequence relation will be called *iterative* if it satisfies the rule

$$(Cut) \qquad \frac{a : b \vdash A \qquad a, A : b \vdash B}{a : b \vdash B}$$

The Cut rule reflects a kind of cumulativity of default reasoning: it allows provable conclusions to be used as positive premises in subsequent inferences. It turns out that having this rule, we can avoid the use of iterative constructions commonly employed in defining logic programming models.

The next result shows that iterative consequence relations are determined by theories and their consequences:

Lemma 1.3 *If $\vdash$ is an iterative consequence relation, then $a : b \vdash A$ if and only if $u : v \vdash A$, for any theory (u, v) such that $a \subseteq u$ and $b \subseteq v$.*

Let $\vdash^i$ be the least iterative consequence relation containing $\vdash$, that is, a consequence relation obtained from $\vdash$ by adding the Cut rule. We show first that $\mathbb{C}\mathrm{n}^i(u : v)$ can be obtained as a result of an iterative construction, which is actually quite common in logic programming literature.

Lemma 1.4 *For any pair (u, v), let $w_0 = \mathbb{C}\mathrm{n}(u : v)$ and $w_{j+1} = w_j \cup \mathbb{C}\mathrm{n}(u, w_j : v)$, for any $j \geq 0$. Then*

$$\mathbb{C}\mathrm{n}^i(u : v) = \bigcup_j w_j.$$

The above construction is similar to a usual construction of proofs in monotonic logical systems, when propositions provable on earlier stages are used as premises of subsequent rules. It is interesting to note that also in our case the construction admits an equivalent description in terms of sets that are 'closed' with respect to the rules. We say that a set w is (u, v)-*closed* in $\vdash$ if $\mathbb{C}\mathrm{n}(u, w : v) \subseteq w$. It is easy to show that an intersection of (u, v)-closed sets is also (u, v)-closed, and hence there exists a least such set. The following lemma shows that this set is exactly $\mathbb{C}\mathrm{n}^i(u : v)$.

Lemma 1.5 $\mathbb{C}n^i(u : v)$ *coincides with the least* (u, v)-*closed set in* $\vdash$.

Note that if (u, v) is a theory, the least (u, v)-closed set is just $\mathbb{C}n(u, v)$. In other words, theories has the same consequences in $\vdash^i$ as in $\vdash$. This means, in turn, that $\vdash$ and $\vdash^i$ have the same t-theories, supported t-theories and stable theories. In addition, the least $(\emptyset, v)$-closed set coincides with the least v-closed set. Hence we have

Lemma 1.6 *A pair* (u, v) *is strongly supported in* $\vdash$ *iff* $u \subseteq \mathbb{C}n^i(\emptyset : v)$.

This lemma gives us an alternative characterization of a strong support: if (u, v) is such a supported pair, then any proposition from u is *provable* in $\vdash^i$ from v as a set of negative premises. We can use this result to give an alternative description of stable theories.

Theorem 1.7 (i) (u, v) *is a stable theory in* $\vdash$ *if and only if*

$$u = \mathbb{C}n^i(\emptyset : v) \quad and \quad \bar{v} = \mathbb{C}n^i(\emptyset : \bar{u})$$

(ii) u *is a stable 2-theory in* $\vdash$ *if and only if* $u = \mathbb{C}n^i(\emptyset : \bar{u})$.

Remark. The above characterization of stable 2-theories makes them equivalent to what we called *extensions* in [1]. The latter were shown to characterize extensions in Reiter's default logic [17]. As a result, we have a kind of equivalence between stable models of Gelfond and Lifschitz and Reiter's extensions. This can be seen as our reformulation of the well known results about translation of logic programs into default logic (see, e.g., [12]).

Many of the properties of stable theories become immediate consequences of the above conditions. Thus, it is clear that positive and negative components of a stable theory are mutually definable, and consequently each of them is sufficient for determining the theory. This connection can be expressed as follows. Let us define the following operator on sets of propositions:

$$f(u) = \mathbb{C}n^i(\emptyset : \bar{u}).$$

This operator is antimonotonic and hence, taken twice, gives a monotonic operator. The next proposition says that stable theories can be identified with fixed points of f^2, while stable 2-theories are just fixed points of f itself (cf. [5]).

Proposition 1.8 (i) (u, v) *is a stable theory iff* $u = f^2(u)$ *and* $v = \overline{f(u)}$. (ii) u *is a stable 2-theory if and only if* $u = f(u)$.

Note that the well-founded model corresponds to the least fixed point of f^2. This immediately implies that such models always exist.

Recall that partial stable models coincide with consistent stable theories. The following proposition gives another characterization of consistent stable theories which is in fact equivalent to that provided in [9].

Proposition 1.9 *A set u is a positive part of a consistent stable theory if and only if $u = f(u \cup f(u))$.*

An important additional constraint that can be imposed on default consequence relations is the following condition:

(*Reflexivity*) $A : \emptyset \vdash A$

Iterative consequence relations that satisfy Reflexivity will be called *reflexive*[2]. The following lemma shows that such consequence relations are uniquely determined by their theories.

Lemma 1.10 *If $\vdash$ is a reflexive consequence relation, then $a : b \vdash A$ if and only if $A \in u$, for any theory (u, v) such that $a \subseteq u$ and $b \subseteq v$.*

An important feature of reflexive consequence relations is that the corresponding provability operator $\mathbb{C}n(u, v)$ has all the usual properties of a deductive closure operator with respect to its first argument (that is, when v is fixed):

C1. If $u \subseteq u'$, then $\mathbb{C}n(u : v) \subseteq \mathbb{C}n(u' : v)$.
C2. $u \subseteq \mathbb{C}n(u : v)$.
C3. $\mathbb{C}n(\mathbb{C}n(u : v) : v) \subseteq \mathbb{C}n(u : v)$.

Let $\vdash^r$ denote the least reflexive consequence relation containing $\vdash$.

Lemma 1.11 $\mathbb{C}n^r(u : v)$ *is the least v-closed set in $\vdash$ that contains u.*

The following lemma describes the connection between the least iterative and the least reflexive consequence relation containing $\vdash$.

Lemma 1.12   (i) $\mathbb{C}n^r(u : v) = u \cup \mathbb{C}n^i(u : v)$;
(ii) $\mathbb{C}n^i(u : v) = \mathbb{C}n(\mathbb{C}n^r(u : v) : v)$

As is well known, models of Comp are sensitive to adding trivial rules $A \leftarrow A$ to a program (see, e.g., [14]). Now the above lemma implies that, for any v, $\mathbb{C}n^r(\emptyset : v) = \mathbb{C}n^i(\emptyset : v)$. Thus, pairs of the form $(\emptyset : v)$ has the same consequences in $\vdash^r$ as in $\vdash^i$. An immediate consequence of this fact is that Reflexivity preserves stable theories (cf. below).

However, Reflexivity makes the notion of a support trivial. As a consequence, for reflexive consequence relations theories coincide with supported theories. This means, in particular, that reflexive consequence relations are already inappropriate for representing models of Comp. This fact can be seen as an example of 'proliferation' of different forms of default reasoning in the sense that different models usually admit different reasoning paradigms.

The above results can serve as an evidence of the usefulness of default consequence relations in studying semantics for logic programs. In the next section we will consider how various rules imposed on default consequence relations influence these models and what consequence relation can be taken as an appropriate logical framework for them.

---
[2] Such consequence relations are in fact equivalent to *clones* of default rules from [19].

2 Kinds of Default Reasoning

2.1 S-consequence Relations and Classical Inference

S-consequence relations[3], defined below, can be seen as an upper bound on any reasonable default system for logic programming.

Definition 2.1 A reflexive consequence relation will be called an *S-consequence relation* if it satisfies the following conditions:

(*Consistency*) $$A : A \vdash B$$

(*Factoring*) $$\frac{a, B : b \vdash A \qquad a : b, B \vdash A}{a : b \vdash A}$$

Consistency excludes, in effect, inconsistent theories from consideration, while Factoring can be seen as a formalization of reasoning by cases in the context of normal programs (see, e.g., [18,20]). It has an effect of restricting admissible theories to complete ones. As a result, it is easy to show that S-consequence relations are determined by their 2-theories. Moreover, it is shown in [2] that the least S-consequence relation containing $\vdash$ is actually the greatest default consequence relation having the same 2-theories as $\vdash$.

The following result shows that an S-consequence relation corresponding to a logic program treats program rules as ordinary formulas of classical logic.

For any program rule r, let $\hat{r}$ denote a logical clause corresponding to r, that is, a logical formula obtained from r by replacing **not** with a classical negation connective and $\leftarrow$ with a classical implication. For a program P, we will denote by $\hat{P}$ the set of such logical formulas corresponding to rules of P. Taking into account our identification of program rules with default sequents, we will also denote by $\hat{s}$ the logical formula corresponding to a sequent s in an obvious way.

Proposition 2.1 *Let $\vdash_P^{sc}$ be the least S-consequence relation corresponding to P and s a default sequent. Then $s \in \vdash_P^{sc}$ iff $\hat{s}$ is a logical consequence of $\hat{P}$.*

Thus, S-consequence relations can be seen as a formalization of a classical inference for normal programs. Furthermore, the following theorem shows that such consequence relations are already monotonic. To be more exact, they are equivalent to (monotonic) Scott consequence relations. A binary relation $a \Vdash b$ between *sets* of propositions is called a *Scott consequence relation* (see [6]) if it satisfies the following conditions:

(*Reflexivity*) $$A \Vdash A;$$

(*Monotonicity*) If $a \Vdash b$ and $a \subseteq a'$, $b \subseteq b'$, then $a' \Vdash b'$;

(*Cut*) $$\frac{a \Vdash b, A \qquad a, A \Vdash b}{a \Vdash b}.$$

---

[3] Such consequence relations were called *stable* in [1,2]. Unfortunately, the name bears unnecessary associations in the present context.

Theorem 2.2 *Let* ⊢ *be an S-consequence relation. Define the following consequence relation between sets of propositions:*

$$a \Vdash_{\vdash} b \equiv a : b \vdash B,$$

for some B ∈ *b. Then* ⊩⊢ *is a Scott consequence relation and a : b* ⊢ *A if and only if a* ⊩⊢ *b, A.*

Remark. The above 'monotonic reduction' of S-consequence relations allows us to take a new look on the results from [10] about the applicability of nonmonotonic rule-based systems to solving 'monotonic' mathematical problems (e.g., the marriage problem). Under suggested formalizations of these problems in terms of default-type rules, the solutions were shown to coincide with extensions (that is, stable 2-theories) of the corresponding nonmonotonic system. However, it turns out that in all examples given in [10] the context makes valid all the conditions of an S-consequence relation! Moreover, it is easy to show that stable 2-theories in such consequence relations coincide with minimal 2-theories. Thus, despite the appearance, the relevant contexts of reasoning in these examples remain thoroughly monotonic.

Classical inference is clearly inappropriate if applied to normal logic programs; it misses the intended meaning of **not** as negation by failure. However, S-consequence relations constitute an important upper bound on reasonable host default systems: if we want to preserve classical logical validity, they should contain rules that are also valid for S-consequence relations. Thus, the main lesson from the above results is that in order to obtain appropriate consequence relations, we must reject, or weaken, some of the conditions characterizing S-consequence relations.

2.2 Consequence Relations Based on Three-Valued Inference

We consider now the question whether an appropriate logical framework for normal logic programs can be based on three-valued inference. As is well-known, three-valued extensions of logic programming semantics (see, e.g., [4,16]) provide a natural generalization of previous approaches while avoiding many of their shortcomings. Still, this does not answer the question whether we can use a three-valued inference in reasoning about such models.

In an attempt to answer this latter question, we define first when a sequent can be considered *valid* with respect to a 3-valued interpretation. It turns out that we have a significant degree of freedom here.

Let ν be a 3-valued interpretation, that is, an assignment of exactly one of the truth-values $\{\mathbf{t}, \mathbf{u}, \mathbf{f}\}$ to all propositions. We extend such interpretations to sets of propositions by taking $\nu(a)$ to be a minimum of all $\nu(A_i)$, for $A_i \in a$, in the t-order of the truth-values: $\mathbf{f} <_t \mathbf{u} <_t \mathbf{t}$. Let $\neg$ be the usual Kleene negation[4] and $\neg b$ denote the set of all $\neg B$ for $B \in b$.

---

[4] That is, $\neg \mathbf{t} = \mathbf{f}, \neg \mathbf{f} = \mathbf{t}$ and $\neg \mathbf{u} = \mathbf{u}$.

Definition 2.2 Let ν be a 3-valued interpretation.

(i) A sequent $a : b \vdash A$ will be said to be *A-valid* with respect to ν iff $\nu(a) \neq \mathbf{f}$ and $\nu(\neg b) = \mathbf{t}$ imply $\nu(A) = \mathbf{t}$.

(ii) $a : b \vdash A$ will be said to be *T-valid* with respect to ν iff $\nu(a \cup \neg b) = \mathbf{t}$ only if $\nu(A) = \mathbf{t}$.

(iii) $a : b \vdash A$ will be said to be *F-valid* with respect to ν iff $\nu(A) = \mathbf{f}$ only if $\nu(a \cup \neg b) = \mathbf{f}$.

(iv) $a : b \vdash A$ will be said to be *t-valid* with respect to ν iff $\nu(a \cup \neg b) \leq_t \nu(A)$.

A-validity reflects an autoepistemic reasoning (see below), T-validity corresponds to a requirement that valid sequents must preserve truth, F-validity corresponds to preservation of 'nonfalsity', while t-validity preserves the 'truth content' (cf. [14]), that is, both truth and nonfalsity. These notions of validity give rise to the corresponding notions of semantic entailment among default sequents. Thus, we will say that a set of sequents S *A-entails* a sequent s (notation $S \models^A s$) if s is A-valid for any interpretation ν in which all of S are A-valid. Similar definitions can be given for T-entailment, F-entailment and t-entailment.

Now we will turn to describing default consequence relations that correspond to these four notions of entailment.

Definition 2.3 An iterative consequence relation will be called

– an *A-consequence relation* if it satisfies Consistency and Factoring;
– a *T-consequence relation* if it satisfies Reflexivity and Consistency;
– an *F-consequence relation* if it satisfies Reflexivity and Factoring;
– a *t-consequence relation* if it satisfies Reflexivity and the following two rules:

(Negative Cut)
$$\frac{c : d \vdash A \quad \{a : b, C_i \vdash B\} \quad \{a, D_j : b \vdash B\} \quad \forall C_i \in c \; \forall D_j \in d}{a : b, A \vdash B}$$

(C-Factoring)
$$\frac{A, a : b \vdash B \qquad a : b, A \vdash B}{a, C : b, C \vdash B}$$

For $x = A, T, F$ or t, let $\vdash^x_S$ denote a least x-consequence relation containing a set of sequents S and $\vdash^x$ a least x-consequence relation containing $\vdash$. The following result shows that the above consequence relations are complete for the corresponding notions of semantic entailment in the sense that $S \models^x s$ for each kind of entailment if and only if s is provable from S in the corresponding default consequence relation.

Theorem 2.3 (3-Completeness Theorems) *For any sequent s,*

$$S \models^x s \quad \text{if and only if} \quad s \in \vdash^x_S .$$

Thus, all the above default consequence relations turn out to be based on 'three-valued reasoning'. We will give below some considerations that can be used in choosing (or rejecting) each of them as a logical paradigm for reasoning about logic programs.

A-consequence Relations. A-consequence relations are in fact structural subsystems of *autoepistemic consequence relations* from [1,2]. The following result shows that they are adequate for semantics based on Clark's completion.

Theorem 2.4 *For any default consequence relation* $\vdash$, $\vdash^A$ *has the same 2-theories and supported 2-theories as* $\vdash$.

Thus, A-consequence relations provide an adequate logical framework for reasoning about ordinary models of Comp. Note, however, that we do not reach completeness here; A-consequence relations are not determined uniquely by their sets of 2-theories and supported 2-theories. As we will explain later, this incompleteness unavoidable as far as essentially nonmonotonic objects are concerned.

Remark. 2-theories and supported 2-theories correspond, respectively, to what we called *stable sets* and *expansions* in [1,2][5]. The latter were shown to correspond to modal stable sets and stable expansions in autoepistemic logic [13]. Using the technique from [1], this result can be reformulated in the form of embedding of logic programs into autoepistemic theories that transform 2-theories to (objective kernels of) stable sets and supported 2-theories into stable expansions. Note, however, that this embedding is not adequate, e.g., for t-theories.

T-consequence Relations. T-consequence relations are reflexive consequence relations that are determined by their *consistent* theories. Though the consistency constraint on interpretations looks plausible and even desirable, it turns out that it preserves neither consistent t-theories nor consistent stable theories. Still, it follows from the results obtained in [2] that such consequence relations preserve stable 2-theories. Moreover, there are additional rules that are appropriate for the latter. An interesting example is the following rule from [3][6]:

$$(Semi\text{-}Factoring) \qquad \frac{A, a : b \vdash B \qquad a : b, A \vdash C}{a : b, B \vdash C}$$

An interesting feature of such consequence relations is that a sequent of the form $a : b, A \vdash A$ 'represents' in some sense the logical clause corresponding to $a : b \vdash A$ (cf. Proposition 2.1).

Proposition 2.5 *If* $\vdash$ *is a T-consequence relation satisfying Semi-Factoring, then* $a : b, A \vdash A$ *iff* $a : b \vdash^{sc} A$.

F-consequence relations. F-consequence relations can be shown to be determined by their *complete* theories. Such consequence relations preserve neither partial theories nor stable theories, nor even stable 2-theories. However, that does not necessarily make them implausible. Moreover, an essentially equivalent system was suggested by Schlipf [18] as a 'logical paradigm' for defining the well-founded semantics. Thus, 3-valued models of stable-by-case completion

---

[5] These objects are called deductively closed sets and weak extensions of $\emptyset$ in [10].

[6] This rule is in fact equivalent to the (Negative Factoring) rule from [1].

from [18] coincide with consistent stable theories of $\vdash_P^F$. As is shown by Schlipf, such models satisfy many of the desiderata for an adequate semantics of logic programs.

t-consequence relations. It can be shown that t-consequence relations are determined by their t-theories that are either complete or consistent. Unfortunately, it turns out that Negative Cut and Reflexivity imply the rule

$$\frac{: A \vdash B}{: B \vdash A}$$

that is clearly inappropriate. For example, a program consisting of a single clause $A \leftarrow \mathbf{not}\, B$ has a unique stable model $(\{A\}, \{B\})$, that is also its well-founded model. However, adding a clause $B \leftarrow \mathbf{not}\, A$ will give us a program with an empty well-founded model and two stable models $(\{A\}, \{B\})$ and $(\{B\}, \{A\})$.

Fortunately, there is an important subsystem of t-consequence relations that turns out to preserve consistent stable theories.

Definition 2.4 A reflexive consequence relation will be called a *TF-consequence relation* if it satisfies C-Factoring.

It can be shown that TF-consequence relations are determined by theories that are either consistent or complete. The following theorem shows that such consequence relations are adequate for reasoning about partial stable models of logic programs. Let $\vdash^{TF}$ denote the least TF-consequence relation containing $\vdash$.

Theorem 2.6 *For any default consequence relation* $\vdash$, $\vdash^{TF}$ *has the same consistent stable theories as* $\vdash$.

Moreover, the following result shows that any TF-consequence relation can be seen as an intersection of a T-consequence relation and an F-consequence relation.

Proposition 2.7 $\vdash$ *is a TF-consequence relation iff* $\vdash = \vdash^T \cap \vdash^F$.

As an immediate consequence of this result we obtain that, in computing partial stable models, we can use T-entailment for computing positive literals and F-entailment for computing negative literals.

Corollary 2.8 *A consistent pair* (u, v) *is a stable theory in* $\vdash$ *if and only if*

$$u = \mathbb{C}\mathrm{n}^T(\emptyset : v) \qquad \overline{v} = \mathbb{C}\mathrm{n}^F(\emptyset : \overline{u}).$$

The main conclusion that can be made from the results of this section is that all the models of normal logic programs considered in this paper admit some kind of three-valued reasoning, though different models give rise to different kinds of reasoning.

3 Conclusions and Further Issues

As we have seen, default consequence relation provide a powerful and efficient framework for studying logic programs and their semantics. In addition, using the results and methods from [1,2], all the constructions and objects considered in this paper can be immediately 'transferred' to default logic and modal nonmonotonic logics.

These results can be seen as a justification of the claim that default consequence relations can serve as a proper logical basis, or framework, for logic programming in which many of the reasoning principles used can be explicitly stated and studied. Moreover, the results of the last section actually bridge the gap between logic programming and common logical theories in showing that major semantics for logic programs admit reasoning based on full-fledged (three-valued) logical inference.

However, a note of caution is in order here. Default consequence relations as such do not provide (and are not intended to be) a formalization of the *nonmonotonic* reasoning behind logic programming semantics, embodied in our case in the understanding of **not** as a negation by failure. In this sense they are systems *for* but not *of* nonmonotonic reasoning. Moreover, as far as provability is concerned, default consequence relations are as monotonic as ordinary deductive systems: proved propositions remain such after addition of new sequents. Negation as failure principle, however, should involve demonstration that certain propositions are *not* provable in a given system. This principle provides us with additional negative premises for subsequent default inference. Clearly, such proofs will be nonmonotonic, but only because addition of new facts may lead to rejection of some of the negative assumptions made earlier. Summing up these remarks, it can be said that default consequence relations are intended to provide no more (but no less) than a proper logical basis for a nonmonotonic reasoning employed in logic programming constructs.

The above considerations have one important implication, namely an essential *incompleteness* of default consequence relations with respect to nontrivial nonmonotonic objects, such as supported and stable theories in our case. This is precisely because if completeness were possible, the corresponding provable propositions would remain provable in all extensions of a default consequence relation, contrary to the nonmonotonic character of these objects.

The study of default consequence relations that is of relevance for logic programming can be extended in many directions. Here we mention only a couple of possibilities. The first direction that suggests itself consists in an extension of the underlying language to include, e.g., a classical negation or implication. The corresponding systems are situated in between pure structural consequence relations considered here and 'full' systems described in [1,2]. Another direction consists in generalizing the very notion of a default sequent to include, e.g., disjunctive rules and negations in the heads. In fact, it can be shown that many of the above constructions and results can be generalized to such systems, though this is an issue for another paper.

Acknowledgments. We would like to thank David Makinson for his detailed comments on an earlier version of the paper.

References

1. A. Bochman (1994) On the relation between default and modal consequence relations *Proc. Fourth International Conference on Principles of Knowledge Representation and Reasoning, KR'94*, Morgan Kaufmann, San Mateo, CA., pp. 63–74.
2. A. Bochman (1993) Modal nonmonotonic logics demodalized. *Annals of Mathematics and Artificial Intelligence* (to appear).
3. J. Chen and S. Kundu (1991) The strong semantics for logic programs. *Methodologies for Intelligent Systems (LNAI 542)*, Z. W. Ras and M. Zamenkova (eds.), Springer Verlag, pp. 490–499.
4. M. C. Fitting (1985) A Kripke-Kleene semantics for logic programs. *JLP* **2**: 295–312.
5. M. C. Fitting (1993) The family of stable models. *JLP* **17**: 197–225.
6. D. M. Gabbay (1976). *Investigations in Modal and Tense Logics*. Dordrecht: D. Reidel.
7. M. Gelfond and V. Lifschitz (1988) The stable model semantics for logic programming *Proc. 5th International Conf./Symp. on Logic Programming*, R. Kowalski and K. Bowen (eds.), MIT Press: Cambridge, MA, pp. 1070–1080.
8. S. Kraus, D. Lehmann and M. Magidor (1990) Nonmonotonic reasoning, peferential models and cumulative logics. *Artificial Intelligence*, **44**: 167–207.
9. E. Laenens, D. Vermeir and C. Zaniolo (1992) Logic programming semantics made easy. *Automata, Languages and Programming (LNCS 623)*, Springer, pp. 499–508.
10. W. Marek, A. Nerode and J. Remmel (1990) A theory of nonmonotonic rule systems I *Annals of Mathematics and Artificial Intelligence*, **1**: 241–273.
11. W. Marek and V.S. Subrahmanian (1992) The relationship between stable, supported, default and autoepistemic semantics for general logic programs. *Theoretical Computer Science*, **103**: 365–386.
12. W. Marek and M. Truszczyński (1989) Stable semantics for logic programs and default theories. *Proc. 1989 North Amer. Conf. on Logic Programming*, MIT Press: Cambridge, MA, pp. 243–256.
13. R. C. Moore (1985). Semantical considerations on non-monotonic logic. *Artificial Intelligence* 25:75–94.
14. H. Przymusinska and T. Przymusinski (1990) Semantic issues in deductive databases and logic programs, in R. Benerji (ed.) *Formal Techniques in Artificial Intelligence*, North-Holland: Amsterdam, pp. 321–367.
15. T.C. Przymusinski (1990) Well-founded semantics coincides with three-valued stable semantics. *Fundamenta Informaticae*, **13**: 445–463.
16. T. Przymusinski (1991) Three-valued nonmonotonic formalisms and semantics of logic programs. *Artificial Intelligence*, **49**: 309–343.
17. R. Reiter (1980). A logic for default reasoning. *Artificial Intelligence* 13:81–132.
18. J. S. Schlipf (1992). Formalizing a logic for logic programming. *Annals of Mathematics and Artificial Intelligence* 5:279–302.
19. H. Thiele On generation of cumulative inference operators by default deduction rules. *LNAI 543*, Springer Verlag, pp. 100–137.
20. A. Van Gelder, K.A. Ross and J.S. Schlipf (1991) The well-founded semantics for general logic programs. *J. ACM*, **38**: 620–650.

Skeptical Rational Extensions

Artur Mikitiuk and Mirosław Truszczyński

University of Kentucky, Department of Computer Science,
Lexington, KY 40506-0046, {artur|mirek}@cs.engr.uky.edu

Abstract. In this paper we propose a version of default logic with the
following two properties: (1) defaults with mutually inconsistent justifications are never used together in constructing a set of default consequences of a theory; (2) the reasoning formalized by our logic is related
to the traditional skeptical mode of default reasoning. Our logic is based
on the concept of a *skeptical rational extension*. We give characterization
results for skeptical rational extensions and an algorithm to compute
them. We present some complexity results. Our main goal is to characterize cases when the class of skeptical rational extensions is closed
under intersection. In the case of normal default theories our logic coincides with the standard skeptical reasoning with extensions. In the case
of seminormal default theories our formalism provides a description of
the standard skeptical reasoning with rational extensions.

1 Introduction

In this paper we investigate a version of default logic with the following two main
properties. First, defaults with mutually inconsistent justifications are never used
together in constructing a set of default consequences of a theory. This has
implications for the adequacy of our system to handle situations with disjunctive
information. Second, the reasoning formalized by our logic is closely related to
the traditional skeptical mode of default reasoning. In the case of normal default
theories it coincides with the standard skeptical reasoning with extensions. In
the case of seminormal default theories our formalism provides a description of
the standard skeptical reasoning with rational extensions. Our logic is defined
by means of a fixpoint construction and not as the intersection of extensions,
as is usually the case with the skeptical reasoning. Hence, our results provide
a fixpoint description of the standard skeptical reasoning from normal default
theories and, in the case of rational extensions, from seminormal default theories.

Default logic, introduced by Reiter [10], is one of the most extensively studied
nonmonotonic systems. Several recent research monographs offer a comprehen-
sive presentation of theoretical and practical aspects of default logic [1, 3, 6].
Default logic assigns to a default theory a collection of theories called *extensions*. Extensions model all possible "realities" described by a default theory
and are used as the basis for two modes of reasoning: *brave* and *skeptical*. In the
brave mode, an arbitrarily selected extension defines the set of consequences for
a default theory. In the skeptical one, the intersection of all extensions serves

as the set of consequences. Skeptical consequences are more robust in the sense that they hold in all possible realities described by a default theory.

All its desirable properties notwithstanding, there are situations where default logic of Reiter is not easily applicable. In particular, default logic does not handle well incomplete information given in the form of disjunctive clauses [9, 2, 4, 7]. To remedy this, several modifications of default logic were proposed: disjunctive default logic [4], cumulative default logic [2], constrained default logic [11] and rational default logic [7]. The first system introduces a new disjunction operator to handle "effective" disjunction. The latter three take into account, in one way or another, the requirement that defaults with mutually inconsistent justifications must not be used in the construction of the same extension. Not surprisingly then, they are somewhat related. Connections between cumulative default logic and constrained default logic are studied in [11]. Relations between constrained default logic and rational default logic are discussed in [8].

In this paper we continue our investigation of rational default logic introduced in [7]. The key idea behind the concept of a rational extension of a default theory (D, W) is that of a maximal set of defaults in D *active* with respect to theories W and S. The collection of all such sets (it is always nonempty) is denoted by $\mathcal{MA}(D, W, S)$. Intuitively, it contains every group of defaults the reasoner can select to justify that S is a rational extension of (D, W) (if none works, S is not a rational extension). That is, S is a rational extension if S can be derived from W by means of some set of defaults $A \in \mathcal{MA}(D, W, S)$. In this paper, we strengthen the requirements for a hypothetical context S to be a rational extension. As a result we obtain a new fixpoint construction and a new class of extensions — *skeptical rational extensions*. (The word "extension" is being used here in a broader sense. A rational or a skeptical rational extension of a default theory is not, in general, an extension of the theory in Reiter's sense — see Examples 1 and 2.) Specifically, for a theory S to be a skeptical rational extension, S must be exactly the set of formulas that can be derived from W by means of *every* set of defaults $A \in \mathcal{MA}(D, W, S)$. In other words, S consists of those formulas the reasoner can justify no matter which element from $\mathcal{MA}(D, W, S)$ is selected for reasoning. This motivates the term *skeptical* used to designate these extensions.

The class of skeptical rational extensions has several desirable properties. For many default theories, it contains a least element with respect to inclusion. In such a case, this least skeptical rational extension can be used as a formal model of skeptical default reasoning (sometimes identical with and sometimes different from the traditional model of skeptical default reasoning).

In this paper we investigate properties of skeptical rational extensions. We restrict ourselves to the propositional case only. We give characterization results for skeptical rational extensions and an algorithm to compute them. We present some complexity results. Our main goal is to characterize cases when the class of skeptical rational extensions is closed under intersection. We obtain the strongest results for normal and seminormal default theories. We show that the intersection of all rational extensions of a seminormal default theory is its *least* skeptical rational extension. In particular, it means that the intersection of all extensions of a normal default theory is, in fact, its least skeptical rational extension.

2 Definitions and Examples

Let $\mathcal{L}$ be a language of propositional logic. A *default* is any expression of the form

$$\frac{\alpha: M\beta_1, \ldots, M\beta_k}{\gamma},$$

where α, β_i, $1 \leq i \leq k$ and γ are propositional formulas from $\mathcal{L}$. The formula α is called the *prerequisite* of d, $p(d)$ in symbols. The formulas β_i, $1 \leq i \leq k$, are called the *justifications* of d. The set of justifications is denoted by $j(d)$. Finally, the formula γ is called the *consequent* of d and is denoted $c(d)$. For a collection D of defaults by $p(D)$, $j(D)$ and $c(D)$ we denote, respectively, the sets of all prerequisites, justifications and consequents of the defaults in D. A default of the form $\frac{\alpha: M\beta}{\beta}$ ($\frac{\alpha: M(\beta \wedge \gamma)}{\gamma}$, resp.) is called *normal* (*seminormal*, resp.).

A default theory is a pair (D, W), where D is a set of defaults and W is a set of propositional formulas. A default theory (D, W) is *normal* (*seminormal*, resp.) if all defaults in D are normal (seminormal, resp.). A default theory (D, W) is *finite* if both D and W are finite.

For a set D of defaults and for a propositional theory S, we define

$$D_S = \left\{ \frac{\alpha}{\gamma} : \frac{\alpha: M\beta_1, \ldots, M\beta_k}{\gamma} \in D, \text{ and } S \nvdash \neg\beta_i, \ 1 \leq i \leq k \right\}$$

and

$$Mon(D) = \left\{ \frac{p(d)}{c(d)} : d \in D \right\}.$$

Given a set of inference rules A, by $Cn^A(\cdot)$ we mean the consequence operator of the formal proof system consisting of propositional calculus and the rules in A (it is defined for all theories in the language).

The key notion of (standard) default logic is the notion of an extension [1]. A theory S is an *extension* for a default theory (D, W) if $S = Cn^{D_S}(W)$.

For a detailed presentation of default logic the reader is referred to [6].

In [7] we introduced the notions of an active set of defaults and a rational extension of a default theory. A set A of defaults is *active* with respect to sets of formulas W and S if it satisfies the following conditions:

AS1 $j(A) = \emptyset$, or $j(A) \cup S$ is consistent,
AS2 $p(A) \subseteq Cn^{A_S}(W)$.

The set of all subsets of a set of defaults D which are active with respect to W and S will be denoted by $\mathcal{A}(D, W, S)$.

Observe, that $\emptyset$ is active with respect to every W and S. Hence, $\mathcal{A}(D, W, S)$ is always nonempty. By the Kuratowski-Zorn Lemma, every $A \in \mathcal{A}(D, W, S)$ is

---

[1] Our definition is different from but equivalent to the original definition of Reiter [10].

contained in a maximal (with respect to inclusion) element of $\mathcal{A}(D, W, S)$ (see [7]). Define $\mathcal{MA}(D, W, S)$ to be the set of all maximal elements in $\mathcal{A}(D, W, S)$.

In [7], we defined S to be a *rational extension* for a default theory (D, W) if $S = Cn^{As}(W)$ for some $A \in \mathcal{MA}(D, W, S)$. We will now define the notion of a skeptical rational extension.

Definition 1. A theory S is a *skeptical rational extension* for a default theory (D, W) if

$$S = \bigcap_{A \in \mathcal{MA}(D, W, S)} Cn^{As}(W). \qquad \square$$

We will illustrate the notions defined above with several examples. The first example exhibits a default theory which does not have an extension or a rational extension but has a skeptical rational extension. In all examples a, b, c and d stand for distinct propositional atoms.

Example 1. Let $D = \{\frac{:M\neg a}{a}, \frac{:M\neg b}{b}\}$. The default theory $(D, \emptyset)$ has no extension and no rational extension. On the other hand, $S = Cn(\{a \vee b\})$ is its skeptical rational extension. Indeed, we have

$$\mathcal{MA}(D, \emptyset, S) = \left\{ \left\{ \frac{:M\neg a}{a} \right\}, \left\{ \frac{:M\neg b}{b} \right\} \right\}$$

and

$$\bigcap_{A \in \mathcal{MA}(D, \emptyset, S)} Cn^{As}(\emptyset) = Cn(\{a\}) \cap Cn(\{b\}) = Cn(\{a \vee b\}) = S. \qquad \square$$

The default theory $(\{\frac{:M\neg a}{a}\}, \emptyset)$ is a classical example of a theory without extensions. More generally, a default theory containing the default $\frac{:M\neg a}{a}$, where a is an atom that does not appear in any other default or formula, does not have an extension. Hence, the fact that $Cn(\{a \vee b\})$ is a skeptical rational extension of the default theory of Example 1 may seem counterintuitive. However, the meaning of the defaults in D is: if $\neg a$ ($\neg b$, resp.) is possible, then conclude a (b, resp.). In the context of $Cn(\{a \vee b\})$, any of the two defaults can fire (but not together). Hence, no matter what is the choice, $a \vee b$ follows.

The next example shows that there are also default theories which have extensions but do not have skeptical rational extensions.

Example 2. Let us consider the default theory (D, W), where $W = \{a \vee b\}$ and

$$D = \left\{ \frac{:M\neg a}{c}, \frac{:M\neg b}{d}, \frac{:M(\neg c \vee \neg d)}{\neg c \wedge \neg d} \right\}.$$

This theory has a unique extension $Cn(\{a \vee b, c, d\})$. We proved in [7] that (D, W) does not have rational extensions.

Assume that S is a skeptical rational extension for (D, W). Then $a \vee b \in S$. If $c \wedge d \notin S$ then

$$\mathcal{MA}(D, W, S) = \left\{ \left\{ \frac{:M\neg a}{c}, \frac{:M(\neg c \vee \neg d)}{\neg c \wedge \neg d} \right\}, \left\{ \frac{:M\neg b}{d}, \frac{:M(\neg c \vee \neg d)}{\neg c \wedge \neg d} \right\} \right\}.$$

Thus, $\bigcap_{A \in \mathcal{MA}(D,W,S)} Cn^{As}(W) = \mathcal{L}$ and $S \neq \mathcal{L}$ (because $c \wedge d \notin S$). So, assume that $c \wedge d \in S$. Then

$$\mathcal{MA}(D, W, S) = \left\{ \left\{ \frac{:M\neg a}{c} \right\}, \left\{ \frac{:M\neg b}{d} \right\} \right\}$$

and $\bigcap_{A \in \mathcal{MA}(D,W,S)} Cn^{As}(W) = Cn(\{a \vee b, c \vee d\}) \neq S$ (because $c \wedge d \notin Cn(\{a \vee b, c \vee d\})$). Hence, (D, W) does not have skeptical rational extensions. $\quad\square$

One of the properties we are especially interested in here is closure under intersection of the family of skeptical rational extensions. The following example presents a default theory for which the family of skeptical rational extensions is closed under intersection. This theory is normal. We will later show that this property holds for every normal default theory with a finite number of extensions.

Example 3. Let $W = \emptyset$ and

$$D = \left\{ \frac{a: Mb}{b}, \frac{a: M\neg b}{\neg b}, \frac{:Ma}{a} \right\}.$$

Let $S_1 = Cn(\{a, b\})$, $S_2 = Cn(\{a, \neg b\})$ and $S = S_1 \cap S_2 = Cn(\{a\})$. Then

$$\mathcal{MA}(D, W, S_1) = \left\{ \left\{ \frac{:Ma}{a}, \frac{a: Mb}{b} \right\} \right\}, \quad \mathcal{MA}(D, W, S_2) = \left\{ \left\{ \frac{:Ma}{a}, \frac{a: M\neg b}{\neg b} \right\} \right\}$$

and

$$\mathcal{MA}(D, W, S) = \left\{ \left\{ \frac{:Ma}{a}, \frac{a: Mb}{b} \right\}, \left\{ \frac{:Ma}{a}, \frac{a: M\neg b}{\neg b} \right\} \right\}.$$

Clearly, S_1 and S_2 are extensions, rational extensions and skeptical rational extensions for (D, W) and S is also a skeptical rational extension for (D, W). $\quad\square$

For some default theories the family of their skeptical rational extensions is not closed under finite intersection.

Example 4. Let $W = \emptyset$ and

$$D = \left\{ \frac{:M\neg a, Md}{b}, \frac{:M(\neg b \vee a), Md}{b \wedge c}, \frac{:M\neg d, M\neg b}{c} \right\}.$$

Let $S_1 = Cn(\{b\})$, $S_2 = Cn(\{c\})$ and $S = S_1 \cap S_2 = Cn(\{b \vee c\})$. Then

$$\mathcal{MA}(D, W, S_1) = \left\{ \left\{ \frac{:M\neg a, Md}{b} \right\}, \left\{ \frac{:M(\neg b \vee a), Md}{b \wedge c} \right\} \right\},$$

$$\mathcal{MA}(D, W, S_2) = \left\{ \left\{ \frac{:M\neg a, Md}{b}, \frac{:M(\neg b \vee a), Md}{b \wedge c} \right\}, \left\{ \frac{:M\neg d, M\neg b}{c} \right\} \right\},$$

and $\mathcal{MA}(D, W, S) = \mathcal{MA}(D, W, S_2)$. It is easy to see that S_1 and S_2 are skeptical rational extensions for (D, W) while S is not. This default theory does not have a least skeptical rational extension.

Let us note that S_1 and S_2 are also rational extensions for (D, W). Let $S_3 = Cn(\{b, c\})$. We have $\mathcal{MA}(D, W, S_3) = \mathcal{MA}(D, W, S_1)$. Hence, S_3 is also a rational extension of (D, W). Finally, it is easy to see that S_3 is the only (Reiter's) extension for (D, W). $\quad\square$

We conclude this section with an alternative characterization of active sets.

Proposition 2. *A set A of defaults is active with respect to sets of formulas W and S if and only if it satisfies* **AS1** *and the following condition:*

AS2′ $p(A) \subseteq Cn^{Mon(A)}(W)$. $\qquad\qquad\qquad\qquad\qquad\qquad\qquad$ □

3 General Properties

In this section we present some results (Theorems 8 and 9, Corollary 10) providing sufficient conditions for the intersection of skeptical rational extensions to be a skeptical rational extension too. These results will be used in Sections 5 and 6. We start with several auxiliary observations. (Simple proofs of Lemmas 3, 4 and 6 are omitted due to space restriction.)

Lemma 3. *Let (D, W) be a default theory. Let S be a set of formulas and let $A \in \mathcal{A}(D, W, S)$. Then $A \in \mathcal{A}(D, W, T)$ for every theory T such that A satisfies* **AS1** *for T.* $\qquad\qquad$ □

Lemma 4. *Let (D, W) be a default theory. Let S and T be theories such that $S \subseteq T$. If $A \in \mathcal{MA}(D, W, S)$ and $A \in \mathcal{A}(D, W, T)$, then $A \in \mathcal{MA}(D, W, T)$.* □

Lemma 5. *Let (D, W) be a default theory and let $S = \bigcap_{i=1}^{k} S_i$ $(k \geq 1)$, where each theory S_i is closed under propositional provability. Then $\mathcal{MA}(D, W, S) \subseteq \bigcup_{i=1}^{k} \mathcal{MA}(D, W, S_i)$.*

Proof. Let $A \in \mathcal{MA}(D, W, S)$. Then A satisfies **AS1** for S. Thus, $j(A) = \emptyset$ or $j(A) \cup S$ is consistent. If $j(A) = \emptyset$ then for every i $(1 \leq i \leq k)$, A satisfies **AS1** for S_i. Let us assume now that $j(A) \cup S$ is consistent. We have

$$Cn(j(A) \cup S) = Cn(j(A) \cup \bigcap_{i=1}^{k} S_i) = Cn(\bigcap_{i=1}^{k}(j(A) \cup S_i)). \qquad (1)$$

We will show that

$$Cn(\bigcap_{i=1}^{k}(j(A) \cup S_i)) = \bigcap_{i=1}^{k} Cn(j(A) \cup S_i). \qquad (2)$$

Clearly, the left-hand side of (2) is contained in the right-hand side. So, we need to prove only the converse inclusion. Consider a formula $\varphi \in \bigcap_{i=1}^{k} Cn(j(A) \cup S_i)$. For every i $(1 \leq i \leq k)$, φ is provable from $j(A) \cup S_i$. By the Compactness Theorem, for every i, there is a finite subset S_i' of S_i such that φ is provable from $j(A) \cup S_i'$. Let φ_i be the conjunction of all formulas from S_i' $(1 \leq i \leq k)$. Then φ is provable from $j(A) \cup \{\varphi_i\}$. Since S_i is closed under propositional consequences, $\varphi_i \in S_i$. Consequently, $\varphi_1 \vee \ldots \vee \varphi_k \in \bigcap_{i=1}^{k} S_i$. Let v be a valuation satisfying $\bigcap_{i=1}^{k}(j(A) \cup S_i)$. Since $\bigcap_{i=1}^{k}(j(A) \cup S_i) = j(A) \cup \bigcap_{i=1}^{k} S_i$, it follows that v satisfies $j(A)$ and v satisfies $\varphi_1 \vee \ldots \vee \varphi_k$. Hence, v satisfies $j(A) \cup \{\varphi_i\}$ for some i $(1 \leq i \leq k)$. Since φ is provable from $j(A) \cup \{\varphi_i\}$, v satisfies φ. Thus, $\varphi \in Cn(\bigcap_{i=1}^{k}(j(A) \cup S_i))$.

It follows from (1) and (2) that if $j(A) \cup S$ is consistent then for some i $(1 \leq i \leq k)$ $j(A) \cup S_i$ is consistent. Hence, in both cases $(j(A) = \emptyset$, or $j(A) \cup S$ is consistent) A satisfies **AS1** for some S_i $(1 \leq i \leq k)$. By Lemma 3, $A \in \mathcal{A}(D, W, S_i)$ for some i $(1 \leq i \leq k)$. Since $S \subseteq S_i$, then by Lemma 4, $A \in \mathcal{MA}(D, W, S_i)$ for some i $(1 \leq i \leq k)$ and we are done. $\square$

We will denote by $GD(D, S)$ the set of *generating* defaults from D with respect to S, that is,

$$GD(D, S) = \left\{ \frac{\alpha: M\beta_1, \ldots, M\beta_k}{\gamma} \in D : S \vdash \alpha \text{ and } S \not\vdash \neg\beta_i, 1 \leq i \leq k \right\}.$$

Lemma 6. *Let a theory S be an extension of a default theory (D, W) and let $A \in \mathcal{A}(D, W, S)$. Then $A \subseteq GD(D, S)$. In particular, if $GD(D, S) \in \mathcal{A}(D, W, S)$ then $\mathcal{MA}(D, W, S) = \{GD(D, S)\}$.* $\square$

Example 4 indicates that the notions of an extension, a rational extension and a skeptical rational extension are, in general, different. However, under some conditions they coincide. One such situation is described in our first theorem (the proof is omitted due to space restriction).

Theorem 7. *Let (D, W) be a default theory and let S be a propositional theory such that $\mathcal{MA}(D, W, S) = \{GD(D, S)\}$. Then S is an extension of (D, W) if and only if S is a rational extension of (D, W) if and only if S is a skeptical rational extension of (D, W).* $\square$

The next several results describe conditions which guarantee that the intersection of skeptical rational extensions is also a skeptical rational extension.

Theorem 8. *Let $\{S_i : i \in I\}$ be a set of skeptical rational extensions for a default theory (D, W), let $S = \bigcap_{i \in I} S_i$ and $\mathcal{MA}(D, W, S) = \bigcup_{i \in I} \mathcal{MA}(D, W, S_i)$. Then S is a skeptical rational extension for (D, W).*

Proof. We have

$$\bigcap_{A \in \mathcal{MA}(D, W, S)} Cn^{A_S}(W) = \bigcap_{A \in \mathcal{MA}(D, W, S)} Cn^{Mon(A)}(W) =$$

$$\bigcap_{i \in I} \bigcap_{A \in \mathcal{MA}(D, W, S_i)} Cn^{Mon(A)}(W) = \bigcap_{i \in I} \bigcap_{A \in \mathcal{MA}(D, W, S_i)} Cn^{A_{S_i}}(W) = \bigcap_{i \in I} S_i = S.$$

$\square$

Theorem 9. *Let $\{S_i : i \in I\}$ be a set of extensions for a default theory (D, W) such that for every $i \in I$, $\mathcal{MA}(D, W, S_i) = \{GD(D, S_i)\}$. Let $S = \bigcap_{i \in I} S_i$ and $\mathcal{MA}(D, W, S) \subseteq \bigcup_{i \in I} \mathcal{MA}(D, W, S_i)$. Then S is a skeptical rational extension for (D, W).*

Proof. Every S_i is a skeptical rational extension for (D, W) (Theorem 7). By Theorem 8, to prove the assertion it suffices to show that $\bigcup_{i \in I} \mathcal{MA}(D, W, S_i) \subseteq \mathcal{MA}(D, W, S)$. Let $A \in \mathcal{MA}(D, W, S_i)$ for some $i \in I$. Hence, $A = GD(D, S_i)$. Since $S \subseteq S_i$, A satisfies **AS1** for S. By Lemma 3, $A \in \mathcal{A}(D, W, S)$. There is $B \in \mathcal{MA}(D, W, S)$ such that $A \subseteq B$. According to the assumption, $B \in \mathcal{MA}(D, W, S_j)$ for some $j \in I$, that is, $B = GD(D, S_j)$. Since $A = GD(D, S_i)$, we have $GD(D, S_i) \subseteq GD(D, S_j)$. Thus, $Cn(W \cup c(GD(D, S_i))) \subseteq Cn(W \cup c(GD(D, S_j)))$. By Theorem 3.57 in [6], we have $S_i = Cn(W \cup c(GD(D, S_i)))$ and $S_j = Cn(W \cup c(GD(D, S_j)))$, so we get $S_i \subseteq S_j$. Since S_i and S_j are extensions of the same default theory, $S_i = S_j$ and $A = B$. Hence, $A \in \mathcal{MA}(D, W, S)$. $\quad\square$

Lemma 5 and Theorem 9 imply the following corollary.

Corollary 10. *Let* $S_1, \ldots, S_k$ *($k \geq 1$) be extensions of a default theory* (D, W) *such that for every* i *($1 \leq i \leq k$),* $\mathcal{MA}(D, W, S_i) = \{GD(D, S_i)\}$. *Then* $S = \bigcap_{i=1}^{k} S_i$ *is a skeptical rational extension for* (D, W). $\quad\square$

Observe that even though in Theorem 9 and Corollary 10 we assume that sets S_i are extensions, by Theorem 7 every S_i is also a rational extension and a skeptical rational extension for (D, W).

4 Algorithmic Issues

Proposition 11. *Let* S *be a skeptical rational extension for a default theory* (D, W) *such that* D *is finite. Then* $S = Cn(W \cup \{\varphi_1 \vee \ldots \vee \varphi_k\})$, *where every* $\varphi_i = \bigwedge c(A_i)$ *for some* $A_i \in \mathcal{MA}(D, W, S)$.

Proof. Since D is finite, $\mathcal{MA}(D, W, S)$ is finite, as well. Let us assume that $\mathcal{MA}(D, W, S) = \{A_1, \ldots, A_k\}$. For each $A_i \in \mathcal{MA}(D, W, S)$ define $\varphi_i = \bigwedge c(A_i)$ (since each A_i is finite, φ_i is well-defined). Since $A_i \in \mathcal{A}(D, W, S)$, $Cn^{(A_i)s}(W) = Cn(W \cup c(A_i)) = Cn(W \cup \{\varphi_i\})$. Hence,

$$S = \bigcap_{i=1}^{k} Cn^{(A_i)s}(W) = \bigcap_{i=1}^{k} Cn(W \cup \{\varphi_i\}) = Cn(W \cup \{\varphi_1 \vee \ldots \vee \varphi_k\}).$$

$\square$

If (D, W) is finite then the number of sets of the form $Cn(W \cup \{\varphi_1 \vee \ldots \vee \varphi_k\})$, where every φ_i is of the form $\bigwedge c(A)$ for some $A \subseteq D$, is also finite. For every such set S, one can compute $\mathcal{MA}(D, W, S)$ and check whether $S = \bigcap_{A \in \mathcal{MA}(D, W, S)} Cn^{As}(W)$. Thus, we have the following algorithm for computing skeptical rational extensions.

1. For every $A \subseteq D$, compute $\varphi_A = \bigwedge c(A)$ ($\varphi_\emptyset = \top$). Let $\Phi = \{\varphi_A : A \subseteq D\}$.
2. For every $\Psi \subseteq \Phi$
   (a) compute $\psi = \bigvee \Psi$,
   (b) for every $A \subseteq D$, verify whether $A \in \mathcal{MA}(D, W, W \cup \{\psi\})$,
   (c) compute $\varphi = \bigvee_{A \in \mathcal{MA}(D, W, W \cup \{\psi\})} \varphi_A$,

(d) check whether $W \cup \{\varphi\} \vdash \psi$ and $W \cup \{\psi\} \vdash \varphi$; if so, output $Cn(W \cup \{\psi\})$ as a skeptical rational extension for (D, W).

The following example shows that there are default theories (D, W) and sets S such that the size of $\mathcal{MA}(D, W, S)$ is exponential in the size of D. It follows that an algorithm for verifying whether S is a skeptical rational extension of (D, W) must have in the worst case an exponential complexity.

Example 5. Let us consider the default theory (D, W) where

$$D = \left\{ \frac{:Mp_1}{p_1}, \frac{:M\neg p_1}{\neg p_1}, \dots, \frac{:Mp_n}{p_n}, \frac{:M\neg p_n}{\neg p_n} \right\},$$

$p_1, \dots, p_n$ are distinct propositional atoms and $W = \emptyset$. Then $\mathcal{MA}(D, W, Cn(\emptyset))$ has 2^n elements, each of them obtained by selecting exactly one default from each pair $\frac{:Mp_i}{p_i}, \frac{:M\neg p_i}{\neg p_i}$. □

The complexity of reasoning with skeptical rational extensions in the general case remains an open problem.

5 Seminormal Default Theories

In this section we study skeptical rational extensions of seminormal default theories. We show that every seminormal default theory has a least skeptical rational extension and that it coincides with the intersection of all rational extensions.

Our first main result of this section shows that every skeptical rational extension of a seminormal default theory can be represented as the intersection of a certain number (possibly infinitely many) of rational extensions.

Theorem 12. *For every skeptical rational extension S of a seminormal default theory (D, W) there is a set $\{S_i : i \in I\}$ of rational extensions for (D, W) such that $S = \bigcap_{i \in I} S_i$.*

Proof. Let S be a skeptical rational extension for (D, W). Consequently, we have $S = \bigcap_{A \in \mathcal{MA}(D, W, S)} Cn^{A_S}(W)$. Let $\mathcal{MA}(D, W, S) = \{A_i : i \in I\}$ and let us denote $S_i = Cn^{(A_i)_S}(W)$ $(i \in I)$. Then $S = \bigcap_{i \in I} S_i$. We will show that each S_i is a rational extension for (D, W).

Since $A_i \in \mathcal{MA}(D, W, S)$, A_i satisfies **AS1** for S. Hence, $j(A_i) = \emptyset$ or $j(A_i) \cup S$ is consistent. If $j(A_i) = \emptyset$ then A_i satisfies **AS1** for S_i. If $j(A_i) \cup S$ is consistent then, since $W \subseteq S$, $j(A_i) \cup W$ is consistent. Since all defaults in A_i are seminormal, $j(A_i)$ implies $c(A_i)$. It follows that $j(A_i) \cup c(A_i) \cup W$ is consistent. Since $A_i \in \mathcal{MA}(D, W, S)$, $Cn^{(A_i)_S}(W) = Cn(W \cup c(A_i))$. Hence, $S_i = Cn(W \cup c(A_i))$. It follows that $j(A_i) \cup S_i$ is consistent. Consequently, A_i satisfies **AS1** for S_i. Thus, in both cases $(j(A_i) = \emptyset$, or $j(A_i) \cup S$ is consistent) A_i satisfies **AS1** for S_i. By Lemma 3, $A_i \in \mathcal{A}(D, W, S_i)$. Since $S \subseteq S_i$, then by Lemma 4, $A_i \in \mathcal{MA}(D, W, S_i)$. It follows that $(A_i)_{S_i} = Mon(A_i) = (A_i)_S$. Since

$S_i = Cn^{(A_i)s}(W)$, we have $S_i = Cn^{(A_i)s_i}(W)$. Hence, S_i is a rational extension of (D, W). □

In [6] a technique for constructing an extension of a default theory from an ordering of defaults was presented and thoroughly studied. In [7] we adapted this technique to the case of rational extensions. We will use some properties of this construction in the proof of the second main result of this section. The reader is referred to [6, 7] for details.

We assume that the set of the atoms of our language $\mathcal{L}$ is denumerable. Consequently, the set of all defaults over the language $\mathcal{L}$ is denumerable.

Let (D, W) be a default theory and $\prec$ a well-ordering of D. We define an ordinal $\eta_\prec$. For every ordinal $\xi < \eta_\prec$ we define a set of defaults AD_ξ and a default d_ξ. We also define a set of defaults $AD_\prec$. We proceed as follows:
If the sets AD_ξ, $\xi < \alpha$, have been defined but $\eta_\prec$ has not been defined then

1. If there is no default $d \in D \setminus \bigcup_{\xi < \alpha} AD_\xi$ such that:
   (a) $j(d) = \emptyset$ or $W \cup c(\bigcup_{\xi < \alpha} AD_\xi) \cup j(\bigcup_{\xi < \alpha} AD_\xi) \cup j(d)$ is consistent, and
   (b) $W \cup c(\bigcup_{\xi < \alpha} AD_\xi) \vdash p(d)$,
   then $\eta_\prec = \alpha$.
2. Otherwise, define d_α to be the $\prec$-least default $d \in D \setminus \bigcup_{\xi < \alpha} AD_\xi$ such that the conditions (a) and (b) above hold. Then set $AD_\alpha = \bigcup_{\xi < \alpha} AD_\xi \cup \{d_\alpha\}$.

When the construction terminates, put $AD_\prec = \bigcup_{\xi < \eta_\prec} AD_\xi$. The theory $Cn(W \cup c(AD_\prec))$ will be called *generated by the well-ordering* $\prec$.

We will need the following property of this construction.

Theorem 13. (extended version of [7]) *Let* (D, W) *be a seminormal default theory and let* $\prec$ *be a well-ordering of* D. *Then* $T_\prec = Cn(W \cup c(AD_\prec))$ *is a rational extension for* (D, W). *Moreover,* $AD_\prec \in \mathcal{MA}(D, W, T_\prec)$. □

It follows from this theorem that every seminormal default theory has a rational extension. In the proof of our next result we will also need the following proposition.

Proposition 14. (extended version of [7]) *Let* (D, W) *be a default theory and let* S *and* T *be rational extensions of* (D, W) *such that* $S = Cn^{As}(W)$ *for some* $A \in \mathcal{MA}(D, W, S)$, $T = Cn^{B_T}(W)$ *for some* $B \in \mathcal{MA}(D, W, T)$ *and* $A \subseteq B$. *Then* $A = B$ *and* $S = T$. □

Now we are ready to present the second main result of this section.

Theorem 15. *The intersection of all rational extensions of a seminormal default theory is the least skeptical rational extension for this theory.*

Proof. Let $\{S_i : i \in I\}$ be the set of all rational extensions of a seminormal default theory (D, W) and let $S = \bigcap_{i \in I} S_i$. By Theorem 12, we need only to prove that S is a skeptical rational extension for (D, W).

Let $A \in \mathcal{MA}(D, W, S)$. Let us consider any well-ordering $\prec$ of D in which the defaults in A precede all other defaults. Assume also that the defaults of A

are ordered by $\prec$ according to the order in which their corresponding monotonic inference rules are applied in the process of computing $Cn^{As}(W)$. It is easy to see that $A \subseteq AD_\prec$. Since the theory (D, W) is seminormal, the theory generated by $\prec$, $Cn(W \cup c(AD_\prec))$, is a rational extension for (D, W) (Theorem 13), that is, $Cn(W \cup c(AD_\prec)) = S_i$ for some $i \in I$. Moreover, $AD_\prec \in \mathcal{MA}(D, W, S_i)$.

$AD_\prec$ satisfies **AS1** for S_i. Hence, $j(AD_\prec) = \emptyset$ or $j(AD_\prec) \cup S_i$ is consistent. If $j(AD_\prec) = \emptyset$ then $AD_\prec = \emptyset$ (every default in D has a justification). Since $A \subseteq AD_\prec = \emptyset$, $A = AD_\prec$. If $j(AD_\prec) \cup S_i$ is consistent then, since $S \subseteq S_i$, $j(AD_\prec) \cup S$ is consistent. By Lemma 3, $AD_\prec \in \mathcal{A}(D, W, S)$. By the maximality of A, we get $A = AD_\prec$. Hence, in both cases $(j(AD_\prec) = \emptyset$, or $j(AD_\prec) \cup S_i$ is consistent) $A = AD_\prec$, that is, $A \in \mathcal{MA}(D, W, S_i)$ for some $i \in I$. Thus, $\mathcal{MA}(D, W, S) \subseteq \bigcup_{i \in I} \mathcal{MA}(D, W, S_i)$.

Moreover, we proved that for every $A \in \mathcal{MA}(D, W, S)$ there is $i \in I$ such that

$$Cn^{As}(W) = Cn(W \cup c(A)) = S_i \text{ and } A \in \mathcal{MA}(D, W, S_i). \tag{3}$$

Thus,

$$\bigcap_{A \in \mathcal{MA}(D,W,S)} Cn^{As}(W) = \bigcap_{i \in I'} S_i$$

for some $I' \subseteq I$. We will show that $I' = I$, that is, that for every $i \in I$, there is $B \in \mathcal{MA}(D, W, S)$ such that $S_i = Cn^{Bs}(W)$.

Since S_i is a rational extension for (D, W), there is $B \in \mathcal{MA}(D, W, S_i)$ such that $S_i = Cn^{Bs_i}(W)$. Since $S \subseteq S_i$, then by Lemma 3, $B \in \mathcal{A}(D, W, S)$. Hence, there is $C \in \mathcal{MA}(D, W, S)$ such that $B \subseteq C$. By (3), there is $j \in I$ such that $C \in \mathcal{MA}(D, W, S_j)$ and $Cn^{Cs}(W) = S_j$. It is easy to see that $C_{S_j} = Mon(C) = C_S$. Hence, $S_j = Cn^{Cs_j}(W)$. By Proposition 14, $B = C$, that is, $B \in \mathcal{MA}(D, W, S)$. Moreover, $B_S = Mon(B) = B_{S_i}$. Thus, $S_i = Cn^{Bs_i}(W) = Cn^{Bs}(W)$. Hence, we have shown that $I' = I$, that is, $\bigcap_{A \in \mathcal{MA}(D,W,S)} Cn^{As}(W) = S$. Thus, S is a skeptical rational extension for (D, W). $\qquad \square$

Corollary 16. *Every seminormal default theory has a skeptical rational extension.* $\qquad \square$

Example 2 shows that Corollary 16 is not true for general default theories.

Theorem 15 shows that the intersection of all rational extensions is a skeptical rational extension. This is not true for an arbitrary family of rational extensions of a seminormal default theory, even if the theory is finite (cf. Theorem 20).

Example 6. Let

$$D = \left\{ \frac{: M(a \wedge \neg b)}{a}, \frac{: M(b \wedge \neg c)}{b}, \frac{: M(c \wedge \neg a)}{c} \right\}.$$

The default theory $(D, \emptyset)$ is a classical example of a seminormal default theory without extensions. This theory has three rational extensions: $S_1 = Cn(\{a\})$, $S_2 = Cn(\{b\})$ and $S_3 = Cn(\{c\})$. According to Theorem 15, their intersection $S = S_1 \cap S_2 \cap S_3 = Cn(\{a \vee b \vee c\})$ is a skeptical rational extension for $(D, \emptyset)$.

However, the intersections of any two rational extensions, $S_{12} = S_1 \cap S_2 = Cn(\{a \vee b\})$, $S_{13} = S_1 \cap S_3 = Cn(\{a \vee c\})$ and $S_{23} = S_2 \cap S_3 = Cn(\{b \vee c\})$, are not skeptical rational extensions. Indeed, it is easy to see that for any i, j ($1 \leq i < j \leq 3$),

$$\mathcal{MA}(D, \emptyset, S_{ij}) = \left\{ \left\{ \frac{: M(a \wedge \neg b)}{a} \right\}, \left\{ \frac{: M(b \wedge \neg c)}{b} \right\}, \left\{ \frac{: M(c \wedge \neg a)}{c} \right\} \right\}.$$

Hence,

$$\bigcap_{A \in \mathcal{MA}(D, \emptyset, S_{ij})} Cn^{A_{S_{ij}}}(\emptyset) = Cn(\{a \vee b \vee c\}) = S \neq S_{ij}.$$

Let us also observe that none of S_1, S_2, S_3 is a skeptical rational extension. $\square$

Theorem 15 implies the following corollary.

Corollary 17. *A formula φ belongs to all skeptical rational extensions of a seminormal default theory (D, W) if and only if φ belongs to all rational extensions of (D, W).* $\square$

The complexity of reasoning with rational extensions was studied in [7]. In particular, we proved that the problem of deciding whether a formula belongs to all rational extensions of a finite default theory is Π_2^P-complete. It remains Π_2^P-complete even under the restriction to the class of normal default theories. Since every normal default theory is seminormal, we obtain the following result.

Corollary 18. *The problem IN-ALL: Given a finite seminormal default theory (D, W) and a formula φ, decide if φ is in all skeptical rational extensions of (D, W), is Π_2^P-complete.* $\square$

The complexity of the problem of deciding whether a formula belongs to at least one skeptical rational extension of a seminormal default theory remains open. The argument we used for the problem IN-ALL does not work here.

6 Normal Default Theories

The results obtained in the previous section for seminormal default theories clearly extend to the case of normal default theories. In this case, however, we can still strengthen some of them. We start this section with a simple proposition.

Proposition 19. *Let S be an extension of a normal default theory (D, W). Then $\mathcal{MA}(D, W, S) = \{GD(D, S)\}$.*

Proof. If S is inconsistent then $\mathcal{MA}(D, W, S) = \{\emptyset\}$ and $GD(D, S) = \emptyset$, so the assertion holds. Assume now that S is consistent. According to Lemma 6, it is sufficient to prove that $GD(D, S) \in \mathcal{A}(D, W, S)$ and this fact is proven in the proof of Theorem 3.1 in [7]. $\square$

The main result of this section shows that finite intersections of extensions (or rational extensions - for normal default theories these notions coincide, see [7]) are skeptical rational extensions. In particular, (rational) extensions are also skeptical rational extensions for normal default theories (unlike in the case of seminormal ones).

Theorem 20. *Let* (D, W) *be a normal default theory.*

1. *Let* $S_1, \ldots, S_k$ *(*$k \geq 1$*) be extensions of* (D, W). *Then* $S = \bigcap_{i=1}^{k} S_i$ *is a skeptical rational extension for* (D, W).
2. *Every skeptical rational extension of* (D, W) *can be represented as the intersection of a certain number (possibly infinitely many) of extensions.*

Proof. The first assertion follows from Proposition 19 and Corollary 10. The second assertion follows from Theorem 12 and from the fact that for normal default theories extensions and rational extensions coincide. □

Corollary 21. *Let* (D, W) *be a normal default theory with a finite number of extensions. Then the family of all skeptical rational extensions of* (D, W) *is closed under intersection. In particular, the family of all skeptical rational extensions of a finite normal default theory is closed under intersection.* □

The question whether the intersection of an arbitrary collection of extensions of a normal default theory is a skeptical rational extension remains open. However, Theorem 15 and the fact that for normal default theories extensions coincide with rational extensions imply a weaker result.

Theorem 22. *The intersection of all extensions of a normal default theory is the least skeptical rational extension for this theory.* □

The following example shows that Theorem 22 is not true for seminormal default theories.

Example 7. Let

$$D = \left\{ \frac{: Ma}{a}, \frac{: M \neg b}{\neg b}, \frac{: M (b \wedge c)}{c} \right\}.$$

The default theory $(D, \emptyset)$ has one extension: $S_1 = Cn(\{a, \neg b\})$. This theory has two rational extensions: S_1 and $S_2 = Cn(\{a, c\})$. It has also two skeptical rational extensions: S_1 and $S_3 = Cn(\{a, \neg b \vee c\})$. Hence, S_1 is the intersection of all extensions while S_3 is the least skeptical rational extension (which, according to Theorem 15, coincides with the intersection of all rational extensions) and $S_1 \neq S_3$. □

Theorems 20 and 22 imply the following corollary.

Corollary 23. *A formula* φ *belongs to some (resp. all) skeptical rational extension(s) of a normal default theory* (D, W) *if and only if* φ *belongs to some (resp. all) extension(s) of* (D, W). □

Using Corollary 23 and the results from [5] on the complexity of the problems IN-SOME and IN-ALL for extensions of normal default theories we get the following complexity result.

Corollary 24. *The problem* IN-SOME*: Given a finite normal default theory* (D, W) *and a formula* φ, *decide if* φ *is in some skeptical rational extension of* (D, W), *is* Σ_2^P-*complete. The problem* IN-ALL*: Given a finite normal default theory* (D, W) *and a formula* φ, *decide if* φ *is in all skeptical rational extensions of* (D, W), *is* Π_2^P-*complete.* □

7 Conclusions

In this paper we proposed a new version of default logic. It is based on the concept of a skeptical rational extension. We showed that in the case of normal default theories our version of default logic coincides with the standard skeptical reasoning with extensions. In the case of seminormal default theories it coincides with the standard skeptical reasoning with rational extensions. We presented some general properties of skeptical rational extensions, an algorithm to compute them and some complexity results. However, the complexity of reasoning with skeptical rational extensions from arbitrary default theories is an open problem.

References

1. P. Besnard. *An introduction to default logic.* Springer-Verlag, Berlin, 1989.
2. G. Brewka. Cumulative default logic: in defense of nonmonotonic inference rules. *Artificial Intelligence*, 50:183–205, 1991.
3. G. Brewka. *Nonmonotonic reasoning: logical foundations of commonsense.* Cambridge University Press, Cambridge, UK, 1991.
4. M. Gelfond, V. Lifschitz, H. Przymusinska, and M. Truszczyński. Disjunctive defaults. In *Second international conference on principles of knowledge representation and reasoning, KR '91*, Cambridge, MA, 1991.
5. G. Gottlob. Complexity results for nonmonotonic logics. *Journal of Logic and Computation*, 2:397–425, 1992.
6. W. Marek and M. Truszczyński. *Nonmonotonic logics; context-dependent reasoning.* Berlin: Springer-Verlag, 1993.
7. A. Mikitiuk and M. Truszczyński. Rational default logic and disjunctive logic programming. In A. Nerode and L. Pereira, editors, *Logic programming and nonmonotonic reasoning*, pages 283–299. MIT Press, 1993.
8. A. Mikitiuk and M. Truszczyński. Constrained and rational default logic. In preparation, 1995.
9. D. Poole. What the lottery paradox tells us about default reasoning. In *Proceedings of the 2nd conference on principles of knowledge representation and reasoning, KR '89*, pages 333–340, San Mateo, CA., 1989. Morgan Kaufmann.
10. R. Reiter. A logic for default reasoning. *Artificial Intelligence*, 13:81–132, 1980.
11. T. Schaub. *Considerations on Default Logics.* PhD thesis, Technischen Hochschule Darmstadt, 1992.

Reasoning with Stratified Default Theories

Paweł Cholewiński

Department of Computer Science, University of Kentucky, Lexington, KY 40506

Abstract. Default logic is one of the principal formalisms for nonmonotonic reasoning. In this paper, we study algorithms for computing extensions for a class of general propositional default theories. We focus on the problem of partitioning a given set of defaults into a family of its subsets. Then we investigate how the results obtained for these subsets can be put together to achieve the extensions of the original theory. The method we propose is designed to prune the search space and reduce the number of calls to propositional provability procedure. It also constitutes a simple and uniform framework for the design of parallel algorithms for computing extensions.

1 Introduction

In this paper, we investigate computational aspects of default logic of Reiter [15]. Default logic is one of the most commonly accepted formalisms for knowledge representation and commonsense reasoning where conclusions must be drawn despite the lack of total knowledge about the world. However, default reasoning problems have high computational complexity. The problem of deciding whether a given formula φ belongs to at least one extension of a finite default theory is Σ_2^P-complete and deciding whether φ belongs to all extensions is Π_2^P-complete [14], [10]. Related questions of existence of an extension and of computing the intersection or the union of all extensions are NP-complete or NP-hard even for very simple classes of default theories [11], [12].

One common technique to overcome difficulties of extensive computations is to decompose a given theory into several subtheories. This approach, commonly referred to as *stratification*, was widely described for logic programming and autoepistemic logic [2], [1], [4], [9], [16], [3]. Etherington [8] and Kautz and Selman [11] applied this approach to some classes of default theories. Their results were based on restricting the syntactic forms of formulas used in default theories. Finally, in [6] and [7] another way of stratification was proposed for arbitrary default theories. In the latter approach no syntactic restrictions on formulas appearing in defaults were imposed, but more restrictive conditions on dependencies between defaults were required. In this approach stratification was designed to speed up the reasoning for a wide class of default theories and not to guarantee the existence or uniqueness of extensions.

In this paper, we focus on the problem of partitioning a given set of defaults D into a family of its subsets $D_1, D_2, \ldots, D_K$. Then we investigate how extensions of a default theory $\Delta = (D, W)$ can be constructed from extensions for

$(D_1, I_1), (D_2, I_2), \ldots, (D_K, I_K)$ where $I_1, I_2, \ldots, I_K$ are suitably chosen propositional theories. This approach to generating extensions can be viewed as follows. Let Ext be a procedure (oracle) which, for a given default theory $\Delta = (D, W)$, returns a list of its extensions. The problem we consider in this paper is: how to use Ext efficiently. It turns out that instead of applying Ext directly to (D, W) we can detect extensions with less computational effort by

1. partitioning D,
2. applying Ext to the subtheories implied by the partition of D,
3. linking the results obtained in step 2.

Algorithms which can serve as Ext were proposed in [13]. In this paper, we investigate the aspects of partitioning D and linking partial results together.

The paper consists of five sections including an appendix containing proofs. In Section 2, we introduce basic notions and give a short survey of previously obtained results. In particular, we recall the definition of stratification used in this paper. Also, we define the concept of a *characteristic poset* P_D for a set of defaults D. In Section 3, we define reasoning procedures for stratified default theories. We discuss the case of an arbitrary poset P_D as well as the extremal cases of chain and antichain. In Section 4, we discuss implications of the results from Section 3 on the complexity of reasoning, scope of their applicability and possible parallelization techniques.

2   Preliminaries

In this section we introduce the basic notions and review the related results. We study finite default theories over propositional language $\mathcal{L}$. For any propositional formula φ by $Var(\varphi)$ we denote the set of all propositional variables which appear in φ. In this paper we assume that conclusion of each default contains at least one propositional variable. An important characteristic of a default d is its set of *conflict variables* $Var^*(d)$, defined as follows. Let $d = \frac{\alpha : \beta_1, \beta_2, \ldots, \beta_k}{\gamma}$. First, we define the set of conflict variables $Var^*(\beta_i)$ for every justification β_i of d. We distinguish the following three cases:

1. if $\beta_i = \gamma$ (normal justification) then $Var^*(\beta_i) = \emptyset$;
2. if $\beta_i = \beta_i' \wedge \gamma$ (seminormal justification) then $Var^*(\beta_i) = Var(\beta_i')$;
3. in all other cases $Var^*(\beta_i) = Var(\beta_i)$.

Finally, by the *set of conflict variables of d* we mean

$$Var^*(d) = \bigcup_{i=1}^{k} Var^*(\beta_i).$$

By a *stratifiable* default theory we mean a default theory $\Delta = (D, W)$ such that:

1. W is consistent;
2. no variable appearing in W appears in a conclusion of any default from D, that is, $Var(W) \cap Var(c(D)) = \emptyset$;
3. every default from D has at least one justification.

In order to investigate extensions of a finite default theory $\Delta = (D, W)$ we construct a graph $G_D = (D, A_D)$ [7] (see also [9], [12], [3]).. The vertices of G_D are the defaults from D. The set of directed edges of G_D, denoted as A_D, contains an arc (d_i, d_j) if:

1. the conclusions of d_i, d_j have a common variable $(Var(c(d_i)) \cap Var(c(d_j)) \neq \emptyset)$, or
2. the conclusion of d_i contains a variable which appears in the set of conflict variables for d_j $(Var(c(d_i)) \cap Var^*(d_j) \neq \emptyset)$, or
3. the conclusion of d_i contains a variable which appears in the prerequisite for d_j $(Var(c(d_i)) \cap Var(p(d_j)) \neq \emptyset)$.

Also, an arc $(d_i, d_j) \in A_D$ for which the condition (2) is satisfied is called *strong*.

Remark. The concept of *strong* arcs was introduced in [6] to prove results on existence of extensions. In this paper we are not interested in existence of extensions but only in pruning search space while computing extensions. Only strong components of G_D will be important to us. Hence, a simpler definition of G_D can be used: $(d_i, d_j) \in A_D$ if and only if $Var(c(d_i)) \cap Var(d_j) \neq \emptyset$. Both definitions yield same graph G_D. Also, by the same reason, self-loops can be ignored and will not be shown in figures.

Example 1. Consider the default theory $\Delta = (D, W)$, where $W = \emptyset$, and D consists of the following default rules:

$$d_1 = \frac{:p}{p \wedge q}, \qquad d_2 = \frac{:\neg p}{\neg p \wedge \neg q}, \qquad d_3 = \frac{:r}{r}, \quad d_4 = \frac{:\neg r}{\neg r}$$

$$d_5 = \frac{:\neg r, s}{s}, \qquad d_6 = \frac{r \vee q : \neg t, p \wedge u}{u}, \ d_7 = \frac{:\neg u}{\neg u}, \ d_8 = \frac{r \vee q : \neg u, p \wedge t}{t}$$

$$d_9 = \frac{p \wedge q : r, \neg s}{\neg v}, \ d_{10} = \frac{p \wedge q : r, s \vee v}{v}.$$

The resulting characteristic graph G_D for D is shown in Figure 1 (self-loops omitted). $\qquad\qquad\qquad\qquad\qquad\qquad\qquad\qquad\qquad\qquad\qquad\qquad\triangle$

The strong components of G_D can be considered as basic building blocks of the theory (D, W). They will be called *strata*. Assume that for a given theory (D, W) the strong components of G_D are $D_1, D_2, \ldots, D_K$. In many cases sizes of these strata are small enough so that known algorithms for computing extensions can deal with them in a reasonable time. For example, problems on graphs such as coloring, hamiltonicity or existence of kernels have encodings as stratifiable theories which admit nontrivial partitioning [5]. The goal of this paper is to investigate how such partial results can be put together to obtain the extensions for (D, W). To this end, we define a *characteristic poset* of a default theory.

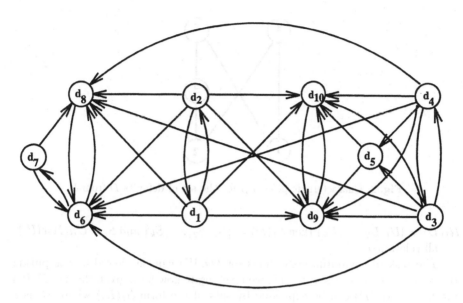

Fig. 1. G_D for the set D from Example 1 (self-loops not shown).

Definition 1. Let D be a set of defaults and $\{D_i : i \in X\}$ be the set of all strong components of G_D. The *characteristic poset* of D is a partially ordered set $P_D = (X, \preceq)$ where $i \preceq j$ if and only if there is a path in G_D from a node of D_i to a node of D_j. $\triangle$

Observe that since D_i's are strong components of G_D the relation $\preceq$ is reflexive, antisymmetric and transitive. Hence $\preceq$ is a partial order.

Example 2. Let D be the set of defaults from Example 1. The graph G_D, shown in Figure 1, has the following strong components:

$$D_1 = \{d_1, d_2\},\ D_2 = \{d_3, d_4\},\ D_3 = \{d_5\},\ D_4 = \{d_6, d_7, d_8\},\ D_5 = \{d_9, d_{10}.\}$$

Thus $P_D = (X, \preceq)$ where $X = \{1, 2, 3, 4, 5\}$ and the partial order relation contains the following pairs: $1 \preceq 4$, $1 \preceq 5$, $2 \preceq 3$, $2 \preceq 4$, $2 \preceq 5$ and $3 \preceq 5$ and $i \preceq i$ for $i \in X$. The Hasse diagram of the poset P_D is shown in Figure 2. $\triangle$

Finally, for any set of defaults D and any propositional theory I we define an operator $\mathcal{U}_I$ as:

$$\mathcal{U}_I(D) = \{U : U \text{ is a set of generating defaults for an extension of } (D, I)\}.$$

That is, for any $U \subseteq D$, $U \in \mathcal{U}_I(D)$ if and only if $S = Cn(I \cup c(U))$ is an extension for (D, I) and U is the set of generating defaults for S. Similarly, by $\mathcal{S}_I(D)$ we denote the set $\{Cn(I \cup c(U)) : U \in \mathcal{U}_I(D)\}$. That is, $\mathcal{S}_I(D)$ is the family of all extensions for (D, W). We adopt the convention that whenever

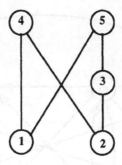

Fig. 2. Characteristic poset P_D for the set of defaults D from 1.

$\mathcal{U}_I(D) = \{U_1, U_2, \ldots, U_k\}$ then $\mathcal{S}_I(D) = \{S_1, S_2, \ldots, S_k\}$ and $S_i = Cn(I \cup c(U_i))$ for all relevant i.

The task of computing extensions for (D, W) can be viewed as computing the set $\mathcal{U}_W(D)$. In the next section we investigate how for a given theory (D, W) the family $\mathcal{U}_W(D)$ can be expressed by sets of the form $\mathcal{U}_I(D_i)$ where D_i is a strong component of G_D and I is a suitably chosen propositional theory.

3 Algorithms for Stratifiable Theories

Let $\Delta = (D, W)$ be a default theory. One method of computing the extensions for Δ was proposed in [7] and relies on computing a topological sort of P_D and then dealing with consecutive strata in a piecewise manner. This way it is possible to find extensions by considering the theory stratum by stratum which often yields significant speedups in reasoning algorithms. We recall this result now.

Theorem 2. *[7] Let $\Delta = (D, W)$ be a stratifiable default theory and let D_i, $i = 1, 2, \ldots, K$ be the all strong components of G_D. Assume that the linear ordering $1, 2, \ldots, K$ is a topological sort of P_D. Then for any set of defaults $U \subseteq D$, $U \in \mathcal{U}_W(D)$ if and only if there exist sets $U_1, U_2, \ldots, U_K$ such that for any i, $i = 1, 2, \ldots, K$:*

1. *$U_i \in \mathcal{U}_{I_i}(D_i)$, where $I_1 = W$ and $I_i = I_{i-1} \cup c(U_{i-1})$ for $i \geq 2$;*

2. *$U = \bigcup_{i=1}^{K} U_i$.* □

The claim of Theorem 2 can be also stated as

$$\mathcal{U}_W(D) = \mathcal{U}_{I_K}(D_K).$$

Theorem 2 defines a direct method of generating extensions for a stratifiable theory. It shows that in this case it is enough to consider one stratum at a time. That is, first find extension S_1 of W by the defaults of the first stratum and append conclusions of defaults generating S_1 to W. Then find extension of "updated" W by the defaults from second stratum and so on. After processing the last stratum the extensions of the original theory are computed.

Example 3. Consider the default theory $\Delta = (D, W)$ from Example 1. The sequence D_1, D_2, D_3, D_4, D_5 corresponds to a topological sort of P_D. An extension for Δ can be computed as follows:

1. $D_1 = \{\frac{:p}{p \wedge q}, \frac{:\neg p}{\neg p \wedge \neg q}\}$ and $I_1 = W = \emptyset$. Let $U_1 = \{\frac{:\neg p}{\neg p \wedge \neg q}\}$. Then $U_1 \in \mathcal{U}_{I_1}(D_1)$. It corresponds to the extension $Cn(\{\neg p \wedge \neg q\})$ of (D_1, I_1).

2. Now, we consider $I_2 = \{\neg p \wedge \neg q\}$ and $D_2 = \{\frac{:r}{r}, \frac{:\neg r}{\neg r}\}$. The theory (D_2, I_2) has an extension $Cn(\{\neg p \wedge \neg q, \neg r\}$ which is generated by the set $U_2 = \{\frac{:\neg r}{\neg r}\}$.

3. $D_3 = \{\frac{:\neg r, s}{s}\}$ and $I_3 = \{\neg p \wedge \neg q, \neg r\}$. The theory (D_3, I_3) has a unique extension $Cn(\{\neg p \wedge \neg q, \neg r, s\}$ generated by the set $U_3 = D_3$.

4. So far, we have computed that $I_4 = \{\neg p \wedge \neg q, \neg r, s\}$ and $D_4 = \{\frac{r \vee q : \neg t, p \wedge u}{u}, \frac{:\neg u}{\neg u}, \frac{r \vee q : \neg u, p \wedge t}{t}\}$. We can select $U_4 = \{\frac{:\neg u}{\neg u}\}$. $U_4 \in \mathcal{U}_{I_4}(D_4)$ and it generates the extension $Cn(\{\neg p \wedge \neg q, \neg r, s, \neg u\}$.

5. Finally, for $I_5 = \{\neg p \wedge \neg q, \neg r, s, \neg u\}$ and $D_5 = \{\frac{p \wedge q : r, \neg s}{\neg v}, \frac{p \wedge q : r, s \vee v}{v}\}$ we have that $\mathcal{U}_{I_5}(D_5) = \{\emptyset\}$ that is $U_5 = \emptyset$.

Consequently, the set $U = \bigcup_{i=1}^{5} U_i$ is a set of generating defaults for an extension of Δ. The extension generated by U has the form $S = Cn(\{\neg p \wedge \neg q, \neg r, s, \neg u\})$.

$\triangle$

Theorem 2 guarantees that if at each step of such reasoning as presented in Example 3 we consider all extensions for (D_i, I_i), then we will find all extensions for (D, W). It is known that if a given default theory can be partitioned into several strata then the number of calls to a propositional provability procedure can be reduced [6], [7]. This often yields significant speedups in reasoning algorithms. However, this method has several disadvantages. It creates a topological sort of the underlying poset P_D what leads to redundant computations. In fact, if strata D_i and D_j are not comparable with respect to P_D then the outcome of processing one of these strata has no effect on applicability of defaults from the other one. But if we impose a linear ordering, in which for example $i < j$, then the method from Theorem 2 will process D_j repeatedly for every output of D_i. Moreover, in such case the strata D_j will be processed with unnecessary large underlying propositional theory I. This is due to the fact that I will contain the consequents of defaults from D_i which were fired. On the other hand, these consequents have no effect on the applicability of defaults from D_j. Finally, this approach leads to unnecessary sequentialization of the whole process.

In this paper we present a different approach. If D_i and D_j are incomparable in P_D then it is possible to process them simultaneously. But this needs a more detailed analysis of and a different algorithm for computing extensions. Now, we will describe such method and prove its correctness. Method described in Theorem 2 can be viewed as a special case of the general approach, namely, the case when P_D is a chain. Before we deal with the most general case of P_D we describe the case of an antichain.

Theorem 3. *Let* $\Delta = (D, W)$ *be a stratifiable default theory. Let* $D_1, \ldots, D_K$ *be a stratification for* D. *If* P_D *is an antichain then*

$$\mathcal{U}_W(D) = \left\{ \bigcup_{i=1}^{K} U_i : \text{for every } i, i = 1, 2, \ldots, K, \quad U_i \in \mathcal{U}_W(D_i) \right\}.$$

$\square$

Theorem 3 says that if P_D is an antichain then the sets of generating defaults for (D, W) are just the unions of sets of generating defaults. Such union contains exactly one set of generating defaults from each $\mathcal{U}_W(D_i)$. The case of an arbitrary poset P_D is more complicated. In this case we need to know the sets of the form $\mathcal{U}_W(D_{\preceq i})$, where $D_{\preceq i}$ is the set of all defaults from the strata D_j such that $j \preceq i$. Consider a stratum D_i which has two immediate predecessors D_{j_1} and D_{j_2}. The set $\mathcal{U}_W(D_{\preceq i})$ depends on the sets $\mathcal{U}_W(D_{\preceq j_1})$ and $\mathcal{U}_W(D_{\preceq j_2})$. We will show that $\mathcal{U}_W(D_{\preceq i})$ contains members of sets $\mathcal{U}_I(D_i)$ where $I = W \cup c(U_1) \cup c(U_2)$ for some $U_1 \in \mathcal{U}_W(D_{\preceq j_1})$ and $U_2 \in \mathcal{U}_W(D_{\preceq j_2})$. However, U_1 and U_2 cannot be arbitrary. The reason is that U_1 and U_2 could have been computed for different contexts in earlier strata.

To describe the case of a general poset P_D we introduce the following notation.

Definition 4. Let $P_D = (X, \preceq)$ be the characteristic poset for a stratifiable default theory $\Delta = (D, W)$. Let $A = \{i_1, i_2, \ldots, i_k\}$ be an antichain in P_D. A sequence $(U_{i_1}, U_{i_2}, \ldots, U_{i_k})$ is a *uniform output sequence* for A if

1. for every j, $j = 1, 2, \ldots, k$, $U_{i_j} \in \mathcal{U}_W(D_{\preceq i_j})$, and
2. for every j, j', $j = 1, 2, \ldots, k$, $j' = 1, 2, \ldots, k$, and for every $l \in X$, if $l \preceq j$ and $l \preceq j'$ then $U_j \cap D_l = U_{j'} \cap D_l$.

Also, we define the *joint output* of A (notation $\mathcal{U}_W(D_{\preceq i_1}) \otimes \mathcal{U}_W(D_{\preceq i_2}) \otimes \ldots \otimes \mathcal{U}_W(D_{\preceq i_k})$) as the set

$$\left\{ \bigcup_{j=1}^{k} U_{i_j} : (U_{i_1}, U_{i_2}, \ldots, U_{i_k}) \text{ is a uniform output sequence for } A \right\}.$$

$\triangle$

Now, we are in a position to describe $\mathcal{U}_W(D_{\preceq i})$ for an arbitrary stratum D_i in P_D in terms of the joint output of immediate predecessors of i.

Theorem 5. *Let* $\Delta = (D, W)$ *be a stratifiable default theory, let* $P_D = (X, \preceq)$, $i \in X$ *and let* $A(i) = \{i_1, i_2, \ldots, i_k\}$ *be the set of all immediate predecessors of* i *in* P_D. *Then*

$$\mathcal{U}_W(D_{\preceq i}) = \bigcup_{V \in \mathcal{F}} \{U \cup V : U \in \mathcal{U}_{W \cup c(V)}(D_i)\}$$

where $\mathcal{F} = \{\emptyset\}$ *if* i *is minimal in* P_D *and for all other vertices* i,
$$\mathcal{F} = \mathcal{U}_W(D_{\preceq i_1}) \otimes \mathcal{U}_W(D_{\preceq i_2}) \otimes \ldots \otimes \mathcal{U}_W(D_{\preceq i_k}).$$

$\square$

Finally, we describe the set $\mathcal{U}_W(D)$. If P_D contains the largest element i_{top} then clearly $\mathcal{U}_W(D) = \mathcal{U}_W(D_{i_{top}})$. Otherwise, we have to take the joint output of the antichain of maximal vertices in P_D.

Theorem 6. *Let* $\Delta = (D, W)$ *be a stratifiable default theory. Let* $Max(P_D) = \{i_1, i_2, \ldots, i_k\}$ *be the set of all maximal vertices in* P_D. *The family of all sets of generating defaults for extensions of* Δ *has the form*

$$\mathcal{U}_W(D) = \mathcal{U}_W(D_{\preceq i_1}) \otimes \mathcal{U}_W(D_{\preceq i_2}) \otimes \ldots \otimes \mathcal{U}_W(D_{\preceq i_k}).$$

$\square$

In the next section, we will discuss implications of these results on complexity of reasoning and possible parallelization techniques for computing extensions. Also, we will show how these results can be adopted to deal with arbitrary default theories that is, how any default theory with consistent W can be converted into a stratifiable default theory.

4 Discussion

We have designed a framework for reasoning with stratifiable default theories. We focused on the problem of partitioning a set of defaults, computing extensions separately for each stratum and linking the obtained results. In our approach the whole process of building extensions for a default theory $\Delta = (D, W)$ is guided by the characteristic poset P_D. One can think of the a whole task of finding an extension for Δ as picking out a successful strategy of searching P_D or determining that there are none. In general, the structure of P_D suggests how theorem provers may be most efficiently used in default reasoning tasks. Also, it suggests how stratified default theories can be processed on parallel machines. In contrast to stratification techniques which use a linearly ordered sequence of strata, P_D allows to process whole antichain of strata concurrently. Moreover, the decomposition of D into a partially ordered set of strata P_D can be done in polynomial time. To be more specific, checking whether W and $c(D)$ have a common variable can be done in $O(l \log l)$ time where l is the length of the default theory (D, W). Clearly, checking whether W is consistent is NP-complete. Graph G_D, its strong components and the characteristic poset P_D can be computed in $O(l^2)$ time. This method of computing stratification can be improved. It is not necessary to use an algorithm for finding strong components of an arbitrary directed graph. One fine algorithm for computing stratification in the case of autoepistemic theories was described in [12]. This algorithm works in $O(l)$ time and uses the specific properties of a characteristic graph. It can be adopted to the case of default theories.

The results we presented in Sections 2 and 3 were stated for stratifiable default theories. That is, we assumed that, in a given default theory $\Delta = (D, W)$, W was consistent, every default had at least one justification and $c(D) \cap W = \emptyset$. In practice, the assumption of consistency of W is not a problem. In fact, if W is inconsistent then for any D the default theory (D, W) has exactly one extension – the set $\mathcal{L}$. The set of generating defaults for this extension consists of all justification-free defaults from D. Thus to detect this case and describe the extensions all we need to do is to perform the consistency check of W.

The second constraint, saying that D may not contain monotonic rules, causes much more problem in practical applications. However, one can use the following theorem to attempt to overcome this difficulty.

Theorem 7. *Let* $\Delta = (D, W)$ *be a default theory and let* W *be consistent. Assume that* $D = D_{mon} \cup D_0$ *where* D_{mon} *is the set of all monotonic defaults from* D. *Let* $\overline{D}_{mon} = \{\frac{\alpha : \gamma}{\gamma} : \frac{\alpha :}{\gamma} \in D_{mon}\}$ *and* $\overline{\Delta} = (\overline{D}_{mon} \cup D_0, W)$. *For any theory* $S \neq \mathcal{L}$, *if* S *is an extension for* Δ *then* S *is an extension for* $\overline{\Delta}$. $\qquad\square$

The method presented in Theorem 7 does not completely overcome the above mentioned difficulty. An extension of $\overline{\Delta}$ is not necessarily an extension of Δ. Thus, each found extension of $\overline{\Delta}$ must be checked whether it is an extension for Δ. On the other hand, testing testing whether a given propositional theory is an extension for Δ is a much simpler task then finding extensions.

The third constraint, $W \cap c(D) = \emptyset$, is the most difficult to satisfy without loosing computational advantages provided by the original partitioning and P_D. As shown in [13] we can replace any formula $\varphi \in W$ with a new default $\frac{\cdot}{\varphi}$ without affecting the set of extensions. Thus, for a given default theory $\Delta = (D, W)$ we can always find a theory $\Delta' = (D', W')$ for which $W' \cap c(D') = \emptyset$ and which has the same extensions as Δ. Then, if we apply Theorem 7 we get a stratifiable theory which contains all extensions of Δ among its own extensions. However, such transformation can give us a very different poset $P_{D'}$ which need not preserve properties of P_D. In fact, each operation of replacing a formula from W with a default adds new default to one of original strata and also can force us to concatenate original strata. Therefore, the granularity of the original partitioning may decrease.

5 Appendix

In this section we present proofs of the results from this paper. In the proofs we use a simple fact which we state as a lemma.

Lemma 8. *Let* T_1 *and* T_2 *be two consistent propositional theories such that* $Var(T_1) \cap Var(T_2) = \emptyset$. *For any propositional formula* φ, *such that* $Var(\varphi) \cap Var(T_2) = \emptyset$,
$$T_1 \cup T_2 \vdash \varphi \quad \text{if and only if} \quad T_1 \vdash \varphi. \qquad\qquad \triangle$$

We will use the following characterization of extensions [13]. For a default theory (D, W) and an arbitrary propositional theory S we define the *reduct of* D *with respect to* S as the set of inference rules
$$Red(D, S) = \left\{ \frac{\alpha}{\gamma} : \frac{\alpha : \beta_1, \ldots, \beta_k}{\gamma} \in D \text{ and } \neg\beta_1 \notin S, \ldots, \neg\beta_k \notin S \right\}.$$

For a set of inference rules R and a theory W, the *base operator* B^R is defined as

$$B^R(W) = W \cup \left\{ \gamma : \frac{\alpha}{\gamma} \in R \text{ and } \alpha \in Cn(W) \right\}.$$

It is known (see [13] for details) that a theory S is an extension for a default theory (D, W) if and only if there is a set $U \subseteq D$ such that $S = Cn(W \cup c(U))$ and $W \cup c(U) = B^{Red(D,S)} \!\uparrow\! \omega(W)$. The following lemma follows by induction from Lemma 8 and definition of operator B.

Lemma 9. *Let R_1 and R_2 be two sets of inference rules and W be a propositional theory such that $Var(W) \cap Var(c(R_1 \cup R_2)) = \emptyset$:*

1. *if $Var(c(R_1)) \cap Var(p(R_2)) = \emptyset$ then $B^{R_1 \cup R_2} \!\uparrow\! \omega(W) = B^{R_1} \!\uparrow\! \omega(B^{R_2} \!\uparrow\! \omega(W))$;*
2. *if $Var(c(R_1)) \cap Var(p(R_2)) = \emptyset$ and $Var(c(R_2)) \cap Var(p(R_1)) = \emptyset$ then $B^{R_1 \cup R_2} \!\uparrow\! \omega(W) = B^{R_1} \!\uparrow\! \omega(W) \cup B^{R_2} \!\uparrow\! \omega(W)$.*

$$\triangle$$

In the proofs we will use the following notation. Let S be an extension for $\Delta = (D, W)$ and P_D be the characteristic poset for Δ. Let U be the set of generating defaults for S. For any vertex i of P_D, by $U_{\preceq i}$ we mean the set $U \cap D_{\preceq i}$ and by $S_{\preceq i}$ we mean the theory $Cn(W \cup c(U_{\preceq i}))$. For a set of defaults D we define a set of inference rules $Mon(D)$ as $Mon(D) = \{ \frac{p(d)}{c(d)} : d \in D \}$.

Lemma 10. *Let $\Delta = (D, W)$ be a stratifiable default theory and $P_D = (X, \preceq)$ be a characteristic poset for D. If S is an extension for Δ and U is a set of generating defaults for S then for any $i \in X$:*

1. *$Red(D, S) \cap Mon(D_{\preceq i}) = Red(D_{\preceq i}, S_{\preceq i})$, and*
2. *$B^{Red(D,S)} \!\uparrow\! \omega(W) \cap c(D_{\preceq i}) = B^{Red(D_{\preceq i}, S_{\preceq i})} \!\uparrow\! \omega(W) \cap c(D_{\preceq i})$, and*
3. *$U_{\preceq i} \in \mathcal{U}_W(D_{\preceq i})$, that is, $S_{\preceq i}$ is an extension for $(D_{\preceq i}, W)$.*

Proof: Let S be an extension for Δ and let U be the set set of generating defaults for S. Then $S = Cn(W \cup c(U))$. Assume that $\frac{\alpha}{\gamma} \in Red(D, S) \cap Mon(D_{\preceq i})$. Then there must be a default $d = \frac{\alpha : \beta_1, \dots, \beta_k}{\gamma} \in D_{\preceq i}$ such that for every $\beta \in J(d)$, $W \cup c(U) \not\vdash \neg \beta$. So, for every $\beta \in J(d)$, $W \cup c(U_{\preceq i}) \not\vdash \neg \beta$ and consequently $\frac{\alpha}{\gamma} \in Red(D_{\preceq i}, S_{\preceq i})$. Assume now, that $\frac{\alpha}{\gamma} \in Red(D_{\preceq i}, S_{\preceq i})$. There must be default $d = \frac{\alpha : \beta_1, \dots, \beta_k}{\gamma} \in D_{\preceq i}$ such that $W \cup c(U_{\preceq i}) \not\vdash \neg \beta$ for any $\beta \in J(d)$. Since for every such β, $Var(\beta) \cap Var(c(U \setminus U_{\preceq i})) = \emptyset$ and $W \cup c(U) = W \cup c(U_{\preceq i}) \cup c(U \setminus U_{\preceq i})$ is consistent, Lemma 8 implies that $W \cup c(U) \not\vdash \neg \beta$. Hence, $\frac{\alpha}{\gamma} \in Red(D, S)$ and since $d \in D_{\preceq i}$ we have that $\frac{\alpha}{\gamma} \in Red(D, S) \cap Mon(D_{\preceq i})$. Therefore the first equality is proven.

To prove that $B^{Red(D,S)} \!\uparrow\! \omega(W) \cap c(D_{\preceq i}) = B^{Red(D_{\preceq i}, S_{\preceq i})} \!\uparrow\! \omega(W) \cap c(D_{\preceq i})$ observe that from part (1) it follows that $Red(D, S) = Red(D_{\preceq i}, S_{\preceq i}) \cup R$ where

$R \subseteq Mon(D \setminus D_{\prec i})$. Hence, $Var(c(R)) \cap Var(p(Red(D_{\prec i}, S_{\prec i})) = \emptyset$. From Lemma 9 (1) we have that

$$B^{Red(D,S)} \!\uparrow\! \omega(W) = B^R \!\uparrow\! \omega(B^{Red(D_{\preceq i}, S_{\preceq i})} \!\uparrow\! \omega(W)) = B^{Red(D_{\preceq i}, S_{\preceq i})} \!\uparrow\! \omega(W) \cup c(R')$$

for some $R' \subseteq R$. Since $c(R') \cap c(D_{\prec i}) = \emptyset$ the second equality is true.

To prove that $S_{\prec i} = Cn(W \cup c(U_{\prec i}))$ is an extension for $(D_{\prec i}, W)$ it is enough to show that $W \cup c(U_{\prec i}) = B^{Red(D_{\preceq i}, S_{\preceq i})} \!\uparrow\! \omega(W)$. Since we have shown that $Red(D_{\prec i}, S_{\prec i}) = Red(D, S) \cap Mon(D_{\prec i})$, from the definition of the base operator B it follows that $B^{Red(D_{\preceq i}, S_{\preceq i})} \!\uparrow\! \omega(W) \subseteq B^{Red(D,S)} \!\uparrow\! \omega(W) = W \cup c(U)$. Also by the definition of B we have that $B^{Red(D_{\preceq i}, S_{\preceq i})} \!\uparrow\! \omega(W) \subseteq W \cup c(D_{\prec i})$. From these two inclusions it follows that $B^{Red(D_{\preceq i}, S_{\preceq i})} \!\uparrow\! \omega(W) \subseteq W \cup c(U_{\prec i})$. On the other hand, if $\gamma \in W \cup c(U_{\prec i})$ then either $\gamma \in W \subseteq B^{Red(D_{\preceq i}, S_{\preceq i})} \!\uparrow\! \omega(W)$, or $\gamma \in c(U_{\prec i})$ In the latter case, $\gamma \in c(U)$ and $\gamma \in c(D_{\prec i})$. So, $\gamma \in B^{Red(D,S)} \!\uparrow\! \omega(W) \cap c(D_{\prec i})$. Thus from point (2) of this lemma it follows that $\gamma \in B^{Red(D_{\preceq i}, S_{\preceq i})} \!\uparrow\! \omega(W)$. This means that $W \cup c(U_{\prec i}) \subseteq B^{Red(D_{\preceq i}, S_{\preceq i})} \!\uparrow\! \omega(W)$. Hence, the equality $W \cup c(U_{\prec i}) = B^{Red(D_{\preceq i}, S_{\preceq i})} \!\uparrow\! \omega(W)$ is proven an $S_{\prec i}$ is an extension for $(D_{\prec i}, W)$. $\qquad\square$

Proof of Theorem 3. Let $U \in \mathcal{U}_W(D)$ and let $S = Cn(W \cup c(U))$ be the extension generated by U. For each i, $i = 1, 2, \ldots, K$, define $U_i = U_{\prec i} = U \cap D_i$. We have that $U = \bigcup_{i=1}^K U_i$ and since in this case P_D is an antichain, it follows from Lemma 10 that for every i, $i = 1, 2, \ldots, K$, $S_i = S_{\prec i} = Cn(W \cup c(U_i))$ is an extension for (D_i, W), that is, $U_i \in \mathcal{U}_W(D_i)$.

Assume now, that there are U_i, $i = 1, 2, \ldots, K$, such that $U_i \in \mathcal{U}_W(D_i)$. That is, each U_i generates an extension $S_i = Cn(W \cup c(U_i))$ for (D_i, W). Let $U = \bigcup_{i=1}^K U_i$. Since W is consistent and every default in D has a justification, all sets $c(U_i)$ are consistent. Since any two different sets from the list $W, U_1, U_2, \ldots, U_K$ have disjoint sets of variables that appear in them, the set $W \cup c(U)$ is consistent. We will show that $U \in \mathcal{U}_W(D)$. Let S be the theory generated by U, that is $S = Cn(W \cup c(U))$. Consider any default $d \in D_i$ and let $\beta \in J(d)$. If $W \cup c(U_i) \vdash \neg\beta$ then $W \cup c(U) \vdash \neg\beta$. On the other hand, if $W \cup c(U_i) \not\vdash \neg\beta$ then Lemma 8 implies that $W \cup c(U) \not\vdash \neg\beta$. So, $Red(D, S) \cap Mon(D_i) = Red(D_i, S_i)$. Hence, $Red(D, S) = \bigcup_{i=1}^K Red(D_i, S_i)$. Also, since P_D is an antichain, for any i, j, $i = 1, \ldots, K$, $j = 1, \ldots, K$, if $i \neq j$ then $Var(p(Red(D_i, S_i)) \cap Var(c(Red(D_j, S_j))) = \emptyset$. Thus from Lemma 9 (2) we have that

$$B^{Red(D,S)} \!\uparrow\! \omega(W) = \bigcup_{i=1}^K B^{Red(D_i, S_i)} \!\uparrow\! \omega(W) = W \cup c(U).$$

So, S is an extension for (D, W). This means that $U \in \mathcal{U}_W(D)$ and the theorem is proven. $\qquad\square$

Proof of Theorem 5 If i is minimal in P_D then $D_{\prec i} = D_i$ and $\mathcal{U}_W(D_{\prec i}) = \bigcup_{V \in \{\emptyset\}} \{U \cup V : U \in \mathcal{U}_{W \cup c(V)}(D_i)\} = \{U : U \in \mathcal{U}_W(D_i)\} = \mathcal{U}_W(D_i)$. Thus the claim holds for all minimal vertices of P_D.

Let now, i be a vertex of P_D and $A(i) = \{i_1, i_2, \ldots, i_k\}$ be the set of all immediate predecessors of i in P_D. Let $U \in \mathcal{U}_W(D_{\preceq i})$ and $S = Cn(W \cup c(U))$. We define U_{i_j}, $j = 1, \ldots, k$, as $U_{i_j} = U \cap D_{\preceq i_j}$. Lemma 10 (3) applied to the theory $(D_{\prec i}, W)$ implies that for all j, $j = 1, \ldots, k$, $U_{i_j} \in \mathcal{U}_W(D_{\preceq i_j})$. Also, from the definition of U_{i_j}, $j = 1, \ldots, k$, it follows that $(U_{i_1}, U_{i_2}, \ldots, U_{i_k})$ is a uniform output sequence for $A(i)$. Let $V = \bigcup_{j=1}^{k} U_{i_j}$. Thus, $V \in \mathcal{F} = \mathcal{U}_W(D_{\preceq i_1}) \otimes \mathcal{U}_W(D_{\preceq i_2}) \otimes \ldots \otimes \mathcal{U}_W(D_{\preceq i_k})$. We will show that $U_i = U \cap D_i \in \mathcal{U}_{W \cup c(V)}(D_i)$. Since $D_i \subseteq D_{\preceq i}$, we have that $Red(D_i, S) \subseteq Red(D_{\preceq i}, S)$. Since $c(V) \subseteq B^{Red(D_{\preceq i}, S)} \uparrow_\omega(W)$, from the definition of the base operator it follows that $B^{Red(D_{\preceq i}, S)} \uparrow_\omega(W) = B^{Red(D_{\preceq i}, S)} \uparrow_\omega(W \cup c(V)) = W \cup c(U)$. Consequently, since $Red(D_i, S) \subseteq Red(D_{\preceq i}, S)$ it follows that $B^{Red(D_i, S)} \uparrow_\omega(W \cup c(V)) \subseteq W \cup c(U)$. On the other hand, since conclusions of the rules from $Red(D_i, S)$ have no common variables with formulas from W and from $p(D_{\preceq i} \setminus D_i)$, it follows from the definition of operator B that $B^{Red(D_{\preceq i}, S)} \uparrow_\omega(W) \subseteq B^{Red(D_i, S)} \uparrow_\omega(W \cup c(V))$. Thus, $W \cup c(U) \subseteq B^{Red(D_i, S)} \uparrow_\omega(W \cup c(V))$. Therefore $W \cup c(U) = W \cup c(V) \cup c(U_i) = B^{Red(D_i, S)} \uparrow_\omega(W \cup c(V))$. So, $S = Cn((W \cup c(V) \cup c(U_i))$ is an extension for $(D_i, W \cup c(V))$. So, $U_i \in \mathcal{U}_{W \cup c(V)}(D_i)$. Thus we proved that $\mathcal{U}_W(D_{\preceq i}) \subseteq \bigcup_{V \in \mathcal{F}} \{U \cup V : U \in \mathcal{U}_{W \cup c(V)}(D_i)\}$.

Assume now, that $(U_{i_1}, U_{i_2}, \ldots, U_{i_k})$ is a uniform output sequence for $A(i)$, $V = \bigcup_{j=1}^{k} U_{i_j}$ and $U_i \in \mathcal{U}_{W \cup c(V)}(D_i)$. To compete the proof it is enough to show that $U_i \cup V \in \mathcal{U}_W(D_{\preceq i})$. For any j, $j = 1, \ldots, k$, U_{i_j} generates an extension for the default theory $(D_{\preceq i_j}, W)$. Since W is consistent and every default from $D_{\preceq i_j}$ has a justification, it follows that the set $c(U_{i_j})$ is consistent [15]. Because $(U_{i_1}, U_{i_2}, \ldots, U_{i_k})$ is a uniform output sequence the theory $c(V)$ must be consistent too. Otherwise, there would be a formula $\varphi = c(d)$, for some $d \in I$ such that $c(V) \vdash \neg c(d)$. Let $d \in U_{i_j}$ for some j. From the definition of uniform output sequence it follows that $c(U_{i_j}) \vdash \neg c(d)$, a contradiction. Since W is consistent and $Var(W) \cap Var(c(V)) = \emptyset$ we have that $W \cup c(V)$ is consistent and same argument can be used to prove that $W \cup c(V) \cup c(U_i)$ is consistent.

Let $S = Cn(W \cup c(V) \cup c(U_i))$. The inequality $Red(D_{\preceq i_j}, S_{\preceq i_j}) \subseteq Red(D_{\preceq i}, S)$ follows from Lemma 8. Thus, we have that $c(U_{i_j}) \subseteq B^{Red(D_{\preceq i_j}, S)} \uparrow_\omega(W) \subseteq B^{Red(D_{\preceq i}, S)} \uparrow_\omega(W)$. Hence $c(V) \subseteq B^{Red(D_{\preceq i}, S)} \uparrow_\omega(W)$. Therefore, $B^{Red(D_{\preceq i}, S)} \uparrow_\omega(W) = B^{Red(D_{\preceq i}, S)} \uparrow_\omega(W \cup c(V))$. Also, by Lemma 8, $Red(D_i, Cn(W \cup c(V))) \subseteq Red(D_{\preceq i}, S)$ and therefore we have that $W \cup c(V) \cup c(U_i) \subseteq B^{Red(D_{\preceq i}, S)} \uparrow_\omega(W)$.

By induction on n one can show that $B^{Red(D_{\preceq i}, S)} \uparrow_\omega(W) \subseteq W \cup c(V) \cup c(U_i)$. Therefore it follows that $W \cup c(U \cup V) = B^{Red(D_{\preceq i}, S)} \uparrow_\omega(W)$ what means that $U \cup V \in \mathcal{U}_W(D_{\preceq i})$. $\qquad \square$

Proof of Theorem 6. If P_D has the largest element i_{top} then $Max(P_D) = \{i_{top}\}$ so $\mathcal{U}_W(D) = \mathcal{U}_W(D_{\preceq i_{top}})$ and the claim is true.

Let us assume then that $Max(P_D) = \{i_1, \ldots, i_k\}$, $k > 1$. Let C be the set of all propositional variables that appear in $\bigcup_{j=1}^{k} D_{i_j}$ and let φ be any tautology such that $Var(\varphi) = C$. Let's define a default $d' = \frac{\varphi:a}{a}$, where a is a propositional variable not appearing in (D, W). Consider the theory (D', W) where $D' = D \cup \{d'\}$. P'_D can be obtained from P_D by adding a new element i_{top} such

that $Max(P_D)$ is the set of all immediate predecessors of i_{top} in P'_D. For any consistent propositional theory $W \cup I$, such that $a \notin Var(W \cup I)$ the default theory $(\{d'\}, I)$ has the unique extension $Cn(W \cup I \cup \{a\})$. Thus, from Theorem 5 it follows that every extension of (D', W) is generated by a set of defaults $\{d'\} \cup V$ where $V \in \mathcal{U}_W(D_{\preceq i_1}) \otimes \mathcal{U}_W(D_{\preceq i_2}) \otimes \mathcal{U}_W(D_{\preceq i_k})$. Since it is easy to see that the mapping $f : 2^D \to 2^{D'}$, defined as $f(U) = U \cup \{d'\}$ establishes one-to-one correspondence between $\mathcal{U}_W(D)$ and $\mathcal{U}_W(D')$ the theorem is proven.
□

Proof of Theorem 7. Let S be a consistent extension for $(D_{mon} \cup D_0, W)$. Let U be the set of generating defaults for S. There are sets $U_{mon} \subseteq D_{mon}$ and $U_0 \subseteq D_0$ such that $U = U_{mon} \cup U_0$. We show that S is an extension for $(\overline{D}_{mon} \cup D_0, W)$. Let $R_1 = Red(D_{mon} \cup D_0, S)$ and $R_2 = Red(\overline{D}_{mon} \cup D_0, S)$. Observe that $R_2 \subseteq R_1$. Hence, $B^{R_2} \uparrow \omega(W) \subseteq B^{R_1} \uparrow \omega(W)$. Suppose that for some default $d \in D$, $Mon(d) \in R_1$ and $Mon(d) \notin R_2$. It must be the case that $d \in D_{mon}$. Let $d = \frac{\alpha :}{\gamma}$. Since $Mon(d) \notin R_2$, we have that $S \vdash \neg\gamma$. Since S is consistent $\gamma \notin S$. Thus, the rule $\frac{\alpha}{\gamma}$ is immaterial in computing $B^{R_1} \uparrow \omega(W)$. That is, $B^{R_1} \uparrow \omega(W) = B^{R_1 \setminus \{\frac{\alpha}{\gamma}\}} \uparrow \omega(W)$. Repeating this argument for each rule $r \in R_1 \setminus R_2$ we eventually receive that $B^{R_1} \uparrow \omega(W) = B^{R_2} \uparrow \omega(W)$ and therefore S is an extension for $\overline{\Delta}$.
□

References

1. K. Apt and H.A. Blair. Arithmetical classification of perfect models of stratified programs. *Fundamenta Informaticae*, 12:1–17, 1990.
2. K. Apt, H.A. Blair, and A. Walker. Towards a theory of declarative knowledge. In J. Minker, editor, *Foundations of Deductive Databases and Logic Programming*, pages 89–142, Los Altos, CA, 1987. Morgan Kaufmann.
3. R. Ben-Eliyahu and R. Dechter. Propositional semantics for disjunctive logic programs. In K. Apt, editor, *Proceedings of International Joint Conference and Symposium on Logic Programming*, pages 813–827, Cambridge, MA, 1992, MIT Press.
4. N. Bidoit and Ch. Froidevaux. General logical databases and programs: Default logic semantics and stratification. *Information and Computation*, 91:15–54, 1991.
5. P. Cholewiński, W. Marek, A. Mikitiuk and M. Truszczyński. Experimenting with default logic In *Proceedings of ICLP-95*, MIT Press, to appear.
6. P. Cholewiński. Seminormal stratified default theories. Technical Report 238-93, University of Kentucky, Lexington, 1993.
7. P. Cholewiński. Stratified default theories. In *Proceedings of CSL-94*, Springer Verlag, to appear.
8. D. W. Etherington. *Reasoning with Incomplete Information*. Pitman, London, 1988.
9. M. Gelfond. On stratified autoepistemic theories. In *Proceedings of AAAI-87*, pages 207–211, Los Altos, CA., 1987. American Association for Artificial Intelligence, Morgan Kaufmann.
10. G. Gottlob. Complexity results for nonmonotonic logics. *Journal of Logic and Computation*, 2:397–425, 1992.

11. H.A. Kautz and B. Selman. Hard problems for simple default logics. In *Principles of Knowledge Representation and Reasoning*, pages 189–197, San Mateo, CA., 1989. Morgan Kaufmann.
12. W. Marek and M. Truszczyński. Autoepistemic logic. *Journal of the ACM*, 38:588 – 619, 1991.
13. W. Marek and M. Truszczyński. *Nonmonotonic Logics; Context-Dependent Reasoning*. Springer-Verlag, 1993.
14. I. Niemelä. On the decidability and complexity of autoepistemic reasoning. *Fundamenta Informaticae*, 1992.
15. R. Reiter. A logic for default reasoning. *Artificial Intelligence*, 13:81–132, 1980.
16. M. Truszczyński. Stratified modal theories and iterative expansions. Technical Report 159-90, Department of Computer Science, University of Kentucky, 1990.

Incremental Methods for Optimizing Partial Instantiation *

Raymond T. Ng** and Xiaomei Tian

Department of Computer Science
University of British Columbia
Vancouver, B.C., V6T 1Z4,
Canada.

Abstract. It has been shown that mixed integer programming methods can effectively support minimal model, stable model and well-founded model semantics for ground deductive databases. Recently, a novel approach called partial instantiation has been developed which, when integrated with mixed integer programming methods, can handle non-ground logic programs. The goal of this paper is to explore how this integrated framework based on partial instantiation can be optimized. In particular, we develop an incremental algorithm that minimizes repetitive computations. We also develop optimization techniques to further enhance the efficiency of our incremental algorithm. Experimental results indicate that our algorithm and optimization techniques can bring about very significant improvement in run-time performance.

1 Introduction

Very active research in the past decade has led to the development of numerous methods for evaluating deductive databases and logic programs. Algorithms, such as magic sets and counting methods, have proven to be very successful for definite and stratified deductive databases [1]. During the past few years, however, several new semantics for disjunctive programs and programs with negations, such as minimal models, stable models and well-founded models [9, 5, 12], have been proposed and widely studied. Recently, it has been shown that mixed integer programming methods can be used to provide a general and rather effective computational paradigm for those semantics [2, 3].

However, like other methods that use linear or integer programming methods for logic deduction [4, 6], the algorithms proposed in [2, 3] is in effect propositional, and can only deal with the ground versions of deductive databases, which are normally much larger in sizes than their non-ground versions. To solve this problem, [7, 8] very recently propose a novel approach, called partial instantiation, which combines unification with mixed integer programming (or with

---

* Research partially sponsored by NSERC Grants OGP0138055 and STR0134419, and NCE IRIS Grants IC-5 and HMI-5. Proofs of all the lemmas and theorems presented in this paper are contained in [11].

** Email: rng@cs.ubc.ca

any other propositional deduction techniques), and which can directly solve a non-ground version of a program. Equally importantly, the approach can handle function symbols, thus making it a true logic programming computational paradigm. While we will discuss partial instantiation in greater details in Section 2, the general strategy is to alternate iteratively between two phases:

evaluation (of propositional program) $\rightarrow$ partial instantiation $\rightarrow$ evaluation ...

More specifically, the initial step begins with evaluating a given non-ground logic program P that may contain disjunctive heads and negations in the bodies as a propositional program using mixed integer programming. This generates a set of true propositional atoms and a set of false propositional atoms. The partial instantiation phase then begins by checking whether unification or "conflict resolution" is possible between atoms in the two sets. If A is an atom in the true set and B an atom in the false set, the most general unifier for A and B is called a *conflict-set* unifier. Then for each conflict-set unifier θ (there can be multiple), clauses in P are instantiated with θ and added to P for further evaluation. In other words, in the next iteration, the (propositional) program to be evaluated is $P \cup P\theta$. This process continues, until either no more conflict-set unifier is found, or the time taken has gone beyond a certain time limit [3].

The main focus of this paper is on how to optimize the run-time performance of the evaluation phase. In particular, as described in [2, 3], the evaluation of program P comprises of two steps: a step to reduce the size of P, followed by the mixed integer programming step to find the models. Let us represent the operations symbolically as $model(sizeopt(P))$. As shown in [2, 3], the operation *sizeopt* to reduce the size of programs is highly beneficial to the subsequent operation of finding the models. Thus, as far as the partial instantiation approach is concerned, if $\theta_1, \ldots, \theta_n$ are all the conflict-set unifiers, an obvious strategy will be to compute $model(sizeopt(P \cup P\theta_1)), \ldots, model(sizeopt(P \cup P\theta_n))$ one by one. The major problem tackled in this paper is how to compute $sizeopt(P \cup P\theta_i)$ incrementally. That is, we try to optimize the evaluation phase by reusing $sizeopt(P)$ to compute $sizeopt(P \cup P\theta_1), \ldots, sizeopt(P \cup P\theta_n)$. As will be shown in Example 2, our task is complicated by the fact that *sizeopt* is not a monotonic operation. The principal contributions of this paper are:

- the development of an algorithm, called Incr, which will be formally proved to be incremental;
- the development of optimizations which may further reduce the size of a program, and save time in computing least models; and
- the implementation and experimental evidence showing that these algorithms and optimizations can lead to significant improvement in run-time efficiency.

---

[3] Partial instantiation may be infinite in the presence of function symbols.

2 Preliminaries

2.1 Review: Partial Instantiation

As described in [7, 8], computing minimal models of logic programs by partial instantiation can be viewed as expanding and processing nodes of *partial instantiation trees*. Given a non-ground logic program P with disjunctive heads and negations in the bodies, the root node of the partial instantiation tree corresponding to P solves P directly as a propositional program. Consider an example presented in [7] where P is the program consisting of the clauses: $P = \{p(X_1, Y_1) \leftarrow q(X_1, Y_1); \ q(a, Y_2) \leftarrow; \ q(X_2, b) \leftarrow \}$ In the root node, P is solved as the program $\{A \leftarrow B; C \leftarrow; D \leftarrow \}$, where A, B, C, D denote $p(X_1, Y_1)$, $q(X_1, Y_1)$, $q(a, Y_2)$ and $q(X_2, b)$ respectively. For this propositional program, the set of true atoms is $T = \{C, D\}$, and the set of false atoms is $F = \{A, B\}$. "Conflict resolution" then looks for unification between an atom in T with an atom in F. For our example, there are two conflict-set unifiers: a) $\theta_1 = \{X_1 = a, Y_1 = Y_2\}$, and b) $\theta_2 = \{X_1 = X_2, Y_1 = b\}$. Now for each conflict-set unifier θ_i, a child node is created which is responsible for the processing of the instantiated program $P \cup P\theta_i$. In other words, the root node of the tree for our example has two child nodes. One corresponds to the program $P_1 = P \cup \{p(a, Y_2) \leftarrow q(a, Y_2)\}$. The other child node corresponds to $P_2 = P \cup \{p(X_2, b) \leftarrow q(X_2, b)\}$. In the evaluation phase of P_2, P_2 again is treated as a propositional program whose true and false sets are $T_2 = \{q(a, Y_2), q(X_2, b), p(X_2, b)\}$ and $F_2 = F$. For T_2 and F_2, there are two conflict-set unifiers which are identical to θ_1, θ_2. Thus, the node for P_2 has two child nodes. Similarly, it is not difficult to verify that the node for P_1 also has two child nodes. This process of expanding child nodes, and alternating between evaluation and partial instantiation continues. A node is a leaf node if its true and false set of atoms cannot be unified. For our example, the partial instantiation tree is finite and has 11 nodes in total.

2.2 Review: Algorithm *SizeOpt*

The following algorithm intends to reduce the size of a given program by deleting clauses whose bodies cannot possibly be satisfied. Since as far as minimal model computation is concerned, a negative literal in the body of a clause can be moved to become a positive literal in the head, hereafter we only consider clauses possibly with disjunctive heads, but no negation in the bodies.

Algorithm SizeOpt ([3]) Input P, a ground disjunctive program, and S_0, the set of atoms that do not appear in the head of any clause in P.

1. Initialize Q to P, Q_d to $\emptyset$ and i to 0.
2. Set R to $\emptyset$.
3. For each clause $Cl \equiv A_1 \vee \ldots \vee A_m \leftarrow B_1 \wedge \ldots \wedge B_n$ in Q, and for some B_j such that $B_j \in S_i$:
   - delete Cl from Q, add Cl to Q_d, and add $A_1, \ldots, A_m$ to R.

4. Increment i by 1, and set S_i to R.

5. For all A in S_i, if A occurs in the head of some clause in Q, delete A from S_i.

6. If S_i is empty, then return Q and Q_d, and halt. Otherwise, go back to Step 2. □

Hereafter, we use the notation $sizeopt(P) = \langle Q, Q_d \rangle$ to denote the application of the above algorithm on P, where Q is the set of retained clauses, and Q_d is the set of deleted clauses.

Example 1 Let P be the following program:

$$A \leftarrow B \wedge C \tag{1}$$
$$B \vee D \leftarrow A \wedge E \tag{2}$$
$$B \leftarrow E \wedge F \tag{3}$$
$$D \leftarrow A \tag{4}$$

Initially, S_0 is the set $\{C, E, F\}$. Thus, after Step 3 in the first iteration of Algorithm SizeOpt, Q_d consists of Clauses 1, 2 and 3, and the only clause remained in Q is Clause 4. After Step 5, S_1 is $\{A, B\}$. In the second iteration of Algorithm SizeOpt, the clause $D \leftarrow A$ is deleted from Q and added to Q_d in Step 3. S_2 is the set $\{D\}$. In the third iteration of Algorithm SizeOpt, execution halts as Q becomes empty. □

Example 2 Let P' be the program obtained by adding the following two clauses to P introduced in the previous example:

$$C \vee G \leftarrow \tag{5}$$
$$E \leftarrow C \tag{6}$$

When Algorithm SizeOpt is applied to P', the situation changes drastically. S_0 is now $\{F\}$. In the first iteration, Clause 3 is the only clause added to Q_d, and S_1 is empty after Step 5. Thus, the algorithm halts in Step 6 without another iteration. □

The above example demonstrates that Algorithm SizeOpt is neither monotonic nor anti-monotonic. The following lemma, proved in [3], shows that Algorithm SizeOpt preserves minimal models.

Lemma 1 ([3]) Let P be a disjunctive deductive database such that $sizeopt(P) = \langle Q, Q_d \rangle$. M is a minimal model of P iff M is a minimal model of Q. □

3 An Incremental Algorithm

Suppose P is the program considered in a node N of a partial instantiation tree, and $\theta_1, \ldots, \theta_m$ are all the conflict-set unifiers. As discussed in Section 2.1, Node N has m children, the j-th of which processes the instantiated program

$P \cup P\theta_j$ (where $1 \leq j \leq m$). As described above, Algorithm SizeOpt can be applied to $P \cup P\theta_j$ to reduce the number of clauses that need to be processed. However, this approach of applying Algorithm SizeOpt directly may lead to a lot of repeated computations, as Algorithm SizeOpt has already been applied to P in Node N (and similarly, the programs in the ancestors of N). To avoid repetitive computations as much as possible, we develop Algorithm Incr that reuses $sizeopt(P)$ to produce $sizeopt(P \cup P\theta_j)$, as shown in Figure 1.

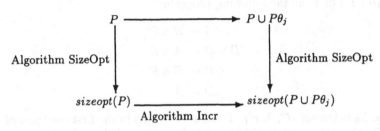

Fig. 1. Incremental Maintenance

3.1 Graphs for Maintaining Deleted Clauses

Recall from Section 2 that $sizeopt(P)$ produces the pair $\langle Q, Q_d \rangle$, where Q_d consists of clauses deleted from P. To facilitate incremental processing, Algorithm Incr uses a directed graph G, called a *DC-graph*, to organize the deleted clauses. The intended properties of a DC-graph are as follows.

- Nodes represent atoms that do not appear in the head of any clause in Q.
- If there is an arc from node B_i to A, then the arc is labeled by a clause $Cl \in Q_d$ such that A appears in the head of Cl and B_i occurs in the body of Cl.

The only exceptions to the above properties are the special *root* node and the arcs originated from this root node. As will be shown later, the root node is the place where a graph traversal begins. Arcs originated from the root node are not labeled, as those arcs do not correspond to any clause in Q_d.

Example 3 Consider the program P discussed in Example 1. Q_d consists of all 4 clauses in P. Figure 2 shows the DC-graph G_1 corresponding to Q_d. For convenience, arcs are labeled by the clause numbers used in Example 1. Furthermore, the label 2,3 of the arc from E to B is a shorthand notation that represents two arcs from E to B with labels 2 and 3 respectively. Notice that G_1 contains a cycle between A and B. □

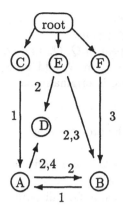

Fig. 2. DC-graph G_1

3.2 Self-sustaining Cycles

Definition 1 Let $A_1 \xrightarrow{Cl_1} A_2 \xrightarrow{Cl_2} \ldots \xrightarrow{Cl_{i-1}} A_i \ldots A_n \xrightarrow{Cl_n} A_1$ be a cycle in DC-graph G, where $A \xrightarrow{Cl} B$ denotes an arc from A to B with label Cl. If there does not exist any arc from outside the cycle to some A_i with label Cl_{i-1} (i.e. $\not\exists B \xrightarrow{Cl_{i-1}} A_i$ for some $B \notin \{A_1, \ldots, A_n\}$), then the cycle is called *self-sustaining*.
□

As shown in the above example, G_1 contains the cycle $A \xrightarrow{Cl_2} B \xrightarrow{Cl_1} A$. This cycle is not self-sustaining because of the arc $C \xrightarrow{Cl_1} A$ (or the arc $E \xrightarrow{Cl_2} B$). The existence of this arc justifies why Clause 1 should be deleted, and why A should remain a node in the graph. On the other hand, if the arcs $C \xrightarrow{Cl_1} A$ and $E \xrightarrow{Cl_2} B$ were removed, the cycle $A \xrightarrow{Cl_2} B \xrightarrow{Cl_1} A$ became self-sustaining. Then for the sake of achieving the kind of incrementality depicted in Figure 1, Clause 2 should be restored (i.e. no longer be kept in Q_d). This would cause node B to disappear from the graph, which in turn leads to the restoration of Clause 1 and the disappearance of node A. Example 5 below will give further details as to why all these actions are necessary. In general, if there exists a self-sustaining cycle in a DC-graph, all the clauses involved in the cycle need to be restored, and all the nodes of the cycle need to be removed. We are now in a position to present Algorithm Incr.

3.3 Algorithm Incr

Algorithm Incr Input $P = \langle Q, Q_d \rangle$, the DC-graph G corresponding to Q_d, and a clause $Cl \equiv A_1 \vee \ldots \vee A_m \leftarrow B_1 \wedge \ldots \wedge B_n$ to be added to P.

1. For each B_i that does not appear in Q and Q_d (i.e. appearing in P the first time), add to graph G a node B_i and an arc from the root to node B_i.

2. For each B_i that is a node,
   (a) For each A_j where $1 \leq j \leq m$,
       (i) If A_j does not appear in Q, Q_d and G, add node A_j to G.
       (ii) If there is a node A_j in G, add an arc from node B_i to node A_j labeled Cl. If there is originally an arc from the root to node A_j, remove that arc.
   (b) Add Cl to Q_d.
3. If there is no such B_i in the previous step,
   (a) Add Cl to Q.
   (b) For each A_j that appears as a node in G where $1 \leq j \leq m$, call Subroutine Remove(A_j).
4. For each self-sustaining cycle in G, call Subroutine Remove(D), where D is some atom in the cycle. □

Subroutine Remove Input atom (node) A.

1. Remove from graph G node A and all the arcs pointing to A.
2. For each arc initially originating from A in G (i.e. $A \xrightarrow{Cl} B$),
   (a) Remove the arc from G.
   (b) If there does not exist another arc pointing to B with label Cl (i.e. $D \xrightarrow{Cl} B$ for some D),
       (i) Remove Cl from Q_d, and add it to Q.
       (ii) Call Subroutine Remove(B) recursively.
3. For each clause Cl in Q_d such that A appears in the body of Cl, if all atoms in the body of Cl do not appear as nodes in G, remove Cl from Q_d, and add it to Q. □

Hereafter, we use the notation $incr(\langle Q, Q_d, G \rangle, Cl) = \langle Q^{out}, Q_d^{out}, G^{out} \rangle$ to denote the fact that when Algorithm Incr is applied to inputs Q (the original set of retained clauses), Q_d (the original set of delete clauses), G (the DC-graph corresponding to Q_d), and Cl (the clause to be inserted), the outputs are Q^{out} (new set of retained clauses), Q_d^{out} (new set of deleted clauses), and G^{out} (new DC-graph). Moreover, we abuse notation by using $\emptyset$ to denote an empty DC-graph, i.e. the DC-graph with the root node only.

Example 4 Apply Algorithm Incr to the 4 clauses in the program P discussed in Example 1. In Figure 3, the first DC-graph (labeled (i)) is graph Gr_1 where $incr(\langle \emptyset, \emptyset, \emptyset \rangle, Cl_1) = \langle \emptyset, \{Cl_1\}, Gr_1 \rangle$. This is the case because nodes B and C are added in Step 1 of Algorithm Incr, node A and the two arcs pointing to A are added in Step 2a. Steps 3 and 4 are not needed in this case. Similarly, the second graph in Figure 3 is DC-graph Gr_2 where $incr(\langle \emptyset, \{Cl_1\}, Gr_1 \rangle, Cl_2) = \langle \emptyset, \{Cl_1, Cl_2\}, Gr_2 \rangle$. This time, node E is added in Step 1 of Algorithm Incr, and the four arcs pointing from A and E to B and D are added in Step 2a. Notice that even though there is a cycle in Gr_2, the cycle is not self-sustaining. It is also not difficult to verify that $sizeopt(\{Cl_1, Cl_2\}) = \langle \emptyset, \{Cl_1, Cl_2\} \rangle$. Similarly, the third graph in Figure 3 is produced by applying Algorithm Incr to add Cl_3

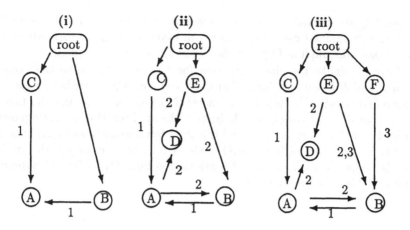

Fig. 3. Applying Algorithm Incr to Add Clauses 1, 2, and 3

to Gr_2. Finally, the graph in Figure 2 is produced by applying Incr to Cl_4 and the third graph. □

The above example only demonstrates the situation when an inserted clause ends up being added to the set Q_d (i.e. Q_d keeps growing). Obviously, this is not always the case, as an inserted clause may indeed end up being added to the set Q. This addition may trigger a series of node removals and the shrinkage of Q_d.

Example 5 Now consider program P', that is by adding Clauses 5 and 6 discussed in Example 2. Let us add Clause 5 first. Steps 1 and 2 of Algorithm Incr are not invoked. But in Step 3a, the clause is added to Q, and Subroutine Remove(C) is called. In Step 1 of Subroutine Remove, node C and the arc from the root to C are removed. As for the arc from C to A labeled Cl_1, this arc is removed. But because of the existence of the arc from B to A labeled Cl_1, Subroutine Remove is *not* called recursively. Furthermore, Step 3 of Remove does not cause any change, and control returns to Algorithm Incr. As for Step 4 of Algorithm Incr, even though there is a cycle from between A and B, this cycle is not self-sustaining because of the arc from E to B with label Cl_2. Thus, Algorithm Incr halts. In functional terms, we have $incr(\langle \emptyset, \{Cl_1, \ldots, Cl_4\}, G_1 \rangle, Cl_5) = \langle \{Cl_5\}, \{Cl_1, \ldots, Cl_4\}, Gr_5 \rangle$, where Gr_5 is the first DC-graph shown in Figure 4. Before we proceed, note that it is not difficult to verify that $sizeopt(\{Cl_1, \ldots, Cl_5\}) = \langle \{Cl_5\}, \{Cl_1, \ldots, Cl_4\} \rangle$.

Now let us add Clause 6. Steps 1 and 2 of Algorithm Incr are not invoked. But in Step 3a, the clause is added to Q, and Subroutine Remove(E) is called. In Step 1 of Subroutine Remove, node E and the arc from the root to E are removed. As for the arc from E to B labeled Cl_2, this arc is removed. But because of the existence of the arc from A to B labeled Cl_2, Subroutine Remove is *not* called recursively. Similarly, the arc from E to B labeled Cl_3 and the arc from E to D

labeled Cl_2 are deleted without recursively calling Remove. Furthermore, Step 3 of Remove does not cause any change, and control returns to Algorithm Incr. The second DC-graph in Figure 4 shows the situation at this point.

However, unlike the above situation for Clause 5, this time the cycle between A and B is self-sustaining. Thus, in Step 4 of Algorithm Incr, Subroutine Remove(B) is called [4]. Step 1 of Remove(B) causes node B and the two arcs from F and A to B to be deleted. In Step 2, the arc from B to A is also removed; Clause 1 is moved from Q_d to Q; and this time Subroutine Remove(A) is invoked recursively. In Step 1 of Remove(A), node A is erased. In Step 2, the arc from A to D is removed; Clauses 2 and 4 are moved from Q_d to Q; and Subroutine Remove(D) is called recursively.

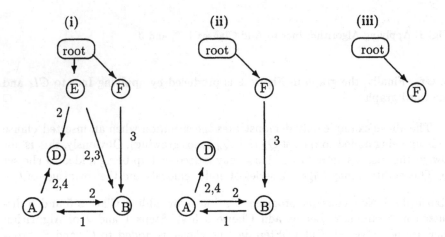

Fig. 4. Applying Algorithm Incr to Add Clauses 5 and 6

Step 1 of Remove(D) erases node D, and Step 3 causes no change. Control now returns to Step 3 of Remove(A). As there is no longer any clause in Q_d with A in the body, control returns to Step 3 of Remove(B). Again as there is no longer any clause in Q_d with B in the body, the executions of Remove(B) and Algorithm Incr are now completed. In functional terms, we have $incr((\{Cl_5\}, \{Cl_1, \ldots, Cl_4\}, Gr_5), Cl_6) = (\{Cl_1, Cl_2, Cl_4, Cl_5, Cl_6\}, \{Cl_3\}, Gr_6)$, where Gr_6 is the last DC-graph shown in Figure 4. □

3.4 Correctness Proof: Incrementality of Algorithm Incr

In the remainder of this section, we present one of the key results of this paper – the theorem proving the incremental property of Algorithm Incr (cf. Theorem 1).

---

[4] It is not difficult to verify that the result is the same, if Remove(A) is called first.

This property has been verified several times in the previous examples. But before we can prove the theorem, we need the following concept and lemma.

Definition 2 Let A be a node in a DC-graph G. The *rank* of A in G, denoted by $rank(A)$, is defined recursively as follows:

1. If there is an arc from the root to A, $rank(A) = 0$.
2. Let $B_{1,1}, \ldots, B_{1,u_1}, \ldots, B_{m,1}, \ldots, B_{m,u_m}$ be all the nodes that have arcs pointing to A, such that: a) $\{Cl_1, \ldots, Cl_m\}$ are all the labels of these arcs, and b) for all $1 \leq j \leq m$, $B_{j,1}, \ldots, B_{j,u_j}$ are all the nodes that have arcs pointing to A with label Cl_j. Then $rank(A) = 1 + max_{j=1}^{m}(min_{i=1}^{u_j} rank(B_{j,i}))$. □

Example 6 Consider the DC-graph G_1 introduced in Figure 3. The nodes with $rank = 0$ are C, E and F. Now consider $rank(A)$. There are the arcs from C and B pointing to A, both with label Cl_1. Thus, $rank(A) = 1 + min\{rank(C), rank(B)\}$. Since $rank(C) = 0$, it is obvious that $rank(A) = 1 + rank(C) = 1$. Now consider $rank(B)$ and all the arcs pointing to B. This time there are two different labels: Cl_2 and Cl_3. For Cl_2, there are the arcs from A and E to B. Based on an analysis similarly to the one for $rank(A)$, the minimum corresponding to Cl_2 is $rank(E) = 0$. For Cl_3, there are the arcs from E and F to B. Thus, the minimum based on Cl_3 is $min\{rank(E), rank(F)\} = 0$. Hence, $rank(B) = 1 + max\{0, 0\} = 1$, where the two zeros correspond to Cl_2 and Cl_3 respectively. Similarly, it is not difficult to verify that $rank(D) = 1 + rank(A) = 2$. Now compare the ranks with the sets S_0, S_1 and S_2 discussed in Example 1. The interesting thing here is that for all atoms A, $rank(A) = k$ iff $A \in S_k$. This property will be proved formally in the lemma below. □

Notice that if a DC-graph contains a self-sustaining cycle, rank assignments to atoms in the cycle are not well-defined. For example, consider the self-sustaining cycle between A and B in the second DC-graph in Figure 4. Then $rank(B)$ depends on $rank(A)$ which in turn depends on $rank(B)$. Thus, both ranks are not well-defined because of the cyclic dependency. Fortunately, since Step 4 of Algorithm Incr removes all self-sustaining cycles, all DC-graphs produced by Incr do not contain any self-sustaining cycle. Then by Definition 1, for the non self-sustaining cycle $A_1 \xrightarrow{Cl_1} A_2 \xrightarrow{Cl_2} \ldots \xrightarrow{Cl_{i-1}} A_i \ldots A_n \xrightarrow{Cl_n} A_1$, there must exist atom A_i such that there exists arc $B \xrightarrow{Cl_{i-1}} A_i$ for some atom $B \notin \{A_1, \ldots, A_n\}$. Thus, in determining $rank(A_i)$, for Clause Cl_{i-1}, $min\{rank(B), rank(A_{i-1})\}$ is always well-defined (cf. the previous example). Thus, there is no cyclic dependency on rank assignments.

Lemma 2 Let $incr(\langle Q, Q_d, G \rangle, Cl) = \langle Q^{out}, Q_d^{out}, G^{out} \rangle$. Then for all nodes $A \in G^{out}$, $rank(A) = n$ iff $A \in S_n$, where the sets $S_0, \ldots, S_n, \ldots$ are the ones produced by applying Algorithm SizeOpt directly on $Q^{out} \cup Q_d^{out}$. □

Now we are in a position to present the theorem that proves the incremental property of Algorithm Incr.

Theorem 1 Let P be a program consisting of clauses $Cl_1, \ldots, Cl_n$. Let $sizeopt(P) = \langle Q, Q_d \rangle$, and let $incr(\ldots incr(\langle \emptyset, \emptyset, \emptyset \rangle, Cl_1), \ldots, Cl_n) = \langle P_n, P_{n,d}, G_n \rangle$. Then: $Q = P_n$ and $Q_d = P_{n,d}$. □

Corollary 1 Given clauses $Cl_1, \ldots, Cl_n$, Algorithm Incr produces the same end result regardless of the order $Cl_1, \ldots, Cl_n$ are inserted. □

4 Further Optimization: Algorithm IncrOpt

A complexity analysis on Algorithm Incr reveals that Step 4 plays a considerable role in determining the efficiency of Incr. It involves finding each and every self-sustaining cycle that may exist in the DC-graph. As shown in Example 5, this is the crucial step that leads to the incremental property of Algorithm Incr. However, the following lemma shows that from the point of view of computing minimal models, self-sustaining cycles need not be detected, and can be left in the graph.

Lemma 3 Let Q be a set of retained clauses and Q_d be a set of deleted clauses maintained in the DC-graph G. Let $A_1 \xrightarrow{Cl_1} A_2 \xrightarrow{Cl_2} \ldots \xrightarrow{Cl_i} A_{i+1} \ldots A_n \xrightarrow{Cl_n} A_1$ be a self-sustaining cycle in G. M is a minimal model of $Q \cup \{Cl_1, \ldots, Cl_n\}$ iff M is a minimal model of Q. □

The above lemma motivates the following algorithm.

Algorithm IncrOpt Exactly the same as Algorithm Incr, but without Step 4 of Incr. □

Hereafter we use the notation $incropt(\langle Q, Q_d, G \rangle, Cl) = \langle Q^{out}, Q_d^{out}, G^{out} \rangle$ for Algorithm IncrOpt in exactly the same way as we use $incr(\langle Q, Q_d, G \rangle, Cl) = \langle Q^{out}, Q_d^{out}, G^{out} \rangle$ for Incr. The corollary below follows directly from Lemma 1, Theorem 1 and Lemma 3.

Corollary 2 Let P be a program consisting of clauses $Cl_1, \ldots, Cl_n$, and let $incropt(\ldots incropt(\langle \emptyset, \emptyset, \emptyset \rangle, Cl_1), \ldots, Cl_n) = \langle P_n, P_{n,d}, G_n \rangle$. M is a minimal model of P iff M is a minimal model of P_n. □

As far as supporting minimal model computation is concerned, Algorithm IncrOpt is more preferable than Algorithm Incr. The reasons are threefold.

- First, as discussed above, IncrOpt does not check for self-sustaining cycles. While cycle detection takes time linear to the number to edges in the graph, checking *all* cycles to see whether they are self-sustaining takes considerably more time. Thus, by not checking self-sustaining cycles, IncrOpt is more efficient than Incr.
- Second, it is easy to see if $incropt(\langle Q, Q_d, G \rangle, Cl) = \langle Q_{opt}^{out}, \_, G_{opt}^{out} \rangle$ and $incr(\langle Q, Q_d, G \rangle, Cl) = \langle Q^{out}, \_, \_ \rangle$, then it is necessary that $Q_{opt}^{out} \subseteq Q^{out}$. More precisely, IncrOpt keeps all clauses in self-sustaining cycles deleted.

Thus, the size of the program Q_{opt}^{out} may be much smaller than that of Q^{out}. The implication is that finding the minimal models based on Q_{opt}^{out} may take considerably less time than finding the minimal models based on Q^{out}.

- The third reason why Algorithm IncrOpt is more preferred applies only to programs P that are definite (i.e. no disjunctive heads). The following lemma shows that for such programs P, Algorithm IncrOpt directly finds the least model of P.

Lemma 4 Let P be a definite program consisting of clauses $Cl_1, \ldots, Cl_n$, and let $incropt(\ldots incropt(\langle \emptyset, \emptyset, \emptyset \rangle, Cl_1), \ldots, Cl_n) = \langle P_n, P_{n,d}, G_n \rangle$. The least model of P is the set $\{A|A$ is the head of a clause in $P_n\}$. □

According to Corollary 1 and Lemma 3, when using Algorithm IncrOpt, different orders of inserting the same collection of clauses do not affect the final DC-graph, and the final sets of retained and deleted clauses. However, different orders may require different execution times – depending largely upon how many times Subroutine Remove is invoked. If Remove is not called at all when inserting a clause $A_1 \vee \ldots \vee A_m \leftarrow B_1 \wedge \ldots \wedge B_l$, the complexity of Algorithm IncrOpt is $O(ml)$. Otherwise, if a is the number of nodes (atoms) in the current graph, then the worst case complexity of recursively calling Remove is $O(alN)$, and that of IncrOpt is $O(ml + alN)$. It is then tempting to conclude that the complexity of IncrOpt for inserting n clauses is $O(n(ml + alN))$. However, this is incorrect because during the process of inserting the n clauses, $Remove(A)$ for all atoms A can only occur at most once. Thus, for inserting n clauses, the complexity of IncrOpt should be $O(nml + al(N + n))$.

On the other hand, if Algorithm SizeOpt is used directly, then there are $(N + n)$ clauses [5]. The worst case complexity of Algorithm SizeOpt for $(N + n)$ clauses is $O(ml(N + n)^2)$. Thus, comparing the complexity figures of Algorithm SizeOpt and IncrOpt does not provide any clear conclusion, as the comparison depends on the magnitude of a, the number of atoms in a DC-graph, relative to the magnitudes of N, n, l and m. Below we will present experimental results evaluating the effectiveness of Algorithm IncrOpt.

5 Experimental Evaluation

5.1 Implementation Overview

For our experimentation, we implemented Algorithm IncrOpt in C. We also implemented two versions of Algorithm SizeOpt. One is a straightforward encoding of the algorithm presented in Section 2.2. The other one tries to minimize searching by extensive indexing. Unfortunately, in all the experiments we have carried out so far, the version with extensive indexing requires so much overhead to set up the indices that the straightforward version takes much less time.

---

[5] Based on Figure 1, the analysis here assumes that P consists of N clauses, and $P\theta_j$ consists of n clauses.

Thus, for all the experimental results reported later for Algorithm SizeOpt, the straightforward version was used.

Recall that in our incremental algorithms, a DC-graph is used to organize the deleted clauses. Each arc in the graph represents a deleted clause. However, not every deleted clause has a corresponding arc in the graph. Given a deleted clause $Cl \equiv A_1 \vee \ldots \vee A_m \leftarrow B_1 \wedge \ldots \wedge B_n$, if all of $A_1, \ldots, A_m$ do not appear in the graph, then this clause would not appear as a label of an arc. In our implementation of the incremental algorithms, we set up a virtual node so that there is an arc from the appropriate node of an atom appearing in the body to the virtual node. More precisely, a virtual node is an atom that appears both in the heads of some clauses in Q and in the heads of some clauses in Q_d. In this way, each deleted clause has a corresponding arc in the DC-graph. This simplifies the construction and maintenance of DC-graph, and makes the implementation more efficient. This is because with the use of virtual nodes, Step 3 of Subroutine Remove can be skipped. Finally, to further speed up the maintenance of DC-graphs, a counter is kept for each clause which records the number of times the clause appears as an arc in the graph. If this counter decreases to zero, the clause is removed from Q_d, and put back to Q.

In the remainder of this section, we will report experimental results evaluating the effectiveness of our proposed algorithms and optimizations. All run-times are in milliseconds, and were obtained by running the experiments in a SPARC-LX Unix time-sharing environment.

5.2 Same Number of Disjunctive Clauses: IncrOpt vs SizeOpt

In this series of experiments, we compared the effectiveness of our incremental algorithm IncrOpt with the original algorithm SizeOpt. For each algorithm, we report i) the total time taken to process 20 randomly generated clauses, ii) the number of clauses deleted, and iii) the time taken to find the minimal models.

	IncrOpt	SizeOpt
processing time for 20 clauses (ms)	3.54	0.33
clauses deleted	19	0
time to find minimal models (ms)	49.17	83.61
total time taken (ms)	52.71	83.94

For just the time taken to process the 20 clauses, our incremental algorithm IncrOpt takes more time than SizeOpt, primarily for maintaining DC-graphs. But as shown above, the extra time is worth spending because IncrOpt manages to delete 19 more clauses than SizeOpt. This is all due to the fact that, as described in Section 4, IncrOpt deletes all the clauses in self-sustaining cycles. Consequently, the times taken for the two algorithms to find the (same collection of) minimal models differ by a wide margin. This clearly demonstrates the importance of deleting more rules, whose impact is multiplied in model computations. At the end, the total time taken by IncrOpt is only about 60% of the time taken by SizeOpt.

5.3 Same Number of Definite Clauses: IncrOpt vs SizeOpt

Based on the results of the previous set of experiments for disjunctive clauses, we surely can predict that for definite clauses, IncrOpt again outperforms SizeOpt. Moreover, Lemma 4 presents a stronger reason for us to believe that IncrOpt will perform even better. The lemma shows that for definite clauses, our incremental algorithms can obtain the least model by simply obtaining the heads of all the clauses not deleted. Indeed, our belief is confirmed by this series of experiments, in which each test program contains 100 randomly generated definite clauses. The following table reports the run-times for a typical program.

	IncrOpt	SizeOpt
processing time for 100 clauses (ms)	9.22	0.73
clauses deleted	89	17
time to find least model (ms)	5.76	580.06
total time taken (ms)	14.98	580.79

The processing time taken by IncrOpt is longer than that by SizeOpt. But again IncrOpt deletes many more clauses, and requires a minimal amount of time to obtain the least model. In contrast, SizeOpt is much less effective in deleting clauses, and requires the invocation of the least model solver whose run-time dominates the entire process.

5.4 Partial Instantiation Trees: IncrOpt vs SizeOpt

Thus far, we have only compared IncrOpt with SizeOpt in those situations where both algorithms are required to process the same number of clauses. But recall that our incremental algorithms are designed for a slightly different purpose: to expand partial instantiation trees efficiently. As described in Section 2.1, if program P in a node N gives rise to conflict-set unifiers $\theta_1, \ldots, \theta_m$, then N has m child nodes, each corresponding to $P \cup P\theta_j$. Thus, as shown in Figure 1, the acid test of the effectiveness of our incremental algorithms is between the time taken for our incremental algorithms to process the clauses in $P\theta_j$ and the time taken for SizeOpt to process all the clauses in $P \cup P\theta_j$. Given the results of the previous series of experiments, we expect IncrOpt to outperform SizeOpt even more in the expansion of partial instantiation trees. This conjecture is confirmed by the following experiment that fully expands the instantiation tree of the program discussed in Section 2.1. Each entry in the table below gives two run-times: i) the time taken to process the clauses in $P\theta_j$ by IncrOpt, or in $P \cup P\theta_j$ by SizeOpt; and ii) the time taken to find the least model.

	IncrOpt	SizeOpt
Node 1 (ms)	0.67/5.47	0.33/45.88
remaining nodes (ms)	0.08/22.2	1.36/201.88
total (processing time/model solving time)	0.75/27.67	1.69/247.76
total (processing time + model solving time)	28.42	249.45

As expected, the processing time of IncrOpt for the first node is relatively long (i.e. 0.67ms), whereas the processing time for each of the subsequent nodes is much shorter (i.e. 0.01ms). This reflects the benefit of being incremental. At the end, the total processing time of IncrOpt is 0.75ms, less than 50% of that of SizeOpt. Furthermore, as shown in previous experiments, IncrOpt requires much less time in finding least models.

6 Conclusions

The objective of this paper is to study how to optimize the expansion of partial instantiation trees for computing minimal and least models. Towards this goal, we have developed Algorithm Incr which is formally proved to be incremental. We have further optimized Incr to delete clauses in self-sustaining cycles. Our experimental results indicate that IncrOpt gives the best performance. More importantly, when compared with the original algorithm SizeOpt, IncrOpt can give very significant improvement in run-time efficiency.

References

1. F. Bancilhon, D. Maier, Y. Sagiv and J. Ullman. (1986) *Magic Sets and Other Strange Ways to Implement Logic Programs*, Proc. ACM-PODS, pp 1–15.
2. C. Bell, A. Nerode, R. Ng and V.S. Subrahmanian. (1992) *Implementing Deductive Databases by Linear Programming*, Proc. ACM-PODS, pp 283–291.
3. C. Bell, A. Nerode, R. Ng and V.S. Subrahmanian. (1992) *Mixed Integer Programming Methods for Computing Nonmonotonic Deductive Databases*, to appear in: Journal of ACM.
4. V. Chandru and J. Hooker. (1991) *Extended Horn Sets in Propositional Logic*, Journal of the ACM, 38, 1, pp 205–221.
5. M. Gelfond and V. Lifschitz. (1988) *The Stable Model Semantics for Logic Programming*, in: Proc. 5th International Conference and Symposium on Logic Programming, ed R. A. Kowalski and K. A. Bowen, pp 1070–1080.
6. R. E. Jeroslow. (1988) *Computation-Oriented Reductions of Predicate to Propositional Logic*, Decision Support Systems, 4, pps 183–187.
7. V.Kagan, A. Nerode and V.S. Subrahmanian (1993) *Computing Definite Logic Programs by Partial Instantiation*, to appear in: Annals of Pure and Applied Logic.
8. V.Kagan, A. Nerode and V.S. Subrahmanian (1994) *Computing Minimal Models by Partial Instantiation*, draft manuscript, submitted to a technical journal for publication.
9. J. Lobo, J. Minker and A. Rajasekar. (1992) *Foundations of Disjunctive Logic Programming*, MIT Press.
10. A. Martelli and U. Montanari. (1982) *An Efficient Unification Algorithm*, ACM Trans. on Programming Languages and Systems, 4, 2, pp 258–282.
11. X. Tian. (1994) *Optimizations for Model Computation Based on Partial Instantiation*, MSc Thesis, University of British Columbia.
12. A. van Gelder, K. Ross and J. Schlipf. (1988) *Unfounded Sets and Well-founded Semantics for General Logic Programs*, in Proc. ACM-PODS, pp 221–230.

A Transformation of Propositional Prolog Programs into Classical Logic

Robert F. Stärk*

Department of Mathematics, Stanford University
Stanford, CA 94305, USA, ⟨staerk@math.stanford.edu⟩

Abstract. We transform a propositional Prolog program P into a set of propositional formulas $prl(P)$ and show that Prolog, using its depth-first left-to-right search, is sound and complete with respect to $prl(P)$. This means that a goal succeeds in Prolog if and only if it follows from $prl(P)$ in classical propositional logic. The generalization of $prl(P)$ to predicate logic leads to a system for which Prolog is still sound but unfortunately not complete. If one changes, however, the definition of the termination operator, then one obtains a theory that allows to prove termination of arbitrary non-floundering goals under Prolog.

1 Introduction

The idea of transforming a propositional Prolog program into a logical theory that reflects the procedural behavior of the program is not new. Several transformations have been proposed for this purpose. Most of them are based on non-classical logics.

Andrews [1] uses folding/unfolding transformations. Van Benthem [18] and Kalsbeek [11] capture computation mechanisms like standard Prolog with substructural Gentzen type calculi. Substructural means that certain classical structural rules are omitted or modified. Cerrito [5] transforms logic programs into sets of linear logic sequents. Elbl [7] investigates the semantics of non-propositional Prolog programs in a version of linear logic.

Our method is to transform a Prolog program into a set of formulas of classical propositional logic. The approach is based on Mints type calculi for pure Prolog in [13] extended by Shepherdson in [14]. A typical rule in these calculi can be described as follows. Assume that $(A:\!-G_i)_{1\le i\le m}$ is a sequence of clauses defining the atom A. Then, for each i there is a rule in the calculus:

$$\frac{\mathsf{F}(G_1 * H) \quad \cdots \quad \mathsf{F}(G_{i-1} * H) \quad \mathsf{S}(G_i * H)}{\mathsf{S}(A * H)} \quad (\mathrm{M}_i)$$

In this rule, H and G_i are sequences of atoms and $*$ denotes concatenation of sequences. The operator F means "fails" and S means "succeeds". Rule (M_i) models what Prolog does when it evaluates a goal of the form $A * H$. Prolog

---

* Supported by the Swiss National Science Foundation.

selects the left-most atom and replaces it with the body G_1 of the first matching clause. The new goal is $G_1 * H$. If it fails, then backtracking starts and the second clause is tried. If finally for some G_i the goal $G_i * H$ succeeds, then the original goal $A * H$ succeeds. If one omits the side atoms H then one can translate the rules (M_1)–(M_m) into the following formula:

$$\bigvee_{i=1}^{m} \left(\bigwedge_{j=1}^{i-1} \mathbf{F} G_j \wedge \mathbf{S} G_i \right) \rightarrow \mathbf{S} A$$

The expressions $\mathbf{F} G_j$, $\mathbf{S} G_i$ and $\mathbf{S} A$ are abbreviations. For an atom B, the expression $\mathbf{S} B$ denotes just a new propositional atom, say B^s. For a sequence of atoms $B_1, \ldots, B_n$, the expression $\mathbf{S}(B_1, \ldots, B_n)$ is an abbreviation for the formula $\mathbf{S} B_1 \wedge \ldots \wedge \mathbf{S} B_n$. This reflects the fact that a goal $B_1, \ldots, B_n$ succeeds if and only if all its subgoals B_i succeed.

But what does $\mathbf{F}(B_1, \ldots, B_n)$ mean? A first attempt is to define $\mathbf{F}(B_1, B_2)$ as $\mathbf{F} B_1 \vee (\mathbf{S} B_1 \wedge \mathbf{F} B_2)$ expressing the failure of B_1, B_2 as: either B_1 fails or B_1 succeeds and B_2 fails. That this is not correct can be seen in the following program:

B_1.
$B_1 :- B_1$.

In this program the atom B_1 succeeds, thus $\mathbf{S} B_1$ is true. The atom B_2 fails trivially, since there are no clauses defining B_2. Thus $\mathbf{F} B_2$ is true. Together we obtain $\mathbf{F} B_1 \vee (\mathbf{S} B_1 \wedge \mathbf{F} B_2)$. But we do not have $\mathbf{F}(B_1, B_2)$, since the goal B_1, B_2 loops in Prolog. (In Prolog the atom B_1 is called "loop" and the atom B_2 is called "fail".) This example has also been studied in [1].

The failure of B_1, B_2 can correctly be expressed as follows: either B_1 fails or B_1 terminates and B_2 fails. This corresponds to the formula $\mathbf{F} B_1 \vee (\mathbf{T} B_1 \wedge \mathbf{F} B_2)$, where a new operator $\mathbf{T}$ is used describing termination of atoms.

In rule (M_i), we see that in order to express the *success* of a goal one has to express what it means that a goal *fails* in Prolog. Thus the notion of *failure* and *negation-as-failure* is built-in to Prolog and we cannot do without it. But in order to express the failure of a goal one has also to express what it means that a goal *terminates* in Prolog.

2 Basic Notions

We use the following metasymbols: A, B for propositional atoms; L, M for literals; G, H for goals; C for program clauses; P for logic programs; φ, ψ for propositional formulas. A *literal* is an atom A or a negated atom of the form $\neg A$. *Goals* are finite sequences of literals; $\varnothing$ is the empty goal. We write $G * H$ for the concatenation of G and H; we write $L * G$ for $\langle L \rangle * G$. A *clause* is an expression of the form $A :- G$. The atom A is the *head* of the clause and G is the *body* of the clause. The body of a clause is a goal. Clauses with empty

```
function eval(P, G): boolean;
begin
  case G of
    ∅: return(true);
    A * H:
      assume that P(A) = (Gᵢ)₁≤ᵢ≤ₘ;
      for i := 1 to m do
        if eval(P, Gᵢ * H) then return(true) end
      end;
      return(false);
    ¬A * H: if eval(P, A) then return(false) else return(eval(P, H)) end
  end
end eval.
```

Fig. 1. The Prolog depth-first search.

bodies are often called *facts* and clauses with non-empty bodies are called *rules*. In examples we identify facts $A : - \varnothing$ with A.

A *logic program* P is a finite sequence of clauses. For an atom A we define $P(A)$ to be the sequence of bodies of clauses of P with head A (in the same order as they appear in P). If there are no clauses with head A then $P(A)$ is the empty sequence. $P(A)$ is called the *definition* of the atom A in P. In this way, a logic program P can be considered as a finite function that assigns to every atom A a finite sequence of goals.

Given a logic program, a goal is evaluated in standard Prolog manner, top-down from left to right as shown in Figure 1. In every step of the evaluation the leftmost literal of the goal G is selected. If it is a positive atom A then it is replaced in the for-loop with the body of the first clause of the program with head A which has not been tried before. If it is the negative literal $\neg A$ then A is evaluated. If A returns *true* then G returns false. If A returns *false* then $\neg A$ is dropped from G and the rest of the goal is evaluated.

The aim of this article is to describe this dynamic procedure by static formulas of classical propositional logic. For this purpose we first analyze the set of those goals G such that $eval(P, G)$ returns *true* and the set of those goals G such that $eval(P, G)$ returns *false*.

Definition 1. Let P be a logic program. The sets $\mathbb{S}(P)$ and $\mathbb{F}(P)$ are inductively generated by the following six rules:

S1. $\varnothing \in \mathbb{S}(P)$.

S2. If $P(A) = (G_i)_{1 \le i \le m}$, $1 \le i \le m$, $G_j * H \in \mathbb{F}(P)$ for $1 \le j < i$, and $G_i * H \in \mathbb{S}(P)$, then $A * H \in \mathbb{S}(P)$.

S3. If $A \in \mathbb{F}(P)$ and $H \in \mathbb{S}(P)$, then $\neg A * H \in \mathbb{S}(P)$.

F1. If $P(A) = (G_i)_{1 \leq i \leq m}$ and $G_i * H \in \mathbb{F}(P)$ for $1 \leq i \leq m$, then $A * H \in \mathbb{F}(P)$.

F2. If $A \in \mathbb{S}(P)$, then $\neg A * H \in \mathbb{F}(P)$.

F3. If $A \in \mathbb{F}(P)$ and $H \in \mathbb{F}(P)$, then $\neg A * H \in \mathbb{F}(P)$.

$\mathbb{S}(P)$ is called the *success* set of P under Prolog and $\mathbb{F}(P)$ is called the *failure* set of P under Prolog. If a goal is in $\mathbb{S}(P)$ then we say that it *succeeds* in Prolog; if a goal is in $\mathbb{F}(P)$ then we say that it *fails* in Prolog. It is easy to see that a goal G belongs to $\mathbb{S}(P)$ if and only if $eval(P, G)$ returns *true*. A goal G belongs to $\mathbb{F}(P)$ if and only if $eval(P, G)$ returns *false*. The definition of $\mathbb{S}(P)$ and $\mathbb{F}(P)$ is more or less the restriction of the Mints type calculus in [14] to propositional programs.

Definition 1 allows us to prove properties about the success set $\mathbb{S}(P)$ and the failure set $\mathbb{F}(P)$ by induction on the definition of these sets. The properties we are interested in are the relations between the success or the failure of a compound goal $G * H$ and the success or the failure of its subgoals G and H. The first obvious result is that a goal $G * H$ succeeds if and only if both subgoals G and H succeed. This gives us the possibility to express the success of a compound goal in terms of of the success of its subgoals. In the case of failure this is not so trivial.

Lemma 2. *Let P be a logic program and let G and H be goals.*

1. *If $G \in \mathbb{F}(P)$ then $G * H \in \mathbb{F}(P)$.*
2. *If $G \in \mathbb{S}(P)$ and $H \in \mathbb{S}(P)$ then $G * H \in \mathbb{S}(P)$.*
3. *It is not possible that G belongs to $\mathbb{S}(P)$ and $\mathbb{F}(P)$.*
4. *If $G * H \in \mathbb{F}(P)$ and $H \in \mathbb{S}(P)$ then $G \in \mathbb{F}(P)$.*
5. *If $G * H \in \mathbb{S}(P)$ then $G \in \mathbb{S}(P)$ and $H \in \mathbb{S}(P)$.*

Proof. By induction on the definition of $\mathbb{S}(P)$ and $\mathbb{F}(P)$. □

As mentioned in the introduction the two terms success and failure do not suffice to express the failure of a compound goal in terms of similar properties of its subgoals. For this purpose we have to introduce the set $\mathbb{T}(P)$ of those goals that terminate in Prolog.

Definition 3. Let P be a logic program. The set $\mathbb{T}(P)$ is inductively generated by the following four rules:

T1. $\varnothing \in \mathbb{T}(P)$.

T2. If $P(A) = (G_i)_{1 \leq i \leq m}$ and $G_i * H \in \mathbb{T}(P)$ for $1 \leq i \leq m$, then $A * H \in \mathbb{T}(P)$.

T3. If $A \in \mathbb{S}(P)$, then $\neg A * H \in \mathbb{T}(P)$.

T4. If $A \in \mathbb{F}(P)$ and $H \in \mathbb{T}(P)$, then $\neg A * H \in \mathbb{T}(P)$.

$\mathbb{T}(P)$ is called the *termination* set of P under Prolog. If a goal is in $\mathbb{T}(P)$ then we say that it *terminates* in Prolog. The definition of the termination set $\mathbb{T}(P)$ looks almost like the definition of the failure set $\mathbb{F}(P)$. The only difference is that we require in T1 that the empty goal $\varnothing$ belongs to $\mathbb{T}(P)$, since the empty goal obviously terminates. The other clauses T2, T3 and T4 are like in the definition of $\mathbb{F}(P)$. Thus it is obvious that $\mathbb{F}(P)$ is a subset of $\mathbb{T}(P)$. The set $\mathbb{T}(P)$ is *not* the set of goals G such that $eval(P, G)$ terminates. This set would simply be the union of $\mathbb{S}(P)$ and $\mathbb{F}(P)$.

With the notion of termination we are able to state the exact relations between the failure or the termination of a compound goal $G * H$ and the failure or the termination of its subgoals G and H.

Lemma 4. *Let P be a logic program and let G and H be goals.*

1. *If $G \in \mathbb{F}(P)$ then $G \in \mathbb{T}(P)$.*
2. *If $G \in \mathbb{T}(P)$ then $G \in \mathbb{S}(P)$ or $G \in \mathbb{F}(P)$.*
3. *If $G \in \mathbb{T}(P)$ and $H \in \mathbb{F}(P)$ then $G * H \in \mathbb{F}(P)$.*
4. *If $G \in \mathbb{T}(P)$ and $H \in \mathbb{T}(P)$ then $G * H \in \mathbb{T}(P)$.*
5. *If $G * H \in \mathbb{F}(P)$ then $G \in \mathbb{F}(P)$ or $(G \in \mathbb{T}(P)$ and $H \in \mathbb{F}(P))$.*
6. *If $G * H \in \mathbb{T}(P)$ then $G \in \mathbb{T}(P)$ and $(G \in \mathbb{F}(P)$ or $H \in \mathbb{T}(P))$.*

Proof. By induction on the definition of $\mathbb{T}(P)$ or $\mathbb{F}(P)$ respectively. □

This lemma tells us that a goal $G * H$ fails if and only if G fails or (G terminates and H fails). The goal $G * H$ terminates if and only if G terminates and (G fails or H terminates). Both laws are used in the next section for axiomatizing the Prolog depth-first search.

3 The axiomatization of the Prolog depth first search

The axiomatization is formulated in an extended language with new, distinct propositional atoms A^s, A^f and A^t for every atom A. The intended interpretation of A^s, A^f and A^t is: A succeeds, A fails and A terminates.

A propositional *valuation* is a function I that assigns to every propositional atom of the new language the value *true* or *false*. A valuation I is extended to propositional formulas in the usual way. We write $I \models \varphi$ for $I(\varphi) = true$. We say that I is a *model* of a set of formulas T (written $I \models T$) if $I \models \varphi$ for all $\varphi \in T$. We say that φ is a *consequence* of T (written $T \models \varphi$) if $I \models \varphi$ for all I such that $I \models T$.

We define three operators S, F and T that transform goals of the original language of the program into formulas of the new extended language. For the empty goal we simply define $\mathsf{S}\varnothing := \top$, $\mathsf{F}\varnothing := \bot$ and $\mathsf{T}\varnothing := \top$. The symbols $\top$ and $\bot$ are the propositional constants for *true* and *false*. For non-empty sequences of literals the operators S, F and T are defined by recursion on the length of the sequence in the following way:

$$\mathsf{S}A := A^s, \qquad \mathsf{S}\neg A := \mathsf{F}A, \qquad \mathsf{S}(L * G) := \mathsf{S}L \wedge \mathsf{S}G,$$
$$\mathsf{F}A := A^f, \qquad \mathsf{F}\neg A := \mathsf{S}A, \qquad \mathsf{F}(L * G) := \mathsf{F}L \vee (\mathsf{T}L \wedge \mathsf{F}G),$$
$$\mathsf{T}A := A^t, \qquad \mathsf{T}\neg A := \mathsf{S}A \vee \mathsf{F}A, \qquad \mathsf{T}(L * G) := \mathsf{T}L \wedge (\mathsf{F}L \vee \mathsf{T}G).$$

The idea is simple. $\mathsf{S}G$ means that the goal G succeeds; $\mathsf{F}G$ means that G fails; $\mathsf{T}G$ means that G terminates. This obvious interpretation will be justified later.

Definition 5. Let P be a logic program. Then $prl(P)$ consists of the following five formulas for every propositional atom A so that $P(A) = (G_i)_{1 \leq i \leq m}$:

1. $\bigvee\limits_{i=1}^{m} \left(\bigwedge\limits_{j=1}^{i-1} \mathbf{F} G_j \wedge \mathbf{S} G_i \right) \rightarrow \mathbf{S} A$

2. $\bigwedge\limits_{i=1}^{m} \mathbf{F} G_i \rightarrow \mathbf{F} A$

3. $\bigwedge\limits_{i=1}^{m} \mathbf{T} G_i \rightarrow \mathbf{T} A$

4. $\neg(\mathbf{S} A \wedge \mathbf{F} A)$

5. $\mathbf{F} A \rightarrow \mathbf{T} A$

It is clear what these formulas express. Formula 1, for example, says that A succeeds provided that there exists a clause $A \leftarrow G_i$ in P so that G_i succeeds and for all clauses $A \leftarrow G_j$ that precede $A \leftarrow G_i$ the body G_j fails. Note, that the left-hand sides of the implications 1–3 are positive formulas, i.e. formulas built up from atoms using conjunction and disjunction only.

The propositional theory $prl(P)$ of a logic program P is in some sense a "completion" of P. It is, however, not a complete theory in the logical sense. This means that it does not decide every proposition. The function prl is a default operator in the sense of Jäger [9].

4 Soundness of Prolog

By soundness we mean that if a goal G succeeds in Prolog, then the formula $\mathbf{S} G$ is a consequence of $prl(P)$, i.e. is true under all valuations which are models of $prl(P)$. In the proof of this theorem the following technical lemma is needed.

Lemma 6. Let P be a logic program and G and H be goals.

1. $\mathbf{F}(G * H)$ is equivalent to $\mathbf{F} G \vee (\mathbf{T} G \wedge \mathbf{F} H)$.
2. $\mathbf{T}(G * H)$ is equivalent to $\mathbf{T} G \wedge (\mathbf{F} G \vee \mathbf{T} H)$.
3. $prl(P) \models \neg(\mathbf{S} G \wedge \mathbf{F} G)$.
4. $prl(P) \models \mathbf{F} G \rightarrow \mathbf{T} G$.

Proof. By induction on the length of the goal G. $\qquad \Box$

Theorem 7 (Soundness). Let P be a logic program and G be a goal.

1. If $G \in \mathbb{S}(P)$ then $prl(P) \models \mathbf{S} G$.
2. If $G \in \mathbb{F}(P)$ then $prl(P) \models \mathbf{F} G$.
3. If $G \in \mathbb{T}(P)$ then $prl(P) \models \mathbf{T} G$.

Proof. By induction on the definition of the sets $\mathbb{S}(P)$, $\mathbb{F}(P)$ and $\mathbb{T}(P)$. $\qquad \Box$

5 Completeness of Prolog

By completeness we mean that if the formula $\mathbf{S}\,G$ is a consequence of $prl(P)$ then the goal G succeeds in Prolog. The proof of the completeness theorem is simple, since $prl(P)$ has a canonical model in which a formula $\mathbf{S}\,G$ is true if and only if the goal G succeeds. For every logic program P we define a canonical valuation I_P. We set $I_P(A^s) := true$ if and only if $A \in \mathbb{S}(P)$; $I_P(A^f) := true$ if and only if $A \in \mathbb{F}(P)$; $I_P(A^t) := true$ if and only if $A \in \mathbb{T}(P)$. The defining properties of I_P are inherited by arbitrary goals:

Lemma 8. *Let P be a logic program and G be a goal. Then we have:*

1. $I_P \models \mathbf{S}\,G \iff G \in \mathbb{S}(P)$.
2. $I_P \models \mathbf{F}\,G \iff G \in \mathbb{F}(P)$.
3. $I_P \models \mathbf{T}\,G \iff G \in \mathbb{T}(P)$.

Proof. By induction on the length of the goal G. $\qquad\square$

Using these properties one can easily prove that the valuation I_P is a model of P.

Theorem 9. *Let P be a logic program. Then I_P is a model of $prl(P)$.*

The valuation I_P is called the *standard model* of $prl(P)$. In general, $prl(P)$ can have several models. Assume that P consists of the clause $A:-\neg A$. Then $prl(P)$ is $\{A^f \to A^s, A^s \to A^f, A^s \vee A^f \to A^t, \neg(A^s \wedge A^f), A^f \to A^t\}$. This theory has two models.

Theorem 10 (Completeness). *Let P be a logic program and G be a goal.*

1. *If $prl(P) \models \mathbf{S}\,G$ then $G \in \mathbb{S}(P)$.*
2. *If $prl(P) \models \mathbf{F}\,G$ then $G \in \mathbb{F}(P)$.*
3. *If $prl(P) \models \mathbf{T}\,G$ then $G \in \mathbb{T}(P)$.*

Proof. If $prl(P) \models \mathbf{S}\,G$ then $\mathbf{S}\,G$ is true in I_P, since I_P is a model of $prl(P)$. By Lemma 8, we obtain $G \in \mathbb{S}(P)$. $\qquad\square$

If a logic program P is *terminating*, i.e. every atom A belongs to $\mathbb{T}(P)$, then I_P is the unique model of $prl(P)$. This can be seen as follows. Assume that J is another model of $prl(P)$. Take an atom A. Since P is terminating, we know that A is in $\mathbb{T}(P)$. Therefore A is either in $\mathbb{S}(P)$ or in $\mathbb{F}(P)$. If A is in $\mathbb{S}(P)$ then by soundness of Prolog we know that A^s is true in J. Thus $J(A^s) = true = I_P(A^s)$ and $J(A^f) = false = I_P(A^f)$. If A is in $\mathbb{F}(P)$ then by soundness of Prolog we know that A^f is true in J. Thus we have $J(A^s) = false = I_P(A^s)$ and $J(A^f) = true = I_P(A^f)$. In both cases we have $J(A^t) = true = I_P(A^t)$. Hence $J = I_P$.

Remark. One could formulate $prl(P)$ in a different way. One could add, for instance, the formula $A^t \to A^s \vee A^f$. One could also replace the implications in the axioms of $prl(P)$ by equivalences.

6 Prolog with function symbols

In this section we present an obvious generalization of $prl(P)$ to predicate logic. Prolog is still sound with respect to the generalized $prl(P)$ but no longer complete. One reason for the loss of completeness is that the canonical valuation I_P is not a model of the predicate version of $prl(P)$; another reason is that the lifting-lemma is not true for Prolog with function symbols. The lifting-lemma is usually the main tool for transforming a propositional completeness proof into a predicate completeness proof.

Let $\mathcal{L}$ be a first order language. $\mathcal{L}$ is given by a set of function symbols and a set of predicate symbols. The terms s, t of $\mathcal{L}$ are built up from variables x, y and function symbols; terms are of the form x or $f(t_1, \ldots, t_n)$. The atoms of $\mathcal{L}$ are of the form $R(t_1, \ldots, t_n)$. Goals, clauses and programs are defined as in the propositional case. We assume that the equality symbol does not appear in programs.

We write $\varphi[\vec{x}]$ to indicate that all free variables of φ are from the list $\vec{x}$. The notion $T \vdash \varphi$ means that the formula φ is derivable from the theory T using the axioms and rules of the classical first-order predicate calculus with equality.

A new language $\hat{\mathcal{L}}$ is defined as follows. $\hat{\mathcal{L}}$ has the same function symbols as $\mathcal{L}$. For every predicate symbol R of $\mathcal{L}$ we take in $\hat{\mathcal{L}}$ three new predicate symbols R^s, R^f and R^t of the same arity as R. The operators $\mathbf{S}$, $\mathbf{F}$ and $\mathbf{T}$ transform goals of the language $\mathcal{L}$ into formulas of $\hat{\mathcal{L}}$. For atoms, they are defined in the following way: $\mathbf{S}\,R(\vec{t}) := R^s(\vec{t})$, $\mathbf{F}\,R(\vec{t}) := R^f(\vec{t})$ and $\mathbf{T}\,R(\vec{t}) := R^t(\vec{t})$. Otherwise the operators are defined as in the propositional case.

Let now P be a program and R be an n-ary predicate symbol. We assume that there are m clauses in P whose heads are of the form $R(\ldots)$ and that the ith clause is of the form $R(t_{i,1}[\vec{y}], \ldots, t_{i,n}[\vec{y}]) :- G_i[\vec{y}]$. For each i, $1 \le i \le m$, we define three $\hat{\mathcal{L}}$-formulas $S_R^{P,i}$, $F_R^{P,i}$ and $T_R^{P,i}$ in the following way:

$$S_R^{P,i}[x_1, \ldots, x_n] := \exists \vec{y} \left(\bigwedge_{k=1}^{n} x_k = t_{i,k}[\vec{y}] \wedge \mathbf{S}\,G_i[\vec{y}] \right),$$

$$F_R^{P,i}[x_1, \ldots, x_n] := \forall \vec{y} \left(\bigwedge_{k=1}^{n} x_k = t_{i,k}[\vec{y}] \to \mathbf{F}\,G_i[\vec{y}] \right),$$

$$T_R^{P,i}[x_1, \ldots, x_n] := \forall \vec{y} \left(\bigwedge_{k=1}^{n} x_k = t_{i,k}[\vec{y}] \to \mathbf{T}\,G_i[\vec{y}] \right).$$

Definition 11. Let P be a logic program. Then $prl(P)$ consists of Clark's equational theory CET (see [6]) and the universal closure of the following five formulas for every predicate symbol R of $\mathcal{L}$ with m clauses in P:

1. $\displaystyle \bigvee_{i=1}^{m} \left(\bigwedge_{j=1}^{i-1} F_R^{P,j}[\vec{x}] \wedge S_R^{P,i}[\vec{x}] \right) \to R^s(\vec{x})$

2. $\displaystyle \bigwedge_{i=1}^{m} F_R^{P,i}[\vec{x}] \to R^f(\vec{x})$

3. $\bigwedge\limits_{i=1}^{m} T_R^{P,i}[\vec{x}] \to R^t(\vec{x})$

4. $\neg(R^s(\vec{x}) \wedge R^f(\vec{x}))$

5. $R^f(\vec{x}) \to R^t(\vec{x})$

It can be shown that Prolog is sound with respect to the predicate version of $prl(P)$. If the goal G returns the answer substitution σ in Prolog, then $prl(P) \vdash \mathbf{S}\, G\sigma$. If the goal G fails in Prolog, i.e. if Prolog returns the answer **no**, then $prl(P) \vdash \mathbf{F}\, G$. Prolog is however not complete with respect to $prl(P)$.

Example 1. Let P be the following logic program (c and d are constants):

> $Q :- R(x)$.
> $R(c) :- R(c)$.
> $R(d)$.

It is easy to see that $prl(P) \vdash \mathbf{S}\, Q$. But the goal Q does not succeed in Prolog. It loops. One reason for this incompleteness is that the goal $R(d)$ succeeds in Prolog, but the goal $R(x)$ loops. In other words, the lifting lemma is not true for Prolog.

Let $\mathfrak{J}_P$ be the Herbrand interpretation in which a ground atom $R^s(\vec{t})$ is true if and only if $R(\vec{t})$ succeeds in Prolog, a ground atom $R^f(\vec{t})$ is true if and only if $R(\vec{t})$ fails in Prolog, and a ground atom $R^t(\vec{t})$ is true if and only if $R(\vec{t})$ terminates in Prolog. In general, the structure $\mathfrak{J}_P$ is *not* a model of $prl(P)$.

Example 2. Let P be the program consisting of the following clauses:

> $Q :- R(x)$.
> $R(f(x)) :- R(x)$.

The formula $(\forall x\ \mathbf{F}\, R(x)) \to \mathbf{F}\, Q$ of $prl(P)$ is not true in $\mathfrak{J}_P$. The reason is that the goal $R(t)$ fails in Prolog for every ground term t but he goal $R(x)$ loops. This example has also been studied in [1].

7 Prolog and LDNF-resolution

In this section we present another generalization to predicate logic. The generalization is based on an observation made by Apt and Pedreschi in [3] which is further substantiated in Apt [2]. The observation is that for most Prolog programs the order of the clauses is not important at least not for the intended inputs. One can change the order of the clauses in most programs and the answers computed by Prolog remain the same. The order of the literals in the body of the clauses, however, is important. Apt and Pedreschi justify their thesis with several example programs.

The idea is to interpret the operator $\mathbf{T}$ differently. In the following $\mathbf{T}\,G$ means that G terminates independently of the order of the clauses in the program. To avoid ambiguities we introduce new operators $\tilde{\mathbf{S}}$, $\tilde{\mathbf{F}}$ and $\tilde{\mathbf{T}}$:

$$
\begin{aligned}
\tilde{\mathbf{S}}\,R(\vec{t}) &:= R^{\mathbf{s}}(\vec{t}), & \tilde{\mathbf{S}}\,\neg A &:= \tilde{\mathbf{F}}\,A, & \tilde{\mathbf{S}}(L * G) &:= \tilde{\mathbf{S}}\,L \wedge \tilde{\mathbf{S}}\,G, \\
\tilde{\mathbf{F}}\,R(\vec{t}) &:= R^{\mathbf{f}}(\vec{t}), & \tilde{\mathbf{F}}\,\neg A &:= \tilde{\mathbf{S}}\,A, & \tilde{\mathbf{F}}(L * G) &:= \tilde{\mathbf{F}}\,L \vee \tilde{\mathbf{F}}\,G, \\
\tilde{\mathbf{T}}\,R(\vec{t}) &:= R^{\mathbf{t}}(\vec{t}), & \tilde{\mathbf{T}}\,\neg A &:= \tilde{\mathbf{T}}\,A, & \tilde{\mathbf{T}}(L * G) &:= \tilde{\mathbf{T}}\,L \wedge (\tilde{\mathbf{F}}\,L \vee \tilde{\mathbf{T}}\,G).
\end{aligned}
$$

For the the empty goal $\tilde{\mathbf{S}}$, $\tilde{\mathbf{F}}$ and $\tilde{\mathbf{T}}$ are defined as $\mathbf{S}$, $\mathbf{F}$ and $\mathbf{T}$. The difference between the operator $\mathbf{T}$ and the operator $\tilde{\mathbf{T}}$ is in negative literals. $\mathbf{T}\,\neg A$ is defined as $\mathbf{S}\,A \vee \mathbf{F}\,A$, whereas $\tilde{\mathbf{T}}\,\neg A$ is defined as $\tilde{\mathbf{T}}\,A$. There is also a difference between $\mathbf{F}$ and $\tilde{\mathbf{F}}$. $\mathbf{F}(L * G)$ is defined as $\mathbf{F}\,L \vee (\mathbf{T}\,L \wedge \mathbf{F}\,G)$, but $\tilde{\mathbf{F}}(L * G)$ is simply the disjunction $\tilde{\mathbf{F}}\,L \vee \tilde{\mathbf{F}}\,G$. This may seem unsound. In the following, however, we will consider the operators $\tilde{\mathbf{S}}$ and $\tilde{\mathbf{F}}$ always together with $\tilde{\mathbf{T}}$. The success of a goal will be expressed by the formula $\tilde{\mathbf{S}}\,G \wedge \tilde{\mathbf{T}}\,G$ and the failure of a goal will be expressed by the formula $\tilde{\mathbf{F}}\,G \wedge \tilde{\mathbf{T}}\,G$.

The formulas $\tilde{S}_R^{P,i}$, $\tilde{F}_R^{P,i}$ and $\tilde{T}_R^{P,i}$ are defined as in the last section. Again we assume that R is an n-ary predicate symbol which has m clauses in P. The ith clause is of the form $R(t_{i,1}[\vec{y}], \ldots, t_{i,n}[\vec{y}]) :- G_i[\vec{y}]$.

$$
\tilde{S}_R^{P,i}[x_1, \ldots, x_n] := \exists \vec{y} \left(\bigwedge_{k=1}^{n} x_k = t_{i,k}[\vec{y}] \wedge \tilde{\mathbf{S}}\,G_i[\vec{y}] \right),
$$

$$
\tilde{F}_R^{P,i}[x_1, \ldots, x_n] := \forall \vec{y} \left(\bigwedge_{k=1}^{n} x_k = t_{i,k}[\vec{y}] \rightarrow \tilde{\mathbf{F}}\,G_i[\vec{y}] \right),
$$

$$
\tilde{T}_R^{P,i}[x_1, \ldots, x_n] := \forall \vec{y} \left(\bigwedge_{k=1}^{n} x_k = t_{i,k}[\vec{y}] \rightarrow \tilde{\mathbf{T}}\,G_i[\vec{y}] \right).
$$

The following definition of the ℓ-completion of a logic program was first given in [16] and later refined in [17]. (In [16] the operator $\tilde{\mathbf{T}}$ is called $\mathbf{L}$.)

Definition 12. Let P be a logic program. Then $\ell comp(P)$ consists of Clark's equational theory CET and the universal closure of the following five formulas for every predicate symbol R of $\mathcal{L}$ with m clauses in P:

1. $\displaystyle\bigvee_{i=1}^{m} \tilde{S}_R^{P,i}[\vec{x}] \rightarrow R^{\mathbf{s}}(\vec{x})$

2. $\displaystyle\bigwedge_{i=1}^{m} \tilde{F}_R^{P,i}[\vec{x}] \rightarrow R^{\mathbf{f}}(\vec{x})$

3. $\displaystyle\bigwedge_{i=1}^{m} \tilde{T}_R^{P,i}[\vec{x}] \rightarrow R^{\mathbf{t}}(\vec{x})$

4. $\neg(R^{\mathbf{s}}(\vec{x}) \wedge R^{\mathbf{f}}(\vec{x}))$

5. $R^{\mathbf{t}}(\vec{x}) \rightarrow R^{\mathbf{s}}(\vec{x}) \vee R^{\mathbf{f}}(\vec{x})$

Theorem 13 (Soundness, [17]). *Let P be a logic program and G be a goal.*

1. *If G returns the answer σ using LDNF-resolution, then $\ell comp(P) \vdash \tilde{\mathsf{S}}\, G\sigma$.*
2. *If G fails using LDNF-resolution, then $\ell comp(P) \vdash \tilde{\mathsf{F}}\, G$.*
3. *If G strongly terminates using LDNF-resolution, then $\ell comp(P) \vdash \tilde{\mathsf{T}}\, G$.*

LDNF-resolution here means SLDNF-resolution with the Prolog literal selection rule which always selects the left-most literal in a goal (see [3] or [17] for details). A goal is called *strongly terminating* if every possible LDNF-computation terminates (without floundering). This property is called *left-terminating* in [3].

Theorem 14 (Completeness, [17]). *Let G be a non-floundering goal.*

1. *If $\ell comp(P) \vdash \tilde{\mathsf{T}}\, G$, then G strongly terminates using LDNF-resolution and thus G terminates in Prolog.*
2. *If $\ell comp(P) \vdash \tilde{\mathsf{T}}\, G \wedge \tilde{\mathsf{S}}\, G\theta$, then there exists an LDNF-computed answer σ for G which is more general than θ. The substitution σ is one of the answers of Prolog.*
3. *If $\ell comp(P) \vdash \tilde{\mathsf{T}}\, G \wedge \tilde{\mathsf{F}}\, G$, then the goal G fails using LDNF-resolution and thus G fails in Prolog.*

Full proofs of these theorems are given in [17]. The restriction of the completeness theorem to non-floundering goals is not essential, since the *modes* of predicates usually prevent floundering (cf. [15] and [17]).

There are interesting relationships between the ℓ-completion and other well-known forms of program completion. For example, Clark's completion of a logic program [6] is equivalent to the theory consisting of CET and the following four axioms for every predicate R:

$$\bigvee_{i=1}^{m} \tilde{S}_R^{P,i}[\vec{x}] \rightarrow R^{\mathsf{s}}(\vec{x}), \quad \bigwedge_{i=1}^{m} \tilde{F}_R^{P,i}[\vec{x}] \rightarrow R^{\mathsf{f}}(\vec{x}),$$

$$\neg(R^{\mathsf{s}}(\vec{x}) \wedge R^{\mathsf{f}}(\vec{x})), \quad R^{\mathsf{s}}(\vec{x}) \vee R^{\mathsf{f}}(\vec{x}).$$

The three-valued completion of Fitting [8] and Kunen [12] is equivalent to the theory consisting of CET and the following three axioms for every predicate R:

$$\bigvee_{i=1}^{m} \tilde{S}_R^{P,i}[\vec{x}] \rightarrow R^{\mathsf{s}}(\vec{x}), \quad \bigwedge_{i=1}^{m} \tilde{F}_R^{P,i}[\vec{x}] \rightarrow R^{\mathsf{f}}(\vec{x}), \quad \neg(R^{\mathsf{s}}(\vec{x}) \wedge R^{\mathsf{f}}(\vec{x})).$$

This theory is called *partial completion* in [10] and *doubled program* by van Gelder and Schlipf in [19]. Clark's completion is obtained from the partial completion by adding the axiom $R^{\mathsf{s}}(\vec{x}) \vee R^{\mathsf{f}}(\vec{x})$. The ℓ-completion is obtained by weakening this (obviously unsound) axiom to $R^{\mathsf{t}}(\vec{x}) \rightarrow R^{\mathsf{s}}(\vec{x}) \vee R^{\mathsf{f}}(\vec{x})$ and by adding axioms for the predicates R^{t}.

To make transparent the differences between the operators S, F, T and $\tilde{\mathsf{S}}$, $\tilde{\mathsf{F}}$, $\tilde{\mathsf{T}}$ we introduce sets $\tilde{\mathbb{S}}(P)$, $\tilde{\mathbb{F}}(P)$, and $\tilde{\mathbb{T}}(P)$.

Definition 15. Let P be a propositional logic program. The sets $\tilde{\mathbb{S}}(P)$ and $\tilde{\mathbb{F}}(P)$ are inductively generated by the following five rules:

S1. $\varnothing \in \tilde{\mathbb{S}}(P)$.

S2. If $P(A) = (G_i)_{1 \leq i \leq m}$, and $H * G_i * H' \in \tilde{\mathbb{S}}(P)$ for some i, $1 \leq i \leq m$, then
$H * A * H' \in \tilde{\mathbb{S}}(P)$.

S3. If $A \in \tilde{\mathbb{F}}(P)$ and $H * H' \in \tilde{\mathbb{S}}(P)$, then $H * \neg A * H' \in \tilde{\mathbb{S}}(P)$.

F1. If $P(A) = (G_i)_{1 \leq i \leq m}$ and $H * G_i * H' \in \tilde{\mathbb{F}}(P)$ for all i, $1 \leq i \leq m$, then
$H * A * H' \in \tilde{\mathbb{F}}(P)$.

F2. If $A \in \tilde{\mathbb{S}}(P)$, then $H * \neg A * H' \in \tilde{\mathbb{F}}(P)$.

The sets $\tilde{\mathbb{S}}(P)$ and $\tilde{\mathbb{F}}(P)$ correspond to what is usually called SLDNF-resolution. A goal is in $\tilde{\mathbb{S}}(P)$ if and only if it succeeds in SLDNF-resolution. A goal is in $\tilde{\mathbb{F}}(P)$ if and only if it fails in SLDNF-resolution.

Lemma 16. *Let P be a propositional logic program and let G and H be goals.*

1. *$G * H \in \tilde{\mathbb{S}}(P) \iff G \in \tilde{\mathbb{S}}(P)$ and $H \in \tilde{\mathbb{S}}(P)$.*
2. *$G * H \in \tilde{\mathbb{F}}(P) \iff G \in \tilde{\mathbb{F}}(P)$ or $H \in \tilde{\mathbb{F}}(P)$.*
3. *If $P(A) = (G_i)_{1 \leq i \leq m}$, $1 \leq i \leq m$, and $H * A * H' \in \tilde{\mathbb{F}}(P)$ then
$H * G_i * H' \in \tilde{\mathbb{F}}(P)$ for all i, $1 \leq i \leq m$.*
4. *If $G * H \in \tilde{\mathbb{F}}(P)$ and $G \in \tilde{\mathbb{S}}(P)$ then $H \in \tilde{\mathbb{F}}(P)$.*
5. *It is not possible that G is in $\tilde{\mathbb{S}}(P)$ and $\tilde{\mathbb{F}}(P)$.*

Proof. By induction on the definition of $\tilde{\mathbb{S}}(P)$ or $\tilde{\mathbb{F}}(P)$ respectively. $\square$

Definition 17. Let P be a propositional logic program. The set $\tilde{\mathbb{T}}(P)$ is inductively generated by the following four rules:

T1. $\varnothing \in \tilde{\mathbb{T}}(P)$.

T2. If $P(A) = (G_i)_{1 \leq i \leq m}$ and $G_i * H \in \tilde{\mathbb{T}}(P)$ for $1 \leq i \leq m$, then $A * H \in \tilde{\mathbb{T}}(P)$.

T3. If $A \in \tilde{\mathbb{T}}(P)$ and $A \in \tilde{\mathbb{S}}(P)$, then $\neg A * H \in \tilde{\mathbb{T}}(P)$.

T4. If $A \in \tilde{\mathbb{T}}(P)$, $A \in \tilde{\mathbb{F}}(P)$ and $H \in \tilde{\mathbb{T}}(P)$, then $\neg A * H \in \tilde{\mathbb{T}}(P)$.

The set $\tilde{\mathbb{T}}(P)$ is the set of left-terminating goals in the sense of [3].

Lemma 18. *Let P be a propositional logic program and let G and H be goals.*

1. *If $G \in \tilde{\mathbb{T}}(P)$ and $H \in \tilde{\mathbb{T}}(P)$ then $G * H \in \tilde{\mathbb{T}}(P)$.*
2. *If $G \in \tilde{\mathbb{T}}(P)$ then $G \in \tilde{\mathbb{S}}(P)$ or $G \in \tilde{\mathbb{F}}(P)$.*
3. *If $G \in \tilde{\mathbb{T}}(P)$ and $G \in \tilde{\mathbb{F}}(P)$ then $G * H \in \tilde{\mathbb{T}}(P)$.*
4. *If $G * H \in \tilde{\mathbb{T}}(P)$ then $G \in \tilde{\mathbb{T}}(P)$ and ($G \in \tilde{\mathbb{F}}(P)$ or $H \in \tilde{\mathbb{T}}(P)$).*

Proof. By induction on the definition of $\tilde{\mathbb{T}}(P)$. $\square$

In the same way as we have proved soundness and completeness of Prolog with respect to $prl(P)$ one can prove the following theorem.

Theorem 19. *Let P be a propositional logic program and let G be a goal. Then we have:*

1. $\ell comp(P) \models \tilde{\mathbf{S}}\, G \iff G \in \tilde{\mathbb{S}}(P)$.
2. $\ell comp(P) \models \tilde{\mathbf{F}}\, G \iff G \in \tilde{\mathbb{F}}(P)$.
3. $\ell comp(P) \models \tilde{\mathbf{T}}\, G \iff G \in \tilde{\mathbb{T}}(P)$.

The relationships between the Prolog based sets $\mathbb{S}(P)$, $\mathbb{F}(P)$, $\mathbb{T}(P)$ and the (S)LDNF-sets $\tilde{\mathbb{S}}(P)$, $\tilde{\mathbb{F}}(P)$, $\tilde{\mathbb{T}}(P)$ are stated in the next lemma.

Lemma 20. *Let P be a propositional logic program and let G be a goal.*

1. $\mathbb{S}(P) \subseteq \tilde{\mathbb{S}}(P)$, $\mathbb{F}(P) \subseteq \tilde{\mathbb{F}}(P)$, $\tilde{\mathbb{T}}(P) \subseteq \mathbb{T}(P)$.
2. $\tilde{\mathbb{S}}(P) \cap \tilde{\mathbb{T}}(P) \subseteq \mathbb{S}(P)$, $\tilde{\mathbb{F}}(P) \cap \tilde{\mathbb{T}}(P) \subseteq \mathbb{F}(P)$.

Proof. Assertion 1 is proved by induction on the definitions of $\mathbb{S}(P)$, $\mathbb{F}(P)$ and $\tilde{\mathbb{T}}(P)$ respectively. Assertion 2 follows from 1. Assume that G is in $\tilde{\mathbb{S}}(P)$ and $\tilde{\mathbb{T}}(P)$. By 1, G is also in $\mathbb{T}(P)$. By Lemma 4(2), G is in $\mathbb{S}(P)$ or in $\mathbb{F}(P)$. If it is in $\mathbb{F}(P)$, then by 1 it is in $\tilde{\mathbb{F}}(P)$. This is not possible by Lemma 16(5). Therefore G is in $\mathbb{S}(P)$. □

The second assertion of this lemma is used in Theorem 14 on the level of predicate logic. It says that if G succeeds (resp. fails) in SLDNF-resolution and G is left-terminating (strongly terminating in LDNF-resolution) then G succeeds (resp. fails) in Prolog.

8 Conclusion

We have presented in this article a sound and complete axiomatization of propositional Prolog in terms of classical propositional logic. To every propositional logic program P (with negation) we have associated a propositional theory $prl(P)$ in an extended language where we have for every propositional atom A new atoms A^s, A^f and A^t expressing success, failure and termination of the atom. Three syntactic operators $\mathbf{S}$, $\mathbf{F}$ and $\mathbf{T}$ have been introduced which transform a goal G into formulas $\mathbf{S}\,G$, $\mathbf{F}\,G$ and $\mathbf{T}\,G$ such that $\mathbf{S}\,G$ (resp. $\mathbf{F}\,G$ or $\mathbf{T}\,G$) is a consequence $prl(P)$ if and only if G succeeds (resp. fails or terminates) in Prolog.

Unfortunately, the obvious generalization of this approach to predicate logic yields a theory for which Prolog is not complete. By changing the interpretation of the termination operator $\mathbf{T}$, however, one obtains a theory with has better properties. This theory is called the ℓ-completion of a logic program and is investigated further in [17].

One could formalize Prolog in a language with variables σ, τ for substitutions. One of the basic relations would then be $A\,\mathbf{R}\,\sigma$ with the interpretation that the atom A returns the answer substitution σ. Since one has then to axiomatize also the theory of substitutions, this approach will become rather complicated. We think that the overhead is not necessary, since one can treat a lot of (pure declarative) Prolog programs already using the much simpler ℓ-completion. If one wants to reason about programs that use all features of full Prolog then one has to use other methods like, for instance, the framework of dynamic algebras in [4].

References

1. J. Andrews. A logical semantics for depth-first Prolog with ground negation. Technical Report CSS/LCCR TR93-10, Centre for Systems Science, Simon Fraser University, 1993.
2. K. R. Apt. Declarative programming in Prolog. In D. Miller, editor, *Logic Programming — Proceedings of the 1993 International Symposium*, pages 11–35. MIT Press, 1993.
3. K. R. Apt and D. Pedreschi. Reasoning about termination of pure Prolog programs. *Information and Computation*, 106(1):109–157, 1993.
4. E. Börger and D. Rosenzweig. A mathematical definition of full Prolog. *Science of Computer Programming*, 1993. To appear.
5. S. Cerrito. A linear axiomatization of negation as failure. *J. of Logic Programming*, 12(1):1–24, 1992.
6. K. L. Clark. Negation as failure. In H. Gallaire and J. Minker, editors, *Logic and Data Bases*, pages 293–322. Plenum Press, New York, 1978.
7. B. Elbl. *Deklarative Semantik von Logikprogrammen mit PROLOGs Auswertungsstrategie*. PhD thesis, Universität der Bundeswehr, München, Germany, 1994.
8. M. Fitting. A Kripke-Kleene semantics for logic programs. *J. of Logic Programming*, 2:295–312, 1985.
9. G. Jäger. Non-monotonic reasoning by axiomatic extensions. In J. E. Fenstad, I. T. Frolov, and R. Hilpinen, editors, *Logic, Methodology and Philosophy of Science VIII*, pages 93–110, Amsterdam, 1989. North-Holland.
10. G. Jäger and R. F. Stärk. A proof-theoretic framework for logic programming. In S. Buss, editor, *Handbook of Proof Theory*. 1994. In Preparation.
11. M. Kalsbeek. Gentzen systems for logic programming styles. Technical Report CT-94-12, ILLC, University of Amsterdam, 1994.
12. K. Kunen. Negation in logic programming. *J. of Logic Programming*, 4(4):289–308, 1987.
13. G. E. Mints. Complete calculus for pure Prolog. *Proc. Acad. Sci. Estonian SSR*, 35(4):367–380, 1986. In Russian.
14. J. C. Shepherdson. Mints type deductive calculi for logic programming. *Annals of Pure and Applied Logic*, 56(1–3):7–17, 1992.
15. R. F. Stärk. Input/output dependencies of normal logic programs. *J. of Logic and Computation*, 4(3):249–262, 1994.
16. R. F. Stärk. The declarative semantics of the Prolog selection rule. In *Proceedings of the Ninth Annual IEEE Symposium on Logic in Computer Science, LICS '94*, pages 252–261, Paris, France, July 1994. IEEE Computer Society Press.
17. R. F. Stärk. First-order theories for pure Prolog programs with negation. *Archive for Mathematical Logic*, 199? To appear.
18. J. van Benthem. Logic as programming. *Fundamenta Informaticae*, 17(4):285–317, 1993.
19. A. Van Gelder and J. S. Schlipf. Commonsense axiomatizations for logic programs. *J. of Logic Programming*, 17(2,3,4):161–195, 1993.

Nonmonotonic Inheritance, Argumentation and Logic Programming

Phan Minh Dung and Tran Cao Son

Computer Science Program
School of Advanced Technology
Asian Institute of Technology
G.P.O Box 2754, Bangkok 10501, Thailand
Email: dung@cs.ait.ac.th, tson@emailhost.ait.ac.th

Abstract. We study the conceptual relationship between the semantics of nonmonotonic inheritance reasoning and argumentation. We show that the credulous semantics of nonmonotonic inheritance network can be captured by the stable semantics of argumentation. We present a transformation of nonmonotonic inheritance networks into equivalent extended logic programs.

1 Introduction

Argument-based approaches to nonmonotonic reasoning have been intensively studied and became prominent in AI and Logic Programming [6, 20, 23, 2, 15] just recently. But reasoning based on arguments represented as paths, has been studied in nonmonotonic inheritance reasoning, a specific field of nonmonotonic reasoning, from the very first day [29] and then in [13, 14, 26, 27, 28, 25, 24, 11]. Path-based reasoning approaches to nonmonotonic inheritance networks are widely accepted because they are intuitive and easy to implement.

The interesting and surprising problem here is that the argument-based semantics of nonmonotonic inheritance network [13, 14, 26, 27, 28, 25, 24, 11] and the general argumentation reasoning [6, 20, 23, 2, 15] seem to have conceptually little in common. Touretzky et al. went so far to claim that one of the fundamental principle of argumentation - *the use of reinstater* - can not be applied in nonmonotonic inheritance reasoning [28].

The relation between nonmonotonic inheritance reasoning and more general frameworks to nonmonotonic reasoning like default logic, autoepistemic logic, logic programming has been intensively studied in [3, 4, 12, 9, 10, 22, 16]. The basic idea of these works is to find a way to translate a nonmonotonic inheritance network into a "equivalent" theory of the respected nonmonotonic logic. But all of these transformations do not preserve the original semantics of nonmonotonic inheritance networks. Hence, conceptually, the relationship between the natural path-based semantics of nonmonotonic inheritance networks and other more general nonmonotonic logics such as default logic, autoepistemic logic, etc. remains unclear. The goal of this paper is to address this problem. We do that by studying the relationship between the argumentation framework given in [2] and inheritance networks.

We show that each acyclic, consistent nonmonotonic inheritance network Γ can be viewed as an argumentation framework AF_Γ so that the credulous semantics of Γ "coincides" with the stable semantics of AF_Γ. Further, we prove that the grounded semantics of AF_Γ is contained in the skeptical semantics of Γ [13, 25]. Thus, we can say that grounded semantics provides the baseline of skepticism in inheritance reasoning.

We present a transformation of consistent nonmonotonic networks into extended logic programs and show that the credulous semantics of the former coincides with the answer set semantics [8] of the latter. To our knowledge, this is the first transformation of nonmonotonic inheritance network into other more general nonmonotonic logics preserving the semantics of nonmontonic inheritance networks.

2 Preliminaries

2.1 Inheritance network

A defeasible inheritance network Γ is defined here as a finite collection of positive and negative direct links between nodes. If x, y are nodes then $x \to y$ (resp. $x \not\to y$) represents a positive (resp. negative) *direct link* from x to y. A network Γ is *consistent* if there exist no two nodes x, y such that both $x \to y$ and $x \not\to y$ belong to Γ. A *positive path* from x_1 to x_n through $x_2, ..., x_{n-1}$, denoted by $\pi(x_1, \sigma, x_n)$ or $\pi(x_1, x_2, ..., x_{n-1}, x_n)$, is a sequence of direct links $x_1 \to x_2$, $x_2 \to x_3, ... , x_{n-1} \to x_n$. Similarly, a *negative path* from x_1 to x_n through x_2, ..., x_{n-1}, denoted by $\bar{\pi}(x_1, \sigma, x_n)$ or $\bar{\pi}(x_1, x_2, ..., x_{n-1}, x_n)$, is a sequence of direct links $x_1 \to x_2$, $x_2 \to x_3, ... , x_{n-1} \not\to x_n$. A generalized path is a sequence of direct links (x_1, x_2), $(x_2, x_3), ... , (x_{n-1}, x_n)$, where (x_i, x_{i+1}) denotes a positive or negative direct link. Γ is *acyclic* if there is no generalized path (x_1, x_2), $(x_2, x_3), ... , (x_{n-1}, x_n)$ with $x_1 = x_n$. The degree of the path $\alpha = \pi(x, \sigma, y)$ (resp. $\alpha = \bar{\pi}(x, \sigma, y)$), denoted by $deg_\Gamma(\alpha)$, is defined as the length (number of edges) of the longest generalized path from x to y. Furthermore, we also use the notation $\pi(x_1, \sigma, x_{n-1}) \to x_n$ (resp. $\pi(x_1, \sigma, x_{n-1}) \not\to x_n$) to denote the path $\pi(x_1, x_2, ..., x_{n-1}, x_n)$ (resp. $\bar{\pi}(x_1, x_2, ..., x_{n-1}, x_n)$).

From now on we will use Γ to denote an arbitrary but fixed network and Φ to denote a set of paths in Γ if no confusion is possible. The notion of inheritability presented here relies on three concepts: constructibility, conflict, and preemption. Their definitions are taken from [14, 28, 13, 24, 25].

Definition 1. A positive path $\pi(x, \sigma, u) \to y$ is *constructible* in Φ iff $\pi(x, \sigma, u) \in \Phi$ and $u \to y \in \Gamma$. A negative path $\pi(x, \sigma, u) \not\to y$ is constructible in Φ iff $\pi(x, \sigma, u) \in \Phi$ and $u \not\to y \in \Gamma$.

Definition 2. $\pi(x, \sigma, y)$ *conflicts* with any path of the form $\bar{\pi}(x, \tau, y)$ and vice versa. A path σ is conflicted in Φ iff Φ contains a path that conflicts with σ.

Different ways have been proposed to define defeasible preemption in Φ [10, 14, 13, 27, 25, 24]. Here, we follow the *off-path* preemption given in [13].

Definition 3. A positive path $\pi(x, \sigma, u) \to y$ (see figure 1) is *preempted* in Φ iff there is a node v such that (i) $v \not\to y \in \Gamma$ and (ii) either v=x or there is a path of the form $\pi(x, \tau_1, v, \tau_2, u) \in \Phi$. A negative path $\pi(x, \sigma, u) \not\to y$ is *preempted* in Φ iff there is a node v such that (i) $v \to y \in \Gamma$ and (ii) either v=x or there is a path of the form $\pi(x, \tau_1, v, \tau_2, u) \in \Phi$.

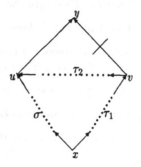

Fig. 1. $\pi(x, \sigma, u) \to y$ is preempted

The credulous semantics of an inheritance network is given in the following definition.

Definition 4. [13] σ is *defeasibly inheritable* in Φ, written as $\Phi \mathrel{\mid\!\sim} \sigma$, iff
either σ is a direct link
or σ is a compound path, $\sigma = \pi(x, \tau, y)$ (likewise for negative path) such that
(i) σ is constructible in Φ, and
(ii) σ is not conflicted in Φ, and
(iii) σ is not preempted in Φ.

Definition 5. A set Φ of paths is a *credulous extension* of the net Γ iff $\Phi = \{ \sigma : \Phi \mathrel{\mid\!\sim} \sigma \}$.

The skeptical semantics for inheritance network is defined by the notion of ideally skeptical extension and is defined as follows.

Definition 6. [25, 24] The intersection of all credulous extensions of Γ is called the *ideally skeptical extension* of Γ.

2.2 Argumentation framework

In the following section, the basics of the abstract theory of argumentation framework of Dung [2] is recalled.

Definition 7. An *argumentation framework* is a pair $AF=<AR, attacks>$, where AR is a set of arguments, and $attacks \in AR \times AR$.

If $(A, B) \in attacks$ we say A attacks B or B is attacked by A. A set of arguments S is said to be attacked by an argument A if there is $B \in S$ such that $(A, B) \in attacks$. Similarly, we say S attacks A if there is $B \in S$ such that $(B, A) \in attacks$.

Definition 8. A set of arguments S is said to be *conflict-free* if there exist no two arguments A,B in S such that $(A, B) \in attacks$.

The stable semantics of AF is defined as follows.

Definition 9. A conflict-free set of arguments S is called a *stable extension* of AF if S attacks every argument which does not belong to S.

It is easy to see that the following proposition holds.

Proposition. *S is stable iff S={A | A is not attacked by S}.*

The stable semantics of argumentation framework captures the semantics of many other mainstream approaches to nonmonotonic reasoning such as extension of Reiter's Default Logic [21], stable expansion of Autoepicstemic Logic [19], and stable model of Logic Programming [7]. We will see in section 3 that the credulous semantics of an inheritance network coincides with the stable semantics of a corresponding argumentation framework.

Often, a more restricted form of skeptical semantics is advocated in many approaches to nonmonotonic reasoning [20, 30]. This form of skeptical semantics is defined in the argumentation framework by the notion of grounded extension defined as the least fixpoint of the following operator.

$F_{AF} : 2^{AR} \to 2^{AR}$, where

$F_{AF}(S) = \{A \mid \forall B,$ if B attacks A, then $\exists C \in S$ such that C attacks B$\}$.

The idea behind this operator will become clear in the next definition.

Definition 10. An argument A is *defendable* wrt S iff for every argument B, if B attacks A, then there is an argument C in S such that C attacks B.

So, we can redefine F_{AF} by $F_{AF}(S) = \{A \mid A$ is defendable wrt S$\}$.

The grounded extension of an argumentation framework is defined next.

Definition 11. The *grounded extension* of an argumentation framework AF denoted by GE_{AF} is the least fixpoint of F_{AF}.

It has been pointed out in [2] that both the semantics, Pollock's Inductive Defeasible Logic [20], and well-founded semantics of Logic Programming [30] are captured by the grounded extension of argumentation framework.

The maximal fixpoint of F_{AF} are called the preferred extension of AF. In general, stable extension are preferred extension but not vice versa. But as we will see later, for any argumentation framework corresponding to inheritance networks, stable semantics and preferred semantics coincide. So it is enough for us to work only with stable semantic.

3 Inheritance Networks as Argumentation Frameworks

Our goal is to clarify the conceptual relationship between the semantics of non-monotonic inheritance networks and the semantics of argumentation frameworks. This will also help to illuminate the conceptual relationship between the semantics of nonmonotonic inheritance networks and other general approaches to nonmonotonic reasoning due to a result of Dung [2] showing that many general approaches to nonmonotonic reasoning [21, 19, 30] can be seen as special cases of argumentation frameworks.

We will show that every nonmonotonic inheritance network Γ can be considered as an argumentation framework $AF_\Gamma = <AR_\Gamma, attacks_\Gamma>$ such that the credulous semantics of Γ coincides with the stable semantics of $AF_\Gamma = <AR_\Gamma, attacks_\Gamma>$ in the sense that every credulous extension of Γ is a stable extension of AF_Γ and vice versa.

A path $\alpha = \pi(x, \sigma, u)$ is called a *prefix* of path $\beta = \pi(x, \sigma, u, \tau, v)$ in Γ. The set of all prefixes of β is denoted by $pre(\beta)$. α is a *proper prefix* of β iff $\alpha \in pre(\beta)$ and $\alpha \neq \beta$.

First, it is intuitive to view any path of Γ as an argument. So, we have $AR_\Gamma = \{\sigma | \sigma$ is a path in $\Gamma\}$.

As next we define the attacks relation of AF_Γ. The underlying principle in defining the attacks relation is that more specific information overrides less specific one. For two conflicted paths, σ and τ, there are following cases:

(i) σ is a direct link. It is clear that we should let σ attack τ, but not vice versa.

(ii) σ and τ are compound paths. In this case, neither σ nor τ are more specific than the other path. Thus, we have: σ attacks τ and vice versa.

Further, it should be also clear that if σ attacks a prefix α of τ then σ attacks τ.

We now consider another kind of attack.

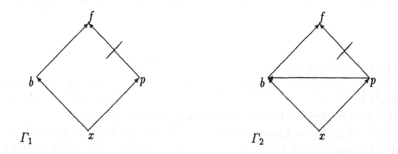

Fig. 2. Motivation of Attack Definition

Example 1. Consider the inheritance network Γ_1 in figure 2. The paths $\sigma = \pi(x, b, f)$ and $\tau = \bar{\pi}(x, p, f)$ are in conflict. Thus, $\pi(x, b, f)$ attacks $\bar{\pi}(x, p, f)$ and vice versa as in case (ii). So, $AF_{\Gamma_1} = <AR_{\Gamma_1}, attacks_{\Gamma_1}>$ with $attacks_{\Gamma_1} = \{(\sigma, \tau), (\tau, \sigma)\}$. Hence AF_{Γ_1} has two stable extensions corresponding to two credulous extensions of $\Gamma_1 : E_1 = \Gamma \cup \{\sigma\}$ and $E_2 = \Gamma \cup \{\tau\}$.

The network Γ_2 in figure 2 is received from the network Γ_1 by adding the positive link $p \rightarrow b$. This is the well-known Penguin-Bird-Fly example. It is clear that in $AF_{\Gamma_2} = <AR_{\Gamma_2}, attacks_{\Gamma_2}>$, σ attacks τ and τ attacks σ as in case of Γ_1. Adding the link $p \rightarrow b$ into Γ_1 makes the argument τ more specific than the argument σ. Thus, due to the principle that more specific information can override less specific one, we can say that adding $p \rightarrow b$ to Γ_1 creates a attack of new kind against $\sigma = \pi(x, b, f)$. We can represent this by viewing the argument $\alpha = \pi(x, p, b)$ in the presence of the link $p \nrightarrow f$ as an attack against the path $\sigma = \pi(x, b, f)$.

These motivations lead to the following definition of attacks.

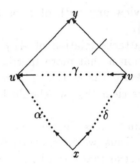

Fig. 3. $\pi(x, \alpha, u) \rightarrow y$ is attacked by $\pi(x, \delta, v, \gamma, u)$ in presence of $v \nrightarrow y$

Definition 12. A path σ attacks path τ iff

(a) σ is a direct link $x \rightarrow y$ (resp. $x \nrightarrow y$) and $\tau = \bar{\pi}(x, \alpha, y)$ (resp. $\tau = \pi(x, \alpha, y)$) or

(b) σ is in conflict with some compound path $\delta \in pre(\tau)$ or

(c) σ, τ are compound paths where there exists a prefix $\beta = \pi(x, \alpha, u) \rightarrow y$ of τ (resp. $\pi(x, \alpha, u) \nrightarrow y$) and $\sigma = \pi(x, \delta, v, \gamma, u)$ with $v \nrightarrow y \in \Gamma$ (resp. $v \rightarrow y \in \Gamma$) (see figure 3).

Remark. From now on we will refer to three types of attacks (a), (b), and (c) defined above as attack by direct link, by conflict, and by preemption, respectively.

So, in our point of view there are two kinds of attacks between two paths σ and τ, symmetric (τ attacks σ and σ attacks τ) and asymmetric (τ attacks σ but not vice versa). Symmetric attacks are equivalent to conflictor in [28] while asymmetric attacks have some similarity to preemptor in [28] but not identical.

Example 2. (Continuation of example 1) For Γ_2 in figure 2 we have $AF_{\Gamma_2}=<AR_{\Gamma_2}, attacks_{\Gamma_2}>$ with

$$AR_{\Gamma_2} = \Gamma_2 \cup \{\sigma, \tau, \alpha, \beta, \delta\} \text{ and}$$

$$attacks_{\Gamma_2} = \{(p \not\rightarrow f, \delta)\} \cup \{(\sigma, \tau), (\tau, \sigma), (\tau, \beta), (\beta, \tau)\} \cup \{(\alpha, \sigma), (\alpha, \beta)\}$$

where $\beta = \pi(x, p, b, f)$, and $\delta = \pi(p, b, f)$. Here, $\{(p \not\rightarrow f, \delta)\}$ is the set of attacks by direct link, $\{(\sigma, \tau), (\tau, \sigma), (\tau, \beta), (\beta, \tau)\}$ is the set of attacks by conflict, and $\{(\alpha, \sigma), (\alpha, \beta)\}$ is the set of attacks by preemption. □

We now prove the coincidence between the credulous extension of Γ and the stable extension of AF_Γ.

Theorem 13. *Let E be a set of paths in Γ. Then, E is a stable extension of AF_Γ iff E is a credulous extension of Γ.*

Proof. See Appendix. □

We give now the definition of the grounded skeptical semantics for an inheritance network.

Definition 14. The grounded skeptical semantics of the inheritance network Γ is defined as the grounded extension GE_{AF_Γ} of the corresponding argumentation framework AF_Γ of Γ.

Since GE_{AF_Γ} is contained in every stable extension of AF_Γ we have the following theorem.

Theorem 15. *The grounded extension GE_{AF_Γ} of AF_Γ is a subset of the ideally skeptical extension of Γ.*

Proof. Since GE_{AF_Γ} is the smallest complete extension of AF_Γ and the complete extensions form a complete semilattice wrt set conclusions [2] we have that GE_{AF_Γ} is contained in every stable extension of AF_Γ. Thus, GE_{AF_Γ} is contained in their intersection which is the ideally skeptical extension. □

4 Transforming Inheritance Network into Logic Program

The coincidence between the credulous semantics of an inheritance network Γ and the stable semantics of the corresponding argumentation framework AF_Γ together with the results in [2] stating that argumentation frameworks in principle can be represented as logic programs, points out that an inheritance network Γ can be transformed into an equivalent logic program P_Γ. Thus, proof procedures based on negation-as-failure can be applied to P_Γ to compute the credulous semantics of Γ.

In this section we transform an inheritance network Γ into an extended logic program P_Γ and show that the credulous semantics of Γ coincides with the answer set semantics of P_Γ[1].

In following we assume that the readers are family with the answer set semantics of Logic Programs [8].

The set of nodes of Γ is the union of two disjoint sets, the set of individuals of Γ, denotes by I_Γ, consists of all nodes x of Γ such that there exists no direct link $y \rightarrow x$ or $y \not\rightarrow x$ in Γ, and the set of predicate (or properties) nodes. In following, a, b, c, ... will represent the individuals of Γ and p, q, r, ... are the predicate nodes if not otherwise specified. For example, in the example 3, a denotes the individual Nixon, and p, r and q denote the predicates Pacifist, Republican and Quaker, respectively.

We first assign an unique natural number $j \in N$ to each direct link $p \rightarrow q \in \Gamma$ (resp. $p \not\rightarrow q \in \Gamma$) of Γ, $p \notin I_\Gamma$, written as $p \rightarrow_j q$ (resp. $p \not\rightarrow_j q$), and introduce a new predicate ab_j representing the abnormal-literal at the edge j. The link $p \rightarrow_j q$ (or $p \not\rightarrow_j q$) is then referred simply as the link j. Based on the attack relationship of AF_Γ the inheritance network Γ can be transformed into an extended logic program P_Γ as follows.

As in case of attack by direct link, any direct link $a \rightarrow p$ (resp. $a \not\rightarrow p$) beginning from a fact node a can be transformed directly into a fact of P_Γ because of there is no arguments which attack $a \rightarrow p$ (resp. $a \not\rightarrow p$). Hence, we have:

(i) For each $a \in I_\Gamma$ if $a \rightarrow p$ (resp. $a \not\rightarrow p$) is in Γ then

$$p(a) \leftarrow \text{(resp. } \neg p(a) \leftarrow \text{)}$$

is a clause of P_Γ.

(ii) For $p \notin I_\Gamma$ and each direct link $p \rightarrow_j q \in \Gamma$, the two clauses

$$q(x) \leftarrow p(x), \; not \; ab_j(x) \quad \text{and}$$
$$ab_j(x) \leftarrow \neg q(x)$$

belong to P_Γ.

---

[1] Grégoire [10] presented an algorithm for transformation of an inheritance network into a logic program. But in our view this could hardly be considered as a transformation because according to the algorithm we first have to compute the extensions of the network and then define a logic program specifying this extension.

Similarly,

(iii) For $p \notin I_\Gamma$ and each direct link $p \not\rightarrow_j q \in \Gamma$

$$\neg q(x) \leftarrow p(x), \ not\ ab_j(x) \quad \text{and}$$
$$ab_j(x) \leftarrow q(x)$$

are clauses of P_Γ.

(iv) For each pair of direct links $p \rightarrow_j q$ and $r \not\rightarrow_k q$ in Γ
(a) If there exists a positive path from p to r over the links $j_1, ..., j_n$ then the clause

$$ab_k(x) \leftarrow p(x), \ not\ ab_{j_1}(x), ..., \ not\ ab_{j_n}(x)$$

belongs to P_Γ.
(b) If there exists a positive path from r to p over the links $j_1, ..., j_n$ then the clause

$$ab_j(x) \leftarrow r(x), \ not\ ab_{j_1}(x), ..., \ not\ ab_{j_n}(x)$$

belongs to P_Γ. □
We illustrate the transformation from Γ into P_Γ in the next two examples.

Example 3. (Nixon-Diamond)

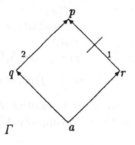

Fig. 4. a=Nixon, p=Pacifist, q=Quaker, r=Republican

The corresponding program P_Γ of Γ in the figure 4 is:

$a \rightarrow r$	$r(a) \leftarrow$
$a \rightarrow q$	$q(a) \leftarrow$
$r \not\rightarrow_1 p$	$\neg p(x) \leftarrow r(x), \ not\ ab_1(x)$
	$ab_1(x) \leftarrow p(x)$
$q \rightarrow_2 p$	$p(x) \leftarrow q(x), \ not\ ab_2(x)$
	$ab_2(x) \leftarrow \neg p(x)$

It is easy to see that P_Γ has only two answer sets $\{r(a), q(a), p(a), ab_1(a)\}$ and $\{r(a), q(a), \neg p(a), ab_2(a)\}$. □

Example 4. (Penguin-Bird-Fly)

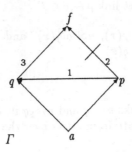

Fig. 5. a=Tweety, p=Penguin, q=Bird, f=Fly

Let consider the net Γ in figure 5. The corresponding program P_Γ consists of

$$
\begin{array}{ll}
a \rightarrow p & p(a) \leftarrow \\
a \rightarrow q & q(a) \leftarrow \\
p \not\rightarrow_2 f & \neg f(x) \leftarrow p(x), \ not\ ab_2(x) \\
& ab_2(x) \leftarrow f(x) \\
q \rightarrow_3 f & f(x) \leftarrow q(x), \ not\ ab_3(x) \\
& ab_3(x) \leftarrow \neg f(x) \\
p \rightarrow_1 q & q(x) \leftarrow p(x), \ not\ ab_1(x) \\
& ab_1(x) \leftarrow \neg q(x) \\
\text{and} & ab_3(x) \leftarrow p(x), \ not\ ab_1(x)
\end{array}
$$

The unique answer set of P_Γ is $\{p(a), q(a), \neg f(a), ab_3(a)\}$. ☐

The relationship between the answer set semantics of P_Γ and the credulous semantics of Γ can be established in the following way. First, for any set of paths E in Γ we define:

$M_E = \{q(a) \mid$ there is a positive path from $a \in I_\Gamma$ to q in E $\} \cup \{\neg q(a) \mid$ there is a negative path from $a \in I_\Gamma$ to q in E $\}$.

The next theorem point out the relationship between the answer set semantics of the program P_Γ and the credulous semantics of the inheritance network Γ.

Theorem 16. *M is an answer set of P_Γ iff there exists a credulous extension E of Γ such that $M_E = M \backslash AB$ where AB denotes the set of all grounded instance of abnormal-predicates in P_Γ.* ☐

5 Conclusion

We have studied the relationship between semantical concepts of nonmonotonic inheritance and of argumentation framework. In chapter 3 we showed that argumentation framework, a general form of argument-based reasoning can be applied to specify nonmonotonic inheritance reasoning in a simple way. We proved that a consistent and acyclic network can be viewed as an argumentation framework so that the credulous semantics of the former coincides with the stable semantics of the latter.

The capturing of credulous semantics of nonmonotonic inheritance by argumentation framework shows that argumentation can be applied successfully to formulize nonmonotonic inheritance reasoning. It is interesting to notice that many new developed approaches to nonmonotonic reasoning [6, 23, 1] can not applied directly to inheritance reasoning as we did for Dung's argumentation framework. Geffner and Pearl's conditional entailment [6] is too weak as we can not draw the conclusion "Tweety is an animal with feather" if we replace the rule "Bird fly" by two rules "Bird are animals with feather" and "Animal with feather fly" in the Penguin-Bird-Fly example. Simary and Loui's defeasible reasoning [23] gives unintuitive answer even in simple cases as in the Penguin-Bird-Fly example given that Tweety is a penguin and a bird. Delgrande and Schaub's general approach [1] cannot give proper answer in the example with $R = \{a \rightarrow r, a \rightarrow p, a \rightarrow \neg q, p \rightarrow q, q \rightarrow \neg s, q \rightarrow r, r \rightarrow s\}$ (figure 18, page 158 [13]).

Acknowledgement

The authors thank Huynh Ngoc Phien for his support during the time in which this paper was written. Thanks also to J.J.Alferes, L.M.Pereira, and P.Mancarella for the many discussions and comments. Many thanks to C.Aravindan who has helped preparing the LATEX version of this paper. The first author is partially supported by EEC Keep in Touch activity KIT011.

References

1. Delgrande, J.P., Schaub, T.H.: A General Approach to Specificity in Default Reasoning, Knowledge and Representation (1994).
2. Dung, P.M.: On the acceptability of arguments and its fundamental role in nonmonotonic reasoning and logic programming and N-person game. AI Vol. **76** 2 (1995) (An extended abstract of this paper can be found in the proceeding of IJCAI, 1993)
3. Etherington, D.: Reasoning with Incomplete Information, Pitman, London. (1988)
4. Etherington, D., Reiter, R.: On Inheritance Hierarchies with Exceptions in Proc. of AAAI-83 (1983) 104–108
5. Geerts P., Vermeir D.: A Nonmonotonic Reasoning Formalism using implicit specificity information. Proceeding of LPNMR'**93**, 380–396
6. Geffner H., Pearl, J.: Conditional entailment: bridging two approaches to default reasoning, Elsevier, AI Vol. **53**, (1992) 209–244
7. Gelfond, M., Lipschitz, V.: The stable model semantics for logic programs, Proceeding of the 5th ICLP, MIT Press, (1988), 1070 – 1079
8. Gelfond, M., Lipschitz, V.: Logic Programs with Classical Negation, Proceeding of the 7th ICLP, MIT Press, (1990), 579 –597

327

9. Gelfond M., Przymusinska H.: Formalization of Inheritance Reasoning in Autoepistemic Logic, Fundamental Informaticae XIII. (1990) 403–443
10. Grégoire, E.: Skeptical Theories of Inheritance and Nonmonotonic Logics, Methodologies for Intelligent System 4, (1989) 430–438
11. Grégoire, E.: Skeptical Inheritance Can Be More Expressive, Proceeding of ECAI (1990) 326–332
12. Haugh, B.: Tractable theories of multiple defeasible Inheritance in Ordinary Nonmonotonic Logics, In Proc. of 7th NCAI. (1988) 421–426
13. Horty, J.F.: Some direct Theories of Non-monotonic Inheritance in Handbook of Logic and Artificial Intelligence and Logic Programming, D.Gabbay and C. Hogger, Oxford University. (1991) 111–187
14. Horty, J.F., Thomason, R.H., Touretzky, D.S.: A skeptical theory of inheritance in non-monotonic semantic networks. AI Vol. **42** (1987) 311–348
15. Kakas, A.C., Kowalski, R.A, Toni, F.: The role of Abduction in Logic Programming. J. Logic and Computations **2** No.6, (1992) 719–770
16. Lin, F.: A study of nonmonotonic reasoning, Ph.D. Dissertation, Standford University. (1991)
17. Matkinson, D., Schlechta K.: Floating conclusion and zombie paths: two deep difficulties in the 'directly skeptical' approach to defeasible inheritance nets. AI Vol. **48** (1980) 99–209
18. McDermott, D., Doyle, J.: Nonmonotonic Logic I. AI, **13** 41–72
19. Moore, R.C.: Semantical Considerations on Nonmonotonic Logics, in Readings in Nonmonotonic Reasoning, Edited by M. L. Ginsberg, Morgan Kaufmann Publishers, Inc., Los Altos, California. (1988) 127–137
20. Pollock, J.L.: Defeasible reasoning. Cognitive Science **17** (1987) 481–518
21. Reiter R.: A Logic for Default Reasoning in Readings in Nonmonotonic Reasoning, Edited by M. L. Ginsberg, Morgan Kaufmann Publishers, Inc., Los Altos, California (1987) 68–93
22. Reiter R., Criscuolo G.: On interacting defaults in Readings in Nonmonotonic Reasoning, Edited by M. L. Ginsberg, Morgan Kaufmann Publishers, Inc., Los Altos, California (1987) 94–100
23. Simary, G.R., Loui R.P.: A mathematical treatment of defeasible reasoning and its implementation. AI **53** (1992) 125–257
24. Simonet, G.: RS theory: a really skeptical theory of inheritance with exceptions, Proceeding of ECAI (1990) 615–626.
25. Stein, L.A.: Resolving ambiguity in non-monotonic inheritance hierarchies. AI **55** (1992) 259–310
26. Thomason, R.H., Horty, J.F: Logics for Inheritance Theory, 2nd Workshop on nonmonotonic reasoning, 220–237
27. Touretzky, D.S., Horty, J.F., Thomason, R.H.: A clash of Intuition: The current state of Non-monotonic Multiple Inheritance Systems. Proceeding of IJCAI (1987) 476–482
28. Touretzky, D.S., Horty, J.F., Thomason, R.H.: A skeptic's Menagerie: Conflictors, Preemptors, Reinstaters, and zombies in nonmonotonic Inheritance. Proceeding of IJCAI, (1991) 478–483
29. Touretzky, D. S.: The mathematics of Inheritance Systems, Morgan Kaufmann Publishers, Inc., Los Altos, California, (1986).
30. Van Gelder A., Ross, K., Schlipf J.S.: Unfounded sets and well-founded semantics for general logic programs. Proceeding of PODS 1988.

Appendix: Proof of selected theorems

In this section we prove that the credulous semantics of Γ and the stable semantics of the corresponding argumentation framework coincide. At first, we prove some general properties of AF_Γ.

Lemma 17. *Let Γ be an acyclic, consistent inheritance network and $AF_\Gamma=<AR_\Gamma, attacks_\Gamma>$ is the corresponding argumentation framework. Then, if σ attacks σ' then $deg_\Gamma(\sigma) \leq deg_\Gamma(\sigma')$.*

Proof. Consider three cases:

1. σ attacks σ' by a direct link then it is clear that $deg_\Gamma(\sigma) = deg_\Gamma(\sigma')$ because σ' and σ have the same begin and the same end nodes.
2. σ attacks σ' by conflict. Then, either σ and σ' have the same begin and end and therefore $deg_\Gamma(\sigma) \leq deg_\Gamma(\sigma')$ or there is some prefix of σ' which is conflict with σ, in this case we have $deg_\Gamma(\sigma) < deg_\Gamma(\sigma')$.
3. σ attacks σ' by preemption. Then, there is a prefix $\alpha = \pi(x, \tau, u) \to y$ (resp. $\pi(x, \tau, u) \not\to y$) of σ' such that $\sigma = \pi(x, \delta, v, \gamma, u)$ with $v \to y$ (resp. $v \not\to y$) in Γ. By definition of deg_Γ we have $deg_\Gamma(\sigma) < deg_\Gamma(\sigma')$.

The lemma is proved from these three cases. □

Lemma 18. *Let S be set of arguments and σ be an argument defendable wrt S. Then, each $\alpha \in pre(\sigma)$ is defendable wrt S.*

Proof. If $\alpha \in pre(\sigma)$ and β is an argument which attacks α then β attacks σ and therefore β is attacked by S. Hence, α is defendable wrt S. □

Lemma 19. *Let S be a set of argument in AF_Γ. If $S \mathrel{|\!\sim} \sigma$ then σ is not attacked by S.*

Proof. Assume that there exists an argument τ in S such that $(\tau, \sigma) \in attacks$. By definition of attack we have three cases:

1. τ attacks σ by direct link. Then, σ is preempted in S.
2. τ attacks σ by conflict. Then, σ is conflicted or non constructible in S.
3. τ attacks σ by preemption. Then, σ is preempted or non constructible in S. From the three cases, we learn that if $(\tau, \sigma) \in attacks$ then $S \mathrel{|\!\not\sim} \sigma$. Contradictory !!! Thus, S does not attack σ. □

The next lemma follows directly from lemma 19, and the definition of the stable extension.

Lemma 20. *If E is a stable extension of AF_Γ and $\sigma \notin E$. Then, $E \mathrel{|\!\not\sim} \sigma$.* □

Further, it is easy to prove the next two lemmae.

Lemma 21. *If E is a stable extension of AF_Γ, then $\Gamma \subset E$, and for every path $\sigma \in E$, σ is not preempted in E.* □

Lemma 22. *Let E be a credulous extension of Γ. Then σ is attacked by E if $\sigma \notin E$.*

Proof. Obviously, σ is a compound path. Without a lost of generality, assume that σ is a positive path $\pi(x, \tau, u) \rightarrow y$. Since σ does not belong to E, $E \not\vdash \sigma$. Hence,

1. σ is conflict in E. It means, there is some path α in E that is conflict with σ. Therefore, σ is attacked by E.
2. σ is not constructible in E. There is a $\alpha \in pre(\sigma)$ such that all proper prefix of α is contained in E but not α. Therefore $E \not\vdash \alpha$. Further, since α is constructible in E. For we have two sub-cases:
   α is conflicted in E. Similarly to the first case, α is attacked by E. Thus, σ is attacked by E.
   α is preempted in E. See next case.
3. σ is preempted in E. That is, there is some node v with $v \not\rightarrow y \in \Gamma$ and there is some $\delta = \pi(x, \alpha, v, \beta, u)$ in E. Thus, δ attacks σ.

From the three cases, we can conclude that σ is attacked by E. $\qquad\square$

From the lemmae 17–22 we can prove the theorem 13:

Proof of Theorem 13

Proof. 1. '$\longleftarrow$' Suppose that E is a stable extension of AF_Γ. Let $\sigma \notin E$. From lemma 19 we have $E \not\vdash \sigma$ (i). Now, let $\sigma \in E$. Any prefix of σ is defendable wrt E (Lemma 18). Thus, all prefixes of σ are containing in E. So, σ is constructible and non-conflicted in E. Further, σ is not preempted in E (Lemma 21). Hence, σ is defeasibly inheritable in E (ii). From (i) and (ii) we can conclude that E is a credulous extensions of Γ.
2. '$\longrightarrow$' Now, if E is a credulous extension of Γ then E is conflict-free (Lemma 18). It is easy to see that E attacks every argument which does not belong to E (Lemma 22). Hence, E is a stable extension of AF_Γ.
The theorem is proved. $\qquad\square$

An Abductive Framework
for Extended Logic Programming *

Antonio Brogi[1], Evelina Lamma[2], Paolo Mancarella[1], Paola Mello[3]

[1] Dipartimento di Informatica, Università di Pisa
Corso Italia 40, 56125 Pisa, Italy. {brogi,paolo}@di.unipi.it
[2] DEIS, Università di Bologna
Viale Risorgimento 2, 40136 Bologna, Italy. elamma@deis.unibo.it
[3] Dipartimento di Informatica, Università di Bari
Via Orabona 4, 70126 Bari, Italy. pmello@deis.unibo.it

Abstract. We provide a simple formulation of a framework where three main extensions of logic programming for non-monotonic reasoning are treated uniformly: Negation-by-default, explicit negation and abduction. The resulting semantics is purely model-theoretic and gives meaning to any consistent abductive logic program. Moreover, it embeds and generalises existing semantics which deal with negation and abduction separately. The abductive framework is equipped with a correct top-down proof procedure.

1 Introduction and Motivations

The need for extending pure logic programming (Horn clauses) to deal with some form of non-monotonic reasoning was recognised since the early years of logic programming. Negation-as-failure was introduced to express negative information, and it has been later recognised as a suitable form of non-monotonic reasoning in Logic Programming (LP). More recently other extensions have been proposed to further enrich the expressive power of LP as a general formalism for knowledge representation. Among others we mention *abductive* LP (e.g., [12, 16, 19]) and LP with *classical* (or better *explicit*) negation (e.g., [18, 22]).

The formalisation of these extensions has called both for new semantics capable of capturing their intended meaning and for new proof procedures to make these extensions usable in practice. For instance, even for the case of negation-as-failure itself (better *negation-by-default*), a number of different semantics characterisations have been defined from different perspectives (e.g., [17, 23, 28]). Something similar is happening for other extensions such as abductive LP (e.g., [12, 16]) and LP with explicit negation (e.g., [18, 23, 24]). In spite of the huge number of proposals, there is no general agreement on what the semantics of each extension should be. Formal comparisons among different semantics are hard to be drawn, mainly because they are often based on different grounds. However some work has been done recently in this direction (e.g., [2, 4, 7, 11, 24]).

Another crucial issue is to understand the relations among different extensions. For instance, the relation between negation-by-default and abduction has been thoroughly studied starting from the work of Eshghi and Kowalski [16]. Dung's work on

---

* The first author has been supported in part by C.N.R. grant 203.01.62.

argumentation [13, 14] represents a remarkable attempt to provide a uniform view of extended LP which encompasses the various extensions proposed so far. The intuitive notions of *argument* and *attack* between arguments may be adopted both for reconstructing a number of Non-Monotonic Reasoning (NMR) formalisms and for understanding different semantics of extended LP in a quite simple way.

The goal of this paper is to provide a simple formulation of a LP-based framework where three main extensions of LP for NMR are treated uniformly, namely negation-by-default, explicit negation and abduction. Our framework can actually be viewed as an argumentation framework in the style of Dung [13], but based on the notion of Herbrand model.

We focus on *abductive logic programs*, that is normal logic programs augmented with abduction and explicit negation. We view an abductive logic program as a positive program where both explicitly negated literals and negation-by-default literals are mapped into new positive atoms. The semantics of an abductive logic program (called *abductive semantics*) is then defined in terms of the Herbrand models of the corresponding positive program. Intuitively speaking, the idea is to restrict, in a step-wise fashion, the set of Herbrand models of the positive version of a program so as to identify the intended models of the original program. We show that our semantics captures in a uniform setting many well-known existing semantics developed both for normal programs and for logic programs with explicit negation.

Though a great deal of work has been done on the semantics of extensions of LP for NMR, relatively little effort has been devoted to the issues related to the computation and implementation of non-monotonic LP semantics. In this paper, we present a query-based proof procedure for abductive logic programs, which is derived from the proof procedure firstly presented by Eshghi and Kowalski [16] and further developed by Kakas and Mancarella [20], and by Dung [12]. The proof procedure is shown to be sound and weakly complete with respect to the abductive semantics. As a by-product, we also prove a weak completeness result for the proof procedure by Eshghi and Kowalski [16] with respect to the scenario semantics by Dung [12].

2 Abductive Logic Programs: Preliminaries

In the sequel, *abductive logic programs* will stand for normal logic programs augmented with abduction and explicit negation. Let us first set up some preliminary notions and notations. We shall use the basic concepts and terminology of logic programming (e.g., as reported by Apt [3]). The standard terminology must be extended to distinguish between negation-by-default, denoted by *not*, and explicit negation, denoted by $\neg$. Literals of the form A or $\neg A$, where A is an atom, will be referred to as *classical literals*, whereas literals of the form *not* L, where L is a classical literal, will be referred to as *default* literals.

An *abductive logic program* is a triple $\langle P, Ab, IC \rangle$ where:

- P is an extended logic program, that is, a set of clauses of the form

$$L_0 \leftarrow L_1, \ldots, L_m, not\ L_{m+1}, \ldots, not\ L_{m+n}$$

where $m, n \geq 0$ and each L_i $(i = 1, \ldots, m + n)$ is a classical literal,

- *Ab* is a set of *abducible predicates*, p, such that neither p nor $\neg p$ occurs in the head of any clause of P,
- *IC* is a set of integrity constraints, that is, a set of closed formulae (where explicit negation $\neg$ applies to atoms only).

In the following, without loss of generality, we consider only (possibly infinite) propositional programs, thus assuming that P (and *IC*) has already been instantiated. Following Eshghi and Kowalski [16], an abductive logic program $\langle P, Ab, IC \rangle$ can be transformed into its *positive version* by transforming P and *IC*. The basic idea is to view negative literals (both classical and default) as new positive atoms (as in [15, 24]).

Given a (ground) program P, let $\mathcal{L}^+$ denote the set containing all the atoms (viz. propositional symbols) A such that A, *not A*, $\neg A$, or *not* $\neg A$ occurs in P. For each $A \in \mathcal{L}^+$ three new propositional symbols are introduced to represent the explicit negation of A ($A^\neg$), the default negation of A (*not_A*), and the default negation of $\neg A$ (*not_A$^\neg$*). We will use the following notation for the sets of new propositional symbols:

$$\mathcal{L}^\neg = \{ A^\neg \mid A \in \mathcal{L}^+ \} \quad and \quad \mathcal{L}^{\mathcal{D}} = \{ not\_L \mid L \in (\mathcal{L}^+ \cup \mathcal{L}^\neg) \}.$$

An abductive logic program $\langle P, Ab, IC \rangle$ is transformed into its *positive version* $\langle P', Ab', IC' \rangle$ as follows:

- P' and IC' are obtained by replacing in P and in IC literals of the form $\neg A$ and *not L* with the corresponding symbols in $\mathcal{L}^\neg$ and $\mathcal{L}^{\mathcal{D}}$ respectively,
- the set of integrity constraints $\{ \leftarrow A, A^\neg \mid A \in \mathcal{L}^+ \}$ is added to IC',
- $Ab' = Ab \cup \{ p^\neg \mid p \in Ab \}$.

From now onwards, we will not distinguish any further between $\langle P, Ab, IC \rangle$ and its positive version $\langle P', Ab', IC' \rangle$, and we will denote directly by $\langle P, Ab, IC \rangle$ the corresponding positive version.

The Herbrand base of an abductive logic program $\langle P, Ab, IC \rangle$ is:

$$\mathcal{B} = \mathcal{L}^+ \cup \mathcal{L}^\neg \cup \mathcal{L}^{\mathcal{D}}.$$

(For the sake of simplicity we assume that the Herbrand base of IC is a subset of $\mathcal{B}$.)

A Herbrand interpretation I of $\langle P, Ab, IC \rangle$ is, as usual, a subset of $\mathcal{B}$. Given an interpretation I, $I^{\mathcal{D}}$ and $I^{\mathcal{A}}$ will stand for the set of default and abducible literals in I respectively. Formally $I^{\mathcal{D}} = I \cap \mathcal{L}^{\mathcal{D}}$ and $I^{\mathcal{A}} = I \cap \{ L \mid pred(L) \in Ab \}$.

In order to treat default literals and integrity constraints, we introduce the notions of *coherency* and *consistency* for an interpretation. Let us first define the natural notion of complement for a literal with respect to negation-by-default.

$$\overline{L} = \begin{cases} \alpha & \text{if } L = not\_\alpha \\ not\_\alpha & \text{otherwise} \end{cases}$$

where α is a classical literal (viz. $\alpha \in (L^+ \cup L^\neg)$). The above definition can be lifted to sets of literals in the obvious way, namely $\overline{X} = \{ \overline{L} \mid L \in X \}$.

Definition 1. Let $\langle P, Ab, IC \rangle$ be an abductive logic program and let $I \subseteq \mathcal{B}$.

(i) I is *coherent* iff $I \cap \overline{I} = \emptyset$. I is *incoherent* otherwise.

(ii) I is *consistent* iff $\not\exists A : A \in I \land A^\neg \in I \land I \models IC$.
I is *inconsistent* otherwise.

A Herbrand interpretation I is *total* if and only if $\forall L \in (\mathcal{L}^+ \cup \mathcal{L}^\neg)$ either $L \in I$ or $\overline{L} \in I$.

3 Model Theory

The semantics of an abductive logic program is defined in terms of the Herbrand models of the positive version of the program. Intuitively the idea is to restrict, in a step-wise fashion, the set of Herbrand models in order to identify the intended meaning of the program.

The Herbrand models of a definite program P can be characterised by means of the notion of *supportedness* introduced by Brogi, Lamma and Mello [6]. Let P be a definite logic program and let H be a subset of the Herbrand base of P. A Herbrand model I is supported by H if and only if I is the least Herbrand model of the program $P \cup H$. It is easy to observe that the least Herbrand model of a program is always supported by the empty set.

We first adapt the definition of supportedness to the case of an abductive logic program $\langle P, Ab, IC \rangle$. Recall that the definite program P is the positive version of the original program and, therefore, any Herbrand model of P can be both incoherent and inconsistent. Thus, it is more convenient to talk about interpretations for the moment.

Definition 2. Let $\langle P, Ab, IC \rangle$ be an abductive logic program, and let H be a set of hypotheses such that $H \subseteq (\mathcal{B}^A \cup \mathcal{B}^D)$. Let T_P be the standard immediate consequence operator.

A Herbrand interpretation $I \subseteq \mathcal{B}$ is *supported* by H iff $I = T_{P \cup H} \uparrow \omega$.

Intuitively, a supported interpretation corresponds to assuming the truth of a set of hypotheses, which may be both default literals and abducibles. In the following, $I(H)$ will denote that I is supported by H.

A supported interpretation may be extended by enlarging the set of assumed hypotheses. In other words, the extensions of a supported interpretation I are all those supported interpretations which assume more hypotheses than I. In the context of abductive logic programs, however, we are interested in considering only those extensions which *conservatively* extend an interpretation I, that is, those extensions which assume only hypotheses coherent with I.

Definition 3. Let $I(H)$ and $J(K)$ be supported interpretations.

$J(K)$ *conservatively extends* $I(H)$ if and only if $K \supseteq H$ and $(\overline{K} \cap I) = \emptyset$.

Intuitively, the conservative extensions of a supported interpretation I are those supported interpretations which assume a larger set of hypotheses ($K \supseteq H$) without contradicting I. Namely, the condition $(\overline{K} \cap I) = \emptyset$ states that there is no (classical) literal L in I whose complement $\overline{L}$ is assumed by the extension J. More precisely, this means that J assumes only default hypotheses which are coherent with I (i.e., $not\_L \in J \implies L \notin I$), and only abducible hypotheses which are coherent with I (i.e., $A \in J^A \implies not\_A \notin I$).

Example 1. Let us consider the following program P:

$$a \leftarrow b$$

where the set of abducible predicates is $Ab = \{b\}$, and the set of integrity constraints IC is $\{\leftarrow b, \quad \leftarrow not\_b\}$. $\{not\_b\}$ and $\{not\_a, not\_b\}$ are the conservative extensions of the supported interpretation $\{not\_b\}$, while the supported interpretation $\{a, b\}$ has no conservative extension (apart from itself). [4]

It is worth noting that the notion of conservative extension can be equivalently expressed in terms of the argumentation-based framework proposed by Dung [13]. Namely, "$J(K)$ conservatively extends $I(H)$" means that the set of hypotheses K is an extended, stronger argument with respect to I which is not attacked by I. The employed notion of attack can be formalised as follows:

$$I(H) \text{ attacks } J(K) \text{ if and only if } I \cap \overline{K} \neq \emptyset.$$

Therefore, the notion of conservative extension can be formulated as follows: $J(K)$ conservatively extends $I(H)$ if and only if J extends I and J is not attacked by I.

The *admissibility* of a supported interpretation depends on its conservative extensions. Namely, an interpretation I is admissible if every conservative extension of I does not contradict the hypotheses of I.

Definition 4. A supported interpretation $I(H)$ is *admissible* if and only if:

$$\forall J : J \text{ conservatively extends } I \implies (\overline{H} \cap J) = \emptyset.$$

Example 2. Let us consider Example 1. The admissible supported interpretations are $I_1 = \{\}$, $I_2 = \{a, b\}$, $I_3 = \{not\_b\}$, $I_4 = \{not\_a, not\_b\}$. Notice that the supported interpretation $\{not\_a\}$ is not admissible since the conservative extension $\{not\_a, a, b\}$ contradicts the hypothesis not_a.

The notion of admissibility can be equivalently expressed in terms of argumentation in the following way: I is an admissible interpretation if and only if none of its conservative extensions attacks I.

Once we have determined the admissible interpretations for an abductive logic program, we wish to identify those admissible interpretations which, intuitively, assume as many default hypotheses as possible. Therefore a *complete* interpretation is an admissible interpretation which assumes all (and only) the default hypotheses not_L such that L does not belong to any of its conservative extensions.

Definition 5. A supported interpretation I is *complete* if and only if:

$$not\_L \in I \iff (\forall J \text{ which conservatively extends } I : L \notin J).$$

---

[4] For the sake of simplicity, from now onwards, we will omit all the default literals of the kind not_A^- in the interpretations.

Example 3. Let us consider Example 1. Among the admissible supported interpretations (Example 2), I_1, I_2 and I_4 are complete. The incompleteness of I_3 is due to the absence of *not_a* among its hypotheses.

The equivalent formulation of complete supported interpretations in terms of argumentation is: I is complete if and only if I assumes all the hypotheses which are not attacked by any of its conservative extensions.

A complete interpretation is a promising candidate for characterising the meaning of an abductive logic program. Complete interpretations, however, do not take into account consistency which concerns the meaning of explicit negation as well as the satisfiability of the integrity constraints of the original program. Actually a supported interpretation may be both incoherent and inconsistent. We will therefore call *pre-model* a complete interpretation which is coherent, and *model* a complete interpretation which is both coherent and consistent.

Example 4. Let us consider Example 1. The pre-models are $I_1 = \{\}$, $I_2 = \{a, b\}$ and $I_4 = \{not\_a, not\_b\}$ (see Example 3). Hence, $\{\}$ is the only model of the program.

Giving meaning to an abductive logic program in terms of its models may be unsatisfactory since it leads to a semantics which is not defined in some cases, even when the program is consistent. Intuitively speaking, an abductive logic program is *consistent* if and only if it has at least one consistent supported interpretation. Obviously, in this case the interpretation supported by the empty set of hypotheses is also consistent. This observation leads to the following natural definition of consistent abductive logic program.

Definition 6. An abductive logic program is *consistent* if and only if the interpretation supported by $\{\}$ is *consistent*.

Example 5. Let us consider the following consistent abductive logic program P:
  $a \leftarrow b$
where $Ab = \{\}$, and the set of integrity constraints IC is $\{\leftarrow not\_b\}$. The only premodel of P, $\{not\_a, not\_b\}$, is inconsistent. Therefore the set of models is empty and no meaning is given to the program.

It is however possible to prove that any abductive logic program has at least one premodel. Thus, in the attempt of assigning a meaning to any *consistent* abductive logic program, we choose to remove contradiction from pre-models, and to identify the semantics in terms of *abductive models*. Abductive models are coherent admissible supported interpretations which – even if not complete – assume as many hypotheses as possible while maintaining the consistency with respect to the integrity constraints. The idea is to consider the *consistent* admissible supported interpretations which assume maximal subsets of the hypotheses of a pre-model M.

Definition 7. Let $\langle P, Ab, IC \rangle$ be a consistent abductive program. An admissible supported interpretation I is an *abductive model* of $\langle P, Ab, IC \rangle$ if and only if it is a maximal (with respect to set inclusion) consistent subset of some pre-model M of $\langle P, Ab, IC \rangle$.

Example 6. Let us consider Example 5. The meaning of the program is the abductive model $\{not\_a\}$. Indeed $\{not\_a\}$ is admissible and it is obtained by removing not_b (the culprit of inconsistency) from the pre-model $\{not\_a, not\_b\}$.

From Definition 7 it immediately follows that each model (i.e., each consistent complete interpretation) is an abductive model, but not vice-versa. We can therefore assign meaning to abductive logic programs in terms of their abductive models. We call the resulting semantics *abductive semantics*. The following proposition states that the abductive semantics gives meaning to any consistent abductive logic program [5].

Proposition 8. *Any consistent abductive logic program has at least one abductive model.*

The abductive semantics embeds the generalised stable model semantics by Kakas and Mancarella [19]. In particular, the following proposition holds.

Proposition 9. *Each generalised stable model corresponds to an abductive model.*

Corollary 10. *Any total abductive model corresponds to a generalised stable model and vice-versa.*

Therefore, in the most general case, that is when $Ab \neq \emptyset$ and IC is a set of closed formulae, the abductive semantics lifts the generalised stable model semantics up to the 3-valued case.

Example 7. Let us consider the following abductive logic program, where P is:
$$a \leftarrow b$$
$$p \leftarrow not\ p$$
the set of abducible predicates is $Ab = \{b\}$, and the set of integrity constraints IC is empty. Consider the sets $\Delta_1 = \{\}$ and $\Delta_2 = \{b\}$. We have that neither $P \cup \Delta_1$ nor $P \cup \Delta_2$ has stable models and thus no generalised stable model exists for P. The problem arises because of the clause $p \leftarrow not\_p$ and the totality requirement of stable model semantics. The abductive semantics, instead, gives meaning to this program in terms of the abductive models $\{a, b\}$ and $\{not\_a, not\_b\}$.

Finally it is worth observing that our abductive framework allows us to reconstruct many well-known semantics of normal and extended logic programming, where abduction is not taken into account.

The case of normal logic programming is recovered by considering an abductive framework where P is a normal logic program viewed as a positive one, $Ab = \emptyset$ and $IC = \emptyset$. In this case, the set of abductive models exactly corresponds to the set of 3-valued stable models [23] (or stationary expansions [24], or complete scenaria [12], being they all equivalent as proved in [7]). All the other results presented in [7] hold. In particular, total abductive models correspond to stable models [17], and the minimal abductive model corresponds to the well-founded model [28].

---

[5] For the lack of space, all the proofs of propositions and theorems are omitted and can be found in [8].

Extended logic programs can be dealt with by considering an abductive logic program where P is an extended logic program viewed as a positive one, $Ab = \emptyset$ and $IC = \{\leftarrow A, A^\neg | A \in \mathcal{L}^+\}$. The set of models captures 3-valued stable semantics [23], whereas abductive models coincide with the complete scenaria semantics for extended logic programs defined in [15]. Other intermediate results can be obtained (see [8]). In particular, total abductive models capture the answer set semantics [18] provided that contradiction is considered with respect to the overall set of literals.

Therefore our framework can be considered a common ground where negation-by-default, explicit negation and abduction can be formally compared and uniformly integrated.

4 Proof Procedure

In this section we present a top-down proof procedure which is correct for the abductive semantics defined in the previous section. We ground our procedure on the proof procedure defined by Kakas and Mancarella [20].

The starting point is to consider an abductive logic program as a positive program (as in Sect. 3) and to include into IC the sets of integrity constraints $\{\leftarrow A, A^\neg | A \in \mathcal{L}^+\}$ and $\{\leftarrow A, not\_A | A \in \mathcal{L}^+\}$. The procedure described in [20] extends Eshghi and Kowalski's procedure [16] so as to manipulate arbitrary abducibles. It deals with ground abductive logic programs and manipulates integrity constraints which are denials containing at least one abducible or default atom.

We adapt this procedure to our framework by:

1. Lifting it to the 3-valued case. In particular, in contrast with [20], we do not consider integrity constraints of the kind $A \vee not\_A$. This allows us to withdraw the totality requirement;
2. Dealing with explicit negation, which is not considered in [20]. Each explicitly negated literal $\neg A$ is viewed as a new positive atom, and an integrity constraint of the form $\leftarrow A, A^\neg$ is added to the program [6].

In the following, we report the main steps of the procedure, which is basically the procedure defined in [20].

Abductive derivation
An abductive derivation from (G_1, Δ_1) to (G_n, Δ_n) in $\langle P, Ab, IC \rangle$ via a selection rule R is a sequence

$$(G_1, \Delta_1), (G_2, \Delta_2), \ldots, (G_n, \Delta_n)$$

such that each G_i has the form $\leftarrow L_1, \ldots, L_k$, and $R(G_i) = L_j$, and (G_{i+1}, Δ_{i+1}) is obtained according to one of the following rules:

(1) If L_j is not abducible or default, then $G_{i+1} = C$ and $\Delta_{i+1} = \Delta_i$ where C is the resolvent of some clause in P with G_i on the selected literal L_j;

---

[6] Notice that these integrity constraints do not satisfy the requirement of [20]. They can be however reduced so as to satisfy this requirement by applying unfolding to the classical literals. An alternative solution is to employ forward evaluation of rules to check any form of integrity constraint, as proposed by Satoh and Iwayama [26].

(2) If L_j is abducible or default and $L_j \in \Delta_i$ then $G_{i+1} =\leftarrow L_1, \ldots, L_{j-1}, L_{j+1}, \ldots, L_k$ and $\Delta_{i+1} = \Delta_i$;

(3) If L_j is abducible or default, $L_j \notin \Delta_i$ and $\overline{L_j} \notin \Delta_i$ and there exists a *consistency derivation* from $(L_j, \Delta_i \cup \{L_j\})$ to $(\{\}, \Delta')$ then $G_{i+1} =\leftarrow L_1, \ldots, L_{j-1}, L_{j+1}, \ldots, L_k$ and $\Delta_{i+1} = \Delta'$.

Steps (1) and (2) are SLD-resolution steps with the rules of P and abductive or default hypotheses, respectively. Step (3) adds a new abductive or default hypothesis to the current set of hypotheses provided that it is consistent.

Consistency derivation

A consistency derivation for an abducible or default literal α from (α, Δ_1) to (F_n, Δ_n) in (P, Ab, IC) is a sequence

$$(\alpha, \Delta_1), (F_1, \Delta_1), (F_2, \Delta_2), \ldots, (F_n, \Delta_n)$$

where:

(i) F_1 is the set of all goals of the form $\leftarrow L_1, \ldots, L_n$ obtained by resolving the abducible or default α with the denials in IC with no such goal been empty ($\leftarrow$);

(ii) For each $i > 1$, F_i has the form $\{\leftarrow L_1, \ldots, L_k\} \cup F_i'$ and for some $j = 1, \ldots, k$ (F_{i+1}, Δ_{i+1}) is obtained according to one of the following rules:

(C1) If L_j is not abducible or default, then $F_{i+1} = C' \cup F_i'$ where C' is the set of all resolvents of clauses in P with $\leftarrow L_1, \ldots, L_k$ on the literal L_j and $\leftarrow \notin C'$, and $\Delta_{i+1} = \Delta_i$;

(C2) If L_j is abducible or default, $L_j \in \Delta_i$ and $k > 1$, then $F_{i+1} = \{\leftarrow L_1, \ldots, L_{j-1}, L_{j+1}, \ldots, L_k\} \cup F_i'$ and $\Delta_{i+1} = \Delta_i$;

(C3) If L_j is abducible or default, $\overline{L_j} \in \Delta_i$ then $F_{i+1} = F_i'$ and $\Delta_{i+1} = \Delta_i$;

(C4) If L_j is abducible or default, $L_j \notin \Delta_i$ and $\overline{L_j} \notin \Delta_i$, and there exists an *abductive derivation* from $(\leftarrow \overline{L_j}, \Delta_i)$ to $(\leftarrow, \Delta')$, then $F_{i+1} = F_i'$ and $\Delta_{i+1} = \Delta'$.

In case (C1) the current branch splits into as many branches as the number of resolvents of $\leftarrow L_1, \ldots, L_k$ with the clauses in P on L_j. If the empty clause is one of such resolvents the whole consistency check fails. In case (C2) the goal under consideration is reduced if literal L_j belongs to the current set of hypotheses Δ_i. In case (C3) the current branch is already consistent under the assumptions in Δ_i, and it is therefore dropped from the consistency check. In case (C4) the current branch of the consistency search space can be dropped provided that $\leftarrow \overline{L_j}$ is abductively provable.

Given a query L (atomic, for the sake of simplicity), the procedure succeeds and returns the set of abducibles Δ if there exists an abductive derivation from $(\leftarrow L, \{\})$ to $(\leftarrow, \Delta)$. The correctness of the proof procedure with respect to our abductive semantics is established by the following two theorems which state the soundness and the weak completeness results, respectively.

Theorem 11. *Consider a consistent abductive logic program and let L be an atomic literal. If the procedure succeeds for L returning a set Δ of abducibles then there exists an abductive model M such that $L \in M$ and $\Delta \subseteq M^A \cup M^D$.*

Theorem 12. *Consider an abductive logic program and let L be an atomic literal. Suppose that every selection of rules in the proof procedure for L terminates with either success or failure. If there exists an abductive model M such that $L \in M$, then there exists a selection of rules such that the procedure succeeds for L returning Δ where $\Delta \subseteq M^A \cup M^D$.*

Therefore the procedure defined here is sound with respect to the abductive semantics, and it is complete when no predicate in the program depends on itself, that is when the program contains no loop.

From Theorem 12, we obtain the following corollary for finite failure.

Corollary 13. *Consider an abductive logic program and let L be an atomic literal. If the procedure finitely fails for L then there exists no abductive model M such that $L \in M$.*

The corollary points out that finite failure can be used to check whether a literal L is *unknown* in all abductive models, since the finite failure of the procedure for both L and $\overline{L}$ means that every abductive model does not contain neither L nor $\overline{L}$.

When considering only normal logic programs, the procedure actually coincides with the abductive proof procedure defined by Eshghi and Kowalski [16], which has been proved sound with respect to admissible scenaria by Dung [12]. Therefore Theorem 12 also states a weak completeness result in the case of finite failure with respect to admissible scenaria.

In [8] we have shown that, for extended logic programs, the abductive semantics embeds the 3-valued stable semantics which, in turns, subsumes the answer set semantics. Therefore, our proof procedure is weakly complete (under the conditions of Theorem 12) with respect to the answer set semantics for extended logic programs, but of course not sound. In fact, the totality requirement of the answer set semantics is not enforced in our procedure.

5 Related Work

In Sect. 3 we have shown how several well-known semantics for normal and extended logic programming can be reconstructed as instances of our abductive framework. In this section we briefly discuss other related work.

Brewka and Konolidge [5] proposed a framework for extended logic programs (without abduction) which gives meaning to any consistent logic program, as in our framework. Differently from Dung [12], from Przymusinski [23] and from our approach, the semantics proposed in [5] is equivalent to the stable model semantics for programs having at least one stable model. This property can be easily enforced in our framework by imposing a preference criterion so as to prefer the total models of a program (if they exist) to the partial models of a program. However, for those programs for which no stable model exists, the semantics proposed in [5] sensibly departs from the 3-valued stable model semantics [23] and from the stationary semantics [24].

Pereira and Alferes [21] introduced an extended stable semantics for logic programs with negation-by-default and explicit negation. The main difference with respect to our approach (as well as with respect to [15, 18, 23]) is that constraints of the form $\neg L \to not\ L$ are enforced for each classical literal L. In order to reconstruct Pereira and Alferes's semantics in our framework, it is necessary both to enforce such constraints and to perform a suitable syntactic transformation of programs, as described by Alferes, Dung and Pereira [1]. Roughly speaking, the transformation consists of replacing each classical literal L occurring in a clause body with the conjunction $L, not\_L^{\neg}$.

Bondarenko, Toni and Kowalski [4], and Dung [13] have recently proposed argumentation frameworks to describe various forms of non-monotonic reasoning. In Sect. 3 we have already shown how our abductive framework can be interpreted as an argumentation framework according to [4, 13]. Intuitively speaking, supported interpretations are the model-theoretic counterpart of sets of assumptions. In contrast with [4], for any interpretation I, we consider only the attacks made by the conservative extensions of I, rather than considering also possible attacks by *any* other supported interpretation. The underlying intuition is that an agent argues only with other agents which assume its hypotheses, and possibly attacks them by assuming more hypotheses. If the agent is not able to counterattack these agents then it is defeated. Conservative extensions are the model-theoretic counterpart of the attacking agents which cannot be counterattacked.

Bondarenko, Toni and Kowalski [4] introduced different definitions of acceptability, which derive from the corresponding notions of counterattack. These definitions may be reconstructed in our framework by suitably modifying the definition of conservative extension.

Dung [14] has recently proposed a new semantics for extended logic programs based on an argumentation framework by extending the work of Dung and Ruamviboonsuk [15]. The main novelty is the introduction of a different notion of *acceptability* of hypotheses which is based on two notions of attack. While we use integrity constraints of the form $\leftarrow A, \neg A$ only to identify the abductive models among the pre-models of a program, Dung maps these constraints into a new notion of attack (called *reductio ad absurdum*), and directly uses them for defining the notion of admissibility.

As far as the proof procedure is concerned, Eshghi and Kowalski [16] firstly introduced a top-down proof procedure to compute negation as failure through abduction. Such a procedure was refined by Dung [12] and extended by Kakas and Mancarella [20] in order to handle arbitrary abducibles. As already pointed out in Sect. 4, the procedure proposed by Kakas and Mancarella [20] is correct with respect to our abductive semantics.

A proof procedure for our abductive framework may be also obtained by adapting the proof procedure proposed by Satoh and Iwayama [26]. The required extension concerns the check for implicit deletions of rules employed in [26]. Indeed there exists an intimate bond between implicit deletion checking and the totality requirement of 2-valued stable models semantics. The check for implicit deletion in [26] actually verifies that the model under consideration assigns an absolute truth value (either true or false) to each atom. Since abductive models are not necessarily total, the removal of the check for implicit deletions from Satoh and Iwayama's procedure would guarantee its correct behaviour with respect to our abductive semantics.

Pereira, Aparicio and Alferes [22] presented a top-down proof procedure for ground normal logic programs which is sound and complete with respect to the 3-valued stable semantics. The main difference with respect to our proof procedure is that a loop check is included so as to guarantee that the execution always terminates.

Teusink [27] presented a proof procedure for extended logic programs which is correct with respect to a proof-theoretic semantics. This semantics guarantees only some form of local consistency, while it does not guarantee global consistency (which is instead ensured in our proof procedure) since it is considered to be too complex.

Denecker and De Schreye [9] introduced a proof procedure for normal abductive logic programs by extending SLDNF resolution. The resulting proof procedure (SLDNFA) is correct with respect to the completion semantics. An important feature of this abductive procedure is its ability of dealing with *non-ground* abductive goals, while our procedure is limited to the ground case. Denecker and De Schreye do not consider general integrity constraints in [9], but only constraints of the kind ← A, not_A. In [10] they overcome this limitation by treating general integrity constraints, though in a rather inefficient way. Actually all the integrity constraints are checked at the end of the proof of a query, that is only after the whole set of abductive hypotheses supporting the query has been computed, while it would be more convenient to check consistency as soon as hypotheses are generated. Our proof procedure checks consistency in such an incremental way, though this is easier to do in our case since hypotheses do not contain variables.

6 Conclusions

We presented a framework for knowledge representation which uniformly integrates abduction and negation (both explicit and by default) in a logic programming setting.

A model-theoretic semantics is defined in terms of Herbrand models (a widely accepted concept in the logic programming community), and it is equipped with an interpretation in terms of argumentations. The distinguishing features of the resulting abductive semantics are that it is defined for any consistent program, and that it is strictly related to well-known semantics for the separate extensions being considered, namely abduction, negation-by-default and explicit negation.

Having in mind a practical use of our abductive framework, we developed a query-based, correct proof procedure in order to provide an effective interpreter for the language. The procedure is basically that defined by Kakas and Mancarella [20] and it coincides with the procedure of Eshghi and Kowalski [16] when only normal logic programs are considered.

References

1. J.J. Alferes, P.M. Dung, and L.M. Pereira. Scenario semantics of extended logic programs. In L. M. Pereira and A. Nerode, editors, *Proc. 2nd Int. Workshop on Logic Programming and Non Monotonic Reasoning*, pages 334–348. The MIT Press, 1993.
2. J.J. Alferes, and L.M. Pereira. On logic program semantics with two kind of negation. In K. Apt, editor, *Proc. Int. Joint Conf. and Symp. on Logic Programming*, pages 574–588. The MIT Press, 1992.
3. K.R. Apt. Logic Programming. In J. van Leeuwen, editor, *Handbook of Theoretical Computer Science*, Vol. B, pages 493–574. Elsevier Science Publisher, 1990.
4. A. Bondarenko, F. Toni, and R.A. Kowalski. An Assumption-based framework for nonmonotonic Reasoning, In L. M. Pereira and A. Nerode, editors, *Proc. 2nd Int. Workshop on Logic Programming and Non Monotonic Reasoning*, pages 171–189. The MIT Press, 1993.

5. G. Brewka, and K. Konolige. An Abductive Framework for General Logic Programs and other Nonmonotonic Systems. *Proc. Int. Joint Conf. on Artificial Intelligence IJ-CAI93*, pages 9–15. AAAI, 1993.

6. A. Brogi, E. Lamma, and P. Mello. Compositional Model-theoretic Semantics for Logic Programs. In *New Generation Computing*, 11(1):1–21. Springer-Verlag, 1992.

7. A. Brogi, E. Lamma, P. Mancarella, and P. Mello. Normal Logic Programs as Open Positive Programs. In K. Apt, editor, *Proc. Int. Joint Conf. and Symp. on Logic Programming*, pages 783-797. The MIT Press, 1992.

8. A. Brogi, E. Lamma, P. Mancarella, and P. Mello. A Unifying View of Logic Programming and Non-monotonic Reasoning. *Tech. Report*, University of Bologna and University of Pisa, December 1994.

9. M. Denecker, and D. De Schreye. SLDNFA: an abductive procedure for normal abductive programs. In K. Apt, editor, *Proc. Int. Joint Conf. and Symp. on Logic Programming*, pages 686–700. The MIT Press, 1992.

10. M Denecker, and D. De Schreye. Representing Incomplete Knowledge in Abductive Logic Programming. In *Proc. International Logic Programming Symposium ILPS93*, pages 147–163. The MIT Press, 1993.

11. J. Dix. Semantics of Logic Programs: Their Intuitions and Formal Properties. In A. Fuhrmann and H. Roth, editors, *Logic, Action and Information*. de Gruyter, Berlin-New York, 1994.

12. P.M. Dung. Negation as Hypothesis: An Abductive Foundation for Logic Programming. In K. Furukawa, editor, *Proc. 8th Int. Conf. on Logic Programming ICLP91*, pages 3–17. The MIT Press, 1991.

13. P.M. Dung. On the Acceptability of Arguments and its Fundamental Role in Nonmonotonic Reasoning. In *Proceedings Int. Joint Conf. on Artificial Intelligence IJ-CAI93*, pages 852–857 . AAAI, 1993.

14. P.M. Dung. An Argumentation Semantics for Logic Programming with Explicit Negation. In D.S. Warren, editor, *Proc. 10th Int. Conf. on Logic Programming ICLP93*, pages 616–630. The MIT Press, 1993.

15. P.M. Dung, and P. Ruamviboonsuk. Well-founded Reasoning with Classical Negation. In *Proc.1st Int. Workshop on Logic Programming and Non-Monotonic Reasoning*, pages 120–132. The MIT Press, 1991.

16. K. Eshghi, and R.A. Kowalski. Abduction Compared with Negation by Failure. In G. Levi and M. Martelli, editors, *Proc. 6th Int. Conf. on Logic Programming ICLP89*, pages 234–254. The MIT Press, 1989.

17. M. Gelfond, and V. Lifschitz. The stable model semantics for logic programming. In R. A. Kowalski and K.A. Bowen, editors, *Proc. 5th Int. Conf. on Logic Programming*, pages 1070–1080. The MIT Press, 1988.

18. M. Gelfond, and V. Lifschitz. Logic Programs with Classical Negation. In D.H.D. Warren and P. Szeredi, editors, *Proc. 7th Int. Conf. on Logic Programming ICLP90*, pages 579–597. The MIT Press, 1990.

19. A.C. Kakas, and P. Mancarella. Generalized stable models: a semantics for abduction. In *Proceedings of 9th European Conference on Artificial Intelligence ECAI90*, pages 385–391. Pitman Publishing, 1990.

20. A.C. Kakas, and P. Mancarella. On the relation between Truth Maintenance and Abduction. In *Proceedings PRICAI90*, 1990.

21. L.M. Pereira, and J.J. Alferes. Well-founded Semantics for Logic Programs with Explicit Negation. In *Proc. ECAI92*, John Wiley & Sons, 1992.

22. L.M. Pereira, J.N. Aparicio, and J.J. Alferes. Derivation Procedures for Extended Stable Models. In *Proc. IJCAI91*, pages 863–868. Morgan Kaufman 1991.

23. T.C. Przymusinski. Extended Stable Semantics for Normal and Disjunctive Programs. In D.H.D. Warren and P. Szeredi, editors, *Proc. 7th Int. Conf. on Logic Programming*, pages 459–477. The MIT Press, 1990.

24. T.C. Przymusinski. Semantics of Disjunctive Logic Programs and Deductive Databases. In *Proc. DOOD'91*, 1991.

25. F. Sadri, and R.A Kowalski. A Theorem-proving Approach to Database Integrity. In J. Minker, editor, *Foundations of Deductive Databases and Logic Programming*, pages 313–362. Morgan-Kaufmann, 1988.

26. K. Satoh, and N. Iwayama. A Query Evaluation Method for Abductive Logic Programming. In K. Apt, editor, *Proc. Int. Joint Conf. and Symp. on Logic Programming*, pages 671–685. The MIT Press, 1992.

27. F. Teusink. A Proof Procedure for Extended Logic Programs. In *Proceedings International Logic Programming Symposium ILPS93*, pages 235–249. The MIT Press, 1993.

28. A. Van Gelder, K.A. Ross, and J.S. Schlipf. Unfounded sets and the well-founded semantics for general logic programs. In *Proc. ACM SIGMOD-SIGACT, Symposium on Principles of Database Systems*, 1988.

Embedding Circumscriptive Theories in General Disjunctive Programs

Chiaki Sakama[1]* and Katsumi Inoue[2]

[1] ASTEM Research Institute of Kyoto
17 Chudoji Minami-machi, Shimogyo, Kyoto 600, Japan
sakama@astem.or.jp
[2] Department of Information and Computer Sciences
Toyohashi University of Technology
Tempaku-cho, Toyohashi 441, Japan
inoue@tutics.tut.ac.jp

Abstract. This paper presents a method of embedding circumscriptive theories in general disjunctive programs. In a general disjunctive program, negation as failure occurs not only in the body but in the head of a rule. In this setting, minimized predicates of a circumscriptive theory are specified using the negation in the body, while fixed and varying predicates are expressed by the negation in the head. Moreover, the translation implies a close relationship between circumscription and abductive logic programming. That is, fixed and varying predicates in a circumscriptive theory are also viewed as abducible predicates in an abductive disjunctive program. Our method of translating circumscription into logic programming is fairly general compared with the existing approaches and exploits new applications of logic programming for representing commonsense knowledge.

1 Introduction

It is well-known that logic programming semantics have close relationships to circumscription. In early studies, Reiter [Rei82] presented that Clark's predicate completion is characterized by circumscription in Horn theories. Lifschitz [Lif85a] showed that the closed world assumption is equivalent to circumscription when the CWA is consistent and circumscription minimizes every predicate in function-free first-order theories. Lifschitz [Lif88] and Przymusinski [Prz88] characterized the perfect model semantics of stratified logic programs and stratified disjunctive logic programs by prioritized circumscription. The results were further generalized by Gelfond *et al.* [GPP89] who introduced various forms of CWAs in terms of circumscription. Recent studies show that the stable model semantics of normal and disjunctive logic programs are also characterized by epistemic circumscription [Lif89, LS92, YY93, SI93].

---

* Address after April 1995: Department of Computer and Communication Science, Wakayama University, 930 Sakaedani, Wakayama 640, Japan.

On the other side, when we consider representing circumscriptive theories in logic programming, difficulties arise due to their axiomatic differences. According to [GL88a], there are three main differences between circumscription and logic programming as follows:

(a) In logic programming, the unique name assumption and the domain closure assumption are usually presumed, while no corresponding assumption exists in the definition of circumscription.
(b) Circumscription contains not only minimized predicates but fixed and varying predicates, while logic programming minimizes predicates by the CWA but lacks the mechanism of representing fixed and varying predicates.
(c) The meaning of logic programs is syntax-dependent and priorities are specified by negation as failure, while circumscription handles classical first-order theories and priorities are specified by the circumscriptive policy.

Such differences usually impose some restriction on formulas included in a circumscriptive theory. For instance, Gelfond and Lifschitz [GL88a] introduce a method of translating circumscriptive theories into normal logic programs, where theories are restricted to stratifiable ones without fixed predicates.

In this paper, we take the above points into consideration and explore a new method of embedding circumscriptive theories in logic programs. As for the part (a), we consider the Herbrand models of circumscriptive theories instead of incorporating both assumptions into the theories as in [GL88a], since each assumption is automatically satisfied in Herbrand models [BS85]. To fill the gap of (b), we introduce a class of *general disjunctive programs*, in which a rule possibly contains negation-as-failure formulas not only in the body but in the head of the rule. Then we show that fixed and varying predicates of circumscriptive theories are expressed by negation-as-failure formulas appearing in the heads of rules. For the part (c), we translate priorities specified by the circumscriptive policy into negation as failure in the body of a rule, together with the introduction of *characteristic clauses*. Our method is fairly general and faithfully translates a large class of circumscriptive theories into logic programs. Moreover, this translation implies a connection between circumscription and abductive logic programming. That is, from the viewpoint of abduction, fixed and varying predicates in a circumscriptive theory are considered as abducible predicates in an abductive disjunctive program.

The rest of this paper is organized as follows. In Section 2, we introduce a class of general disjunctive programs and present their semantics. In Section 3, we give a translation from circumscriptive theories into general disjunctive programs. It is shown that there is an equivalence relationship between the PZ-minimal Herbrand models of a circumscriptive theory and the stable models of a translated general disjunctive program. In Section 4, we show that circumscriptive theories are also expressed in terms of abductive disjunctive programs. Section 5 discusses related issues and applications to other nonmonotonic formalisms, and Section 6 summarizes the paper.

2   General Disjunctive Programs

Logic programs considering in this paper are disjunctive logic programs, which possibly contain negation-as-failure formulas not only in the body but in the head of a rule.

A *general disjunctive program* (GDP) is a set of rules of the form

$$A_1 \mid \ldots \mid A_k \mid not\, A_{k+1} \mid \ldots \mid not\, A_l$$
$$\leftarrow B_{l+1}, \ldots, B_m, not\, B_{m+1}, \ldots, not\, B_n\,, \tag{1}$$

where A_i's and B_j's are atoms and $n \geq m \geq l \geq k \geq 0$. The disjunction in the left-hand side is the *head* and the conjunction in the right-hand side is the *body* of the rule. Each rule with variables stands for the set of its ground instances as usual. Intuitive reading of the rule (1) is that if each $B_{l+1}, \ldots, B_m$ holds and each $B_{m+1}, \ldots, B_n$ does not hold then some A_i $(1 \leq i \leq k)$ holds or some A_j $(k + 1 \leq j \leq l)$ does not hold.

Logic programs with such positive occurrences of negation as failure were initially introduced by Lifschitz and Woo as a subset of the logic of *minimal belief and negation as failure* (MBNF) [LW92, Lif94a]. Recently, its applications to commonsense reasoning in logic programming were exploited by Inoue and Sakama [IS94]. In [IS94], programs called *general extended disjunctive programs* (GEDPs) are introduced, which generalize *extended disjunctive programs* (EDPs) of [GL91] by introducing positive *not*. GDPs considering here are a subclass of GEDPs, i.e., GEDPs without classical negation. A GDP reduces to a *normal disjunctive program* (NDP) when each rule contains no *not* in the head. An NDP is called a *normal logic program* (NLP) if each rule contains at most one atom in the head, while an NDP is called a *positive disjunctive program* (PDP) if each rule contains no *not* in the body.

An *interpretation* of a program is a subset of the Herbrand base of the program. An interpretation I *satisfies* a ground rule of the form (1) iff $\{B_{l+1}, \ldots, B_m\} \subseteq I$ and $\{B_{m+1}, \ldots, B_n\} \cap I = \emptyset$ imply either $\{A_1, \ldots, A_k\} \cap I \neq \emptyset$ or $\{A_{k+1}, \ldots, A_l\} \not\subseteq I$.

A declarative meaning of a GDP is given by the *stable model semantics*. First, let Π be a PDP and M an interpretation. Then M is a *minimal model* of Π iff M is a minimal set satisfying each rule in Π. Next, let Π be a GDP and M an interpretation. The PDP Π^M is defined as follows: a rule

$$A_1 \mid \ldots \mid A_k \leftarrow B_{l+1}, \ldots, B_m \tag{2}$$

is in Π^M iff there is a ground rule of the form (1) from Π such that

$$\{A_{k+1}, \ldots, A_l\} \subseteq M \text{ and } \{B_{m+1}, \ldots, B_n\} \cap M = \emptyset.$$

Then, M is a *stable model* of Π iff M is a minimal model of Π^M.

By definition, any stable model of a GDP Π satisfies every ground rule from Π. For any ground formula F, we write $\Pi \models F$ iff $M \models F$ holds in every stable model M of Π.

The above definition of stable models is a special case of the definition of answer sets in a GEDP [LW92, IS94], and it reduces to the notion of stable models of Przymusinski [Prz91] when Π is an NDP, and that of Gelfond and Lifschitz [GL88b] when Π is an NLP.

It should be noted that in contrast with the case of NDPs or NLPs, a stable model of a GDP is not always minimal.

Example 2.1 Let Π be a program consisting of the single rule

$$A \mid not\ A \leftarrow .$$

Then Π has two stable models $\emptyset$ and $\{A\}$.

In the next section, we will see that this unique feature of stable models of GDPs is useful to represent fixed and varying predicates in circumscriptive theories.

3 Embedding Circumscriptive Theories in GDPs

Circumscription is a form of nonmonotonic reasoning initially introduced by McCarthy [Mc80] and thoroughly developed by Lifschitz [Lif94b]. We first review the framework of circumscription. The following definition is due to [Lif85b].

Given a first-order theory T, let P and Z be disjoint tuples of predicates from T. Then *circumscription* of P in T with *variable* Z is defined as the second-order formula:

$$Circ(T; P; Z) = T(P, Z) \wedge \neg \exists P'\, Z'(T(P', Z') \wedge P' < P), \qquad (3)$$

where $T(P, Z)$ is a theory containing predicate constants P, Z, and P', Z' are tuples of predicate variables similar to P, Z. The set of all predicates other than P, Z from T is denoted by Q, which is called the *fixed* predicates.

A model of $Circ(T; P; Z)$ is called a *PZ-minimal model* of T, in which extensions of each predicate from P are minimized with those from Z varied and those from Q fixed. In particular, when Z is empty it is called a *P-minimal model*. For any first-order formula F, $Circ(T; P; Z) \models F$ iff $M \models F$ holds for every PZ-minimal model M of T [Lif85b].

We introduce some notations which will be useful. P^+, Z^+, and Q^+ respectively denote the sets of all positive literals with predicates from P, Z, and Q, while P^-, Z^-, and Q^- denote the sets of all negative literals correspondingly. We also use the letters P, Z, and Q to denote the sets of all (positive or negative) literals whose predicates belong to P, Z, and Q, respectively. The small letters $p_1, \ldots, p_l$, $z_1, \ldots, z_m$, and $q_1, \ldots, q_n$ are used to denote atoms from P^+, Z^+, and Q^+, respectively.

In this paper, we consider a first-order theory T as a set of *clauses* of the form:

$$A_1 \vee \ldots \vee A_l \vee \neg B_1 \vee \ldots \vee \neg B_m$$

where each A_i $(1 \leq i \leq l; l \geq 0)$ and B_j $(1 \leq j \leq m; m \geq 0)$ are atoms and every variable in the formula is assumed to be universally quantified at the front. Each clause is also identified with the set of its literals. A first-order theory is simply said a theory hereafter.

Given a theory T we restrict our attention to its *Herbrand models*, since we are interested in a semantic relationship to logic programming. Note that such a restriction has an effect to incorporate both the *domain closure assumption* and the *unique name assumption* into T [BS85].

To express circumscription in terms of logic programming, we first introduce the notion of characteristic clauses. We say a clause C_1 *subsumes* a clause C_2 if $C_1\theta \subseteq C_2$ holds for some substitution θ, and C_1 does not have literals more than C_2.[3]

Definition 3.1 Let T be a theory. Then a *characteristic clause* of T is defined as a clause C such that

(i) $T \vdash C$ where C consists of literals from $P^+ \cup Q$, and
(ii) for any clause C' such that $T \vdash C'$, if C' subsumes C then C subsumes C'.

The set of all characteristic clauses of T is denoted as $cc(T)$.

The notion of characteristic clauses is also introduced in [BS85, GPP89, Prz89, HIP91] and discussed in [Ino92] in a general setting.

Next we define a transformation from a first-order theory to a GDP.

Definition 3.2 Given a theory T, a GDP Π_T is constructed as follows.

1. For any clause in T of the form:

$$p_1 \vee \ldots \vee p_l \vee z_1 \vee \ldots \vee z_m \vee q_1 \vee \ldots \vee q_n$$
$$\vee \neg p_{l+1} \vee \ldots \vee \neg p_s \vee \neg z_{m+1} \vee \ldots \vee \neg z_t \vee \neg q_{n+1} \vee \ldots \vee \neg q_u , \qquad (4)$$

Π_T has the rule:

$$z_1 \mid \ldots \mid z_m \mid q_1 \mid \ldots \mid q_n$$
$$\leftarrow p_{l+1}, \ldots, p_s, z_{m+1}, \ldots, z_t, q_{n+1}, \ldots, q_u, not\, p_1, \ldots, not\, p_l . \qquad (5)$$

2. For every characteristic clause of T of the form:

$$p_1 \vee \ldots \vee p_l \vee q_1 \vee \ldots \vee q_n \vee \neg q_{n+1} \vee \ldots \vee \neg q_u , \qquad (6)$$

Π_T has the rule:

$$p_1 \mid \ldots \mid p_l \mid q_1 \mid \ldots \mid q_n \leftarrow q_{n+1}, \ldots, q_u . \qquad (7)$$

3. For any atom $A \in Q \cup Z$, Π_T has the rule:

$$A \mid not\, A \leftarrow . \qquad (8)$$

---

[3] This subsumption is called the θ-subsumption in [Lov78].

The intuitive meaning of the above transformation is explained as follows. First, for any clause (4) from T, predicates from P have higher priorities to Z for minimizing, hence each p_i $(i = 1, \ldots, l)$ is shifted to the body in its negated form $not\ p_i$ in (5). Second, any characteristic clause (6) from T is added to Π_T as a rule (7). This transformation retains each rule that can make atoms from P true. The rule (8) says that each atom A from Q or Z is either true or not.

The rest of this section shows that the above transformation exactly embeds a circumscriptive theory $Circ(T; P; Z)$ in a GEDP Π_T. In the following, PZ (or P)-minimal models mean PZ (or P)-minimal *Herbrand* models. Also any atom A is identified with the rule $A \leftarrow$ in a program. We begin with preliminary lemmas.

Lemma 3.1 Let T be a theory such that $Q = Z = \emptyset$. Then M is a P-minimal model of T iff M is a stable model of Π_T.

Proof: In this case, there is a characteristic clause $p_1 \vee \ldots \vee p_l$ from T iff there is a corresponding rule of the form $p_1 \mid \ldots \mid p_l \leftarrow$ in $\Pi_T{}^M$. Then, M is a P-minimal model of T iff M is a P-minimal model of $cc(T)$ and satisfies each rule of the form (5) iff M is a stable model of Π_T. □

Lemma 3.2 [Lif94b, (3.2)] Let Ψ be a set of closed formulas containing no predicate from P, Z. Then, $Circ(T \wedge \Psi; P; Z) = Circ(T; P; Z) \wedge \Psi$. □

Lemma 3.3 Let T be a theory such that $Z = \emptyset$. Then M is a P-minimal model of T iff M is a stable model of Π_T.

Proof: M is a P-minimal model of T such that $M \cap Q = \Psi$ iff M is a model of $Circ(T; P) \wedge \Psi$. Here, $Circ(T; P) \wedge \Psi = Circ(T \wedge \Psi; P)$ by Lemma 3.2. Since Ψ is all the true atoms from Q in M, M is a model of $Circ(T \wedge \Psi; P)$ iff M is a model of $Circ(T \wedge \Psi; P, Q)$. In this case, the fixed predicates are considered empty, so that M is a model of $Circ(T \wedge \Psi; P, Q)$ iff M is a stable model of $\Pi_{T \wedge \Psi}$ by Lemma 3.1. For each $A \in \Psi$, there is a rule of the form (8) and $A \in \Pi_T{}^M$. Then, $\Pi_{T \wedge \Psi}{}^M = \Pi_T{}^M \wedge \Psi = \Pi_T{}^M$. Thus, M is a minimal model of $\Pi_{T \wedge \Psi}{}^M$ iff M is a minimal model of $\Pi_T{}^M$ iff M is a stable model of Π_T. □

Let $M_P = M \cap P$, $M_Q = M \cap Q$, $M_Z = M \cap Z$, and Σ a disjunction of literals.

Lemma 3.4 Let T be a theory and M a P-minimal model of $cc(T)$. Then there is a model N of T such that $N_P = M_P$ and $N_Q = M_Q$.

Proof: When M is a P-minimal model of $cc(T)$, M satisfies any clause $\Sigma \subset P^+ \cup Q$ such that $T \vdash \Sigma$. Now we show the result by two steps.

(I) First, suppose that M does not satisfy some ground clause C : $\Sigma_0 \vee \neg p_1 \vee \ldots \vee \neg p_l$ such that $T \vdash C$, $T \nvdash \Sigma_0$, and $\Sigma_0 \subset P^+ \cup Q$. Then $M \nvDash \Sigma_0$ and $M \vDash p_i$ $(i = 1, \ldots, l)$. We first show that for each p_i there is a ground clause C_i : $p_i \vee \Sigma_i$ from $cc(T)$ such that $M \nvDash \Sigma_i$. Suppose to the contrary that there is no such clause for some p_i. Then, for any ground clause Σ_i, $cc(T) \vdash p_i \vee \Sigma_i$ implies $M \models \Sigma_i$. Hence, $M \setminus \{p_i\}$ is a model of $cc(T)$, which contradicts the fact that M is a P-minimal model of $cc(T)$. Next, by C and each C_i, $T \vdash \Sigma_0 \vee \Sigma_1 \vee \ldots \vee \Sigma_l$. However, since $M \not\models \Sigma_i$ for $i = 0, \ldots, l$, $M \not\models \Sigma_0 \vee \Sigma_1 \vee \ldots \vee \Sigma_l$. This contradicts the fact that M satisfies any clause $\Sigma \subset P^+ \cup Q$ such that $T \vdash \Sigma$. Therefore, M satisfies any clause $\Sigma' \subset P \cup Q$ such that $T \vdash \Sigma'$.

(II) Next, let us consider the set $\mathcal{Z}_M(T)$ of ground clauses defined by

$$\mathcal{Z}_M(T) = \{\, l_1 \vee \ldots \vee l_n \mid \Sigma \vee l_1 \vee \ldots \vee l_n \in ground(T),$$
$$\Sigma \subset P \cup Q, \; M \not\models \Sigma, \text{ and } l_i \in Z \text{ for } i = 1, \ldots, n \,\},$$

where $ground(T)$ is the set of ground clauses from T. Suppose that $\mathcal{Z}_M(T)$ is unsatisfiable. By Herbrand's Theorem, there is a finite subset Γ of $\mathcal{Z}_M(T)$ such that Γ is unsatisfiable. Then, let

$$\Gamma_M(T) = \{\, \Sigma \mid \Sigma \vee l_1 \vee \ldots \vee l_n \in ground(T) \text{ and } l_1 \vee \ldots \vee l_n \in \Gamma \,\}.$$

By $\Gamma \vdash \square$ and $T \wedge \bigwedge_{\Sigma_i \in \Gamma_M(T)} \neg \Sigma_i \vdash \Gamma$, $T \wedge \bigwedge_{\Sigma_i \in \Gamma_M(T)} \neg \Sigma_i \vdash \square$ holds, thereby $T \vdash \bigvee_{\Sigma_i \in \Gamma_M(T)} \Sigma_i$. However, $M \not\models \Sigma_i$ for each $\Sigma_i \in \Gamma_M(T)$ by the construction of $\Gamma_M(T)$ and $\mathcal{Z}_M(T)$. Hence, $M \not\models \bigvee_{\Sigma_i \in \Gamma_M(T)} \Sigma_i$, which contradicts the result of (I) that M satisfies every clause $\Sigma' \subset P \cup Q$ such that $T \vdash \Sigma'$. Therefore, $\mathcal{Z}_M(T)$ is satisfiable. Let N_Z be a model of $\mathcal{Z}_M(T)$, and put $N = M \cup N_Z$. Then, N satisfies every clause of T, and $N_P = M_P$ and $N_Q = M_Q$. $\square$

Lemma 3.5 Let T be a theory and M a model of T. Then, M is a PZ-minimal model of T iff $M_P \cup M_Q$ is a P-minimal model of $cc(T)$.

Proof: ($\Rightarrow$) Suppose that M is a PZ-minimal model of T. Since $T \vdash C$ for any $C \in cc(T)$, M is also a model of $cc(T)$. Also $cc(T)$ does not include any literal from Z, $M_P \cup M_Q$ is a model of $cc(T)$. Suppose that $M_P \cup M_Q$ is not a P-minimal model of $cc(T)$. Then there is a P-minimal model N of $cc(T)$ such that $N_Q = M_Q$ and $N_P \subset M_P$. By Lemma 3.4, there is a model M' of T such that $M'_P = N_P$ and $M'_Q = N_Q$. However, since $M'_P \subset M_P$ and $M'_Q = M_Q$, this contradicts the fact that M is a PZ-minimal model of T. Therefore, $M_P \cup M_Q$ is a P-minimal model of $cc(T)$.

($\Leftarrow$) Suppose that $M_P \cup M_Q$ is a P-minimal model of $cc(T)$. By Lemma 3.4, there is a model N of T such that $N_P = M_P$ and $N_Q = M_Q$. If N is not a PZ-minimal model of T, there is a PZ-minimal model N' of T such that $N'_P \subset N_P$ and $N'_Q = N_Q$. In this case, however, $N'_P \cup N'_Q$ is a P-minimal model of $cc(T)$ by the ($\Rightarrow$) part, which contradicts the fact that $M_P \cup M_Q = N_P \cup N'_Q$ is a P-minimal model of $cc(T)$. Hence, $N = M_P \cup M_Q \cup N_Z$ is a PZ-minimal model of T. Since N and M differ only on predicates from Z, M is also a PZ-minimal model of T. $\square$

Given a theory T and an interpretation M, let us define a theory T^{M_P} such that the clause

$$z_1 \vee \ldots \vee z_m \vee q_1 \vee \ldots \vee q_n$$
$$\vee \neg p_{l+1} \vee \ldots \vee \neg p_s \vee \neg z_{m+1} \vee \ldots \vee \neg z_t \vee \neg q_{n+1} \vee \ldots \vee \neg q_u \qquad (9)$$

is in T^{M_P} iff a clause

$$p_1 \vee \ldots \vee p_l \vee z_1 \vee \ldots \vee z_m \vee q_1 \vee \ldots \vee q_n$$

$$\vee \neg p_{l+1} \vee \ldots \vee \neg p_s \vee \neg z_{m+1} \vee \ldots \vee \neg z_t \vee \neg q_{n+1} \vee \ldots \vee \neg q_u \qquad (10)$$

is in T and $M_P \not\models p_1 \vee \ldots \vee p_l$. Then the following property holds.

Lemma 3.6 If M is a PZ-minimal model of T, $M_Q \cup M_Z$ is a P-minimal model of T^{M_P}.

Proof: Since T^{M_P} contains no literals from P^+, any model N of T^{M_P} such that $N_P = \emptyset$ is a P-minimal model of T^{M_P}. On the other hand, if M is a model of T, $M_Q \cup M_Z$ is a model of T^{M_P} by definition. Therefore, $M_Q \cup M_Z$ is a P-minimal model of T^{M_P}. $\qquad\square$

Now we are ready to show the main result of this paper.

Theorem 3.7 Let T be a theory. Then M is a PZ-minimal model of T iff M is a stable model of Π_T.

Proof: Put $\Pi_T = \Pi_{sft} \cup \Pi_{cc} \cup \Pi_{pnf}$ where Π_{sft} is the set of rules of the form (5), Π_{cc} is the set of rules of the form (7), and Π_{pnf} is the set of rules of the form (8). Then, $\Pi_T{}^M = \Pi_{sft}{}^{M_P} \cup \Pi_{cc} \cup \Pi_{pnf}{}^{M_Q} \cup \Pi_{pnf}{}^{M_Z}$. Since $\Pi_{pnf}{}^{M_Q} = M_Q$ and $\Pi_{pnf}{}^{M_Z} = M_Z$, it becomes $\Pi_T{}^M = \Pi_{sft}{}^{M_P} \cup \Pi_{cc} \cup M_Q \cup M_Z$. Suppose that M is a PZ-minimal model of T. By Lemma 3.5, $M_P \cup M_Q$ is a P-minimal model of $cc(T)$, thereby a minimal model of $\Pi_{cc} \cup M_Q$ (by Lemma 3.3). Since $\Pi_{cc} \cup M_Q$ contains no atom from Z, $M_P \cup M_Q \cup M_Z$ is a minimal model of $\Pi_{cc} \cup M_Q \cup M_Z$. On the other hand, by Lemma 3.6, $M_Q \cup M_Z$ is a P-minimal model of T^{M_P}. Hence M satisfies T^{M_P}, so does $\Pi_{sft}{}^{M_P}$. Therefore, M is a minimal model of $\Pi_{sft}{}^{M_P} \cup \Pi_{cc} \cup M_Q \cup M_Z = \Pi_T{}^M$, hence a stable model of Π_T.

Conversely, suppose that M is a stable model of Π_T. Then M is a minimal model of $\Pi_T{}^M$. For each rule of the form:

$$z_1 \mid \ldots \mid z_m \mid q_1 \mid \ldots \mid q_n \leftarrow p_{l+1}, \ldots, p_s, z_{m+1}, \ldots, z_t, q_{n+1}, \ldots, q_u \qquad (11)$$

in $\Pi_{sft}{}^{M_P}$, there is a corresponding clause of the form (9) in T^{M_P}. Then if M satisfies (11), so does (9). Also, for each rule not included in $\Pi_{sft}{}^{M_P}$ but in Π_{sft}, $M_P \models p_1 \vee \ldots \vee p_l$ holds, hence M also satisfies the corresponding clause (10) in T. Thus, M is a model of T. On the other hand, $M_P \cup M_Q$ is a minimal model of $\Pi_{cc} \cup M_Q$, thereby a P-minimal model of $cc(T)$ (by Lemma 3.3). Therefore, M is a PZ-minimal model of T by Lemma 3.5. $\qquad\square$

Corollary 3.8 Let T be a theory and F a ground formula. Then, $Circ(T; P; Z) \models F$ iff $\Pi_T \models F$. $\qquad\square$

Note that $Circ(T; P; Z)$ is satisfiable iff Π_T has a stable model. When $Circ(T; P; Z)$ is unsatisfiable, T is also unsatisfiable [Lif94b, Corollary 6.3.3]. In this case, $cc(T)$ contains the empty clause only [Ino92, Proposition 2.5], then Π_T contains "$\leftarrow$" meaning the falsity and has no stable model.

Example 3.1 Let T be the theory:

$$bird(x) \vee \neg penguin(x), \tag{12}$$
$$\neg fly(x) \vee \neg penguin(x), \tag{13}$$
$$fly(x) \vee ab(x) \vee \neg bird(x), \tag{14}$$
$$bird(Joe), \tag{15}$$

where $P = \{ab\}$, $Z = \{fly\}$, and $Q = \{bird, penguin\}$. Then, $Circ(T; P; Z)$ has two PZ-minimal models: $M_1 = \{bird(Joe), fly(Joe)\}$ and $M_2 = \{bird(Joe), ab(Joe), penguin(Joe)\}$. In this case, Π_T becomes

$$bird(x) \leftarrow penguin(x), \tag{16}$$
$$\leftarrow fly(x), penguin(x), \tag{17}$$
$$fly(x) \leftarrow bird(x), not\, ab(x), \tag{18}$$
$$ab(x) \leftarrow penguin(x), \tag{19}$$
$$bird(Joe) \leftarrow, \tag{20}$$
$$fly(x) \mid not\, fly(x) \leftarrow, \tag{21}$$
$$bird(x) \mid not\, bird(x) \leftarrow, \tag{22}$$
$$penguin(x) \mid not\, penguin(x) \leftarrow, \tag{23}$$

which has the stable models M_1 and M_2.

In the above translation, by (12), (13), and (14) the characteristic clause

$$ab(x) \vee \neg penguin(x)$$

is obtained, which is translated into (19) in Π_T. Note that in this theory the translation of [GL88a] cannot be applied, since T contains the fixed predicates.

Example 3.2 Our translation can also handle theories representing *multiple inheritance*. Consider the theory T consisting of the clauses

$$pacifist(x) \vee ab_1(x) \vee \neg republican(x),$$
$$\neg pacifist(x) \vee ab_2(x) \vee \neg quaker(x),$$

where $P = \{ab_1, ab_2\}$, $Z = \{pacifist\}$, and $Q = \{republican, quaker\}$. Then Π_T becomes

$$pacifist(x) \leftarrow republican(x),\, not\, ab_1(x),$$
$$\leftarrow pacifist(x), quaker(x),\, not\, ab_2(x),$$
$$ab_1(x) \mid ab_2(x) \leftarrow republican(x), quaker(x),$$
$$pacifist(x) \mid not\, pacifist(x) \leftarrow,$$
$$republican(x) \mid not\, republican(x) \leftarrow,$$
$$quaker(x) \mid not\, quaker(x) \leftarrow.$$

Then, given the facts

$$republican(Nixon) \leftarrow, \quad quaker(Nixon) \leftarrow,$$

Π_T has two stable models: $\{\, republican(Nixon), quaker(Nixon), ab_1(Nixon) \,\}$ and $\{\, republican(Nixon), quaker(Nixon), pacifist(Nixon), ab_2(Nixon) \,\}$, which are exactly the PZ-minimal models of T. Note that the above theory is not represented by a stratified NLP, hence cannot be handled by [GL88a].

4 Connection with Abductive Disjunctive Programs

Abductive logic programming [KKT92] is an extension of logic programming, which realizes abductive reasoning in AI by incorporating a set of hypotheses into programs. In this section, we show that circumscriptive theories are expressed by abductive disjunctive programs using the result of the previous section.

An *abductive disjunctive program* (ADP) is a pair $\langle \Pi, \Gamma \rangle$ where Π is an NDP and Γ is a finite set of predicates called the *abducible predicates*. The set of all ground atoms $\mathcal{A}_\Gamma$ having abducible predicates from Γ is called the *abducibles*. Let $\langle \Pi, \Gamma \rangle$ be an ADP. An interpretation M is called a *belief set* of $\langle \Pi, \Gamma \rangle$ iff M is a stable model of the NDP $\Pi \cup E$ for some set $E \subseteq \mathcal{A}_\Gamma$. By definition, stable models of an NDP are belief sets of an ADP with $E = \emptyset$.

Inoue and Sakama [IS94] observe that a rule of the form:

$$A \mid not\, A \leftarrow$$

in a GDP is viewed as a rule expressing an abductive hypothesis, i.e., an abducible A is true or not. The correspondence is formally presented as follows.

Lemma 4.1 [IS94] Let $\langle \Pi, \Gamma \rangle$ be an ADP. Suppose that $gdp(\Pi, \Gamma)$ is a GDP such that

$$gdp(\Pi, \Gamma) = \Pi \cup \{\, A \mid not\, A \leftarrow \mid A \in \mathcal{A}_\Gamma \,\}.$$

Then, M is a belief set of $\langle \Pi, \Gamma \rangle$ iff M is a stable model of $gdp(\Pi, \Gamma)$. □

Since GDPs Π_T introduced in the previous section are of the form $gdp(\Pi, \Gamma)$, the above result implies that circumscriptive theories can be expressed in terms of ADPs by identifying fixed and varying predicates as abducible predicates.

Given a theory T, let $adp(T; P; Z) = \langle \Pi, \Gamma \rangle$ where (i) for each clause in T of the form (4), Π contains the rule of the form (5); (ii) for every characteristic clause of T of the form (6), Π contains the rule of the form (7); and $\Gamma = Q \cup Z$. Then the following theorem holds by Theorem 3.7 and Lemma 4.1.

Theorem 4.2 Let T be a theory. Then M is a PZ-minimal model of T iff M is a belief set of $adp(T; P; Z)$. □

Example 4.1 Let T be the theory of Example 3.1. Then $adp(T; P; Z)$ becomes $\langle \Pi, \Gamma \rangle$ where Π consists of rules (16) – (20), and $\Gamma = \{\, fly, bird, penguin \,\}$. Then it is easy to see that M_1 and M_2 become the belief sets of $\langle \Pi, \Gamma \rangle$.

5   Discussion

In this section, we discuss related issues and further applications to nonmonotonic reasoning.

1. (*Comparison with GL-translation*)
Gelfond and Lifschitz [GL88a] introduce a translation from circumscriptive theories into NLPs. Comparing both translations, the GL-translation is restricted to stratifiable theories, and every clause is assumed to contain at most one variable predicate and no fixed predicates. By contrast, we have no such restriction and consider GDPs instead of NLPs. On the other hand, under the above condition they provide a translation from *prioritized circumscription* into NLPs, while we do not consider the issue here. The possibility of extending our translation to prioritized circumscription is currently under investigation.

It is known that circumscription with fixed and varying predicates can be reduced to circumscription without them [dKK89, CEG92]. Using the techniques, it might be possible to translate circumscriptive theories into disjunctive logic programs without handling fixed and varying predicates directly. However, these reductions introduce extra predicates which are not included in the original theory. In this sense, our method faithfully translates the original circumscriptive theory into a logic program by representing fixed and varying predicates in an appropriate manner.

2. (*Connection with EDPs*)
In this paper, we have used the rule of the form (8) for each atom A having fixed or varying predicates. On the other hand, if we use the *CWA-rule*

$$\neg A \leftarrow not\, A$$

for each atom A with a minimizing predicate, it is possible to use the rule

$$A \mid \neg A \leftarrow \qquad (24)$$

instead of (8). Then there will be a one-to-one correspondence between the PZ-minimal Herbrand models of a first-order theory T and the answer sets of the translated EDP. Note that in this case an answer set contains negative literals explicitly, which are false in the corresponding PZ-minimal model. The rule of the form (24) is called a *completeness rule* in [LT94].

3. (*Proof procedure*)
Some proof procedures are known for computing circumscription directly [Prz89, Gin89, HIP91]. On the other hand, the translation presented in this paper enables us to use proof procedures for GDPs also for circumscriptive theories. For example, a simple model generation procedure for GEDPs in [IS94] is used for this purpose.

Note that our translation is different from the approach of [Lif85b, Lif94b], which reduces the second-order circumscription formula (3) to an equivalent first-order formula. In our translation, computation of characteristic clauses is achieved by first-order deduction and always possible as long as the number of characteristic clauses is finite. Procedures for computing characteristic clauses are given in [Prz89, Ino92].

4. (*Relation to other NMR formalisms*)

Circumscription is related to other nonmonotonic formalisms. According to [GLPS94], circumscription $Circ(T; P; Z)$ is represented by *autoepistemic logic* (AEL) as follows:

$$T \land (P \supset BP) \land (Q \supset BQ) \land (BQ \supset Q).$$

Also, by analogy with the above translation, circumscription is represented by the MBNF formula[4]

$$BT \land (not\, P \supset B\neg P) \land (not\, Q \supset B\neg Q) \land (not\, Q \lor BQ).$$

Comparing these AEL/MBNF translations with our GDP translations, some interesting correspondences are observed. First, the last formulas in the AEL/MBNF translations correspond to the rule (8) in our GDP translation.[5] Second, the AEL/MBNF translations include the CWA-formulas for each atom from both P and Q, while we have no such rules but include rules (7) for handling P. Third, in the AEL/MBNF translations no care is taken for each Z, while our GDP translation have rules (5) and (8) to cope with Z properly.

The result of this paper is also extended to another translation from circumscription into AEL/MBNF by combining the translation from GDPs into AEL/MBNF in [IS94, LW92]. These correspondences will generalize the result of this paper to the non-Herbrand model case.

5. (*Application to parametric knowledge*)

Lifschitz [Lif93] argues that in nonmonotonic systems some additional assumption makes the domain under consideration smaller, and consequently the class of conclusions true in all the domains becomes larger. He calls such assumptions as *parameters*, and the monotonicity property as *restricted monotonicity*. In circumscription, fixed predicates have the property of such parameters. For instance, in Example 3.1, $bird(Joe)$ and $penguin(Joe)$ are considered as parameters, therefore all theorems of T are included in the theorems of $T \cup \{penguin(Joe)\}$. Lifschitz provides a way to realize restricted monotonicity in EDPs. According to [Lif93], T is translated into the EDP Π'_T:

$$bird(x) \leftarrow penguin(x), \tag{16}$$

$$\neg fly(x) \leftarrow penguin(x), \tag{25}$$

$$fly(x) \leftarrow bird(x),\, not\, ab(x), \tag{18}$$

$$ab(x) \leftarrow not\, \neg penguin(x), \tag{26}$$

$$bird(Joe) \leftarrow . \tag{20}$$

---

[4] Lifschitz [Lif94a] gives a similar translation without Q.

[5] Using a different formalization [Kon89], the formula $BQ \supset Q$ in the AEL translation can be replaced with $\neg Q \supset B\neg Q$.

The differences between Π'_T and Π_T are the existence of the rule (25) instead of (17), the rule (26) instead of (19), and the absence of (21), (22) and (23). Note here that the above translation realizes restricted monotonicity, while there is no correspondence between the answer sets of Π'_T and the PZ-minimal models of the original theory. In fact, Π'_T has the unique answer set $\{\,bird(Joe),\ ab(Joe)\,\}$, while no corresponding PZ-minimal model of T exists. On the other hand, in our translation there is an exact matching between the the stable models of Π_T and the PZ-minimal models of T by the presence of (21) – (23). Moreover, Lifschitz does not provide a method of encoding parametric knowledge in logic programming in general, while our translation illustrates that such knowledge is represented using the positive occurrences of negation as failure. Since parametric knowledge plays an important role in theories of action [Lif93], our translation will contribute to those theories.

6 Summary

This paper has presented a method of embedding circumscriptive theories in general disjunctive programs. We introduced a translation from circumscriptive theories into general disjunctive programs, and showed that the PZ-minimal Herbrand models of a circumscriptive theory are exactly the stable models of the transformed GDP. The result is also applied to abductive characterization of circumscription. That is, by identifying fixed and varying predicates with abducible predicates, the PZ-minimal Herbrand models of a circumscriptive theory are expressed by the belief sets of an abductive disjunctive program. We finally addressed extensions and applications to other nonmonotonic formalisms.

The result of this paper implies that general disjunctive programs are as expressive as circumscriptive theories under the Herbrand model semantics. Moreover, the positive occurrences of negation as failure increase the expressive power of logic programming as a nonmonotonic formalism, and exploit new applications of logic programming as a knowledge representation tool.

References

[BS85] Bossu, G. and Siegel, P., Saturation, Nonmonotonic Reasoning and the Closed World Assumption, *Artificial Intelligence* 25, 13-63, 1985.

[CEG92] Cadoli, M., Eiter, T. and Gottlob, G., An Efficient Method for Eliminating Varying Predicates from a Circumscription, *Artificial Intelligence* 54, 397-410, 1992.

[dKK89] de Kleer, J. and Konolige, K., Eliminating the Fixed Predicates from a Circumscription, *Artificial Intelligence* 39, 391-398, 1989.

[GL88a] Gelfond, M. and Lifschitz, V., Compiling Circumscriptive Theories into Logic Programs, *Proc. AAAI-88*, 455-459. Extended version in: *Proc. 2nd Int. Workshop on Nonmonotonic Reasoning*, LNAI 346, 74-99, 1988.

[GL88b] Gelfond, M. and Lifschitz, V., The Stable Model Semantics for Logic Programming, *Proc. ICLP'88*, 1070-1080.

[GPP89] Gelfond, M., Przymusinska, H. and Przymusinski, T., On the Relation between Circumscription and Negation as Failure, *Artificial Intelligence* 38, 75-94, 1989.

[GL91] Gelfond, M. and Lifschitz, V., Classical Negation in Logic Programs and Disjunctive Databases, *New Generation Computing* 9, 365-385, 1991.

[GLPS94] Gelfond, M., Lifschitz, V., Przymusinska, H. and Schwarz, G., Autoepistemic Logic and Introspective Circumscription, *Proc. TARK'94*, 197-207.

[Gin89] Ginsberg, M. L., A Circumscriptive Theorem Prover, *Artificial Intelligence* 39, 209-230, 1989.

[HIP91] Helft, N., Inoue, K. and Poole, D., Query Answering in Circumscription, *Proc. IJCAI-91*, 426-431.

[Ino92] Inoue, K., Linear Resolution for Consequence Finding, *Artificial Intelligence* 56, 301-353, 1992.

[IS94] Inoue, K. and Sakama, C., On Positive Occurrences of Negation as Failure, *Proc. KR'94*, 293-304.

[KKT92] Kakas, A. C., Kowalski, R. A. and Toni, F., Abductive Logic Programming, *J. Logic and Computation* 2, 719-770, 1992.

[Kon89] Konolige, K., On the Relation between Autoepistemic Logic and Circumscription, *Proc. IJCAI-89*, 1213-1218.

[Lif85a] Lifschitz, V., Closed World Databases and Circumscription, *Artificial Intelligence* 27, 229-235, 1985.

[Lif85b] Lifschitz, V., Computing Circumscription, *Proc. IJCAI-85*, 121-127.

[Lif88] Lifschitz, V., On the Declarative Semantics of Logic Programs with Negation, in *Foundations of Deductive Databases and Logic Programming* (J. Minker ed.), Morgan Kaufmann, 177-192, 1988.

[Lif89] Lifschitz, V., Between Circumscription and Autoepistemic Logic, *Proc. KR'89*, 235-244.

[LW92] Lifschitz, V. and Woo, T. Y. C., Answer Sets in General Nonmonotonic Reasoning (preliminary report), *Proc. KR'92*, 603-614.

[Lif93] Lifschitz, V., Restricted Monotonicity, *Proc. AAAI-93*, 432-437.

[Lif94a] Lifschitz, V., Minimal Belief and Negation as Failure, *Artificial Intelligence* 70, 53-72, 1994.

[Lif94b] Lifschitz, V., Circumscription, in *Handbook of Logic in Artificial Intelligence and Logic Programming* (D. M. Gabbay, *et al.* eds.), Clarendon Press, 297-352, 1994.

[LT94] Lifschitz, V. and Turner, H., From Disjunctive Programs to Abduction, *Proc. ICLP'94 Workshop on Nonmonotonic Extensions of Logic Programming*, 111-125.

[LS92] Lin F. and Shoham, Y., A Logic of Knowledge and Justified Assumptions, *Artificial Intelligence* 57, 271-289, 1992.

[Lov78] Loveland, D. W., *Automated Theorem Proving: A Logical Basis*, North-Holland, 1978.

[Mc80] McCarthy, J., Circumscription – A Form of Nonmonotonic Reasoning, *Artificial Intelligence* 13, 27-39, 1980.

[Prz88] Przymusinski, T. C., On the Declarative Semantics of Deductive Databases and Logic Programs, in the same source book as [Lif88], 193-216, 1988.

[Prz89] Przymusinski, T., An Algorithm to Compute Circumscription, *Artificial Intelligence* 38, 49-73, 1989.

[Prz91] Przymusinski, T. C., Stable Semantics for Disjunctive Programs, *New Generation Computing* 9, 401-424, 1991.

[Rei82] Reiter, R., Circumscription implies Predicate Completion (sometimes), *Proc. AAAI-82*, 418-420.

[SI93] Sakama, C. and Inoue, K., Relating Disjunctive Logic Programs to Default Theories, *Proc. LPNMR'93*, 266-282.

[YY93] Yuan, L. Y. and You, J-H., Autoepistemic Circumscription and Logic Programming, *J. Automated Reasoning* 10, 143-160, 1993.

Stable Classes and Operator Pairs
for Disjunctive Programs

Jürgen Kalinski

Institute of Computer Science III
University of Bonn, Römerstr. 164
53117 Bonn, Germany
email: cully@cs.uni-bonn.de

Abstract. Baral and Subrahmanian introduced the notion of stable classes for normal logic programs. In contrast to stable models stable classes always exist and can be given a constructive characterization. We generalize the Baral-Subrahmanian approach to disjunctive programs and propose mf-stable classes for different functions mf. Such mf-stable classes always exist and are sound with respect to stable model semantics. Operationalizations for approximate but efficient query evaluation are defined in terms of three-valued interpretations and their relation with mf-stable classes is analyzed. Finally, analogous concepts are given for an approach based on states instead of models.

1 Introduction

Stable model semantics as proposed by Gelfond and Lifschitz [4] is one of the most elegant approaches concerning the semantics of normal logic programs. It generalizes the perfect model semantics and is closely related with Autoepistemic Logic [12] as a major formalization of nonmonotonic reasoning. Furthermore, stable model semantics could subsequently be extended to disjunctive logic programs (Gelfond and Lifschitz [5], by Przymusinski [15]).

Unfortunately, stable models do not necessarily exist, and even when they exist they are hard to compute. Baral and Subrahmanian therefore introduced the notion of stable classes for normal programs (cf. [1] and [2]). Essentially, a stable class is a set $\mathcal{I}$ of interpretations such that

$$\mathcal{I} = \{ S_P(I) \mid I \in \mathcal{I} \}$$

where S_P is an antimonotonic operator. They prove that normal programs always have stable classes and that

$$\{ \operatorname{lfp}(S_P^2), \operatorname{gfp}(S_P^2) \}$$

is the smallest (with respect to a Hoare-ordering) stable class, thereby providing a constructive characterization. They furthermore show that this stable class is equivalent to the well-founded semantics, another (by now) standard formalization of the meaning of normal logic programs [20] which is more efficient than the stable model semantics.

The goal of this paper is to provide a generalization of stable classes for disjunctive programs and constructive characterizations for query evaluation procedures. The main results are:

1. We propose a generalization of the Baral-Subrahmanian approach defining mf-stable classes for certain model filters mf. Every mf-stable class is sound with respect to stable model semantics. For every model filter and every disjunctive program there is an mf-stable class. Stable models are exactly characterizable by the information-greatest model filter.

2. We have a closer look at so-called complete model filters mf and show that they are isomorphic to three-valued interpretations.

3. We propose query evaluation procedures for disjunctive programs in terms of standard operators on three-valued interpretations: the van Emden-Kowalski and the Minker-Rajasekar consequence operators as well as Weak Generalized Closed World Assumption. Existing implementations of these operators can thus be used for query evaluation. It is even possible to evaluate queries for disjunctive programs by exclusively referring to normal programs at runtime.

4. Finally, an analogous approach is presented where the central concepts are based on states instead of models.

2 Preliminaries

A *disjunctive logic program* is a set of rules $A_1 \vee \ldots \vee A_m \leftarrow L_1, \ldots, L_n$ where every A_i is an atom and every L_i an atom or an atom negated by '$\neg$'. Function symbols are not allowed such that the Herbrand universe and Herbrand base B_P of P are finite. When $m = 1$, the program is called *normal*. Programs without any occurrence of '$\neg$' are *positive*. Disjunctions denoted by '$A \vee B$' are regarded as sets of atoms. A logic program P is tacitly identified with the finite set of all ground instantiations of its rules in its Herbrand universe. The set of minimal models of a positive program will be denoted by $\mathcal{MIN}(P)$.

The reader is assumed to be familiar with van Emden and Kowalski's *immediate consequence operator* T_P (cf. [19]). For positive disjunctive programs an appropriate monotonic operator on disjunctions of atoms has been proposed by Minker and Rajasekar (cf. [11]):

$$T_P^{\vee}(I) = \{ \delta \mid \delta \text{ a disjunction of atoms}, \beta \leftarrow A_1, \ldots, A_n \in P,$$
$$A_i \vee \gamma_i \in I \text{ for } 1 \leq i \leq n, \delta = \beta \vee \gamma_1 \vee \ldots \vee \gamma_n \}$$

Theorem 1. (Minker/Rajasekar [11]) Let P be a positive disjunctive program and δ a disjunction of atoms. Then the following are equivalent:

1. δ *is a logical consequence of P.*
2. $\delta' \in \text{lfp}(T_P^{\vee})$ *for some subclause $\delta' \subseteq \delta$.*

Example 1. In a database application disjunctions can be used for the representation of uncertain values:

```
diagnosis(patient1,disease1)
diagnosis(patient2,disease2) V diagnosis(patient2,disease3)
allergic(patient2,ingredient)
```

Rules describe interdependencies between facts. Positive rules specify which facts can be derived from others:

```
prescribe(P,medicine1) ← diagnosis(P,disease1)
```

Negation allows the derivation of facts based on the absence of information about other facts:

```
prescribe(P,medicine2) ← diagnosis(P,disease2),
                         ¬allergic(P,ingredient)
```

Definition 2. (Gelfond/Lifschitz [4]) Let P be a disjunctive program and I a Herbrand interpretation. The *Gelfond-Lifschitz transformation P/I* of P modulo I is defined by the following procedure:

– Remove every rule whose body contains some $\neg A$ with $A \in I$.
– Delete every expression $\neg A$ in the remaining rules.

Definition 3. (Gelfond/Lifschitz [4] [5], Przymusinski [15]) Let P be a disjunctive program. A Herbrand interpretation I is called a *stable model* of P, if it is a minimal model of P/I.

The *stable model semantics* ascribes a logic program its set of stable models.

3 Model Filters

3.1 Basic Concepts

Stable models must not necessarily exist and it is hard to decide whether there is one (cf. [9] and [10]). When they exist, there might be an exponential number of them. Baral and Subrahmanian have therefore introduced the alternative concept of stable classes for *normal* programs (cf. [1] and [2]). Whereas stable model semantics uses a single assumptive interpretation for the Gelfond-Lifschitz transformation and checks whether this guess can be verified, Baral and Subrahmanian start from a whole set of assumptions yielding a set of program transformations and correspondingly a smallest model for each of them. A set $\mathcal{I}$ is called a stable class, if it is 'verifiable', i.e. a set which is identical with the collection of smallest models of P/I for every $I \in \mathcal{I}$:

$$\mathcal{I} = \{ J \mid J \subseteq B_P \text{ and there is } I \in \mathcal{I} \text{ such that}$$
$$J \text{ is the smallest model of } P/I \}$$

Baral and Subrahmanian concentrate on normal logic programs and define stable classes in terms of an antimonotonic operator. We regard the model-oriented formulation as advantageous, as it allows the following generalization for disjunctive programs:

$$\mathcal{I} = \{ J \mid J \subseteq B_P \text{ and there is } I \in \mathcal{I} \text{ such that}$$
$$J \text{ is a minimal model of } P/I \}$$

In most general terms we are given a set $\mathcal{I}$ of interpretations which are used as an assumptive background for Gelfond-Lifschitz transformations of P, and we have a mapping mf which derives another set of interpretations from $\mathcal{I}$. Every superset $\mathcal{J}$ of $\mathcal{I}$ is a less precise assumption set, so that $mf(\mathcal{J})$ should be a superset of $mf(\mathcal{I})$. Furthermore, we hold on to the idea that positive programs are descriptions of their minimal models. Hence, $mf(\mathcal{I})$, i.e. the worlds described by P under the assumptions given by $\mathcal{I}$, should contain the minimal models of P/I for every $I \in \mathcal{I}$.

Definition 4. Let P be a disjunctive program. A monotonic mapping

$$mf : \wp(\wp(B_P)) \to \wp(\wp(B_P))$$

with $mf(\mathcal{I}) \supseteq \mathcal{MIN}(P/I)$ for every $I \in \mathcal{I}$ is called a *model filter* for P.

Definition 5. Let P be a program and mf a model filter. A set of interpretations $\mathcal{I}$ is called an *mf-stable class*, if and only if

$$\mathcal{I} = mf(\mathcal{I})$$

At first sight it seems that we have complicated the whole thing by the use of sets of assumptions instead of single assumptions. All of the following sections are to prove and demonstrate that rather the opposite is the case.

Example 2. The Baral-Subrahmanian filter has already been mentioned above:

$$mf_\top(\mathcal{I}) := \{ J \mid J \subseteq B_P \text{ and there is } I \in \mathcal{I} \text{ such that}$$
$$J \text{ is a minimal model of } P/I \}$$

For normal programs every fixpoint of this filter is a stable class in the sense of [1] and every singleton fixpoint is a stable model.

Another filter is defined by $mf_\bot(\mathcal{I}) := \wp(B_P)$.

3.2 Properties

Every set $\mathcal{I}$ of interpretations represents a certain amount of information: It specifies a number of true statements — namely those which are true in every element of $\mathcal{I}$ — and a number of false statements — namely those which are false in every element of $\mathcal{I}$. When speaking about orders on sets of interpretations, we will from now on always mean the information order defined by

$$\mathcal{I}_1 \sqsubseteq \mathcal{I}_2 \quad \text{iff} \quad \mathcal{I}_1 \supseteq \mathcal{I}_2$$

Proposition 6. *For every disjunctive program*

$$\text{lub}_{\sqsubseteq} \{ mf^n(\wp(B_P)) \mid n \in \mathbb{N} \}$$

is the smallest mf-stable class for every model filter mf.

This proposition justifies the name 'model filter': Initially we are given the entire set $\wp(B_P)$ of all interpretations. But iterative applications of mf will stepwise remove members from its input: $\wp(B_P) \supseteq mf(\wp(B_P)) \supseteq mf^2(\wp(B_P)) \supseteq \ldots$.

Theorem 7. *Let P be a disjunctive program and mf a model filter. Then* lfp (mf) *contains every stable model of P.*

Note that the theorem also holds for the trivial filter $mf_\perp$. Obviously, there are 'good' and 'bad' filters and Sections 5 and 6 will present operationalizations with different quality/efficiency-ratios. The quality of model filters can be expressed in terms of the information order

$$mf_1 \sqsubseteq mf_2 \quad \text{iff} \quad mf_1(\mathcal{I}) \sqsubseteq mf_2(\mathcal{I}) \quad \forall \mathcal{I} \subseteq \wp(B_P)$$

where $mf_\perp$ and $mf_\top$ are the smallest (worst) and greatest (best) model filter. Stable models can be characterized in terms of $mf_\top$.

Proposition 8. *An interpretation I is a stable model of a normal program, if and only if* $\{I\} = mf_\top(\{I\})$. *An interpretation I is a stable model of a disjunctive program, if and only if* $I \in mf_\top(\{I\})$.

Example 3. The relation between an assumptive interpretation I and the minimal models of P/I can be visualized by a graph with interpretations as nodes and where there is an arc from I to J, if J is a minimal model of P/I.

$P_1:$	$P_2:$	$P_3:$
$p \lor q$	$p \leftarrow \neg p$	$p \leftarrow \neg q$
$p \leftarrow \neg p$		$q \leftarrow \neg p$

For P_1 we obtain Figure 1.[1] P_1 has the stable model $\{\{p\}\}$, and $\{\{p\}, \{q\}\}$ is an $mf_\top$-stable class: In the $P_1/\{p\}$-transformation the rule is removed and the remaining disjunction has two minimal models.

P_2 has no stable model but an $mf_\top$-stable class $\{\emptyset, \{p\}\}$.

P_3 has two stable models $\{p\}$ and $\{q\}$. Both $\{\{p\}\}$ and $\{\{q\}\}$ are fixpoints of $mf_\top$, but also $\{\{p\}, \{q\}\}$, $\{\emptyset, \{p, q\}\}$ and $\{\emptyset, \{p\}, \{q\}, \{p, q\}\}$ are $mf_\top$-stable classes (see Figure 1).

Example 4. Consider the programs:

$P_4:$	$P_5:$
$p \lor q$	$p \lor q$
	$p \leftarrow \neg r$

---

[1] Stable models are indicated by loops in this graph.

Fig. 1. Minimal Model Relations for P_1 and P_3

P_4 has two stable models $\{p\}$ and $\{q\}$ and $\{\{p\},\{q\}\}$ is its unique mf_T-stable class.

P_5 has the unique stable model $\{p\}$ and:

$$mf_T(\wp(B_{P_5})) = \{\,\{p\},\{q\}\,\}$$
$$mf_T^2(\wp(B_{P_5})) = \{\,\{p\}\,\} = \mathrm{lfp}\,(mf_T)$$

3.3 Complete Model Filters

We will now propose a model filter which is distinctly different from the Baral-Subrahmanian style. Let us assume that we are given a 'reasonable' set $\mathcal{I}$ of interpretations (e.g. a set which contains all stable models). Once again each of its members yields some P/I. We now derive a new reasonable set from the given one:

- If an atom is true under every reasonable assumption, i.e. if it is true in every minimal model of P/I for every $I \in \mathcal{I}$, then it is regarded to be certainly true.
- If an atom is false under every reasonable assumption, i.e. if it is in no minimal model of P/I for any $I \in \mathcal{I}$, then it is regarded to be certainly false.

We will then derive from $\mathcal{I}$ every interpretation where every atom of the first kind is true and every atom of the second kind false. In formal terms this is expressible as:

$$mf_C(\mathcal{I}) := \{\, J \mid J \subseteq B_P,\ \bigcap_{I \in \mathcal{I}} \mathcal{MIN}(P/I) \subseteq J \subseteq \bigcup_{I \in \mathcal{I}} \mathcal{MIN}(P/I) \,\}$$

The mf_C-filter has an attractive property: It is closed under union and intersection and for $I_1, I_2 \in \mathcal{I}$ and $I_1 \subseteq J \subseteq I_2$ we have $J \in \mathcal{I}$ as well. Let such sets be called *complete*. A model filter mf is called *complete*, if $mf(\mathcal{I})$ is complete for every $\mathcal{I} \subseteq \wp(B_P)$. The collection $\mathcal{C}$ of all complete sets of interpretations and the set of all complete model filters are lattices once again; mf_C is the information-greatest complete model filter.

Example 5. For program P_5 the smallest fixpoints of mf_C and $mf_\top$ are identical although reached via different iterations. P_1 demonstrates that mf_C is strictly weaker than $mf_\top$:

$$\text{lfp}\,(mf_C) = \{\emptyset, \{p\}, \{q\}, \{p, q\}\} \sqsubseteq \{\{p\}, \{q\}\} = \text{lfp}\,(mf_\top)$$

The difference is that $P_1/\{p\}$ has two minimal models which constitute the fixpoint of $mf_\top$. As mf_C generates the intersection and union of these two models, the fixpoint contains the additional members $\emptyset$ and $\{p, q\}$.

4   Three-Valued Interpretations and Operator Pairs

4.1   Three-Valued Interpretations

Stable classes have been motivated by the facts that stable models do not always exist and that they are hard to compute. As an alternative way out three-valued instead of two-valued interpretations have been proposed. We will first analyze the relation between three-valued interpretations and sets of interpretations and then show, in how far three-valued interpretations support a constructive characterization of stable classes.[2]

Definition 9. A pair $\langle I_1, I_2 \rangle$ of interpretations is called a *three-valued interpretation*, if $I_1 \subseteq I_2$ or if $\langle I_1, I_2 \rangle = \langle B_P, \emptyset \rangle$.

A three-valued interpretation describes incomplete knowledge about the world. Its first component tells us which atoms must at least be true, and its second component which atoms can at most be true. Everything else must be false.

Definition 10. (Przymusinski [13]) Let P be a disjunctive program. A three-valued interpretation $\langle I_1, I_2 \rangle$ is called a *three-valued stable model* of P, if I_1 is a minimal model of P/I_2 and I_2 a minimal model of P/I_1.

While in Section 3 we have constantly been referring to minimal models of P/I for a given assumptive interpretation I, we will from now on make use of standard operational terms like:

$$S_P(I) := \text{lfp}\,(T_{P/I}) \quad \text{and} \quad S_P^\vee(I) := \text{lfp}\,(T_{P/I}^\vee) \cap B_P$$

Stable models of normal programs can then alternatively be characterized by the fixpoint equation $I = S_P(I)$.

Lemma 11. *The operators S_P and $S_P^\vee$ are anti-monotonic. Therefore S_P^2 and $S_P^{\vee^2}$ are monotonic.*

---

[2] Three-valued interpretations are usually denoted by a pair of true and false ground atoms. The subsequently chosen representation is more convenient for our purposes.

It has been observed by Baral and Subrahmanian that lfp (S_P^2) and gfp (S_P^2) provide a constructive characterization of certain stable classes for normal programs (see [1] and [2]). In the next section Baral and Subrahmanian's constructive characterization will be generalized for disjunctive programs. It will be formulated in terms of standard operators, such that existing implementations and algorithms can be exploited within query evaluation procedures for disjunctive programs (e.g. [17]).

Let $\mathcal{P}$ denote the set of all three-valued interpretations and define

$$\langle I_1, I_2 \rangle \preceq \langle J_1, J_2 \rangle \quad \text{iff} \quad I_1 \subseteq J_1 \text{ and } J_2 \subseteq I_2$$

and $\langle I_1, I_2 \rangle \preceq \langle B_P, \emptyset \rangle$.[3] Three-valued interpretations form a lattice isomorphic to complete sets of interpretations.

Theorem 12. *The set $\langle \mathcal{P}, \preceq \rangle$ is a complete lattice and there is a lattice isomorphism $h_{\mathcal{P} \to \mathcal{C}} : \mathcal{P} \to \mathcal{C}$.*

4.2 Three-Valued Operator Pairs

Model filters are transformation mappings on sets of interpretations. Correspondingly, a transformation for three-valued interpretations will now be defined. Every three-valued interpretation $\langle I_1, I_2 \rangle$ consists of two components: I_1 represents what must at least be true and I_2 what can at most be true. The transformation mapping will consist of two components as well: one for the derivation of necessary and another one for the derivation of potential truths.

Definition 13. Let g_1 and g_2 be two antimonotonic mappings on $\wp(B_P)$ which satisfy the following conditions for every $I \subseteq B_P$:

1. $g_1(I)$ is sound with respect to every minimal model of P/I
   (i.e. $g_1(I) \subseteq J \quad \forall J \in \mathcal{MIN}(P/I)$).
2. $g_2(I)$ is complete with respect to every minimal model of P/I
   (i.e. $g_2(I) \supseteq J \quad \forall J \in \mathcal{MIN}(P/I)$).

Then a mapping $p_{\langle g_1, g_2 \rangle}$ on three-valued interpretations with

$$p_{\langle g_1, g_2 \rangle} \langle I_1, I_2 \rangle := \langle g_1(I_2), g_2(I_1) \rangle$$

for $I_1 \subseteq I_2$ is called a *three-valued operator pair*.

The crucial feature and advantage of three-valued operator pairs is that both of its constituents are mappings on single interpretations. For practical applications these mapppings should be 'simple'.

For every operator pair $p_{\langle g_1, g_2 \rangle}$ there is a model filter $mf_{\langle g_1, g_2 \rangle}$ such that the following diagram commutes:

$$
\begin{array}{ccc}
\mathcal{P} & \xrightarrow{h_{\mathcal{P} \to \mathcal{C}}} & \mathcal{C} \\
{\scriptstyle p_{\langle g_1, g_2 \rangle}} \downarrow & & \downarrow {\scriptstyle mf_{\langle g_1, g_2 \rangle}} \\
\mathcal{P} & \xrightarrow{h_{\mathcal{P} \to \mathcal{C}}} & \mathcal{C}
\end{array}
$$

---

[3] In [13] this is called the Fitting-ordering in contrast to the standard ordering.

Hence, lfp $(p_{\langle g_1, g_2 \rangle}) \preceq \langle I_1, I_2 \rangle$ for every three-valued stable model $\langle I_1, I_2 \rangle$. For normal programs the well-founded semantics $p_{\langle S_P, S_P \rangle}$ (see e.g. [13]) commutes with mf_C.[4]

5 Operator Pair Based on WGCWA

The best complete model filter mf_C refers to the set of all atoms that are true in some minimal model of P/I. Note that this is nothing but the complement of the so-called *generalized closed world assumption* GCWA. As even for positive programs GCWA is hard to compute, the *weak generalized closed world assumption* WGCWA has been proposed as an efficient alternative (cf. [7] and [16] where it is called disjunctive database rule). Weak generalized closed world assumption also turns out to be an appropriate candidate for operator pairs.

Definition 14. Every normal rule $A_i \leftarrow L_1, \ldots, L_n$ $(1 \le i \le m)$ is a *normalization* of the disjunctive rule

$$A_1 \vee \ldots \vee A_m \leftarrow L_1, \ldots, L_n$$

Let P be a disjunctive program. Then P^{N+} denotes the *strong normal transformation* of P which is obtained by replacing every disjunctive rule by the set of all of its normalizations. A *weak normal transformation* of P is obtained by replacing every disjunctive rule by one of its normalizations.

Lemma 15. Let P be a disjunctive program. If A is true in some minimal model of P/I, then $A \in S_{P^{N+}}(I)$.

Theorem 16. The mapping $p_{\langle S_P^\vee, S_{P^{N+}} \rangle}$ defined by

$$p_{\langle S_P^\vee, S_{P^{N+}} \rangle} \langle I_1, I_2 \rangle := \langle S_P^\vee(I_2), S_{P^{N+}}(I_1) \rangle$$

is a three-valued operator pair.

As our approach refers to the weak generalized closed world assumption in the second component, disjunctions are read inclusively in the derivation of false atoms. As a consequence, lfp $\langle S_P^\vee, S_{P^{N+}} \rangle$ need not coincide with unique stable models (provided they exist), not even for stratified programs.

Example 6. P_5 is a stratified program and $\{p\}$ its unique stable (perfect) model. The least fixpoint of $p_{\langle S_P^\vee, S_{P^{N+}} \rangle}$ is $\langle \{p\}, \{p, q\} \rangle$. It yields the information that p is true and r false whereas the status of q remains undetermined. ∎

---

[4] This result is conceptually simpler than Baral and Subrahmanian's who have to define a Hoare ordering between stable classes in order to establish a correspondence between $p_{\langle S_P, S_P \rangle}$ and mf_T.

Of course, WGCWA provides only a rough estimation of what must be false. Its adequateness and limits will be demonstrated by the next example. It demonstrates that a database or knowledge-based system which implements operator pairs should support queries with respect to both the sound and the complete component of the smallest fixpoint: the former is referred to by queries of the "Which — do certainly —"-type and the latter by "For which — is there any evidence that —"-queries.

Example 7. The following zoological program has 16 stable models:

```
sparrow(a) ∨ bat(a)              sparrow(b) ∨ hawk(b)
sparrow(c) ∨ penguin(c)          ostrich(d) ∨ penguin(d)
bird(X) ← penguin(X)             bird(X) ← ostrich(X)
bird(X) ← sparrow(X)             bird(X) ← hawk(X)
ab1(X) ← penguin(X)              ab1(X) ← ostrich(X)
mammal(X) ← bat(X)               ab2(X) ← bat(X)
flyer(X) ← bat(X)                flyer(X) ← bird(X),¬ab1(X)
walker(X) ← penguin(X)           walker(X) ← mammal(X),¬ab2(X)
```

In our approach a query like "Which animals do certainly fly?" is answered with a and b and "For which animals is there any evidence that they fly?" is additionally answered with c. d does certainly not fly. Note that walker(c) and walker(d) are left undetermined, as penguin(c) and penguin(d) are true in a possible world of the upper viewpoint. The undefined truth value of walker(a) has a different reason: From a cautious viewpoint bat(a) is not known for sure. Hence ab2(a) is not known for sure and the upper viewpoint is based on the assumption that ¬ab2(a) is true.

6 Operator Pair Based on Normal Programs

The three-valued operator $p_{\langle S_P^{\vee}, S_{PN+}\rangle}$ is quite an unbalanced pair. Its second component is much more efficient than its first, as it refers to normal programs. The $S_P^{\vee}$-part cares for derivations from disjunctive data. Now as $p_{\langle S_P^{\vee}, S_{PN+}\rangle}$ has already been a compromise between efficiency and expressiveness, one might be interested in even more efficient alternatives. A normal program P^{N+} has been proposed for the complete component of the operator pair. We are now looking for another normal program P^{N-} which can be used for the sound component. A first candidate is easily found, namely the normal fragment of P — i.e. P without all of its non-normal rules.

Theorem 17. Let P be a disjunctive program with strong normalization P^{N+} and normal fragment P^{N-}. Then

$$p_{\langle S_{PN-}, S_{PN+}\rangle}\langle I_1, I_2\rangle := \langle S_{PN-}(I_2), S_{PN+}(I_1)\rangle$$

is a three-valued operator pair of P.

Example 8. For P_5 we obtain:

$$p_{\langle S_{pN-},S_{pN+}\rangle}(\emptyset, B_P) = \langle \emptyset, \{p,q\}\rangle = \mathrm{lfp}\,(p_{\langle S_{pN-},S_{pN+}\rangle})$$

Note that although all the disjunctive information is completely discarded in P^{N-}, this is still different from restricting oneself to normal programs: Disjunctions block the derivation of falsehoods (like p or q). What concerns positive inferences the normal fragment of a program is too rough an estimation. ∎

But when it is our goal to have efficient runtime query evaluation procedures, why should we not spend additional efforts during 'compile time'? We can especially perform derivations from disjunctive data at this early stage and enrich the normal fragment of P with derivable normal rules and facts which are true in every stable model. We thus obtain a two-phase approach:

- Disjunctive programs are used for data modeling and knowledge representation. Additional facts and normal rules can be derived from P offline in a preprocessing stage. By a kind of view materialization P^{N-} is the augmentation of the normal fragment of P by these implications.
- Normal programs are used for efficient query evaluation and runtime reasoning. The evaluation procedure refers to P^{N-} and P^{N+} in mutual recursion.

A more detailed description of the preprocessing phase can be found in [6].

7 State-Based Filters and Pairs

The formalism presented above can easily be adapted to alternative concepts. While the definition of model filters refers to models, one might as well consider disjunctive states. We will shortly sketch, how this view can be expressed within our framework.

Definition 18. A *disjunctive state* is a set S of ground disjunctions such that $\delta \in S$ implies $\delta' \in S$ for every clause $\delta' \supseteq \delta$.

For positive programs P we define $ST(P) := \{\delta \mid P \vdash \delta\}$ to be the set of all ground disjunctions that can be inferred from P. The Gelfond-Lifschitz transformation is extended from interpretations (i.e. sets of atoms) to sets of disjunctions.

Definition 19. A disjunctive state S is called a *stable state* of a disjunctive program P, if it is the set of all ground disjunctions implied by P/S:

$$S = ST(P/S)$$

Example 9. Consider the program:

    p ∨ q
    r ← ¬p

It has two stable models {p} and {q, r}, but a unique stable state, namely all disjunctions implied by {p ∨ q, r}. Atom r is true with respect to stable state semantics, but not under stable model semantics. The former interprets the negation as "It is not known that p", whereas the latter reads it as "There is no evidence for p". In fact, stable states are a special case of autoepistemic expansions (see [6]).

Definition 20. A state filter *sf* is a monotonic mapping on sets of states such that

$$sf(\mathcal{S}) \supseteq \{ T \mid T \text{ a disjunctive state,}$$
$$\text{there is } S \in \mathcal{S}, T = \mathcal{ST}(P/S) \}$$

A fixpoint of *sf* is called an *sf-stable class*.

The set of all disjunctive states ordered by information content is a complete lattice.[5]

Definition 21. Let g_1 and g_2 be two antimonotonic mappings on three-valued interpretations where for every I:

$$g_1(I) \subseteq \{ A \mid A \in B_P, P/I \vdash A \} \subseteq g_2(I)$$

Then a mapping $p_{\langle g_1, g_2 \rangle}$ on three-valued interpretations with

$$p_{\langle g_1, g_2 \rangle} \langle I_1, I_2 \rangle := \langle g_1(I_2), g_2(I_1) \rangle$$

is called a *state-based operator pair*.

Theorem 22. Let P be a disjunctive program, P^{N-} an enriched normal fragment and P^{N+} a weak normalization. Then

$$p_{\langle S_P^\vee, S_P^\vee \rangle} \langle I_1, I_2 \rangle := \langle S_P^\vee(I_2), S_P^\vee(I_1) \rangle$$
$$p_{\langle S_{P^{N-}}, S_{P^{N+}} \rangle} \langle I_1, I_2 \rangle := \langle S_{P^{N-}}(I_2), S_{P^{N+}}(I_1) \rangle$$

are state-based operator pairs.

Example 10. Consider the disjunctive statement guilty(a) ∨ guilty(b). In a state-based interpretation the rule innocent(X) ← ¬guilty(X) stands for the principle that someone is to be presumed innocent as long as his guilt has not been proved: $p_{\langle S_P^\vee, S_P^\vee \rangle}$ yields the falsehood of both guilty(a) and guilty(b) and the truth of innocent(a) and innocent(b). The model-based approach represents a policeman's perspective. The rule then reads: Only when there is no evidence whatever for someone's guilt, then he can be assumed to be innocent ($p_{\langle S_P^\vee, S_{P^{N+}} \rangle}$ leaves the status of guilty(a), innocent(a), guilty(b) and innocent(b) undetermined).

---

[5] As is the isomorphic set of minimal representations of disjunctive states. In fact, it would be more adequate to study such minimal representations instead of interpretations in the current section (see [6]). We prefer to use interpretations just for the sake of brevity.

8   Related Work

As Baral and Subrahmanian [2] mention, Van Gelder's 'alternating fixpoint construction' [21] is essentially equivalent to their constructive characterization of stable classes. Both approaches confine their attention to normal programs. Fitting [3] has recently studied the connections between extremal models in the broad setting of bilattices. All these approaches are based a single operator, whereas we pointed out how the application of operator pairs to three-valued interpretations allows to capture disjunctive programs as well.

Przymusinski [14] presents a very general framework for nonmonotonic reasoning which also makes use of two different inference operators. Operator pairs as presented in our paper are a kind of instantiation of Przymusinski's schema. The formulation in terms of standard logic programming operators allows to reuse existing implementation techniques.

Baral and Subrahmanian [1] present a generalization of their approach for autoepistemic reasoning which is, however, not sound with respect to Moore's Autoepistemic Logic [12]. As proved in [6], mf_c can be given a sound generalization in the autoepistemic case.

Knowledge Compilation has been studied by Selman and Kautz in a series of papers (e.g. [18]) where they concentrate on the monotonic case.

9   Conclusion

We have shown that the notion of stable classes can be naturally extended to disjunctive programs and that stable classes exist for a whole lattice of filter operators. Three-valued interpretations are isomorphic to complete sets of interpretations. Three-valued operator pairs as transformations on three-valued interpretations can be seen as operationalizations of stable classes.

Two operator pairs for disjunctive programs have been proposed, namely one based on the Minker-Rajasekar-operator together with WGCWA and another exclusively based on normal programs. By the previous results both are sound with respect to three-valued (and hence two-valued) stable model semantics.

When the concept of models is replaced by disjunctive states, stable states and state-based operator pairs can be defined. From a state-based perspective negation is interpreted distinctly different. We have finally proposed an operator pair for the state-based approach.

References

1. Chitta Baral and V. S. Subrahmanian. Dualities between alternative semantics for logic programming and nonmonotonic reasoning. In LPNMR91 [8], pages 69–86.
2. Chitta R. Baral and V. S. Subrahmanian. Stable and extension class theory for logic programs and default logics. *Journal of Automated Reasoning*, 8:345–366, 1992.

3. Melvin Fitting. The family of stable models. *Journal of Logic Programming*, 17:197–225, 1993.

4. Michael Gelfond and Vladimir Lifschitz. The stable model semantics for logic programming. In *Proc. of the 5th Int. Conf. and Symp. on Logic Programming*, pages 1070–1080. MIT Press, 1988.

5. Michael Gelfond and Vladimir Lifschitz. Classical negation in logic programs and disjunctive databases. *New Generation Computing*, 9:365–385, 1991.

6. Jürgen Kalinski. *Weak Autoepistemic Reasoning — A Study in Autoepistemic Reasoning from a Logic Programming Perspective*. PhD thesis, Universität Bonn, 1994.

7. Jorge Lobo, Jack Minker, and Arcot Rajasekar. *Foundations of Disjunctive Logic Programming*. MIT Press, 1992.

8. *Proc. of the 1st Int. Workshop on Logic Programming and Non-monotonic Reasoning*. MIT Press, 1991.

9. Wiktor Marek and Miroslaw Truszczyński. Autoepistemic logic. *Journal of the ACM*, 38(3):588–619, 1991.

10. Wiktor Marek and Miroslaw Truszczyński. Computing intersection of autoepistemic expansions. In LPNMR91 [8], pages 37–50.

11. Jack Minker and Arcot Rajasekar. A fixpoint semantics for disjunctive logic programs. *Journal of Logic Programming*, 9:45–74, 1990.

12. Robert C. Moore. Semantical considerations on nonmonotonic logic. *Artificial Intelligence*, 25(4):75–94, 1985.

13. Teodor C. Przymusinski. Extended stable semantics for normal and disjunctive programs. In *Proc. of the 7th Int. Conf. on Logic Programming*, pages 459–477. MIT Press, 1990.

14. Teodor C. Przymusinski. Autoepistemic logics of closed beliefs and logic programming. In LPNMR91 [8], pages 3–20.

15. Teodor C. Przymusinski. Stable semantics for disjunctive programs. *New Generation Computing*, 9:401–424, 1991.

16. K. A. Ross and R. W. Topor. Inferring negative information from disjunctive databases. *Journal of Automated Reasoning*, 4:397–424, 1988.

17. Dietmar Seipel. Non-monotonic reasoning based on minimal models and its efficient implementation. In Bernd Wolfinger, editor, *Innovationen bei Rechen- und Kommunikationssystemen, GI-Fachgespräch: Disjunctive logic programming and disjunctive databases*, pages 53–60. Springer, 1994.

18. Bart Selman and Henry Kautz. Knowledge compilation using Horn approximations. In *Proc. of the 9th National Conf. on Artificial Intelligence*, pages 904–909, 1991.

19. M. H. van Emden and R. A. Kowalski. The semantics of predicate logic as a programming language. *Journal of the ACM*, 23(4):733–742, 1976.

20. A. Van Gelder, K. A. Ross, and J. S. Schlipf. The well-founded semantics for general logic programs. *Journal of the ACM*, 38(3):620–650, 1991.

21. Allen Van Gelder. The alternating fixpoint of logic programs with negation. In *Proc. of the 8th Symp. on Principles of Database Systems*, pages 1–10, 1989.

Nonmonotonicity and Answer Set Inference

David Pearce

Deutsches Forschungszentrum
für Künstliche Intelligenz (DFKI)
Stuhlsatzenhausweg 3
66123 Saarbrücken, Germany
pearce@dfki.uni-sb.de

Abstract

The study of abstract properties of nonmonotonic inference has thrown
up a number of general conditions on inference relations that are often
thought to be desirable, and sometimes even essential, for an adequate
system of nonmonotonic reasoning. However, several of the key conditions
on inference that have been proposed in the literature make explicit reference to the *classical* concept of logical consequence, and there is a general
tendency to focus attention on inference operations that are *supraclassical*
in the sense of extending classical consequence. Against this trend I argue
for the importance of systems that are not supraclassical. I suggest that
their inference relations should measure up to adequacy conditions that
are more sensitive to the style of reasoning for which they are intended,
and which take account of the underlying logic of the monotonic subsystem, if such a subsystem can be identified. I illustrate these points by
considering some properties of the inference relation associated with the
answer set semantics of extended disjunctive databases.

1   Introduction

Systems of nonmonotonic reasoning have won wide acceptance, especially in
the artificial intelligence community, for their ability to handle various forms of
commonsense inference. Since the pioneering papers of [7], [14] and [13], logicians
have also studied the abstract properties of inference relations that fail to satisfy
monotonicity, and a broad classification of types of nonmonotonic inference is
now available. A thorough and highly readable survey of these, together with
a discussion of the major systems proposed in AI, can be found in a recent
handbook article by David Makinson, [15].

This paper addresses an issue arising out of studies like those of [13] and
[15]. I shall focus mainly on [15], since it provides a very broad classification
scheme and is likely to become a standard reference work in the area. Makinson, as is customary, distinguishes between inference relations, or consequence

operations, that extend classical logic, and those which do not; the former being known as *supraclassical*. Several well-known formalisations of nonmonotonic reasoning are, indeed, supraclassical; but there is a number of interesting and important systems that are not. They include: (i) general systems of nonmonotonic inference designed to extend a particular standard nonclassical logic, eg. intuitionistic logic; (ii) systems like those of logic programming and deductive databases which embody (explicitly or implicitly) certain constructive forms of reasoning. The latter have become especially important in recent years, due to an increasing interaction between the fields of nonmonotonic reasoning and logic programming. In particular, several approaches to the semantics of extended logic programs have been directly inspired by nonmonotonic formalism developed in AI; and, conversely, insights drawn from logic programming have been applied to tackle certain concrete problems of nonmonotonic reasoning as well as to help design more adequate general formalisms.

In view of this interaction between the two areas it is important to acquire a clear understanding of the general properties of the inference relations arising in logic programming and formalisms inspired by it. With respect to properties like *cumulativity, rationality* and several others Jürgen Dix in [2] and several later works has undertaken a rather exhaustive analysis of the major semantics proposed for *normal* and *disjunctive* logic programs. And, with the help of such properties, he has formulated certain conditions of adequacy that logic programming semantics ought to fulfil. Similarly, Makinson [15] includes a particular system of logic programming, the stable model semantics of Gelfond & Lifschitz [9], in his table of nonmonotonic systems and their properties. However, under classification schemes like Makinson's several key properties of inference relations are well-defined only for systems that are closed under Boolean connectives and are therefore not applicable to formalisms employing a restricted syntax, like that of Horn clauses and certain extensions of them. Thus, while he is able to classify the consequence operation of the stable model semantics as failing cumulativity, some other key properties, like supraclassicality and distribution, are simply non-applicable.

To a large extent, however, the situation changes if one takes account of recent developments in *extended logic programming* that have taken place mainly since [15] was written. For example, the stable model semantics has been generalised to apply to so-called general disjunctive programs or extended disjunctive databases, where a controlled use of disjunction and negation is permitted. Although in their usual presentations they are not fully closed under Boolean connectives, the syntax of such systems is rich enough to allow one to express and to study additional properties of the associated consequence operation. This will be one of my aims in this note.

There is also a second, more philosophical point to be made. Studies like [13], [15] and [2] not only have the descriptive aim to classify properties of nonmonotonic consequence operations, they also have a normative or methodological aim to select, from among a range of different properties, those which, for one purpose or another, might be considered particularly desirable: eg. in

Makinson's case most notably *cumulativity* and *distribution*. Now, in some cases the desirability of such properties seems to be intimately related to the question whether the consequence operation under scrutiny is or is not supraclassical. If, for instance, the monotonic subpart of the system is classical, and the system as a whole supraclassical, one may well expect certain properties of classical consequence (apart, of course, from monotonicity itself) to continue to hold for the nonmonotonic extension; while properties that do not hold classically, like standard constructivity properties, will continue to fail. On the other hand, if the system possesses, say, a constructive monotonic subsystem, and on this ground fails supraclassicality, then one will not expect its consequence operation to satisfy certain overtly classical conditions on inference, whereas one might hope that properties acceptable to constructive inference in general continue to hold for the extended nonmonotonic system. I shall illustrate these points with respect to the inference relation of the answer set semantics of logic programs and disjunctive databases.

2 General Properties of Nonmonotonic Inference

The consequence operation C of any ordinary deductive logic satisfies the well-known closure conditions of Tarski: *inclusion* ($A \subseteq C(A)$), *idempotence* ($C(A) = C(C(A))$) and *monotony* ($B \subseteq A \Rightarrow C(B) \subseteq C(A)$); where A, B are sets of propositions in some language **L**. Expressed in terms of an inference relation $\vdash \subseteq 2^{\mathbf{L}} \times \mathbf{L}$, the three conditions become *reflexivity*, *cut* and *monotony*, formulated respectively as:

$$\varphi \in A \;\;\Rightarrow\;\; A \vdash \varphi$$
$$A \vdash \psi_i, \forall i \in I \text{ and } A \cup \{\psi_i : i \in I\} \vdash \varphi \;\;\Rightarrow\;\; A \vdash \varphi$$
$$B \vdash \varphi \text{ and } B \subseteq A \;\;\Rightarrow\;\; A \vdash \varphi$$

The study of general properties of inference relations $\vdash \subseteq 2^{\mathbf{L}} \times \mathbf{L}$ or operations $C : 2^{\mathbf{L}} \longrightarrow 2^{\mathbf{L}}$ that do not satisfy monotony, but are in other respects 'well-behaved', now has a well-established tradition. Makinson [15], who surveys the field and studies the properties of many nonmonotonic inference relations that have been proposed in the literature, distinguishes between 'pure' conditions that only make reference to $\vdash$ or C, and 'mixed' conditions that involve the connectives of the object language. Besides the pure conditions of reflexivity and cut that may remain on dropping or weakening full monotony, the most discussed has been that of *cautious monotony*:

$$A \subseteq B \subseteq C(A) \;\;\Rightarrow\;\; C(A) \subseteq C(B). \tag{1}$$

Alongside these three pure conditions, Makinson gives special prominence to two mixed conditions on inference: *supraclassicality* and *distribution*. They may be called *core* conditions, since, under suitable assumptions, they entail several other conditions that have been studied in nonmonotonic reasoning. Supraclassicality is simply the condition that C extends the usual consequence operation Cn of classical logic, ie. $Cn \leq C$, or for all $A \subseteq \mathbf{L}$,

$$Cn(A) \subseteq C(A).$$

It is clear that also *distribution* ranks very high on Makinson's list of desirable properties.[1] For a consequence operation C, distribution is formulated as the condition

$$C(A) \cap C(B) \subseteq C(Cn(A) \cap Cn(B))$$

for any $A, B \subseteq \mathbf{L}$, where again Cn is the usual consequence operation of classical logic. He writes:

> The condition of distribution is perhaps too complex to carry an immediate intuitive conviction. Its justification lies in its immense power, its intuitively appealing instances, and the fact that it is satisfied by some important forms of nonmonotonic reasoning... [15, §2.2]

Now, under this formulation, classical consequence is used in expressing distribution. Indeed, all the further implications of distribution are discussed in the context of *supraclassical* inference relations. Moreover, in Makinson's classification table, distribution is only considered with respect to systems closed under Boolean connectives, and, among them, only two systems that are not supraclassical (where distribution also fails). When the corresponding consequence operation *is* supraclassical and satisfies absorption (see §3 below), however, Makinson observes that distribution entails three well-known properties of classical inference, expressed for the corresponding $\vdash$ relation as:

$$A \cup \{\varphi\} \vdash \alpha \quad \text{and} \quad A \cup \{\psi\} \vdash \alpha \;\Rightarrow\; A \cup \{\varphi \vee \psi\} \vdash \alpha \qquad (2)$$

$$A \cup \{\varphi\} \vdash \alpha \quad \text{and} \quad A \cup \{\sim\varphi\} \vdash \alpha \;\Rightarrow\; A \vdash \alpha \qquad (3)$$

$$A \cup \{\varphi\} \vdash \psi \;\Rightarrow\; A \vdash \varphi \to \psi \qquad (4)$$

perhaps these are what Makinson has in mind by the "intuitively appealing instances" of distribution. (2) is called *disjunction in the antecedent*, (3) *proof* or *reasoning by cases*, and (4) *conditionalisation*. Let us analyse them first with respect to some standard (monotonic) inference relations $\vdash$ of deductive logic. Although, '$\to$' in (4) is intended to be material (truth-functional) implication, clearly we can evaluate this property in the context of any logic admitting an implication connective. Since (4) is satisfied by any logic possessing a deduction theorem, it holds in most of the standard systems of constructive and many-valued logic, like Heyting's intuitionistic logic, H, Nelson's constructive logic, N, and Lukasiewicz's 3-valued logic, L.[2] Indeed, the possession of a deduction

---

[1] For instance, of Reiter's default logic, he writes: "Now we pass to the bad news, concerning [the failure of] distribution and cautious monotony." [15, §3.2]

[2] See any standard text on logical systems, eg. [23].

theorem is sometimes regarded as an adequacy condition that an implication connective '$\rightarrow$' should fulfil. It fails, however, for some systems, like strong Kleene 3-valued logic, where although $p \vdash p$ holds, in general $\vdash p \rightarrow p$ does not.

The status of (2) and (3), on the other hand, is intimately tied to the nature of the $\vdash$ relation. In particular, the $\vdash$ relations of constructive systems, like H and N, satisfy (2) but not (3), since reasoning by cases is not a constructively approved principle.[3] Although (2) holds, it cannot, of course, be used to derive (3) (by setting $\psi = \sim \varphi$), since in these logics $\varphi \vee \sim \varphi$ is not a tautology.

As Makinson observes, there are nonmonotonic systems, like Reiter's default logic (DL) [21], whose inference relations are supraclassical and satisfy absorption, but where distribution fails. In fact, more than one 'instance' of distribution fails. Makinson himself provides, for DL, a simple counterexample to (3). Similarly, DL does not support disjunction in the antecedent:

Example 1 *Let p, q, r be distinct atomic propositons and let D consist of two rules $p : /r$ and $q : /r$. Then adding to these rules either the fact p or the fact q leads to a theory that, in each case, has r in its extension. But r is not entailed by the theory comprising just D and the disjunction $p \vee q$.*

So, despite its supraclassicality, Reiter's inference relation fails even some principles that are constructively sound.

3   A Case Study: Answer Set Semantics

In this section I want to consider a particular approach to logic programming semantics, that of *answer sets*, due to Gelfond & Lifschitz [10], and to analyse how its associated nonmonotonic inference relation fares with respect to some of the general conditions on inference discussed by Makinson. There are several reasons to take a special interest in the approach of Gelfond & Lifschitz. First, it is uniformly applicable to logic programs or deductive databases in which, in addition to negation-as-failure, a second, explicit or *strong* negation is present (denoted here by '$\sim$'), and where disjunction is permitted in the 'heads' of database rules. Thus, it generalises both extended and disjunctive logic programming. Moreover, Michael Gelfond [8] has recently further extended the approach to cover the use of epistemic operators, and I shall make use here of part of his extended semantics. Secondly, the semantics relates closely to several well-established general nonmonotonic formalisms, such as default logic and nonmonotonic modal logics, including autoepistemic logic. Thirdly, although answer sets are characterised by a fixed-point construction, and the semantics has a partly 'procedural' flavour, it breaks down very neatly into a monotonic, purely logical part, supplemented by a nonmonotonic, procedural device; moreover, inference on the monotonic subpart admits a clear characterisation in terms of a well-known standard logical system. Finally, unlike some variants of the well-founded semantics for logic programs, the answer set approach does not in general yield unique intended

---

[3](2) holds for these logics, since each has $(\alpha \rightarrow \gamma) \rightarrow ((\beta \rightarrow \gamma) \rightarrow (\alpha \vee \beta \rightarrow \gamma))$ as one of its standard axioms. Whence (2) follows by two applications of the deduction theorem.

models. The presence of a plurality of models or answer sets, like the plurality of extensions in default logic, seems to be responsible for the failure of some conditions on nonmonotonic inference, notably cautious monotony. However, even when inference is defined in the sceptical sense (as truth in all answer sets) the semantics is less 'sceptical' than some based on well-founded models, and yields in some instances intuitively more convincing results.

The formulas of extended disjunctive databases (XDDBs) have the form

$$L_1 \wedge \ldots \wedge L_m \wedge not\ L_{m+1} \wedge \ldots \wedge not\ L_n \rightarrow K_1, \vee \ldots \vee K_k \tag{5}$$

where each L_i, K_j is a literal (atom or strongly negated atom) and we may have $m = n$ and m or n may be zero. An XDDB is a set of such formulas, and an extended logic program or XLP Π is a set of such formulas, with $k = 1$. The set of all ground literals in the language of Π is denoted by Lit. Since answer sets are defined for databases without variables, each formula of form (5) is treated as shorthand for the set of its ground instances. The above is only a notational variant of the formalism of [10], which regards (5) as a rule and adopts different symbols for conjunction, disjunction and implication.

Let Π be an XDDB without '*not*'. Adapting the definition of [10] to the above notation, we recall that an *answer set* of Π is a minimal (under set-theoretic inclusion) subset S of Lit such that

(A1) for each formula $L_1 \wedge \ldots \wedge L_m \rightarrow K_1 \vee \ldots \vee K_k$ of Π, if $L_1, \ldots, L_m \in S$ then, for some $i = 1, \ldots, k$, $K_i \in S$;

(A2) if S contains a pair of complementary literals of the form $A, \sim A$ (where A is an atom), then $S = Lit$.

If a set S of literals fulfils (A1) and (A2) for each formula of Π, we say that S *satisfies* the formulas of Π. Further, we call an answer set S of Π *noncontradictory* if it does not contain a complementary pair of literals.

Now, let Π be an arbitrary XDDB. For any set of ground literals $S \subset Lit$, the database Π^S is the database without '*not*' obtaining from Π by deleting

(i) each formula containing a subformula *not* L with $L \in S$, and

(ii) all subformulas of the form *not* L in the remaining formulas.

Since the transformed database does not contain '*not*', its answer sets are defined (as previously). Then a set S of ground literals is said to be an *answer set* of an extended database Π if and only if S is an answer set of Π^S.

A database is called *inconsistent* if it possesses a contradictory answer set. Gelfond and Lifschitz show that an inconsistent database has Lit as its unique answer set, and that some (consistent) databases possess no answer sets.

It is clear that answer set consequence is normally intended to be understood in the sceptical sense, ie. a database Π entails a literal L iff L belongs to each answer set of Π. In order to extend this relation to Boolean compound formulas, we can make use of an extension of the semantics employed by Gelfond [8]. Adapting Gelfond's approach to the case of XDDBs, truth ($\models$) and falsity ($\dashv$)

of formulas in $(\wedge, \vee, \sim)$ wrt a set S of ground literals, is defined by:

For atomic A, $S \models A$ if $A \in S$; $S \dashv A$ if $\sim A \in S$

$S \models \sim A$ if $S \dashv A$; $S \dashv \sim A$ if $S \models A$

$S \models \varphi \wedge \psi$ if $S \models \varphi$ and $S \models \psi$; $S \dashv \varphi \wedge \psi$ if $S \dashv \varphi$ or $S \dashv \psi$

And $(\varphi \vee \psi)$ is defined as $\sim (\sim \varphi \wedge \sim \psi)$.

In this manner, we can define a nonmonotonic inference relation $\vdash$ between databases and formulas in $(\wedge, \vee, \sim)$, by

$$\Pi \vdash \varphi \quad \text{iff} \quad S \models \varphi \text{ for all answer sets } S \text{ of } \Pi.$$

For the remainder of this section '$\vdash$' will be this inference relation corresponding to answer set semantics. A simple consequence of the above truth conditions for $\models$ that will be useful later is the following. If S is any consistent set of literals and φ an implication-free database formula, ie. of form (5) with $n = 0$, then S satisfies φ in the sense mentioned above if and only if $S \models \varphi$. Similarly, S satisfies a *not*-free database formula $L_1 \wedge \ldots \wedge L_m \rightarrow K_1 \vee \ldots \vee K_k$ if and only if $S \models K_1 \vee \ldots \vee K_k$ whenever $S \models L_1 \wedge \ldots \wedge L_m$.

The answer set semantics conservatively extends the stable model semantics for normal programs. It is readily seen that its inference relation $\vdash$ is not supraclassical. Indeed, several principles of classical reasoning fail, including the *excluded middle*, and *contraposition*. With respect to the above three properties of (supraclassical) distribution, we can note the following. Condition (4) does not apply, since we have no implication connective on the right of $\vdash$. However the remaining conditions are applicable, provided the syntax of the formulas φ and ψ is suitably restricted. The following simple example shows the failure of (3), as well as of the principles just mentioned.

Example 2 *Let Π be the database $\{p \rightarrow q\}, \{\sim q\}$, where p, q are atoms. Then (i) not $\Pi \vdash \sim p$ (failure of contraposition); (ii) $\Pi \cup \{p\} \vdash p \vee \sim p$ and $\Pi \cup \{\sim p\} \vdash p \vee \sim p$, but not $\Pi \vdash p \vee \sim p$ (failure of excluded middle and of proof by cases).*[4]

On the other hand we have:

Observation 1 $\vdash$ *satisfies disjunction in the antecedent.*

Proof. Suppose $\Pi \cup \{\bigvee L_i\} \vdash \varphi$ and $\Pi \cup \{\bigvee K_j\} \vdash \varphi$, for literals L_i, K_j. Let S be any answer set of $\Pi \cup \{\bigvee L_i \vee \bigvee K_j\}$. We have to show that $S \models \varphi$. Since S is an answer set of $\Pi \cup \{\bigvee L_i \vee \bigvee K_j\}$, it is, by definition, an answer set of $(\Pi \cup \{\bigvee L_i \vee \bigvee K_j\})^S$, and the latter, by construction, is the same as $\Pi^S \cup \{\bigvee L_i \vee \bigvee K_j\}$. So S is a minimal subset of literals satisfying the formulas of Π^S together with the condition $L_i \in S$, for some i, or $K_j \in S$, for some j.

---

[4]Actually, (i) shows the failure of *modus tollens*, although it is easily seen that Π is not equivalent under $\vdash$ to the contrapositionally equivalent database $\Pi' = \{\sim q \rightarrow \sim p\}, \{\sim q\}$, since Π' derives $\sim p$.

Suppose $L_i \in S$, for some i. Then S is a minimal set satisfying the formulas of Π^S together with the condition $L_i \in S$. So S is an answer set of $\Pi^S \cup \{\bigvee L_i\}$, which, again by construction, equals $(\Pi \cup \{\bigvee L_i\})^S$. So S is an answer set of $\Pi \cup \{\bigvee L_i\}$, and therefore $S \models \varphi$, by the first hypothesis of the proposition. Similarly, if some $K_j \in S$, then S is an answer set of $\Pi \cup \{\bigvee K_j\}$, and by the second hypothesis $S \models \varphi$. $\square$

So much for the 'instances' of distribution. Let us now consider the 'pure conditions' of nonmonotonic consequence. The principle properties usually singled-out are those of *inclusion* (or *reflexivity*), *cut* and *cautious monotony*; any inference relation satisfying all three being called *cumulative*. These three conditions are applicable, providing that appropriate syntactic restrictions are imposed on the formulas present. Using such restrictions Makinson shows, for example, that the stable model semantics satisfies both reflexivity and cut. For answer set inference on XDDBs, Π, the most general form of these conditions becomes:

$$\varphi \in \Pi \quad \Rightarrow \quad \Pi \mathrel{\vdash\!\!\sim} \varphi \tag{6}$$

$$\Pi \mathrel{\vdash\!\!\sim} \varphi \text{ and } \Pi \cup \{\varphi\} \mathrel{\vdash\!\!\sim} \psi \quad \Rightarrow \quad \Pi \mathrel{\vdash\!\!\sim} \psi \tag{7}$$

where φ is a disjunction of ground literals and ψ any Boolean sentence.[5]

Observation 2 *Answer set inference, $\mathrel{\vdash\!\!\sim}$, satisfies reflexivity and cut.*

Proof. Reflexivity is trivial by the construction of answer sets. To verify cut, suppose that (i) $\Pi \mathrel{\vdash\!\!\sim} \varphi$ and (ii) $\Pi \cup \{\varphi\} \mathrel{\vdash\!\!\sim} \psi$, where φ is a disjunction of ground literals, say $\bigvee L_i$. Let S be any answer set of Π. Then we have to show that $S \models \psi$. By (i), S is a minimal subset of literals satisfying Π^S, and moreover the condition $L_i \in S$, for some i. So S is a minimal set satisfying $\Pi^S \cup \{\bigvee L_i\} = (\Pi \cup \{\bigvee L_i\})^S = (\Pi \cup \{\varphi\})^S$. Therefore S is an answer set of $\Pi \cup \{\varphi\}$, and, by (ii), $S \models \psi$. $\square$

However, $\mathrel{\vdash\!\!\sim}$ fails cautious monotony. In fact, in [10] a correspondence between Reiter's default extensions and answer sets is established for the case of extended logic programs, where no disjunction is present. Since, as Makinson [15] points out, there are examples of default theories not involving disjunction that fail cautious monotony, these carry over to the case of answer sets. It follows that $\mathrel{\vdash\!\!\sim}$ is not a cumulative relation.

Among the 'non-pure' conditions of nonmonotonic inference, Makinson draws particular attention to two properties where negation is essentially involved. They are:

$$A \mathrel{\vdash\!\!\sim} \psi \quad \text{and} \quad A \cup \{\varphi\} \mathrel{\vdash\!\!\sim} \sim\psi \quad \Rightarrow \quad A \mathrel{\vdash\!\!\sim} \sim\varphi \tag{8}$$

$$A \mathrel{\vdash\!\!\sim} \psi \quad \text{and not} \quad A \cup \{\varphi\} \mathrel{\vdash\!\!\sim} \psi \quad \Rightarrow \quad A \mathrel{\vdash\!\!\sim} \sim\varphi \tag{9}$$

---

[5]Actually, we can also allow φ to be any Boolean formula by expressing it (via the de Morgan rules) in conjunctive normal form and adopting the convention that $\bigwedge_{i \in I} \varphi_i \in \Pi$ if $\varphi_i \in \Pi$ for all $i \in I$, and $\Pi \cup \{\bigwedge_i \varphi_i\} = \Pi \cup \{\varphi_i : i \in I\}$. For $\mathrel{\vdash\!\!\sim}$, the slightly more general form of cut, where a set of formulas appears in place of φ, can easily be deduced from (7.

(8), which is not named, is discussed in his (Observation 2.2.5) and given roughly the interpretation that only contra-indicated propositions yield contra-indicated results. (9) is known as *rationality* and is analysed at length by Dix [2] in connection with logic programming semantics.[6] A simple example shows that answer set inference fails both conditions.

Example 3 *Let Π be the database $\{p \to \sim q, not\ p \to q\}$, for distinct atoms, p, q. We have: $\Pi \mathrel{\mid\!\sim} q$, not $\Pi \cup \{p\} \mathrel{\mid\!\sim} q$ and $\Pi \cup \{p\} \mathrel{\mid\!\sim} \sim q$ but not $\Pi \mathrel{\mid\!\sim} \sim p$.*

Notice however that, in translating the above conditions into the language of disjunctive databases the inference in each case is to the *strong* negation of the proposition, ie. $\Pi \mathrel{\mid\!\sim} \sim p$, which, in turn, means that p is explicity false in each answer set S of Π (ie. $S \mathrel{=\mid} p$). This illustrates a difference with respect to the stable model semantics for normal programs containing only one sort of negation. For, in the latter case an atom is regarded as false if it is not true in all models: querying the database yields only the answers *yes* or *no*, rather than *yes, no* or *unknown*. Weakening the conclusion of (8), for instance, one obtains a variant that is satisfied by answer set inference.

Observation 3 *Let Π be a consistent database, φ a disjunction of literals, and ψ a Boolean formula. If $\Pi \mathrel{\mid\!\sim} \psi$ and $\Pi \cup \{\varphi\} \mathrel{\mid\!\sim} \sim \psi$, then φ is not true in any answer set of Π.*

Proof. Assume the hypotheses of the proposition for some $\varphi := \bigvee L_i$, where the L_i are literals. Suppose for the contradiction that Π possesses an answer set S such that $S \models \bigvee L_i$. Since S is an answer set of Π^S satisfying the condition $L_i \in S$ for some i, S is also an answer set of $\Pi^S \cup \{\bigvee L_i\} = (\Pi \cup \{\varphi\})^S$. So S is an answer set of $\Pi \cup \{\varphi\}$ and, by the first hypothesis, $S \models \psi$, by the second $S \models \sim \psi$, contradicting the assumption that Π is consistent. $\square$

The conclusion of this condition, that φ is not true in *any* answer set of Π, might be expressed by saying that Π *weakly* falsifies φ, in contrast to the *strong* falsification of φ expressed in the conclusions of (8) and (9). I shall argue below that the condition expressed in Observation 3 is more reasonable than either (8) or (9).

Let us now return to the matter of supraclassicality. It transpires that answer set inference fails supraclassicality, not in a random or *ad hoc* manner, but in a fully systematic fashion. First, note that on databases or programs without '*not*', the inference relation is monotonic; so nonmonotonicity arises entirely due to the presence of negation-as-failure. In fact, for *not*-free databases, the inference relation can be fully characterised by the following (see the author's ([18]):

Proposition 1 *Let Π be a disjunctive database without 'not' and φ any Boolean formula. Then $\Pi \mathrel{\mid\!\sim} \varphi$ iff $\Pi \vdash_N \varphi$, where $\vdash_N$ is the inference relation of Nelson's constructive logic.*

---

[6] (9) is also called *rational monotony* and is usually formulated as the principle: if $A \mathrel{\mid\!\sim} \psi$ and not $A \mathrel{\mid\!\sim} \sim \varphi$, then $A \cup \{\varphi\} \mathrel{\mid\!\sim} \psi$. As noted in [15], the two formulations are equivalent.

This property can also be extended to open databases and existential queries.

It follows that $\vdash$ can be regarded as a nonmonotonic extension of the constructive inference relation $\vdash_N$. In what sense, however, might one claim that the consequence operation, say C, corresponding to $\vdash$ is *supraconstructive* in that $Cn_N \leq C$, where Cn_N denotes N-consequence? Clearly, one has to take account of the special syntax of databases. First, $\vdash$ takes only Boolean formulas on the right; secondly, on the left it allows databases containing an additional negation, '*not*'. An obvious posssibility is to consider Cn_N restricted to sets of Boolean formulas. Thus, let $\mathcal{L} = \mathcal{L}(\wedge, \vee, \sim)$ be the collection of all Boolean formulas. And define $C_N : 2^{\mathcal{L}} \longrightarrow 2^{\mathcal{L}}$, by $C_N = Cn_N \cap \mathcal{L}$. Notice that C_N is a consequence operation satisfying inclusion, idempotence, monotony and compactness that is closely related to 3-valued or partial logic.[7] Furthermore, if S is a set of literals, $\mathbf{S}$ a collection of sets of literals, and $\models$ the truth relation defined earlier, we set

$$Th(S) \quad := \quad \{\varphi \in \mathcal{L} : S \models \varphi\} \tag{10}$$

$$Th(\mathbf{S}) \quad := \quad \{\varphi \in \mathcal{L} : \forall S \in \mathbf{S}, S \models \varphi\} \tag{11}$$

Thus, if $\mathbf{S}$ is the collection of all answer sets of a database Π, then $\Pi \vdash \varphi$ iff $\varphi \in Th(\mathbf{S})$.

Analogously to the case of default theories, where facts are distinguished from rules, we can regard any database Π as comprising a *disjunctive fact base* A, consisting of the set of formulas (5) where $n = 0$, together with a *rule base*, $\mathcal{R}$, consisting of the remaining formulas (5) with $n > 0$. So any Π can be represented in the form $(A, \mathcal{R})$. Then, with any pure rule base $\mathcal{R}$ we can associate a consequence operation $C^{\mathcal{R}} : 2^{\mathcal{L}} \longrightarrow 2^{\mathcal{L}}$, by putting $C^{\mathcal{R}}(A) = Th(\mathbf{S})$, where $\mathbf{S}$ is the collection of answer sets of the database $(A, \mathcal{R})$. Equivalently,

$$C^{\mathcal{R}}(A) = \{\varphi \in \mathcal{L} : (A, \mathcal{R}) \vdash \varphi\}.$$

When $\mathcal{R}$ is a fixed but arbitrary rule base we sometimes write just C instead of $C^{\mathcal{R}}$. For $C^{\mathcal{R}}$ to be defined for arbitrary sets of $\mathcal{L}$-formulas, we assume that each formula $\psi \in \mathcal{L}$ is written in conjunctive normal form $\psi = \bigwedge_{i \in I} \varphi_i$, where each φ_i is a disjunction of literals. Then $C^{\mathcal{R}}(\{\psi\}) := C^{\mathcal{R}}(\cup\{\varphi_i : i \in I\})$ and $C^{\mathcal{R}}(A) = C^{\mathcal{R}}(\cup\{\psi : \psi \in A\})$.[8]

---

[7]In fact, we could replace C_N by an equivalent operation corresponding to a suitable 3-valued logic, see [18]. However, the notation C_N emphasises that we are dealing with a restriction of Cn_N.

[8]It is easily seen that any Boolean formula is equivalent, under $\models$ or C_N, to a formula in conjunctive normal form. As the reader may have noticed by now, the term 'disjunctive database' is rather too modest. In view of the nature of the $\models$ relation we could equally well consider from the outset database formulas of type (5) where in place of each L_i on the left of the implication an arbitrary Boolean formula φ_i may appear, and in place of the disjunction $\bigvee K_j$ on the right of the implication a single Boolean formula ψ may stand. We could then define the answer sets S of a *not*-free database Π by replacing (A1) with the condition that S be a minimal set of literals satisfying: for every formula $\varphi_1 \wedge \ldots \wedge \varphi_m \rightarrow \psi$ of Π, if $S \models \varphi_1 \wedge \ldots \wedge \varphi_m$, then $S \models \psi$. This generalisation is actually carried through in [8]. I have tried here to stay as close as possible to the original format of [10] since it is better known and it facilitates comparisons with other approaches to disjunctive logic programming.

From Proposition 1 it is easily verified that, for any rule base $\mathcal{R}$, the associated consequence operation $C^{\mathcal{R}}$ satisfies $C_N \leq C^{\mathcal{R}}$, ie. for all $A \subseteq \mathcal{L}$, $C_N(A) \subseteq C^{\mathcal{R}}(A)$. However, in view of a result of [19] a much stronger property can be established. Each XDDB formula of form (5) can be interpreted as a formula of N by reading *not* as intuitionistic negation, '¬'. Intuitionistic negation is definable in N, eg. by $\neg\varphi := \varphi \rightarrow\sim \varphi$, and the resulting system with two negations is a conservative extension of H (see eg. [11]). Using the above interpretation of database formulas as formulas of N we can associate with any rule base $\mathcal{R}$ a monotonic consequence operation, $C_N^{\mathcal{R}}$, on sets A of boolean formulas, defined by

$$C_N^{\mathcal{R}}(A) = \{\varphi \in \mathcal{L} : \{A \cup \mathcal{R}\} \vdash_N \varphi\}.$$

Then we have:

Observation 4 (supraconstructivity) *For any rule base $\mathcal{R}$, $C_N^{\mathcal{R}} \leq C^{\mathcal{R}}$.*

Proof. In [19] it is shown that, under the above interpretation of database formulas, for any XDDB Π and boolean formula φ, if $\Pi \vdash_N \varphi$ then $\Pi \vdash \varphi$. The observation then follows immediately from the definitions. □

Makinson in [15] lays considerable stress on the importance of supraclassicality. As he remarks, any supraclassical inference operation C that is also idempotent, satisfies *left absorption* $(CnC = C)$ and moreover, if cumulative, satisfies *full absorption*, $CnC = C = CCn$. He suggests, furthermore, that an approach to nonmonotonic inference

> deserves the name "logical" only if it leads to an inference operation C satisfying the full absorption principle $CnC = C = CCn$. In other words, only if the propositions we are allowed to infer from a set A form a classical theory $(C = CnC)$, which moreover depends only upon the logical content of A rather than its manner of presentation $(C = CCn)$. [15, §2.2]

Here "logical" is to be contrasted with approaches of a more procedural kind; although it should be noted that under Makinson's treatment of inference in default logic (in much the same style we have defined C for answer set consequence) DL is indeed regarded as a "logical" approach.

Since the operation C is supraconstructive rather than supraclassical, it does not of course fulfil Makinson's strictures on absorption. But there seems to be no reason *a priori* to construe nonmonotonic inference as being "non-logical" if the propositions to be inferred from a set A do not form a *classical* theory, since they may form a 'coherent' set with respect to some other, non-classical logic.

In view of this we can inquire whether answer set consequence C obeys absorption principles, or their instances, defined with respect to C_N rather than Cn. The answer turns out to be positive.

Observation 5 C satisfies left absorption wrt C_N, ie. for any $\mathcal{R}$ and A, $C_N C^{\mathcal{R}}(A) = C^{\mathcal{R}}(A)$.

Proof. Left absorption is a straightforward consequence of the fact that, for any set S of literals, $Th(S) = C_N(Th(S))$ (see the author's [18]). Since, by the monotony of C_N, we have $A \subseteq Th(S) \Rightarrow C_N(A) \subseteq C_N(Th(S))$, it follows from the above closure of $Th(S)$ that $A \subseteq Th(S) \Rightarrow C_N(A) \subseteq Th(S)$. Now, if **S** is the collection of all answer sets of a database $(A, \mathcal{R})$, then for every $S \in \mathbf{S}$, $Th(\mathbf{S}) \subseteq Th(S)$, hence for all S, $C_N(Th(\mathbf{S})) \subseteq Th(S)$. Therefore $C_N(Th(\mathbf{S})) \subseteq Th(\mathbf{S})$, and hence $C_N C \leq C$. Conversely, $C \leq C_N C$ holds by inclusion. $\square$

Observation 6 C *satisfies right absorption wrt* C_N, *ie. for any* $\mathcal{R}$ *and* A, $C^{\mathcal{R}} C_N(A) = C^{\mathcal{R}}(A)$.

Proof. We have to show that for any boolean φ, $(C_N(A), \mathcal{R}) \hspace{-0.5em}\vdash\hspace{-0.5em}\sim \varphi$ if and only if $(A, \mathcal{R}) \hspace{0.1em}\vdash\hspace{-0.5em}\sim \varphi$. It suffices therefore to verify, for any database $(A, \mathcal{R})$, that any subset S of literals is an answer set of $(A, \mathcal{R})$ iff it is an answer set of $(C_N(A), \mathcal{R})$. Thus, consider any answer set S of $(A, \mathcal{R})$. S is an answer set of $(A, \mathcal{R})^S$ and therefore a minimal set satisfying the formulas of $(A, \mathcal{R})^S = (A, \mathcal{R}^S)$. So S is a minimal set satisfying the formulas of $\mathcal{R}^S$ together with the conditions $S \models \varphi$ for each $\varphi \in A$. Moreover, since $C_N(A) \subseteq Th(S)$, $S \models \psi$ for each $\psi \in C_N(A)$. From the truth conditions for $\models$, it follows that S satisfies the formulas of $(C_N(A), \mathcal{R}^S)$, and is clearly a minimal set with this property. Therefore S is an answer set of $(C_N(A), \mathcal{R}^S) = (C_N(A), \mathcal{R})^S$, and hence an answer set of $(C_N(A), \mathcal{R})$, as required. Conversely, any answer set S of $(C_N(A), \mathcal{R})$ is an answer set of $(C_N(A), \mathcal{R})^S = (C_N(A), \mathcal{R}^S)$, satisfying the formulas of $\mathcal{R}^S$ and the conditions $S \models \varphi, \forall \varphi \in A$. S is a minimal set with this property, since, otherwise there would be a proper subset S' of S satisfying the formulas of $\mathcal{R}^S$ and the formulas in A. But then also $S' \models \psi, \forall \psi \in C_N(A)$, contradicting the assumption that S is a minimal set satisfying the formulas of $(C_N(A), \mathcal{R}^S)$. Since the answer sets of A and $C_N(A)$ coincide, we have $C C_N(A) = C(A)$. $\square$

From full absorption the following properties of C, whose analogs for the classical case are proved by Makinson, can be deduced.

$$\varphi, \psi \in C(A) \quad \Rightarrow \quad \varphi \wedge \psi \in C(A)$$
$$\varphi \in C(A), \psi \in C_N(\varphi) \quad \Rightarrow \quad \psi \in C(A)$$
$$C_N(A) = C_N(B) \quad \Rightarrow \quad C(A) = C(B)$$
$$A \subseteq B \subseteq C_N(A) \quad \Rightarrow \quad C(A) = C(B)$$

This concludes our brief study of answer set inference.

4 Related Work and Open Questions

Recently Dietrich and Herre [3, 4, 12] have studied what they call *deductive bases* of nonmonotonic inference operations. Roughly speaking, given a nonmonotonic consequence operation C over some language **L**, a monotonic and compact consequence operation $C_{\mathbf{L}}$ over **L** is said to be a deductive basis for

C if C extends $C_{\mathbf{L}}$ and satisfies full absorption with respect to it. They have investigated this notion mainly in the context of Poole systems and systems of minimal reasoning. An interesting general property, proved in [3], is that if an operation C is cumulative, then it possesses a maximal deductive basis. The content of Observations 4 and 5 comes therefore quite close to the claim that Nelson's constructive logic N forms a deductive basis for answer set inference, assuming the proposed translation of database rules into formulas of N. From the present paper and the work of Dietrich and Herre one can extract the following, more general research programme: to any semantics for extended or disjunctive logic programs associate a corresponding consequence relation C and identify, for a given translation of program rules into logical formulas, the deductive bases for C.[9]

5 Summary and Conclusions

Despite the restricted syntax of extended disjunctive databases, we have seen that most of the key properties of nonmonotonic inference relations are applicable either to the inference relation $\vdash$ or the consequence operation C associated with answer set semantics. Using whichever of $\vdash$ or C is the most convenient, we observed that, of the 'pure' conditions, *reflexivity* (or *inclusion*) and *cut* are satisfied, *cautious monotony* is not. Of the 'mixed' conditions, involving logical connectives, we considered, besides the 'core' properties of *supraclassicality* and *distribution*, also *absorption* and *rationality*. Under the formulation of these conditions given by Makinson [15], all of them fail, suggesting at first sight that answer sets do not determine a particularly well-behaved or 'coherent' notion of nonmonotonic inference.[10]

On further analysis, however, such a conclusion would seem unwarranted. Besides supraclassicality, also absorption and distribution are conditions involving the interaction of C with the classical consequence operation Cn. None of these three conditions seems appropriate, therefore, in a case like that of answer sets where C has a well-defined monotonic 'part' that does not correspond to classical inference. However, once we have identified the inference relation of the monotonic subsystem, in our case as a fragment of Nelson's constructive logic N, we can formulate more appropriate conditions involving the two notions of inference in question. In this manner, we could observe that C is *supraconstructive* in the sense of extending C_N, and it satisfies full absorption with respect to C_N. In the case of distribution, we discussed with respect to $\vdash$ the three conditions of *disjunction in the antecedent*, *proof by cases* and *conditionalisation*

---

[9]The version of right absorption proved here is somewhat weaker than that required to show that N is a deductive basis for answer set inference. In work still in progress, however, we investigate more fully the deductive base problem for answer sets and stable models and extend the results given here.

[10]I am not suggesting here that David Makinson himself would necessarily endorse this conclusion. My point is rather that if one should choose to use the principle conditions in his classification table as a 'check-list' of desirable properties of nonmonotonic inference, then the approach of answers sets would seem to fare rather badly.

which follow from it (in the supraclassical, absorbant case), and from which distribution seems to acquire its main attraction. Of these conditions only the first two are applicable, and only the first is constructively sound. As is appropriate for a supraconstructive notion of inference, therefore, the answer set semantics supports disjunction in the antecedent and does not support proof by cases.

Lastly, let us consider the condition of rationality that figures prominently in [2]. While rationality (9) as well as the slightly weaker condition (8) fail, there is a still weaker version of (8) that does hold. This condition, formulated in Observation 3, can be expressed roughly as follows: if a set of premises Π *verifies* ψ, whilst Π together with φ *falsifies* ψ, then Π does not *verify* φ; indeed, as I suggested earlier, we might say that Π *weakly falsifies* φ, since the latter fails in every answer set. I find this weaker version of rationality far more plausible and intuitive than (full) rationality for an inference relation extending constructive logic with strong negation, where, in particular, verification and falsification (or refutation) are considered as separate primitive notions and are interpreted in a strong, constructive sense. For such a system, rationality, which amounts to the condition 'if Π verifies ψ, whilst Π with φ does not verify ψ, then Π falsifies φ', looks far too strong.[11]

What general lessons might be learnt from this case study? The principle one, I believe, is that in most of the general studies of conditions on nonmonotonic inference (of which [15] is an especially clear example) supraclassicality, and other closely related mixed conditions, are allotted an over-prominent position. The privileged place assigned to classical consequence when discussing possible adequacy criteria for nonmonotonic inference 'infects' these criteria to the extent that any nonmonotonic inference relation based on, say, a constructive monotonic logic is likely to fail them. This emphasis on 'classically-based' conditions would be relatively harmless were to it transpire that virtually all interesting and useful systems of nonmonotonic reasoning are in fact supraclassical. But this is certainly not the case, as can be seen not only from the example of logic programming inference, but from many other approaches to nonmonotonic reasoning, eg. [16, 6, 22, 5, 1, 17, 20]. Moreover one of the main motives for studying nonmonotonic reasoning is to acquire logical formalisms suitable for applications in knowledge representation and reasoning (KR). Once we step into the field of KR and consider the kinds of systems that are actually implemented and used in practice, we see that supraclassicality is the exception rather than

---

[11] A mild analogy with the failure of *modus tollens* in the monotonic case of N might help to illustrate this point. Due to the constructive reading of both $\rightarrow$ and of negation $\sim$ (or refutation) in N, the following situation arises. Suppose that a consistent set of premises Π contains the formula $(p \rightarrow q)$. If Π additionally verifies p, then it must verify q by *modus ponens*, because the implication tells us that any method of proving p can be converted into a method for proving q. If instead, however, Π falsifies q (ie. $\Pi \vdash_N \sim q$), we cannot, without further information, conclude (as *modus tollens* would prescribe) that Π *falsifies* p, because we have no constructive proof to that effect. By the consistency of Π, we can draw only the weaker conclusion that p is not verified; alternatively, that any extension of Π by p would be inconsistent. Likewise, if the antecedent of the weak rationality condition is satisfied, we can conclude that φ cannot be true, but an explicit refutation of it (or a verification of $\sim \varphi$) may be lacking.

the rule. Relational and deductive databases, logic programs, concept languages and other KR systems not only restrict the syntax of full first-order logic, but are usually associated with some nonclassical, particularly constructive, forms of inference. If nonmonotonic logics are to make a genuine contribution to problems in KR, they must be sensitive to the special requirements that arise in KR. Moreover, when analysing desirable conditions on nonmonotonic inference it seems important to develop a classification scheme that is at least in touch and in tune with these special requirements, and which is appropriate for studying logic programming and other knowledge representation formalisms that are in use or under development at present. One such system is that of disjunctive databases equipped with an answer set inference relation. I hope to have illustrated how poorly the standard classification of inference conditions applies there.

5.1 Acknowledgements

I am grateful to Michael Gelfond for sketching a simple argument to verify Observation 1. Work on this paper was supported by the Freie Universität Berlin under the research project SLIP: *Systems of Logic as a Theoretical Foundation for Knowledge and Information Processing.*

References

[1] Aiello, L, Amati, G, & Pirri, F, Intuitionistic Autoepistemic Logic, forthcoming in *Studia Logica.*

[2] Dix, J, *Nichtmonotones Schließen und dessen Anwendung auf Semantiken logischer Programme*, Doctoral Disseration, Karlsruhe, 1992.

[3] Dietrich, J, Deductive Bases of Nonmonotonic Inference Operations, NTZ Report, Universität Leipzig, 1994.

[4] Dietrich, J & Herre, H, Outline of Nonmonotonic Model Theory, NTZ Report, Universität Leipzig, 1994.

[5] Fischer Servi, G, Non-Monotonic Consequence Based on Intuionistic Logic, AILA Preprint, Padova, 1991.

[6] Gabbay, D, Intuitionistic Basis for Non-Monotonic Logic, in *Proc 6th Conf on Automated Deduction*, LNCS 138, Springer-Verlag, 1982, 260-273.

[7] Gabbay, D, Theoretical Foundations for Non-Monotonic Reasoning in Expert Systems, in K Apt (ed), *Logics and Models of Concurrent Systems*, Springer-Verlag, 1985.

[8] Gelfond, M, Logic Programming and Reasoning with Incomplete Information, *Ann Math and AI* 12 (1994), 89-116.

[9] Gelfond, M, & Lifschitz, V, The Stable Model Semantics for Logic Programs, in K Bowen & R Kowalski (eds), *Proc 5th Int Conf on Logic Programming 2*, MIT Press, 1070-1080.

[10] Gelfond, M, & Lifschtz, V, Classical Negation in Logic Programs and Disjunctive Databases, *New Generation Computing* (1991), 365-387.

[11] Gurevich, Y, Intuitionistic Logic with Strong Negation, *Studia Logica* 36 (1977), 49-59.

[12] Herre, H, Compactness Properties of Nonmonotonic Inference Operations, in C MacNish, D Pearce & L M Pereira (eds), *Logics in Artificial Intelligence. Proceedings JELIA '94*, LNAI 838, Springer-Verlag, 1994.

[13] Kraus, S, Lehmann, D, & Magidor, M, Nonmonotonic Reasoning, Preferential Models and Cumulative Logics, *Artificial Intelligence* 44 (1990), 167-207.

[14] Makinson, D, General Theory of Cumulative Inference, in M Reinfrank et al (eds), *Non-Monotonic Reasoning*, LNAI 346, Springer-Verlag, 1989.

[15] Makinson, D, General Patterns in Nonmonotonic Reasoning, in D Gabbay, C J Hogger & J A Robinson (eds), *Handbook of Logic in Artificial Intelligence*, Clarendon Press, Oxford, 1994.

[16] McCarty, L T, & van der Meyden, R, An Intuitionistic Interpretation of Finite and Infinite Failure, in Pereira, L M, & Nerode, A, (eds), *Logic Programming and Non-Monotonic Reasoning*, MIT Press, 1993. 417-436.

[17] Pearce, D, Default Logic and Constructive Logic, in B Neumann (ed), *Proc ECAI 92*, John Wiley, 1992.

[18] Pearce, D, Answer Sets and Constructive Logic, I: Monotonic Databases, in A Fuhrmann & H Rott (eds), *Logic, Action and Change*, De Gruyter, Berlin, 1994.

[19] Pearce, D, Answer Sets and Nonmonotonic S4, in R. Dyckhoff (ed), *Extensions of Logic Programming*, LNAI 798, Springer-Verlag, 1994.

[20] Pereira, L M, Alferes, J J, & Aparicio, J N, Default Theory for Well Founded Semantics with Explicit Negation, in D Pearce & G Wagner (eds), *Logics in AI. Proc JELIA 92*, LNAI 475, Springer-Verlag, 1992, 339-356.

[21] Reiter, R, A Logic for Default Reasoning, *Artificial Intelligence* 13 (1980), 81-132.

[22] Turner, R, *Logics for Artificial Intelligence*, Ellis Horwood, 1984.

[23] Wòjcicki, R, *Theory of Logical Calculi*, Kluwer, Dordrecht, 1988.

Trans-Epistemic Semantics for Logic Programs

Arcot Rajasekar

Computer Science Department
University of Kentucky, Lexington, KY 40506
sekar@cs.engr.uky.edu

Abstract. Each stable model of a logic program is computed in isolation. This does not allow one to reason in any stable model with information from other stable models. Such information interchange is needed when computing with full introspection, as performed by Gelfond's epistemic specifications, or when modeling multi-agent reasoning using stable models. In this paper, we define syntactic and semantic structures that allow the use of information from multiple stable models when computing one stable model. Hence a notion of second order stability is introduced and every computed model should be stable at that level. We define a concept of trans-epistemic (te-) logic programs that is reduced to a logic program using information from a trans-epistemic interpretation. The te-interpretation is checked for stability against the set of stable models of the logic program using a consensus function. We discus the properties of trans-epistemic stable models and motivate their use with examples.

1 Motivation

Stable models were introduced by Gelfond and Lifschitz [3] in order to capture the semantics of logic programs with default negation. An important aspect of stable model semantics is that there can be multiple stable models for a normal (or general) logic program. Query processing in such a case becomes difficult; one can ask

(*)   *whether a formula is (is not) in every (some) stable model.*

But a main drawback with the approach was that the answer for such queries cannot be fed back to compute additional consequences in the stable models. *Each stable model is computed in isolation. Moreover, the query process is not compatible with the derivation process.* The last point is a significant departure from the concept in logic programming that the body of a rule is equivalent to a query.

The main aim of this paper is to describe an extension to logic programs that will retain the (body = goal) relationship. For facilitating this extension, we need to find a means to allow for a stable model to utilize information that is derived in other models. Moreover, one can generalize the concept of this information usage beyond the four types described in (*), by allowing arbitrary user-defined functions for testing for truth (falsity) in the canonical models of a program. For this purpose, we introduce the concept of trans-context that is instrumental in communicating information between canonical models. We build upon the concept of 'epistemic specification' defined by Gelfond in [2, 1].

There are several situations where one can use such systems that require sharing of information among canonical models. Gelfond [2] provides several cases where such an approach is required to capture the concept of incomplete information such as *unknown p*. We motivate the need for the increased expressive power for inferring with trans-extensional knowledge using the following examples that capture multi-agent reasoning.

Example 1 *Consider a court case. The prosecuting attorney would like to convince the jury that a suspect is normally guilty if his alibi can be consistently assumed to be false. The defense attorney would like to convince the jury that a suspect is normally innocent if his alibi can be consistently assumed to be true. The two statements can be seen as defeasible that provide a nonmonotonic inference about a person's innocence. A jury has to find a person to be either guilty or innocent. The judge requires the jury to return a verdict of guilty only if they find the suspect to be guilty* beyond any doubt; *otherwise they should give a verdict of innocent. If the jury returns a verdict of guilty the judge will give orders for imprisonment; otherwise the suspect will be freed.*

Assume that in a particular case, person 'a' is suspected and neither of the attorney's were able to prove or disprove the alibi presented by the suspect. All the above statements, except for the judge's requirement, can be encoded in a (any) non-monotonic system leading to the following two belief sets:

$E_1 = \{suspect(a), \neg alibi(a), guilty(a), imprison(a)\}$

$E_2 = \{suspect(a), alibi(a), innocent(a), free(a)\}$

The two beliefs can be viewed as the reasoning used by two jurors in the case, who were convinced by the opposing lawyers. If the two jurors use only their beliefs in their sentencing the suspect, one would be left with a hung jury. But if they are allowed to confer, and asked to return a guilty verdict only if both are convinced of the suspect's guilt then both the juries will return the innocent verdict. To encode the exact ruling required by the judge, that a person is imprisoned (in any belief set) only if found guilty beyond doubt, requires one to consider the knowledge about a suspect's guilt at other belief sets and to use the knowledge to free a person if found innocent even in one belief set. That is, one has to free a suspect in every extension. Hence, a need for migration of information and beliefs across belief sets is required. In such a case, one will have the following belief sets:

$E_1' = \{suspect(a), \neg alibi(a), innocent(a), free(a)\}$

$E_2' = \{suspect(a), alibi(a), innocent(a), free(a)\}$

Even though in E_1' the conclusion of innocence is not supported by the local *belief set (¬alibi), it is forced (railroaded!) by consistency requirement from a* global *belief set. Notice that this was possible because the truth about the suspect's alibi was undecided and hence defeasible; if there was a solid proof for the truth value of the suspect's alibi in either belief set, then the outcome would not be as shown above.* □

The first example allows one to reason internally in each belief set with full introspection (as in Gelfond's epistemic specification). The second example has a similar flavor of reasoning but is not concerned with truth (or falsity) in every intended model.

Example 2 [1] *Next we study a slightly more involved way of using knowledge from other canonical models. A company wants to study the market for a forthcoming product 'a' and wants to make some decisions about further products depending upon the success of the product 'a'. In a test marketing study, it has been shown that the product will succeed if educated and youth like it, or if uneducated, elderly and men like it or if elderly and women like it. Market studies of similar products has shown that if men liked the product the women will hate it and vice versa, and if the old liked the product the youth will hate it and vice versa. It has been found that the uneducated liked a product if the educated hated it. It has also been found that if youth and women liked a product the educated liked it too. To keep ahead of competition, the company wants to start working on a new line of products also if possible. It wants to start on product 'b' if the product 'a' can succeed in at least 3 possible scenarios, it wants to start on product 'c' if the product 'a' can succeed in less than 3 scenarios. If the company wants to make product 'b' it needs to buy widget w_1, otherwise it needs to buy widget w_2 to make product 'c'. Hence the company needs to reason with several scenarios. Part of the reasoning about the products can be encoded as rules and defaults as follows:*

$success(X) \leftarrow young\_like(X) \wedge educated\_like(X),$
$success(X) \leftarrow old\_like(X) \wedge men\_like(X) \wedge uneducated\_like(X),$
$success(X) \leftarrow old\_like(X) \wedge women\_like(X) \wedge educated\_like(X),$
$buy(w_1) \leftarrow make(b),$
$buy(w_2) \leftarrow make(c),$
$old\_like(X) \leftarrow \neg young\_like(X),$
$young\_like(X) \leftarrow \neg old\_like(X),$
$women\_like(X) \leftarrow \neg men\_like(X),$
$men\_like(X) \leftarrow \neg women\_like(X),$
$uneducated\_like(X) \leftarrow \neg educated\_like(X),$
$educated\_like(X) \leftarrow young\_like(X) \wedge women\_like(X)$

As can be seen, there are more than one scenario possible each with an outcome of success or failure for the product as (partially) shown below:

$E_1 = \{old\_like(a), men\_like(a), uneducated\_like(a), success(a)\}$
$E_2 = \{old\_like(a), women\_like(a), uneducated\_like(a), failure(a)\}$
$E_3 = \{young\_like(a), men\_like(a), uneducated\_like(a), failure(a)\}$
$E_4 = \{young\_like(a), women\_like(a), educated\_like(a), success(a)\}$

Since there are two scenarios in which the product 'a' is successful, the company should buy widget w_2 and start making product 'c'. In traditional logic programming (or other nonmonotonic reasoning) systems, even though one is able to conclude that the product 'a' is successful in two extensions, one is unable to feed that information into the theory to further conclude about production of product 'c' or the need to buy widget w_2. □

---

[1] No offense is intended to any group of people by this example.

Notice that the above problem is a departure from epistemic specification, since the reasoning used is not concerned in truth or falsity in every intended model. The examples shown above provide motivation for enhancing the capabilities of default reasoner to reason internally with more than one extension and to share information among models. That is, each intended model should be able to reason with trans-model knowledge.

We introduce the syntax and semantics of an extended logic program called trans-epistemic logic programs (te-logic programs). A central idea in te-logic programs is the concept of a trans-context that plays a role similar to that played by the stable set in normal logic programs. The trans-context can be seen as a global context that is used to check whether a te-logic program rule is allowed or disallowed in deriving conclusions in any canonical model. Once a set of rules are allowed by a trans-context, a set of stable models can be computed for these logic program rules. The set of these models, called trans-epistemic models, are in turn used to compute the trans-context using a user-defined mapping function ϱ that maps a set of theories into a single theory. A trans-stable model can be seen as a stable trans-context, a fixpoint of the mapping ϱ. There can be zero, one or more trans-stable models for te-logic programs (possibly giving rise to another recursion of trans-trans-epistemic specifications!!).

Before, we discuss trans-epistemic logic programs, we provide an overview of Gelfond's epistemic specification.

1.1 Epistemic Specification

Gelfond [2], provides the definition of an *epistemic specification* by augmenting logic programming with two new modalities K and M. Gelfond's epistemic specification are rules of the form:

$$F \leftarrow G_1, \ldots, G_m, not\, G_{m+1}, \ldots, not\, G_k \tag{1}$$

where $F, G_{m+1}, \ldots, G_k$ are formulas in a first order language $\mathcal{L}_0$ and $G_1, \ldots, G_m$ are either formulas in $\mathcal{L}_0$ or modal formulas of the form KF and MF, where KF stands for "F is known to be true", and MF stands for "F is believed to be true". Let $A = \{A_i\}$ be a collection of sets of ground literals and let W be a set of ground literals. (According to Gelfond, A can be thought of as a collection of possible belief sets of a reasoner while W represents his current (working) set of beliefs [2]). The notion of truth ($\models$) and falsity ($\dashv$) of formulae in $\mathcal{L}_0$ is defined with respect to a pair $M = <A, W>$ as follows[2]:

$M \models p(\bar{t})$ iff $p(\bar{t}) \in W$

$M \models KF$ iff $< A, A_k > \models F$ for every A_k from A

$M \models AF$ iff $< A, A_k > \models F$ for some A_k from A

$M \dashv KF$ iff $M \not\models KF$

$M \dashv MF$ iff $M \not\models MF$.

As can be seen, the operator K captures the notion of *true in every set of A* and the operator M captures the notion of *true in some set of A*. For example, the

---

[2] We provide only a subset of the set of definitions. Please refer [3] for a full definition.

following sentence (from [2])

$$interview(X) \leftarrow \neg Keligible(X), \neg K \neg eligible(X)$$

corresponds to the intuitive meaning given by the sentence "the students whose eligibility is not determined by other rules should be interviewed by the scholarship committee". That is, if $eligible(p)$ and $\neg eligible(p)$ is not available in any set of intended models then $interview(p)$ should be in every intended model.

Gelfond defines the concept of a world view A as a collection of belief sets for a given epistemic specification T. He defines a method to compute the world view by first guessing a collection of sets of literals and checking whether one obtains the same set of belief sets after performing a double reduct operation. Gelfond shows that in the case when T is a normal logic program the world view coincides with the set of stable models. In the case of extended disjunctive program he shows that the world view coincides with the set of all answer sets of T. He also shows the applicability of the formalism to various forms of common sense reasoning such as modeling **unknown**, generalization of closed world assumption, modeling unique names axiom, etc.

A second proposal in trans-epistemic reasoning is described by Przymusinski [12] as an extension of auto-epistemic logic of minimal beliefs (AELB). He introduces a belief operator $\mathcal{B}$ with meaning of "true in all minimal models". He defines the concept of a static auto-epistemic expansion of a theory if it satisfies the following fixpoint equation:

$$T^* = \{Con(T \cup \{\mathcal{L}F : T^* \vdash F\} \cup \{\neg \mathcal{L}F : T^* \nvdash F\} \cup \{\mathcal{B}F : T^* \vdash_{min} F\})\},$$

where Con^* is the propositional inference system augmented with a consistency and conjunctive belief axioms. Przymusinski shows that the formalism captures the semantics of several types of logic programs. Proposals to capture 'full introspection' was also described by Levesque [4, 5].

2   Trans-Epistemic Logic Programs

Let $\mathcal{L}$ be a propositional language. A *trans-epistemic logic program (te-logic program)* is a two-tuple (D, W), where W is a collection of formulas in $\mathcal{L}$, and D is a collection of rules, called *trans-epistemic logic programs rules (te-rules)*, of the form:

$$p \leftarrow q_1, \ldots, q_n, \neg r_1, \ldots, \neg r_m, As_1, \ldots, As_k, Bt_1, \ldots, Bt_l \qquad (2)$$

where p, q's, r's, s's and t's are predicate symbols in $\mathcal{L}$. The $\neg$ operator represents default negation[3]. The s_i's (resp. t_i's, r_i's) are called A-predicates (resp. B-predicates, $\neg$-predicates). A $\neg$-free (resp. A-free, B-free) rule is one with $m = 0$ (resp. $k = 0$, $l = 0$). Note that when the rules are both A-free and B-free then they are the same as normal logic program rules.

The semantics of te-logic programs can be captured as a stable semantics. For a given te-logic program P, the Herbrand universe U_P, Herbrand base B_P and Herbrand interpretations are defined in the usual manner.

---

[3] We restrict ourselves to extending normal logic programs; the syntax and semantics for classical negation and exclusive disjunctions can be developed in a similar fashion.

Definition 1 *Let P be a te-logic program. Let $\varrho : 2^{(2^{B_P})} \longrightarrow 2^{B_P}$ be a mapping from a set of interpretations to an interpretation. Let $E, S \subseteq B_P$ be interpretations of P. E is called a* trans-context.
A new program $P^{E,S}$ is obtained from P as follows:

1. *Remove from P a rule of the form 2 if $\exists i, s_j \notin E$ or $\exists j, t_j \in E$ or $\exists j, r_j \in S$.*
2. *In remaining rules, remove all A-predicates, B-predicates and $\neg$-predicates.*

If the least Herbrand model of $P^{E,S}$ is equal to S then S is a te-stable model *of P with respect to E.*
The set of all te-stable models of P with respect to E is denoted as $MM^E(P)$.
The set E is called a trans-stable model *of P under ϱ if*

$$E = \varrho(MM^E(P)) \tag{3}$$

$\square$

The mapping $\varrho : 2^{(2^{B_P})} \longrightarrow 2^{B_P}$ plays an important role in defining the semantics of te-logic programs.. It can be seen as a concentrator of information contained in the te-stable models. In this sense we call the function as a *consensus function*. An intuitive and familiar ϱ function is the intersection function that provides the set of formulas that are true in every te-extensions. Actually, with the intersection function, the operator K of Gelfond is the same as that of A and the operator M is equivalent to that of A with the union function. With extended logic programs (with classical negations) one can mimic Mt with $B \sim t$ since $Kt =\sim M \sim t$ where $\sim$ denotes classical negation. The ϱ function can be viewed as a communication medium between the te-stable models passing information from one to another. By defining complex ϱ functions one can define complex interactions between them. We discuss several interpretations of ϱ in a later section. Below, we may avoid mentioning the term *under ϱ* if ϱ is implicit.

A te-logic program can have zero, one or many sets of trans-stable models with respect to a given mapping ϱ. Consider the te-logic program $P = \{p \leftarrow Bp\}$.
It has no trans-stable models when ϱ is the union function. Since, if one assumes the trans-context to be $E = \{p\}$ then one obtains a singleton set of te-stable models $MM^E(P) = \{\{\}\}$ whose union is not equal to E. If $E = \{\}$ then one obtains a singleton set of extensions $MM^E(P) = \{\{p\}\}$ whose union is again not equal to E.

Example 3 *To illustrate the role played by trans-contexts, consider the normal logic program*
$P = \{r \leftarrow p, \neg q \qquad s \leftarrow p, \neg r$
$\qquad q \leftarrow p, \neg s \qquad p \leftarrow.$
The program P does not have any stable models. But consider that one applies an additional constraint stating that a program rule can be applied only if the conclusion of the rule is in consensus with all intended models. When this constraint is applied one can have four different consensus,

1) agreeing on only p being true.   *3) agreeing on p and r being true.*
2) agreeing on p and q being true.   *4) agreeing on p and s being true.*
The above constraint can be implemented as a te-logic program P':
$r \leftarrow p, \neg q, Ar$ $s \leftarrow p, \neg r, As$
$q \leftarrow p, \neg s, Aq$ $p \leftarrow.$
The four sets of trans-stable models of P' are given as follows:
   1) $\{\{p\}\}$ *3)* $\{\{p, r\}\}$
   2) $\{\{p, q\}\}$ *4)* $\{\{p, s\}\}$
The sets of trans-stable models capture the mutually exclusive nature of q, r and s.

We next discuss the examples introduced in Section 1.

Example 4 *The statements in Example 1 can be encoded as a te-logic program P :*
$guilty(X) \leftarrow suspect(X), bad\_alibi(X), \neg innocent(X)$
$innocent(X) \leftarrow suspect(X), good\_alibi(X)$
$innocent(X) \leftarrow suspect(X), Ainnocent(X)$
$guilty(X) \leftarrow suspect(X), Binnocent(X)$
$good\_alibi(X) \leftarrow \neg bad\_alibi(X)$
$bad\_alibi(X) \leftarrow \neg good\_alibi(X)$
$suspect(a)$ $free(X) \leftarrow innocent(X)$ $imprison(X) \leftarrow guilty(X)$
The mapping ϱ is defined as follows: $\varrho(\{E_1, \ldots E_k\}) = \bigcup_{i=1}^{k} E_i$. *The first two rules capture the idea that a juror uses the suspect's alibi and the juror's (local) belief of innocence to deliver the verdict. The next two te-default rules deals with the trans-model information, and captures the notion that a guilty verdict can be given only when the suspect is judged non-innocent by every juror; otherwise a verdict of innocence is to be delivered.*

Note that the trans-stable models can have contradictory information. We have have two sets of te-stable models for P w.r.t. ϱ:
$E_1 = \{suspect(a), good\_alibi(a), innocent(a), free(a)\},$
$E_2 = \{suspect(a), bad\_alibi(a), innocent(a), free(a)\}.$
The trans-stable model is $E_1 \cup E_2$. One can see the close interaction between local stable sets and the trans-context. Even if a trans-context forces a guilty verdict, a local te-stable model can still have an innocent status, effectively making the trans-context unstable (because of the union function). It can be established that a guilty verdict is pronounced only if every te-extension agrees on it; not otherwise. Of course, each juror can find a suspect guilty because of different reasons.

In the case of Example 2 the mapping ϱ can be defined as follows:
$$\varrho(\{E_1, \ldots, E_k\}) = \{\varphi | \varphi \in E_i, \ IN(\varphi, E_1) + \ldots + IN(\varphi, E_k) \geq 3\}$$
where $IN(\varphi, S)$ is 1 if $\varphi \in S$; otherwise 0. This also shows that the mapping function ϱ can be quite complicated.

Definition 1 provides a fixpoint definition for the trans-stable models. A similar characterization can be provided for the set of te-stable models. (Proofs of results given here can be found in [13].)

Theorem 1 *Let P be a te-logic program and $\overline{E} = \{E_1, \ldots, E_n\}$ be a set of interpretations. $\varrho : 2^{(2^{B_P})} \longrightarrow 2^{B_P}$ be a mapping that takes a set of interpretations and returns an interpretation. Then, $\overline{E}$ is the set of te-stable models of P iff*

$$\overline{E} = MM^{\varrho(\overline{E})}(P). \tag{4}$$

Next we study the monotonicity properties of te-logic programs. To facilitate this study, we apply a concept similar to the S-proof system developed by Marek and Truszczynski [9] for default logic.

Definition 2 *Let P be a te-logic program. Let $S, E \subseteq B_P$. An SE-derivation (or SE-proof) of a predicate p from P is a finite sequence $p_1, \ldots, p_h$ such that $p_h = p$ and for every p_i, $i = 1, \ldots, k$, is derived as follows:*
there is a te-program rule in P

$$p_i \leftarrow q_1, \ldots, q_n, \neg r_1, \ldots, \neg r_m, As_1, \ldots, As_k, Bt_1, \ldots, Bt_l$$

such that $\{q_1, \ldots, q_n\} \subseteq \{p_1, \ldots, p_{i-1}\}$, $\{r_1, \ldots r_m\} \cap S = \emptyset$, $\{s_1, \ldots s_k\} \subseteq E$ and $\{t_1, \ldots t_l\} \cap E = \emptyset$.
By $Cn^{S,E}(P)$ we denote the set of all formulas having an SE-proof from P. □

Given a set $E \subseteq B_P$ and a te-logic program P, it can be seen that

Theorem 2

$$MM^E(P) = \{S | S = Cn^{S,E}(P)\}. \tag{5}$$

The monotonicity property of the operator Cn is as follows:

Lemma 1 *The operator $Cn^{S,E}(P)$ is monotone in P and antimonotone in S.*
□

But the monotonicity with respect to the trans-contexts depends upon whether or not the te-logic program rules are B-free or A-free:

Lemma 2 *The operator $Cn^{S,E}(P)$ is monotone in E if the rules in P are B-free. The operator $Cn^{S,E}$ is antimonotone in E if the rules in P are A-free.*
□

We next provide a means of relating te-logic programs and normal logic programs.

Definition 3 *Given a te-logic program P and set $E \subseteq B_P$, we define a set of trans-consistent rules, P_E, as follows:*
$$P_E = \{p \leftarrow q_1, \ldots, q_n, \neg r_1, \ldots, \neg r_m |$$
$$p \leftarrow q_1, \ldots, q_n, \neg r_1, \ldots, \neg r_m, As_1, Bt_1, \ldots, Bt_l \in P$$
$$\forall i, 1 \leq i \leq k, s_i \in E \text{ and } \forall j, 1 \leq j \leq l, \neg r_j \notin E\}. \qquad \square$$

Note that the trans-consistent program rules can be viewed as a reduct of P with respect to E.

Theorem 3 *For every te-logic program P and for every set $E \subseteq B_P$,*

$$MM^E(P) = Stable\_models(P_E).$$

That is every te-stable model of P with respect to E is a stable model of P_E. □

The above equivalence between the stable models and te-stable models provides a means to derive some monotonicity properties of the set $MM^E(P)$. Let E_1, E_2 be two sets of theories in $\mathcal{L}$. Then, $E_1 \sqsubseteq E_2$ if $\forall s \in E_1 \; \exists t \in E_2, \; s \subseteq t$. If $E_1 = \{s\}$ and $E_2 = \{t\}$ are singletons, then we use $s \subseteq t$ instead of $\{s\} \sqsubseteq \{t\}$. A property $\mathcal{P}$ is $\sqsubseteq$-monotone if $S \sqsubseteq T$ implies that $\mathcal{P}(S) \sqsubseteq \mathcal{P}(T)$. A property $\mathcal{P}$ is anti-$\sqsubseteq$-monotone if $S \sqsubseteq T$ implies that $\mathcal{P}(S) \sqsupseteq \mathcal{P}(T)$. We have the following results:

Lemma 3 *If we consider only $\neg$-free te-logic programs then $MM^E(P)$ is $\sqsubseteq$-monotone in P.* □

Lemma 4 *If P contains only $\neg$-free te-logic program rules that are also B-free, then $MM^E(P)$ is $\sqsubseteq$-monotone in E.* □

Lemma 5 *If D contains only $\neg$-free te-logic program rules that are also A-free, then $MM^E(P)$ is anti-$\sqsubseteq$-monotone in E.* □

The next set of properties are carried over from those of stable models of logic programs.

Corollary 1 *If $\overline{E} = \{E_1, \ldots, E_k\}$ is a set of te-stable models for a te-logic program P with respect to a mapping ϱ then the set E is the set of minimal models of normal logic program $P_\varrho(E)$.* □

Lemma 6 *If $\overline{E} = \{E_1, \ldots, E_k\}$ is a set of te-stable models for a te-logic program P with respect to a mapping ϱ then the set E forms an anti-chain.* □

Lemma 7 *Let $\overline{E} = \{E_1, \ldots, E_k\}$ be a set of te-stable models for a te-logic program P with respect to a mapping ϱ. Then for every theory Z such that $Z \subseteq \bigcap_{i=1}^{k} E_i$, E is also a set of te-stable models for the te-logic program $(P \cup Z)$ with respect to the mapping ϱ.* □

Next we show how one can embed normal logic programs as te-logic programs. One simple embedding is motivated by the fact that A-free and B-free te-logic programs and normal logic programs are equivalent Theorem 3. A second embedding provides more insight into the properties of te- logic programs. Let P be a normal logic program. We define an embedding for P as follows:

$$te\_embed(P) = \{p \leftarrow q_1, \ldots, q_n, Br_1, \ldots, Br_m | p \leftarrow q_1, \ldots, q_n, \neg r_1, \ldots, \neg r_m \in P\}$$
$$(6)$$

Then we have the following property:

Theorem 4 *Let P be a normal logic program in $\mathcal{L}$. Let $\varrho : 2^{(2^{B_P})} \longrightarrow 2^{B_P}$ be a mapping given by $\bigcap$. If E_i is an stable model for P then $\{E_i\}$ is a set of te-stable models of the te-logic program te_embed(P) with respect to mapping ϱ.* $\quad\square$

Theorem 5 *Let P be a te-logic program such that every te-program rule in P is A-free and $\neg$-free. Let $\varrho : 2^{(2^{B_P})} \longrightarrow 2^{B_P}$ be a mapping given by $\bigcap$. If $\{E_i\}$ is a set of te-stable models of P with respect to mapping ϱ, then E_i is a stable model for the logic program P' such that te_embed(P') $= P$.* $\quad\square$

The following property is a consequence of the above embedding.

Theorem 6 *Let P be a te-logic program such that every te-program rule in P is A-free and $\neg$-free. Let $\varrho : 2^{(2^{B_P})} \longrightarrow 2^{B_P}$ be a mapping given by $\bigcap$. Then the set of trans-stable models of P with respect to ϱ form an antichain.* $\quad\square$

But a negative result is that arbitrary te-logic programs need not have te-stable models that form an antichain as can be seen from Example 3.

3   Meanings of ϱ Function

The various meaning that can be attributed to the mapping operator ϱ and the consistency checking process provides a wide range of te-rule logic definitions. The function ϱ can have a simple to quite complex semantics.

1. ϱ can return the intersection of its input set providing the intuitive meaning about formulas that are *true* in every element of the input set. In this case, one can see that every te-stable model for a logic program P with respect to $\varrho = \bigcap$ contains the corresponding trans-stable model.
2. ϱ can return the union of its input set providing the intuitive meaning about formulas that are present at least one element of the input set. In this case, every te-stable model for a logic program P with respect to $\varrho = \bigcup$ is contained in the corresponding trans-stable model.
3. ϱ can return the set of formulas that are in (or at least in) a fixed number of the input set.
4. ϱ can return multi-valued interpretation for the formulas in the input set based on the count of how many input sets contain the formulas.
5. ϱ can be defined as in the above cases but using only a (randomly chosen) subset of the input set in constructing the return set. This can give semantics such as a two-out-of-five random sample of the input set contain the formula φ. Such semantics might be useful in dealing with infinite sets. This case also provides an interesting combination of nonmonotonic and probabilistic reasoning.
6. If the consequences of te-rules themselves contain rules, one can get a rich semantics. Then ϱ can return the $\bigcup(\bigcap, etc.)$ of the input set provided by the rule. If the rules in the consequences are themselves te-rules then one can have a recursion that may be quite bizarre.

7. ϱ can return the $\cup(\cap,$ *etc.*) of the set of elements of the input set that have a certain property. Eg. the union of all sets that contain $guilty(a)$. The 'property' can be quite general: it can be a set of first order formulas, consequences of a logic program, or the intersection of the answer sets of another logic program or the objective consequences of an auto-epistemic logic. Hence, we can capture a concept of two (or more) different systems controlling one another (possibly in a mutual manner). This type of interaction ca be useful in hybrid knowledge systems [6, 11].

8. ϱ can return the empty set possibly giving a meaning similar to normal logic program for some types of consistency checking. Returning a constant set can also do likewise.

Obviously, the above set of different meanings attributable to ϱ is not exhaustive. In full generality, one may have more than one type of consistency checking operator $A_1, \ldots A_r, B_1, \ldots, B_s$, each parametrized by a different ϱ function! That is, each modality operator can have its own semantics and checking against (possibly) different trans-contexts.

The concept of using more than one trans-context for obtaining a trans-stable model can be used to advantage in query processing. This is motivated by the following result.

Theorem 7 *Let P be a te-logic program and let $\varrho : 2^{(2^{B_P})} \longrightarrow 2^{B_P}$ be a mapping that takes a set of interpretations and returns an interpretation. Let $E_1, E_2 \subseteq B_P$ such that $E_1 \subseteq E_2$. Define:*

$$P_{E_1,E_2} = \{p \leftarrow q_1, \ldots, q_n, \neg r_1, \ldots, \neg r_m |$$
$$p \leftarrow q_1, \ldots, q_n, \neg r_1, \ldots, \neg r_m, As_1, \ldots, As_k, Bt_1, \ldots, Bt_l \in P$$
$$\forall i, 1 \le i \le k, s_i \in E_1 \text{ and } \forall j, 1 \le j \le l, \neg r_j \notin E_2 \}.$$

Let $\{S_1, \ldots S_h\}$ be a set of stable models of P_{E_1,E_2}. Then $\{S_1, \ldots S_h\}$ is a set of te-stable models of P if $E_1 \subseteq \varrho(\{S_1, \ldots S_h\}) \subseteq E_2$. $\square$

The set $E = \varrho(\{S_1, \ldots S_h\})$ forms the trans-stable model of P. The advantage provided by this theorem is that to generate the te-stable-models (alternatively the trans-stable models) of a te-logic program, one needs to guess two sets of atoms: one a lower bound and another an upper bound of the actual trans-stable model. In fact E_1 can be a subset of all A-predicates in P and $E_2 = B_P - S$ where S is a subset of all B-predicates in P such that $S \cap E_1 = \emptyset$. Note that the choice of E_1 and E_2 are independent of the ϱ function.

4 Discussion

The trans-epistemic logic programs extend epistemic specifications in two significant manner. The first extension is generalizing the use of information from multiple canonical models beyond those prescribed in (*) (page 1). That is, a user can define the manner in which the information from the canonical models can be combined and applied in a rule. The second extension is the introduction of a single structure of trans-context that not only makes the interaction between the canonical models to be uniform, but also eases query processing.

A side effect of these extensions are the concepts of user-definable consensus function ϱ and trans-stable models. The multiple stable models of rule logic can be viewed as defining the different sets of beliefs that a reasoning agent can construct and support from a given set of facts and defeasible rules. Likewise, multiple sets of te-stable models of te-logic programs can be viewed as defining the sets of beliefs that a co-operating group of reasoning agents can construct, support and agree to from a given set of facts and defeasible rules.

An outcome of this extension is the concept of $\sqsubseteq$-monotonism. Similar to the case of $\neg$-predicates introducing anti-monotonism (under $\subseteq$), B-predicates can be seen to introduce anti-$\sqsubseteq$-monotonism. Properties of $\sqsubseteq$-monotonism needs to be studied in greater detail. Similarly, The properties of different functions ϱ needs to be studied in greater detail. Even though, we described a 'guess-and-check' inference system that utilizes an underlying stable-model generating system, proof-systems that do not require guessing can possibly be constructed. Systems based on forward chaining techniques [14, 7, 8, 10] that compute extensions to default theories need to be explored.

In this paper we have motivated the need for reasoning with multiple extension. We also defined a system that provides a generalized mechanism for such interaction. We also discussed several properties of such systems. We believe that an important application of trans-epistemic programs is in the area of simulating and studying the behavior of co-operating reasoning agents.

References

1. M. Gelfond. Strong Introspection. In *Proc. of AAAI-91*, 1991.
2. M. Gelfond. Logic Programming and Reasoning with Incomplete Information. *Annals of Mathematics and Artificial Intelligence*, 12, 1994.
3. M. Gelfond and V. Lifschitz. The Stable Model Semantics for Logic Programming. In R.A. Kowalski and K.A. Bowen, editors, *Proc. 5th International Conference and Symposium on Logic Programming*, pages 1070–1080, Seattle, Washington, August 15-19 1988.
4. G. Lakemeyer and H.J. Levesque. A Tractable Knowledge Representation Service with Full Introspection. In *Proc. TARK 88*, 1988.
5. H.J. Levesque. All I Know: A Study in Autoepistemic Logic. *Artificial Intelligence*, 42(2-3):263–309, 1990.
6. J.J. Lu, A. Nerode, J.B. Remmel, and V.S. Subrahmanian. Toward A Theory of Hybrid Knowledge Bases. Technical report, MSI, Cornell University, 1993.
7. W. Marek, A. Nerode, and J.B. Remmel. A Contex for Belief Revision: FC-normal Nonmonotonic Programs. In *Proc. Workshop on Defeasible Reasoning and Constraint Solving*, 1992. To appear in Annals of Pure and Applied Logic.
8. W. Marek, A. Nerode, and J.B. Remmel. Rule Systems, Well-ordering and Forward Chaining. Technical report, MSI, Cornell University, 1993.
9. W. Marek and M. Truszczynski. Relating autoepistemic and default logics. In *Proceedings of KR-89*, pages 276–288, 1989.
10. W. Marek and M. Truszczynski. *Non-Monotonic Logics: Context Dependent Reasoning*. Springer-Verlag, 1993.

11. A. Nerode and V.S. Subrahmanian. Hybrid Knowledge Bases. Technical report, University of Maryland, 1992.
12. T. Przymusinski. A Knowledge Representation Framework Based on Autoepistemic Logic of Minimal Beliefs. In *Proc. of Logic Programming and Non-Monotonic Reasoning Retreat, Shaker Village, KY*, 1994.
13. A. Rajasekar. Theories of Trans-Epistemic Defaults. Technical report, University of Kentucky, 1994.
14. R. Reiter. A Logic for Default Reasoning. *Artificial Intelligence*, 13(1 and 2):81–132, April 1980.

Computing the acceptability semantics

Francesca Toni[1] and Antonios C. Kakas[2]

[1] Department of Computing, Imperial College, 180 Queen's Gate,
London SW7 2BZ, UK, ft@doc.ic.ac.uk
[2] Department of Computer Science, University of Cyprus, 75 Kallipoleos Street,
Nicosia P.O. Box 537, Cyprus, antonis@turing.cs.ucy.ac.cy

Abstract. We present a proof theory and a proof procedure for non-monotonic reasoning based on the acceptability semantics for logic programming, formulated in an argumentation framework. These proof theory and procedure are defined as generalisations of corresponding proof theories and procedures for the stable theory and preferred extension semantics. In turn, these can be seen as generalisations of the Eshghi-Kowalski abductive procedure for logic programming.

1 Introduction

Recently, it has been shown [2, 4, 6, 9] that many non-monotonic reasoning formalisms, including logic programming (LP), Reiter's default logic and Moore's autoepistemic logic, can be understood within an argumentation framework. In particular, the argumentation-based understanding of negation as failure provides a uniform framework for capturing many existing semantics for LP. The general aim of this paper is to study the problem of computing the argumentation semantics for non-monotonic reasoning. Our study will concentrate on the special case of LP, but most of our work carries through without major changes to other non-monotonic formalisms.

One way which shows how the argumentation-based view of negation as failure can unify many of the existing semantics for LP is through a particular argumentation semantics, called the acceptability semantics [9]. This is due to the fact that many of these semantics, e.g. stable models, partial stable models, preferred extensions, stationary expansions, complete scenaria, well-founded model and stable theories (and semantics equivalent to these) can be captured as approximations of the full acceptability semantics (see [9]). Thus, the specific aim of the paper is to develop a proof theory and a proof procedure for the acceptability semantics. The computational model we present gives a general, uniform framework for computing other existing semantics for negation as failure that are approximations of the acceptability semantics. Proof theories and procedures for computing these semantics are approximations of the general proof theory and procedure for the full acceptability semantics.

In the argumentation-based approach to the acceptability semantics, non-monotonic reasoning is seen as the formation of theories which are extensions of a given theory T in a monotonic language by suitable subsets of a given set of hypotheses $\mathcal{H}$ (c.f. [10]). A set of hypotheses Δ can be seen as an argument supporting the consequences of the extended theory $T \cup \Delta$ in the monotonic language. In general, $\mathcal{H}$ contains conflicts (or incompatibilities) with respect to the theory T. Acceptability is a criterion for selecting subsets of $\mathcal{H}$ that provide appropriate conflict-free extensions of T. This criterion is defined on the subsets of $\mathcal{H}$ recursively so that its satisfaction by a particular subset (extension) depends on whether other (conflicting) extensions might or might not be acceptable under the same criterion. Hence acceptability is a second-order criterion.

This second-order nature of acceptability makes its computation (from a technical point of view) particularly interesting especially in view of the non-determinism there is in selecting parts of the hypotheses in $\mathcal{H}$. The aim of this paper is to develop a constructive proof theory for acceptability that understands and takes into account these second order and non-determinism features. In doing this, our primary concern will not be the complexity of the computation but rather the understanding of the general issues and problems involved in computing acceptability (and its various approximations).

The proof theory and procedure for acceptability is a generalisation of proof theories and procedures for (sound) approximations of the acceptability semantics such as preferred extensions [3] and stable theories [8], which are in turn a generalisation of the Eshghi-Kowalski (E-K) abductive proof procedure presented in [5]. In fact, for clarity of presentation, we will first give, in sections 3 and 4, the computational framework for the special cases of the preferred extension and stable theory semantics, and then show, in sections 5 and 6, how these need to be extended to capture the full acceptability semantics. The proofs of the results presented are omitted due to lack of space, and can be found in an extended version of this paper.

2 Acceptability semantics

In this section we review the acceptability semantics [9] formulated within an abstract argumentation framework [2, 4, 6, 9], which consists of a *theory* T in a *monotonic logic*, a set of possible *hypotheses* (also called *assumptions* or *abducibles*) $\mathcal{H}$, and a notion of *attack* between sets of hypotheses. Different instances of this abstract argumentation framework correspond to different non-monotonic formalism. For the special case of LP, T is a logic program P, the set of assumptions $\mathcal{H}$ is the set $\{not\, p \,|\, p$ is a variable-free atom in the Herbrand base of $P\}$, and the monotonic background logic is the classical logic of definite Horn programs. These programs are obtained by replacing each negation as failure literal in the program P and in the hypotheses set $\mathcal{H}$ by a new positive atom [5]. We will denote these programs by the same symbol P, and the consequence relation of this logic by $\models$. (Unless otherwise specified, we will assume that programs containing variables represent the set of all their variable-free instances over the given Herbrand universe.) The notion of **attack** between sets of negation as failure abducibles is given as follows [7]: a set of abducibles Δ_1 attacks another set Δ_2 iff $P \cup \Delta_1 \models p$ for some $not\, p$ in Δ_2.

Within the abstract argumentation framework, the notion of acceptability is specified as follows (see [9] for the details of the formal definition). Given a logic program P and two sets of hypotheses (subsets of $\mathcal{H}$) Δ and Δ_0,

(ACC) Δ is acceptable to Δ_0 iff for all sets of hypotheses A,
  if A attacks $\Delta - \Delta_0$, then A is not acceptable to $\Delta_0 \cup \Delta$.

A set of hypotheses is **acceptable** iff it is acceptable to the empty set of hypotheses. Note that a set of hypotheses that attacks itself can not be acceptable, since there is no attack against the empty set of assumptions.

In developing the proof theory for acceptability it is useful to unfold the condition "A is not acceptable to $\Delta_0 \cup \Delta$" in definition (ACC) to obtain the equivalent definition:

(ACC') Δ is acceptable to Δ_0 iff for all sets of hypotheses A,
  if A attacks $\Delta - \Delta_0$, then there exists a set of hypotheses D such that
  D attacks $A - (\Delta_0 \cup \Delta)$ and D is acceptable to $A \cup \Delta_0 \cup \Delta$.

The set D plays the role of a **defence** against the **attack** A. As we will see later on, the proof theory constructs acceptable sets Δ by augmenting in a suitable way an initial subset of Δ with defences.

The base case for both definitions (ACC) and (ACC') is given by

(BC1) Δ is acceptable to Δ_0 if there exists no set $A \subseteq \mathcal{H}$ attacking $(\Delta - \Delta_0)$.

In particular, since there are no attacks against the empty set of hypotheses, (BC1) occurs when $\Delta - \Delta_0 = \emptyset$, giving

(BC2) Δ is acceptable to Δ_0 if $\Delta \subseteq \Delta_0$.

In order to illustrate the acceptability semantics and its main features relevant for the development of the proof theory let us consider a few examples.

Example 1. Consider the program P

$$s \leftarrow not\, r$$
$$r \leftarrow not\, s.$$

$\Delta_1 = \{not\, s\}$ is acceptable, since every attack against Δ_1 is a superset of $\{not\, r\}$, which is attacked by Δ_1 itself. This shows how a set of abducibles, that we want to prove acceptable, can be used in its own defence. Similarly, $\Delta_2 = \{not\, r\}$ is acceptable. However, $\Delta_1 \cup \Delta_2$ cannot be acceptable, since it attacks itself.

Example 2. Consider the program of example 1 extended by the clause

$$q \leftarrow not\, s.$$

$\{not\, q\}$ is not acceptable, since the attack $\{not\, s\}$ against it is acceptable, as we have seen in example 1. However, the set $\{not\, q, not\, r\}$ is acceptable, because $not\, r$ provides an acceptable defence against the attack $\{not\, s\}$.

This example shows that the construction of an acceptable set Δ can be done *incrementally* starting from a given set of hypotheses Δ_1, by adding to Δ_1 defences for it. A property of the acceptability semantics that can be used as a basis for such a construction is

$\Delta = \Delta_1 \cup \Delta_2$ is acceptable to $\Delta_1 \cap \Delta_2$ iff

Δ_1 is acceptable to Δ_2 and Δ_2 is acceptable to Δ_1.

Therefore, to construct an acceptable set $\Delta = \Delta_1 \cup \Delta_2$ we need to find a new set Δ_2, disjoint from Δ_1, such that Δ_2 can be shown to be acceptable to Δ_1 and Δ_2 does not affect the acceptability of Δ_1 to Δ_2. While choosing Δ_2, one has to take into account the non-determinism that arises in the program. In example 2, it was necessary to choose $\Delta_2 = \{not\, r\}$ among $\{not\, s\}$ and $\{not\, r\}$, to render $\{not\, q\} \cup \Delta_2$ acceptable.

Example 3. Given the program P

$$q \leftarrow not\, r$$
$$r \leftarrow not\, s$$
$$s \leftarrow not\, q$$

$\Delta_1 = \{not\, q\}$ is not acceptable, since the attack $\{not\, r\}$ against Δ_1 is acceptable to Δ_1. In fact, any attack against $\{not\, r\}$ must contain $not\, s$, and therefore Δ_1 provides a defence against this attack. Similarly, $\Delta_2 = \{not\, r\}$ and $\Delta_3 = \{not\, s\}$ are not acceptable. If we add to P an extra rule

$$p \leftarrow not\, q$$

then $\Delta = \{not\, p\}$ is acceptable, since the attack Δ_1 is still not acceptable to Δ, for the same reason as above.

Note that, in this example, the defence $\{not\, r\}$ used to show that the attack $\{not\, q\}$ is not acceptable should not be added to $\Delta = \{not\, p\}$. Therefore, the proof theory for constructing acceptable sets should be able to separate which defences or more generally which parts of defences can be included in the final Δ. This adds an extra level of non-determinism in the computation.

As shown in [9], many semantics for LP correspond to approximations of the full acceptability relation, obtained through iterations of the acceptability specification starting from the above base cases (BC1/2). Specifically, preferred extension [3] and stable theory [8] semantics (and other semantics which are equivalent to them) are captured by the approximation at the second iteration:

(A-ACC) Δ is acceptable to Δ_0 if for all sets of hypotheses A,
if A attacks $\Delta - \Delta_0$, then there exists a set of hypotheses D such that
D attacks $A - (\Delta_0 \cup \Delta)$ and $D \subseteq A \cup \Delta \cup \Delta_0$.

If $\Delta_0 = \emptyset$, then Δ is acceptable with respect to this approximation of the notion of acceptability iff Δ is weakly stable [8], where a stable theory corresponds to a maximal weakly stable set. As a special case, if $\Delta_0 = \emptyset$ and the condition

$D \subseteq A \cup \varDelta \cup \varDelta_0$ is replaced by its special case $D \subseteq \varDelta \cup \varDelta_0$, then $\varDelta$ is acceptable with respect to the resulting approximation of definition (ACC′) iff $\varDelta$ is admissible [3], where a preferred extension corresponds to a maximal admissible set. The proof theories and procedures for admissibility and weak stability semantics we will develop next will be based upon the formulation of these semantics via (ACC′).

3 Proof theory for admissibility and weak stability

The proof theory will be based on an understanding of the semantics in terms of trees defined as follows:

Definition 1. **A tree for a set of hypotheses** $\varDelta$ is a labelled and-tree where
 (1) each node is a set of abducibles, labelled either as "defence" or "attack";
 (2) the root is $\varDelta$, labelled as "defence";
 (3) the root has as children all the attacks against $\varDelta$
   and each such child is labelled as "attack";
 (4) each node A labelled as "attack" has at most one child,
   labelled as "defence", chosen amongst all the attacks against $A - \varDelta$.

Note that defences are non-deterministically chosen, so that many trees can be associated to the same $\varDelta$. Note also that a tree for some set of assumptions can be infinite in breadth, since there may be infinitely many attacks against the root of the tree. However, by definition, trees are finite in depth; more precisely, they are at most two levels deep.

Definition 2. A tree for some set of hypotheses is called

 - **admissible** iff all its leaves are "defences" nodes, and each "defence" node is a subset of the root.
 - **weakly stable** iff all its leaves are "defences" nodes, and each "defence" node D which is child of an "attack" node A is a subset of the union of the root together with A.

Theorem 3. *Given a set of hypotheses* $\varDelta$:

 - $\varDelta$ *is admissible iff there exists an admissible tree for* $\varDelta$.
 - $\varDelta$ *is weakly stable iff there exists a weakly stable tree for* $\varDelta$.

The trees as defined above require that *all* attacks against the root are considered. Since every superset of an attack is itself an attack, there might be infinitely many attacks to be considered to build admissible and weakly stable trees. However, any such attack is a superset of a *minimal* (with respect to set inclusion) attack. From the property that if $\varDelta$ attacks $(A - \varDelta)$ then $\varDelta$ attacks $(A' - \varDelta)$, for all $A' \supseteq A$, to check whether a set of assumptions $\varDelta$ is admissible (or weakly stable) we only need to construct the subtree of some admissible (or weakly stable) tree for $\varDelta$ where the set of all "attack" nodes contains at least the set of all minimal attacks.

Definition 4. A restricted tree for a set of assumptions Δ is a tree as in definition 1 but with condition (3) replaced by

(3′) the root has as children at least all *minimal* attacks against Δ and each such child is labelled as "attack".

Restricted admissible and weakly stable trees for sets of assumptions are defined similarly to definition 2.

Theorem 5. *Given a set of hypotheses Δ*

 — Δ *is admissible iff there exists a restricted admissible tree for Δ.*
 — Δ *is weakly stable iff there is a restricted weakly stable tree for Δ.*

We want to compute admissible or weakly stable sets of assumptions Δ that are explanations (in the program P) of a given goal G, i.e. such that $P \cup \Delta \models G$. In this section we will assume that an explanation Δ_1 for G is given. (The computation of Δ_1 will be discussed in the next section.) Δ_1 might not be admissible (weakly stable). The task of the proof procedure is to compute a superset Δ of Δ_1 which is admissible (weakly stable). In the remainder of this section we will concentrate on the computation of weakly stable sets of assumptions. The computation of admissible sets is a special case, as we will see later.

The abstract proof theory for the weak stability semantics is based upon the construction of restricted weakly stable trees. In the sequel, if no confusion arises, restricted trees will be simply referred to as trees. Such trees are constructed via the derivation of **partial trees**. Like in full (restricted) trees, nodes of partial trees are sets of assumptions that are labelled as "attack" or as "defence". However, unlike in full trees, some of the branches of partial trees are only partially built, and the roots of partial trees may grow during the derivation. The growth of the root corresponds to the extension of Δ to Δ_1.

The abducibles that are added to Δ_1 to form Δ are chosen from the abducibles occurring in defences in the subtree below Δ_1. In fact, by adding these abducibles to the root Δ_1, we increase the possibility of the defences to be subsets of the (extended) root together with the parent attack, and therefore we obtain a derived partial tree which is closer to a weakly stable tree. As an example, let us consider the program in example 2 and $\Delta_1 = \{not\, q\}$. Figure 1 (a) shows a tree for Δ_1 which is not weakly stable (and also not admissible), since the defence $D_1 = \{not\, r\}$ is not a subset of $\Delta_1 \cup A_1$. However, the tree for $\Delta = \Delta_1 \cup D_1$ is weakly stable (and admissible), as shown in figure 1 (b). (Arrows from one node to others indicate that the assumptions in the first set are contained in the union of the nodes to which the arrows are pointing.)

The addition of abducibles in defences to the root of partial trees will not always lead to weakly stable trees. For example, consider the situation sketched in figure 2. Here, the addition of $not\, q$ in D to the current root Δ_1 of the given partial tree renders any derivable full tree non-weakly stable, since $not\, q$ can not be possibly selected in A as the hypothesis to be counter attacked in a weakly stable tree when $not\, q$ belongs to the root. In general, given a partial tree, if an

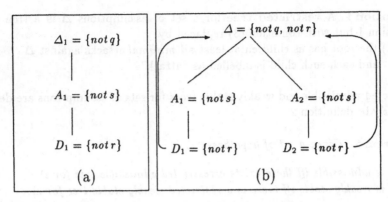

Fig. 1. Δ_1 needs to be extended, for example 2

abducible c has already occurred in some "attack" node A followed by a "defence" node D and c is the single specific abducible in A that is "contradicted" by D, then c should not be promoted to the root. In the sequel, any such abducible c will be referred to as **culprit**.

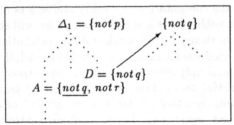

Fig. 2. Abducibles in defences added to Δ_1 should be disjoint from the set of culprits.

In the following definition we assume that nodes of partial trees are marked to indicate that they should not be selected further. Moreover, we assume that, given a node N in a partial tree, **branch**(N) denotes the set of all abducibles occurring in the union of all nodes from the root up and including the parent node of N, if N is not the root, and the empty set, if N is the root.

Definition 6. A derivation of partial trees is a sequence of partial trees $T_0, \ldots, T_n$ such that, given T_i $(i = 0, \ldots, n - 1)$, with $culprits_i$ the set of all culprits in "attack" nodes in T_i, if N is the selected (unmarked) node in T_i, then the next partial tree T_{i+1}, with associated $culprits_{i+1}$, is obtained as follows:

(α) if N is an "attack" node, then
if $(N - branch(N))$ is not empty,
then an abducible c in $(N - branch(N))$ and an attack D against c are chosen, and
$culprits_{i+1}$ is $culprits_i \cup \{c\}$ and T_{i+1} is T_i with
(1) N marked and
(2) D added as a "defence" node child of N;
(δ) if N is a "defence" node, then $culprits_{i+1} = culprits_i$ and
if $(N - branch(N))$ is empty then T_{i+1} is T_i with N marked,
otherwise, if none of the abducibles in $(N - branch(N))$ belongs to $culprits_i$, given

a set of attacks $A_1, \ldots, A_m, \ldots$ against $(N-branch(N))$ that includes all minimal attacks against $(N-branch(N))$,
then T_{i+1} is T_i with N marked, the set of hypotheses in the root extended by $(N-branch(N))$, and $A_1, \ldots, A_m, \ldots$ added as additional "attack" nodes children of the root.

Note that, once a "defence" node is selected, if it does not contain any previously chosen culprit, all abducibles in the node that do not occur in the branch up to the node are added to the root.

Definition 7. Given a set of abducibles Δ_1, a **successful derivation** from Δ_1 is a derivation from the partial tree T_0 consisting only of the (unmarked) root Δ_1 to a partial tree T_n where all nodes are marked and all leaves are defences. If the root of T_n is Δ, then we say that the derivation **computes** Δ from Δ_1.

Theorem 8. *Given a set of abducibles Δ_1. If there is a successful derivation computing Δ from Δ_1, then Δ is weakly stable (soundness). Conversely, if there exists a weakly stable set of abducibles Δ such that $\Delta_1 \subseteq \Delta$, then there exists a successful derivation computing Δ' from Δ_1, where $\Delta_1 \subseteq \Delta' \subseteq \Delta$ (completeness).*

The proof theory for admissibility is a special case of the proof theory for weak stability, where for each node N in a partial tree, if N is different from the root then $branch(N)$ denotes the set of all assumptions in the root, if N is the root then $branch(N)$ denotes the empty set.

4 Procedures for admissibility and weak stability

The proof theory developed in the previous section can be specialised to give procedures for the computation of admissibility and weak stability by incorporating a specific (resolution-based) way of computing the notion of attack. Similarly to the previous section, we will concentrate on the computation of weakly stable sets of assumptions, the computation of admissible sets being a special case. The procedure solves directly the problem of generating a weakly stable explanation Δ for a goal G. In fact, the procedure interleaves the generation of Δ_1 and the extension of Δ_1 to Δ. Moreover, neither the query nor the program are assumed to be variable-free.

The procedure is based on the generation of full trees via the derivation of partial trees. However, the nodes of both full and partial trees are goals, represented as sets of abducibles and atoms, instead of sets of abducibles only, in order to accommodate the computation of Δ_1 from G and the computation of the notion of attack, both via resolution. Note that the set of attacks computed via resolution is a superset of the set of all minimal attacks. The trees used to define the proof procedure are called **goal-trees**. Similarly to the previous sections, we will assume that nodes in goal-trees are possibly marked either as "attack-goal" or "defence-goal". Note that attack-goals and defence-goals represent potential attacks and defences. The actual attacks and defences result from resolving away all non-abducible conditions in the goals. If some of these conditions can not be resolved away, then there are no corresponding attacks and defences for such goals.

In the abstract proof theory, "defence" nodes, which were sets of abducibles only, were marked once selected. In the procedure, instead, defence-goals might be selected many times before they are actually marked, due to the fact that defence-nodes contain atoms as well as abducibles. Moreover, in order to keep the alternation of attack- and defence-goals in branches of goal-trees, nodes are not only added to goal-trees, but are also replaced by new nodes (after resolution takes place).

We will denote the selection rule of conditions in goals by $\Re$. We will assume that $\Re$ is safe, i.e. abducibles (negative literals) are selected by $\Re$ only if they are variable-free. Moreover, we will suppose that each goal G in a goal-tree is divided into two parts, with the first part, referred to as $G_{already}$, a set of variable-free negative literals which is transparent to the selection rule $\Re$. Namely, $\Re(G)$ stands for $\Re(G - G_{already})$. Furthermore, we will suppose that $\Re(G) = \square$ iff $(G - G_{already}) = \emptyset$. Finally, for each goal G in a goal-tree, if G is different from the root then $branch(G)$ denotes the set of all variable-free abducibles occurring in the branch from the root of the given goal-tree up and including the goal parent of G, and if N is the root then $branch(G)$ denotes the empty set.

Definition 9. A derivation of partial goal-trees is a sequence of partial goal-trees $T_0, \ldots, T_n$ where, given a goal-tree T_i ($i = 0, \ldots, n-1$), with $culprits_i$ the set of all culprits in attack-goals in T_i, if G is the selected (unmarked) node in T_i, then T_{i+1}, with associated $culprits_{i+1}$, is obtained as follows:

(α) If G is an attack-goal and $\Re(G) \neq \square$, then

    (α_{atom}) If $\Re(G) = q$, where q is an atom, then $culprits_{i+1} = culprits_i$, and
        if $G_1, \ldots, G_m$ are all the resolvents of G on q (in P) and $m \geq 1$,
        then T_{i+1} is T_i with G replaced by the m attack-goals $G_1, \ldots, G_m$ as children
        of the parent node of G, and $G_{1\,already} = \ldots = G_{m\,already} = G_{already}$.

    (α_{naf}) If $\Re(G) = not\,q$, where q is an atom, then
        if $not\,q$ is in $branch(G)$, then $culprits_{i+1} = culprits_i$, and T_{i+1} is T_i with $G_{already}$
        replaced by $G_{already} \cup \{not\,q\}$,
        otherwise $culprits_{i+1}$ is $culprits_i \cup \{not\,q\}$, and T_{i+1} is T_i with:
        (1) G marked, and
        (2) the goal $\{q\}$ added as a defence-goal child of G;

(δ) If G is a defence-goal, then $culprits_{i+1} = culprits_i$, and

    ($\delta_\square$) If $\Re(G) = \square$, then T_{i+1} is T_i with G marked;

    (δ_{atom}) If $\Re(G) = q$, where q is an atom, then
        if there exists some resolvent G' of G on q (in P),
        then T_{i+1} is T_i with G replaced by the defence-goal G' and $G'_{already} = G_{already}$;

    (δ_{naf}) If $\Re(G) = not\,q$, where q is an atom, then
        if $not\,q$ is in $branch(G)$, then T_{i+1} is T_i with $G_{already}$ replaced by $G_{already} \cup$
        $\{not\,q\}$, otherwise
        if $not\,q$ is not in $branch(G)$ and $not\,q$ does not belong to $culprits_i$, then T_{i+1}
        is T_i with $G_{already}$ replaced by $G_{already} \cup \{not\,q\}$ and the root of T_{i+1} is the

goal $R' = R \cup \{not\,q\}$, where R is the root of T_i, $R'_{already} = R_{already} \cup \{not\,q\}$, and the goal $\{q\}$ is an additional (attack) child of the extended root R'.

Note that a defence-goal is marked only when all the abducibles in it have been generated and dealt with (case $\delta_{\square}$). An attack-goal is marked once a culprit is selected in it (case α_{naf}). Also, note that abducibles in defence-goals are added to the root one by one.

Definition 10. A **successful derivation of goal-trees** from a goal G is a derivation from the goal-tree T_0 consisting only of the (unmarked) root G to a goal-tree T_n where all nodes are marked and all leaves are defence-nodes. If Δ is the root of T_n then we will say that the derivation **computes** Δ from G.

Theorem 11. *Given a goal G, if there exists a successful derivation of goal-trees computing Δ from G, then $P \cup \Delta \models G$ and Δ is weakly stable (soundness).*

However, the proof procedure is incomplete, for the same reasons that SLDNF is incomplete, namely because of the possible occurrence of floundering and loops.

As in the case of the proof theory, the procedure for admissibility is a special case of the procedure for weak stability, where for each node N in a partial tree, if N is different from the root then $branch(N)$ denotes the set of all variable-free abducibles in the root, if N is the root then $branch(N)$ denotes the empty set.

An implementation of the procedure requires some representation of the tree structure. Regarding this issue, note that some strategies for the selection of nodes in the tree might be more convenient than others. For example, we can use a combination of breadth-first and depth-first strategies. The generation of different attack-goals, against some given abducibles in the root or against a given defence-goal, can be interleaved (breadth-first strategy). But once an attack-goal G has been generated and chosen, the treatment of all the other attack-goals at the same level is suspended until a complete subtree from G, whose nodes are all marked, has been constructed (depth-first strategy). This combined strategy allows to record only a single branch at each time, without keeping a record of the whole tree. If we apply this combined strategy to the instance of the procedure to compute admissibility, we obtain a procedure which is equivalent to the E-K proof procedure [5], in the sense that the two procedures compute the same answers to the same goals. Our procedure, however, differs from the E-K procedure because of the use of culprits. This allows a more effective detection of failures.

5. Proof theory for acceptability semantics

The trees defined in section 3 are not sufficient to express the acceptability semantics. The following definition gives a more general notion of tree.

Definition 12. A **tree for sets of hypotheses** Δ and Δ_0 is a labelled and-tree, where

(1) each node is a set of abducibles, labelled either as "defence" or "attack",
(2) the root is Δ, labelled as "defence",
(3) each node D labelled as "defence" has as children all the attacks against $D - (\Delta_0 \cup branch(D))$, and each such child is labelled as "attack",
(4) each node A labelled as "attack" has at most a single child, labelled as "defence", chosen amongst all the attacks against $A - (\Delta_0 \cup branch(A))$.

In each branch of such a tree, not only the root but any "defence" node can be followed by "attack" nodes. This is due to part (3) of the definition. As a result, any branch of the tree can be infinite, and therefore trees of this kind can not only be infinite in breadth, as in the case of simple trees, but also infinite in depth. Note that, as for simple trees, defences are non-deterministically chosen, so that many trees can be associated to the same Δ, Δ_0. In the rest of the paper, we will focus on trees as in definition 12.

In any tree for Δ, Δ_0, a "defence" node D is a leaf iff there are no attacks against $D - (\Delta_0 \cup branch(D))$. This corresponds to the base case (BC1) for the acceptability semantics. As a special case, this can happen if $D \subseteq (\Delta_0 \cup branch(D))$, namely if the base case (BC2) applies.

Definition 13. A tree for some sets of assumptions Δ, Δ_0 is called **acceptable** iff all its leaves are "defence" nodes.

Theorem 14. *Δ is acceptable to Δ_0 iff there is an acceptable tree for Δ, Δ_0.*

Unfortunately, the analogous of theorem 5 does not hold for acceptability, since it is not sufficient to look only at minimal attacks to guarantee for a set to be acceptable. This is illustrated by the program in example 2 extended by the clause

$$p \leftarrow not\, q.$$

$\Delta = \{not\, p\}$ is not acceptable (to $\emptyset$). Figure 3 shows (a subtree of) the only tree for Δ, $\emptyset$. $A_1 = \{not\, q\}$ is the only minimal attack against Δ, and it gives rise to the left branch, terminating with a "defence" node $D_2 = D_1$. However, there is a subset of the set of all attacks that it is sufficient to consider in the construction of the tree in order to guarantee acceptability. Attacks in this subset are built from minimal attacks and the culprits chosen in the branches below the minimal attacks in the tree. Every non-minimal attack constructed in this way needs to be extended via culprits, recursively. For instance, in the example described in figure 3. it is sufficient to consider only the additional, non-minimal attack $A_1^* = \{not\, q, not\, r\}$, constructed from the minimal attack A_1 and the culprit $not\, r$ in A_2, to detect that Δ is not acceptable, since A_1^* gives rise to the right branch of the tree which has the "attack" node A_2^* as leaf.

Definition 15. Given sets of abducibles Δ, Δ_0, a **restricted tree** for Δ, Δ_0 is a tree as in definition 12 but with condition (3) replaced by

(3') each node D labelled as "defence" has as children, labelled as "attack", all sets of assumptions A satisfying the following conditions:

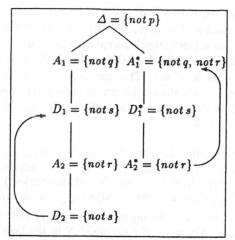

Fig. 3. Minimal attacks are not sufficient for acceptability, for example 2

- A is a minimal attack against $D - (\Delta_0 \cup branch(D))$, or
- A is an **extended attack**, i.e. $A = A' \cup C$, where A' is a (minimal or extended) attack child of D in the tree and C is a non-empty subset of the set of culprits that appear in the subtree below A'.

Theorem 16. Δ *is acceptable to* Δ_0 *iff there exists a* **restricted acceptable tree** *for* Δ, Δ_0, *i.e. a restricted tree whose leaves are all defences.*

Analogously to the admissibility and weak stability case, the proof theory for acceptability is based on the construction of acceptable trees via the derivation of partial trees. However, in the acceptability case, the promotion of abducibles in defences to the root of partial trees is not compulsory, since trees are allowed to grow more that two levels deep. Moreover, there are cases where this promotion is unwanted, because it leads to non-acceptable sets of assumptions. For example, consider the program in example 3, and the tree sketched in figure 4. Here, if the abducible $not\,r$ in the "defence" node D_1 in the left branch is added to Δ_1, we

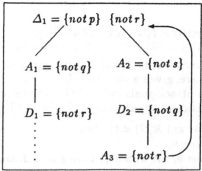

Fig. 4. Not all abducibles in defences should be promoted to Δ_1, for example 3

obtain a non-acceptable tree with a leaf, A_3, which is an "attack" node. Indeed, $\Delta_1 \cup \{not\, r\}$ is not acceptable. However, Δ_1 alone is acceptable. In fact, we can expand the left branch to a (restricted) acceptable tree. As a result, the proof theory should allow the *choice* whether abducibles in "defence" nodes should be added to the current root or not. The non-determinism in choosing which abducibles in a "defence" node should be promoted to Δ_1 is restricted by the fact that these should be disjoint from the culprits in "attack" nodes. In the sequel we will assume $\Delta_0 = \emptyset$.

Definition 17. A derivation of partial trees is a sequence of partial trees $T_0, \ldots, T_n$ such that, given T_i $(i = 0, \ldots, n-1)$, with *culprits*$_i$ the set of all culprits in "attack" nodes in T_i, if N is the selected (unmarked) node in T_i, then the next partial tree T_{i+1}, with associated *culprits*$_{i+1}$, is obtained as follows:

(α) as in definition 6, but with T_{i+1} having in addition to (1) and (2):
   (3) for each defence D' that is not the parent of N in the branch up to N in T_i and for each "attack" node A which is a child of such D' such that A is either the attack A' in the branch with N or is a superset of A' (obtained as an extended attack for an earlier culprit),
   an extended attack $A \cup \{c\}$ is added as an additional "attack" node child of D' in T_{i+1};
(δ) if N is a "defence" node, then *culprits*$_{i+1}$ = *culprits*$_i$ and
   if $(N - branch(N))$ is empty then T_{i+1} is T_i with N marked,
   otherwise, let $(N - branch(N)) = D_1 \cup D_2$ for some sets of assumptions D_1 and D_2 such that $D_2 \cap$ *culprits*$_i = \emptyset$;
   then, given a set of attacks $A_1, \ldots, A_m, \ldots$ against D_1 containing all minimal attacks against D_1 and a set of attacks $B_1, \ldots, B_k, \ldots$ against D_2 containing all minimal attacks against D_2,
   T_{i+1} is T_i with N is marked, $A_1, \ldots, A_m, \ldots$ added as "attack" nodes children of N, the set of hypotheses in the root extended by D_2, and $B_1, \ldots, B_k, \ldots$ added as (additional) "attack" nodes children of the root.

Note that, once a "defence" node is selected, only one subset, D_2, of the set of (new) abducibles in that node is promoted to the root.

Given a set of abducibles Δ_1, the notion of successful derivation of partial reduced trees is given as in definition 7. We then have the same soundness and completeness results as in theorem 8 with weakly stable replaced by acceptable.

6 A proof procedure for acceptability

Definition 18. A derivation of partial goal-trees is a sequence of partial goal-trees $T_0, \ldots, T_n$ where, given a goal-tree T_i $(i = 0, \ldots, n-1)$, with *culprits*$_i$ the set of all culprits in attack-goals in T_i, if G is the selected (unmarked) node in T_i, then T_{i+1}, with associated *culprits*$_{i+1}$, is obtained as follows:

(α) If G is an attack-node and $\Re(G) \neq \square$, then
   (α_{atom}) as in definition 9,
   (α_{naf}) as in definition 9, but with T_{i+1} having in addition to (1) and (2):
      (3) for each defence-goal D different from the parent of G in the branch that

leads to G in T_i such that A is child of D in the branch containing G and $A'_1, \ldots, A'_n$ are all the children of D that are extensions of A in T_i, then the extended attack-goals $A \cup \{not\, q\}, A'_1 \cup \{not\, q\}, \ldots, A'_n \cup \{not\, q\}$ are additional children of D in T_{i+1}. The *already* part of the new extended attack-goals is the same as that of the specific attack they extend.

(δ) If G is a defence-goal, then $culprits_{i+1} = culprits_i$, and

($\delta_\Box$) as in definition 9,

(δ_{atom}) as in definition 9,

(δ_{naf}) If $\Re(G) = not\, q$, where q is an atom, then

if $not\, q$ is in $branch(G)$, then T_{i+1} is T_i where $G_{already}$ is replaced by $G_{already} \cup \{not\, q\}$, otherwise, if $not\, q$ is not in $branch(G)$, then

either: if $not\, q$ does not belong to the subset of $culprits_i$ consisting of all culprits selected in the subtree of T_i with root G, then T_{i+1} is T_i where $G_{already}$ is replaced by $G_{already} \cup \{not\, q\}$ and the goal $\{q\}$ is an attack-goal child of G;

or: if $not\, q$ does not belong to the whole set $culprits_i$, then T_{i+1} is T_i where $G_{already}$ is replaced by $G_{already} \cup \{not\, q\}$ and the root of T_{i+1} is the goal $R' = R \cup \{not\, q\}$, where R is the root of T_i, $R'_{already} = R_{already} \cup \{not\, q\}$, and the goal $\{q\}$ is an additional (attack) child of the extended root R'.

Note that the test in the "either" case in (δ_{naf}), that the selected abducible does not occur as a culprit in the subtree below the defence-goal, is needed because defences are incrementally built, and therefore contribute only partially to the *branch*. The new appearance of *not q* in the defence would invalidate a previously chosen culprit *not q* in the branch below the defence goal. This feature was not present in the proof theory, but is analogous to the test that we need to make when we extend the root.

The notion of successful derivation and the soundness result are formulated similarly to the weak stability case (see definition 10 and theorem 11).

In the case of acceptability, the combined (breadth- depth-first) selection strategy in the actual implementation of the procedure has the advantages we observed in the weakly stability case, as well as the additional advantage that all the extended attacks for each given attack can be constructed at the same time, in case (α_{naf}), and the construction of the required subset of $culprits_i$ in the "either" case in (δ_{naf}) does not need to be separated from the construction of the whole set of culprits.

7 Conclusions

We have presented a proof theory and a proof procedure for the acceptability semantics in LP. These have been defined as extensions of proof theories and procedures for the weak stability (stable theory) and admissibility (preferred extension) semantics. In this paper, we have focused on understanding the main problems underlying the computation of these semantics rather than on the complexity of the computation. The study of more effective procedures for acceptability is an issue for future research.

Proof theories and procedures have been defined in the context of an argumentation-based interpretation of the (LP) semantics. In defining these proof theories, we have abstracted away from the specific features of the argumentation framework for LP, and only assumed that the framework satisfies some general properties, e.g. that the background logic is monotonic and that no set of assumptions can attack the empty set of assumptions. As a result, the given proof theories can be used as a basis to define procedures for acceptability and weak stability/admissibility in any non-monotonic formalism that can be formulated in the argumentation framework, whenever this satisfies the same properties.

Other argumentation semantics for LP have been proposed, e.g. [1]. The use of the computational scheme given in this paper for developing procedures for such semantics is an issue for future research.

Acknowledgements

This research was supported by the Fujitsu Research Laboratories. The authors are grateful to R.A. Kowalski for many helpful suggestions and comments on earlier drafts, and to P.M. Dung and P. Mancarella for many helpful discussions.

References

1. J.J. Alferes, L.M. Pereira, An argumentation-theoretic semantics based on non-refutable falsity. *Proc. Work. on Non-monotonic Extensions of LP, ICLP94*
2. A. Bondarenko, F. Toni, R. A. Kowalski An assumption-based framework for non-monotonic reasoning. *Proc. 2nd International Workshop on Logic Programming and Non-monotonic Reasoning* (1993)
3. P. M. Dung, Negation as hypothesis: an abductive foundation for logic programming. *ICLP'91*
4. P. M. Dung, On the acceptability of arguments and its fundamental role in non-monotonic reasoning and logic programming. *IJCAI '93*
5. K. Eshghi, R. A. Kowalski, Abduction compared with negation as failure. *ICLP'89*
6. A. C. Kakas, Default reasoning via negation as failure. *Foundations of Knowledge Representation and Reasoning"*, LNAI 810 Springer Verlag (1994)
7. A. C. Kakas, R. A. Kowalski, F. Toni, Abductive logic programming. *Journal of Logic and Computation* 2(6) (1993)
8. A. C. Kakas, P. Mancarella, Stable theories for logic programs. *ISLP'91*
9. A. C. Kakas, P. Mancarella, P.M. Dung, The Acceptability Semantics for Logic Programs. *ICLP'94*
10. D. Poole A logical framework for default reasoning. *Artificial Intelligence* 36 (1988)

Index

D. Aquilino 57
P. Asirelli 57
K.A. Berman 113
H.A. Blair 99
A. Bochman 245
S. Brass 85
A. Brogi 330
P. Cholewiński 273
C.V. Damásio 29
J.C.P. da Silva 175
M. Denecker 15
J. Dix 85
P.M. Dung 316
T. Eiter 1
M. Fitting 143
J.V. Franco 113
D. Gabbay 203
L. Giordano 203
G. Gottlob 1
M. Halfeld Ferrari Alves 71
K. Inoue 344
J. Kalinski 358
A.C. Kakas 401
E. Lamma 330
D. Laurent 71
N. Leone 1
V. Lifschitz 127
P. Mancarella 330
V.W. Marek 43
A. Martelli 203
N. McCain 127
P. Mello 330
A. Mikitiuk 259
R. Miller 217
A. Nerode 43
R.T. Ng 287
N. Olivetti 203
D. Pearce 372
L.M. Pereira 29
T.C. Przymusinski 127,156
A. Rajasekar 388
J.B. Remmel 43
C. Renso 57
C. Sakama 344

J.S. Schlipf 113
T.C. Son 316
N. Spyratos 71
R.F. Stärk 127,302
X. Tian 287
F. Toni 401
M. Truszczyński 259
F. Turini 57
H. Turner 156
W. van der Hoek 189
S.R.M. Veloso 175
C. Witteveen 189
J.-H. You 231
L.-Y. Yuan 231

Springer-Verlag
and the Environment

We at Springer-Verlag firmly believe that an international science publisher has a special obligation to the environment, and our corporate policies consistently reflect this conviction.

We also expect our business partners – paper mills, printers, packaging manufacturers, etc. – to commit themselves to using environmentally friendly materials and production processes.

The paper in this book is made from low- or no-chlorine pulp and is acid free, in conformance with international standards for paper permanency.

Lecture Notes in Artificial Intelligence (LNAI)

Vol. 764: G. Wagner, Vivid Logic. XII, 148 pages. 1994.

Vol. 766: P. R. Van Loocke, The Dynamics of Concepts. XI, 340 pages. 1994.

Vol. 770: P. Haddawy, Representing Plans Under Uncertainty. X, 129 pages. 1994.

Vol. 784: F. Bergadano, L. De Raedt (Eds.), Machine Learning: ECML-94. Proceedings, 1994. XI, 439 pages. 1994.

Vol. 795: W. A. Hunt, Jr., FM8501: A Verified Microprocessor. XIII, 333 pages. 1994.

Vol. 798: R. Dyckhoff (Ed.), Extensions of Logic Programming. Proceedings, 1993. VIII, 360 pages. 1994.

Vol. 799: M. P. Singh, Multiagent Systems: Intentions, Know-How, and Communications. XXIII, 168 pages. 1994.

Vol. 804: D. Hernández, Qualitative Representation of Spatial Knowledge. IX, 202 pages. 1994.

Vol. 808: M. Masuch, L. Pólos (Eds.), Knowledge Representation and Reasoning Under Uncertainty. VII, 237 pages. 1994.

Vol. 810: G. Lakemeyer, B. Nebel (Eds.), Foundations of Knowledge Representation and Reasoning. VIII, 355 pages. 1994.

Vol. 814: A. Bundy (Ed.), Automated Deduction — CADE-12. Proceedings, 1994. XVI, 848 pages. 1994.

Vol. 822: F. Pfenning (Ed.), Logic Programming and Automated Reasoning. Proceedings, 1994. X, 345 pages. 1994.

Vol. 827: D. M. Gabbay, H. J. Ohlbach (Eds.), Temporal Logic. Proceedings, 1994. XI, 546 pages. 1994.

Vol. 830: C. Castelfranchi, E. Werner (Eds.), Artificial Social Systems. Proceedings, 1992. XVIII, 337 pages. 1994.

Vol. 833: D. Driankov, P. W. Eklund, A. Ralescu (Eds.), Fuzzy Logic and Fuzzy Control. Proceedings, 1991. XII, 157 pages. 1994.

Vol. 835: W. M. Tepfenhart, J. P. Dick, J. F. Sowa (Eds.), Conceptual Structures: Current Practices. Proceedings, 1994. VIII, 331 pages. 1994.

Vol. 837: S. Wess, K.-D. Althoff, M. M. Richter (Eds.), Topics in Case-Based Reasoning. Proceedings, 1993. IX, 471 pages. 1994.

Vol. 838: C. MacNish, D. Pearce, L. M. Pereira (Eds.), Logics in Artificial Intelligence. Proceedings, 1994. IX, 413 pages. 1994.

Vol. 847: A. Ralescu (Ed.) Fuzzy Logic in Artificial Intelligence. Proceedings, 1993. VII, 128 pages. 1994.

Vol: 861: B. Nebel, L. Dreschler-Fischer (Eds.), KI-94: Advances in Artificial Intelligence. Proceedings, 1994. IX, 401 pages. 1994.

Vol. 862: R. C. Carrasco, J. Oncina (Eds.), Grammatical Inference and Applications. Proceedings, 1994. VIII, 290 pages. 1994.

Vol 867: L. Steels, G. Schreiber, W. Van de Velde (Eds.), A Future for Knowledge Acquisition. Proceedings, 1994. XII, 414 pages. 1994.

Vol. 869: Z. W. Raś, M. Zemankova (Eds.), Methodologies for Intelligent Systems. Proceedings, 1994. X, 613 pages. 1994.

Vol. 872: S Arikawa, K. P. Jantke (Eds.), Algorithmic Learning Theory. Proceedings, 1994. XIV, 575 pages. 1994.

Vol. 878: T. Ishida, Parallel, Distributed and Multiagent Production Systems. XVII, 166 pages. 1994.

Vol. 886: M. M. Veloso, Planning and Learning by Analogical Reasoning. XIII, 181 pages. 1994.

Vol. 890: M. J. Wooldridge, N. R. Jennings (Eds.), Intelligent Agents. Proceedings, 1994. VIII, 407 pages. 1995.

Vol. 897: M. Fisher, R. Owens (Eds.), Executable Modal and Temporal Logics. Proceedings, 1993. VII, 180 pages. 1995.

Vol. 898: P. Steffens (Ed.), Machine Translation and the Lexicon. Proceedings, 1993. X, 251 pages. 1995.

Vol. 904: P. Vitányi (Ed.), Computational Learning Theory. EuroCOLT'95. Proceedings, 1995. XVII, 415 pages. 1995.

Vol. 912: N. Lavrač, S. Wrobel (Eds.), Machine Learning: ECML – 95. Proceedings, 1995. XI, 370 pages. 1995.

Vol. 918: P. Baumgartner, R. Hähnle, J. Posegga (Eds.), Theorem Proving with Analytic Tableaux and Related Methods. Proceedings, 1995. X, 352 pages. 1995.

Vol. 927: J. Dix, L. Moniz Pereira, T.C. Przymusinski (Eds.), Non-Monotonic Extensions of Logic Programming. Proceedings, 1994. IX, 229 pages. 1995.

Vol. 928: V.W. Marek, A. Nerode, M. Truszczyński (Eds.), Logic Programming and Nonmonotonic Reasoning. Proceedings, 1995. VIII, 417 pages. 1995.

Vol. 929: F. Morán, A. Moreno, J.J. Merelo, P.Chacón (Eds.), Advances in Artificial Life. Proceedings, 1995. XIII, 960 pages. 1995.

Vol. 934: P. Barahona, M. Stefanelli, J. Wyatt (Eds.), Artificial Intelligence in Medicine. Proceedings, 1995. XI, 449 pages. 1995.

Vol. 941: M. Cadoli, Tractable Reasoning in Artificial Intelligence. XVII, 247 pages. 1995.

Lecture Notes in Computer Science

Vol. 905: N. Ayache (Ed.), Computer Vision, Virtual Reality and Robotics in Medicine. Proceedings, 1995. XIV, 567 pages. 1995.

Vol. 906: E. Astesiano, G. Reggio, A. Tarlecki (Eds.), Recent Trends in Data Type Specification. Proceedings, 1995. VIII, 523 pages. 1995.

Vol. 907: T. Ito, A. Yonezawa (Eds.), Theory and Practice of Parallel Programming. Proceedings, 1995. VIII, 485 pages. 1995.

Vol. 908: J. R. Rao Extensions of the UNITY Methodology: Compositionality, Fairness and Probability in Parallelism. XI, 178 pages. 1995.

Vol. 909: H. Comon, J.-P. Jouannaud (Eds.), Term Rewriting. Proceedings, 1993. VIII, 221 pages. 1995.

Vol. 910: A. Podelski (Ed.), Constraint Programming: Basics and Trends. Proceedings, 1995. XI, 315 pages. 1995.

Vol. 911: R. Baeza-Yates, E. Goles, P. V. Poblete (Eds.), LATIN '95: Theoretical Informatics. Proceedings, 1995. IX, 525 pages. 1995.

Vol. 912: N. Lavrač, S. Wrobel (Eds.), Machine Learning: ECML – 95. Proceedings, 1995. XI, 370 pages. 1995. (Subseries LNAI).

Vol. 913: W. Schäfer (Ed.), Software Process Technology. Proceedings, 1995. IX, 261 pages. 1995.

Vol. 914: J. Hsiang (Ed.), Rewriting Techniques and Applications. Proceedings, 1995. XII, 473 pages. 1995.

Vol. 915: P. D. Mosses, M. Nielsen, M. I. Schwartzbach (Eds.), TAPSOFT '95: Theory and Practice of Software Development. Proceedings, 1995. XV, 810 pages. 1995.

Vol. 916: N. R. Adam, B. K. Bhargava, Y. Yesha (Eds.), Digital Libraries. Proceedings, 1994. XIII, 321 pages. 1995.

Vol. 917: J. Pieprzyk, R. Safavi-Naini (Eds.), Advances in Cryptology - ASIACRYPT '94. Proceedings, 1994. XII,

Vol. 918: P. Baumgartner, R. Hähnle, J. Posegga (Eds.), Theorem Proving with Analytic Tableaux and Related Methods. Proceedings, 1995. X, 352 pages. 1995. (Subseries LNAI).

Vol. 919: B. Hertzberger, G. Serazzi (Eds.), High-Performance Computing and Networking. Proceedings, 1995. XXIV, 957 pages. 1995.

Vol. 920: E. Balas, J. Clausen (Eds.), Integer Programming and Combinatorial Optimization. Proceedings, 1995. IX, 436 pages. 1995.

Vol. 921: L. C. Guillou, J.-J. Quisquater (Eds.), Advances in Cryptology – EUROCRYPT '95. Proceedings, 1995. XIV, 417 pages. 1995.

Vol. 922: H. Dörr, Efficient Graph Rewriting and Its Implementation. IX, 266 pages. 1995.

Vol. 923: M. Meyer (Ed.), Constraint Processing. IV, 289 pages. 1995.

Vol. 924: P. Ciancarini, O. Nierstrasz, A. Yonezawa (Eds.), Object-Based Models and Languages for Concurrent Systems. Proceedings, 1994. VII, 193 pages. 1995.

Vol. 925: J. Jeuring, E. Meijer (Eds.), Advanced Functional Programming. Proceedings, 1995. VII, 331 pages. 1995.

Vol. 926: P. Nesi (Ed.), Objective Software Quality. Proceedings, 1995. VIII, 249 pages. 1995.

Vol. 927: J. Dix, L. Moniz Pereira, T. C. Przymusinski (Eds.), Non-Monotonic Extensions of Logic Programming. Proceedings, 1994. IX, 229 pages. 1995. (Subseries LNAI).

Vol. 928:V.W. Marek, A. Nerode, M. Truszczyński (Eds.), Logic Programming and Nonmonotonic Reasoning. Proceedings, 1995. VIII, 417 pages. 1995. (Subseries LNAI).

Vol. 929: F. Morán, A. Moreno, J.J. Merelo, P. Chacón (Eds.), Advances in Artificial Life. Proceedings, 1995. XIII, 960 pages. 1995. (Subseries LNAI).

Vol. 930: J. Mira, F. Sandoval (Eds.), From Natural to Artificial Neural Computation. Proceedings, 1995. XVIII, 1150 pages. 1995.

Vol. 931: P.J. Braspenning, F. Thuijsman, A.J.M.M. Weijters (Eds.), Artificial Neural Networks. IX, 295 pages. 1995.

Vol. 932: J. Iivari, K. Lyytinen, M. Rossi (Eds.), Advanced Information Systems Engineering. Proceedings, 1995. XI, 388 pages. 1995.

Vol. 933: L. Pacholski, J. Tiuryn (Eds.), Computer Science Logic. Proceedings, 1994. IX, 543 pages. 1995.

Vol. 934: P. Barahona, M. Stefanelli, J. Wyatt (Eds.), Artificial Intelligence in Medicine. Proceedings, 1995. XI, 449 pages. 1995. (Subseries LNAI).

Vol. 935: G. De Michelis, M. Diaz (Eds.), Application and Theory of Petri Nets 1995. Proceedings, 1995. VIII, 511 pages. 1995.

Vol. 936: V.S. Alagar, M. Nivat (Eds.), Algebraic Methodology and Softwasre Technology. Proceedings, 1995. XIV, 591 pages. 1995.

Vol. 937: Z. Galil, E. Ukkonen (Eds.), Combinatorial Pattern Matching. Proceedings, 1995. VIII, 409 pages. 1995.

Vol. 938: K.P. Birman, F. Mattern, A. Schiper (Eds.), Theory and Practice in Distributed Systems. Proceedings, 1994. X, 263 pages. 1995.

Vol. 941: M. Cadoli, Tractable Reasoning in Artificial Intelligence. XVII, 247 pages. 1995. (Subseries LNAI).